*Circuit Design
Using Personal Computers*

Circuit Design
Using Personal Computers

THOMAS R. CUTHBERT, JR.

Director, Advanced Technology
Collins Transmission Systems Division
Rockwell International Corporation
Dallas, Texas

A Wiley-Interscience Publication

JOHN WILEY & SONS
New York   Chichester   Brisbane   Toronto   Singapore

Library of Congress Cataloging in Publication Data:

Cuthbert, Thomas R. (Thomas Remy), 1928–
  Circuit design using personal computers.

  "A Wiley-Interscience publication."
  Includes bibliographical references and index.
  1. Radio circuits.   2. Electronic circuit design—Data
processing.   3. Microcomputers.   I. Title.
TK6553.C87   1982   621.3841'2   82-16015
ISBN 0-471-87700-X

Printed in the United States of America

10 9 8 7 6 5 4

To my late parents,
Tommy and Brownie

Preface

Circuit design, an essential part of electrical engineering, has become an exciting field because of the availability of responsive personal computers. Productive interaction with the designer's own computer has been possible for several years, but only recently has it become completely respectable through the introduction of a highly touted personal computer by the largest manufacturer of big computers. Modern circuit design usually involves extensive mathematical calculations based on increasingly theoretical concepts to satisfy escalating performance requirements. I wrote this book to show how effective personal computers can be in circuit design.

The first goal is to describe practical radio frequency circuit design techniques that are especially appropriate for personal computers and have one or more fundamental concepts or applications. For example, the polynomial root-finder algorithm can solve as many as 20 complex roots and is based on the important Cauchy–Riemann conditions. It works well, and the underlying principles are worth studying. The second goal is to exploit the interaction between circuit designer and computer to clarify both design techniques and fundamental concepts. It is possible to produce valuable answers rapidly while developing a feel for the procedure and obtaining insight into fundamental processes, such as errors between exact derivatives and their finite-difference estimates.

The most frequently encountered design procedures are appropriate for personal computers, even though there are a few heavily used procedures that must be performed on large computers. This book is based on the premise that most designers are better served by computer programs that they can call their own. Only a few must master large-computer operating procedures and program manuals several inches thick; these procedures are beyond the scope of this book. Rather, I have selected some of the most productive and interesting circuit design techniques, some very old and others quite recent. Many students have recently developed an appreciation and interest in these topics precisely because the techniques have become visible on the personal computer. Excessive theoretical analysis has been avoided by providing references to more detailed explanations; these also provide the interested student with efficient avenues for further investigation.

This book is intended for practicing electrical engineers and for university students with at least senior-class standing. The topics should also interest electronics engineers who design circuits derived in terms of complex variables and functions to provide impedance matching, filtering, and linear amplification. Circuits operating from very low frequency through millimeter waves can be designed by these techniques. The necessary numerical methods should also interest those who do not have specific applications.

The numerical methods include solution of complex linear equations, integration, curve fitting by rational functions, nonlinear optimization, and operations on complex polynomials. These programmed tools are applied to examples of filter synthesis to illustrate the subject as well as the numerical methods. Several powerful direct-design methods for filters are described, and both single-frequency and broadband impedance-matching techniques and limitations are explained. An efficient ladder network analysis method, suitable for hand-held or larger computers, is described and programmed for confirming network design and evaluating various effects, including component sensitivities. Linear-amplifier design theory is based on the concept of bilinear transformations and the popular impedance-mapping technique. This also enables a description of load effects on passive networks and is the design basis of filters that absorb rather than reflect energy.

The methods are supported by seventeen programs in reverse Polish notation (RPN) for Hewlett–Packard HP-67 and HP-97 hand-held programmable calculators and, with minor modifications, for models HP-41C and HP-9815. There are also 28 programs in Microsoft BASIC language for PET and similar desktop computers. PET is a registered trademark of Commodore Business Machines, a division of Commodore International. Microsoft Consumer Products has furnished a consistent and widely accepted BASIC programming language to many prominent personal-computer manufacturers. Some of the BASIC programs are short enough for hand-held computers, but most require a desktop computer having several thousand bytes of random-access memory and appropriate speed. Each chapter, except for the introduction, contains a set of problems, most of which require a hand-held calculator for solution.

The material in this book was and is being used in a two-semester graduate-level course at Southern Methodist University. The first semester covered numerical methods—including optimization, examples of filter synthesis, and ladder network analysis—contained in Chapters Two through Five. The more specialized, second-semester content included impedance matching, linear amplifier design, direct-coupled filters, and the other direct filter design methods in Chapters Six through Nine. The course was taught with several students in the classroom and the majority on a closed-circuit television network that included video output from a desktop, BASIC language personal computer on the instructor's desk. The ability to edit and rerun the programs in this book was a most valuable teaching aid. All students were encouraged to acquire at least a hand-held computer; university desktop personal computers were available, but many industrial students had their own.

This material was taught three times as a 48-hour industrial seminar for practicing engineers who desired a cognitive overview of the field with emphasis on personal computing. Approximately 6 hours of study per chapter should be spent for all but Chapter One, and good visual aids and computer TV monitors are required in the classroom. More limited seminars may also be taught as follows: numerical methods in Chapters Two and Five; numerical methods, filter synthesis, and elliptic filters in Chapters Two, Three, and Sections 9.2–9.4; ladder network analysis and sensitivities in Chapter Four; impedance matching and direct-coupled and stub filters in Sections 6.1–6.5, Chapter Eight, and Section 9.1; and linear amplifiers, impedance mapping, and filter-load effects in Chapter Seven and Sections 9.5 and 9.6. Individual engineers with some familiarity with the subject will find this book a good basis for review and discovery of new design methods and programs. Access to or ownership of a desktop computer is a necessity; the minimum requirement is ownership of a programmable hand-held calculator and access to a desktop computer or a readily accessible, responsive computer terminal to run BASIC language programs.

I wish to express my deep appreciation to colleagues at Collins Radio Company, Texas Instruments, and Rockwell International for their suggestions, constructive criticism, and other contributions to understanding. Special recognition is due to Dr. Kenneth W. Heizer, Southern Methodist University, who endured years of my questions concerning technical relevance and origins. His knowledge and patience satisfied my aspiration for both generality and applicability.

THOMAS R. CUTHBERT, JR.

Plano, Texas
December 1982

Contents

Chapter 1

Introduction

This book describes design and analysis techniques for radio frequency circuits that are particularly well suited to small computers. This chapter presents the rationale and an overview of the book's organization.

Both entering and experienced engineers are addressed for entirely different reasons. Many new electrical engineering graduates have received heavy exposure to digital circuits and systems during recent training. Apparently, the curriculum time limitations have resulted in less thorough treatment of analog topics, especially filter, impedance matching, and similar circuit design techniques. The industrial need has not diminished. Experienced engineers are probably far less aware of the new opportunities available through small-computer design methods. These computing aids are becoming a necessity in this field, if only to meet the competition from those already exploiting the opportunities. This book establishes a level of capability of hand-held and desktop computers for those who have not been following recent applications.

Engineers can now own the computers and programs as well as their technical expertise. It is interesting to estimate the current (1982) costs for the equipment an engineer may consider buying for professional use. The following figures do not account for potential tax advantages. Typical programmable-calculator and peripheral equipment costs range from \$150 to \$800. Typical desktop personal computers and peripheral equipment costs range from \$500 to \$5000, and professional-grade equipment (e.g., Hewlett–Packard, Wang, and Digital Equipment) costs about twice these amounts. The most expensive desktop computing systems cost as much as \$30,000. It is estimated that within five years the same performance may be obtained at half these costs; conversely, twice the performance may be available at the same cost.

Hamming (1973, p. 3) has remarked that computing should be intimately bound with both the source of the problem and the use that is going to be made of the answers; it is not a step to be taken in isolation from reality. Therefore, the art of connecting the specific problem with the computing is important, but it is best taught within a field of application. That is the

1

viewpoint taken in this book. It is now desirable for an engineer with average job assignments to make effective use of small computers. Tasks that were not feasible just 10 years ago can now be performed routinely. Hamming also noted that computing is for insight; modern technology certainly demands a high level of understanding. Design difficulties often can be detected early by computer analysis.

Running the programs in this book will provide many engineers with a convincing display of principles and results that are often elusive as abstractions. For example, calculus literally works before your eyes on the computer screen when an optimizer is reporting its strategy and progress! Program modifications are suggested throughout the text to demonstrate computational behavior for degenerate cases. Most readers will find that using the programs while studying this material will improve comprehension immensely. Many of the suggested extensions have been developed and programmed by the author, and are known to be both feasible and worthwhile.

The computer programs furnished in this text are deliberately unsophisticated. The best program is one written, or at least adapted, by the end user. This avoids a significant amount of computer programming, namely the effort any programmer expends to anticipate all possible reactions of the unknown user. Also, a prudent engineer will be skeptical unless programs written by others are exceptionally well documented, tested, and constructed to deal with degenerate cases and to provide user prompting. Often there is little professional satisfaction in simply running programs others have developed; after all, a clerk could do that job at lower cost.

A valuable feature on many desktop computers is a TRACE program that allows the user to either step through the program execution one line at a time or to execute slowly with current line numbers displayed for correlation with code listings and/or flowcharts. Another recommended computer feature is an EDITOR program that enables a search for the names of variables. Most BASIC languages allow only "global" variable names, which are not private within subroutines. A good EDITOR facilitates the expansion and combination of programs in this book without the risk of conflicting variable names.

Most of the short programs are furnished in Hewlett–Packard's reverse Polish notation (RPN). For Texas Instruments hand-held calculators, such as the TI-59 and others using an algebraic operating system (AOS) language, coding can originate with the equations provided and in the format of the given programs. Differences between RPN and AOS have been discussed by Murdock (1979). Hand-held computers have not been made obsolete by desktop computers; there are many occasions when a completely portable computer is much more effective when used at the place of immediate need.

Numerous geometric illustrations have been employed in place of more abstract explanations. A number of graphs depicting design parameter trends are also presented; use of computers does not diminish the value of graphic displays, sensitivity and similar computations not withstanding.

It is assumed that the reader will pursue interesting topics at his or her own level of sophistication once the possibilities are clear. To that end, extensive references are provided for further inquiry. Many references are quite recent, which is not to overlook some of the older classics—for example, an early direct-coupled-filter article by Dishal (1949). There are some derivations that are crucial to important issues; these have been included in an appendix or outlined in the problem set.

There are several indispensable numerical analysis tools that will be required throughout this book and that are applicable in almost all phases of electrical engineering. Chapter Two begins with the most elemental of these (especially in steady-state sinusoidal network analysis): the complex addition, subtraction, multiplication, and division functions. A hand-held computer program is given for convenient usage, and the reader will need to have ready access to this capability on many occasions. The Gauss–Jordan method for solving real equations in real unknowns is discussed in connection with a BASIC language program; this is used later in Chapter Two for fitting rational polynomials to discrete complex data sets and in Chapter Three in the Gewertz synthesis method. A very convenient extension of this method to solve systems of complex equations is also described; this technique is convenient for solving nodal analysis equations and similar systems.

Chapter Two also describes the trapezoidal rule and its application in the Romberg method of numerical integration; this is used in the broadband impedance–matching methods in Chapter Six. Also, Simpson's rule is derived for later use in time-domain analysis in Chapter Four. Chapter Two concludes with methods for fitting polynomials to data. First, real polynomials are generated to provide a minimax fit to piecewise linear functions using Chebyshev polynomials. Second, complex data are fit by a rational function of a complex variable, especially the frequency-axis variable in the Laplace s plane. This will be applied to broadband matching, and is useful in other ways, such as representing measured antenna impedance data.

Many of the computer aids developed in this book are not only efficient tools, but are based on important principles worth the attention of any network designer. Moore's root finder in chapter three is a good example, because it depends on the Cauchy–Riemann conditions and a powerful but little-known method for evaluating a complex-variable polynomial and its derivatives.

Engineers interested in network synthesis, automatic control, and sampled data systems need many other mathematical aids. Polynomial addition and subtraction of parts, multiplication, long division, and partial and continued fraction expansions of rational polynomials are described in Chapter Three. Their application to network synthesis is used to develop the characteristic and transducer functions in terms of the ABCD (chain) matrix of rational polynomials. These are then realized as doubly terminated ladder networks. Gewertz's singly terminated network synthesis method concludes Chapter Three; this method accomplishes input impedance synthesis, given the network function's real part.

Chapters Four and Five need not be considered together, but the efficient ladder network analysis method in Chapter Four is constructed so as to become a part of the powerful gradient optimizer in Chapter Five. The recursive ladder network analysis method is based on assumed load power and impedance (therefore current) and accommodates flexible interconnection of two-terminal branches. The topological description is very compact, so that the technique can be employed in hand-held as well as in larger computers. Node voltages and branch currents are available for many purposes, including the powerful Tellegen method for sensitivity calculations. Two-port chain matrix parameters are described for use in cases where transmission line, bridged-T, and arbitrary two-port network sections appear in cascade. A node-bridging analysis technique is discussed to avoid the need for nodal or other matrix methods for only slightly more complicated ladder network problems. The input and transfer quantities obtained are related to the terminal voltages and currents. This is developed by introducing the first of several scattering parameter explanations in order to simplify the calculations. Almost all other topics in this book depend on the engineer's ability to check his or her design by means of a ladder network simulation. Simpson's numerical integration is used to evaluate Fourier and convolution integrals so that the frequency samples of network response previously generated can provide time response to an impulse or any arbitrary excitation. Chapter Four concludes with a compact explanation of sensitivities computed by approximate finite differences and by the exact Tellegen method. Applications discussed include establishing tolerances and automatic adjustment of network elements to approach some arbitrary frequency or time response—in other words, optimization.

Chapter Five is a practical application of nonlinear programming (optimization) for adjustment of design variables. It begins with a brief review of essential matrix algebra in terms of geometric concepts. Then the significant properties of conjugate gradient search methods are illustrated by computer demonstration, and the role of linear searches in useful algorithms is illustrated. A Fletcher–Reeves optimizer is furnished in the BASIC language with several practical examples. The creation of sampled objective functions, their component response functions, and gradients are described as related to network optimization. Methods for enforcing simple bounds and for satisfying more complicated constraints conclude Chapter Five. Numerous opportunities are used to point out in passing some mathematical concepts of general interest: for example, the relationship of eigenvalues and eigenvectors to ellipsoidal quadratic forms. Only gradient methods are discussed; the reasoning behind this choice and a few remarks concerning the alternative direct search class of optimizers are provided.

Design methods and computer programs for impedance matching at a frequency and over lowpass and bandpass intervals are contained in Chapter Six. At single frequencies, resistance and phase transformations are obtained by L, T, and pi networks. Complex source and load specifications are accommodated by explanation of programs for the $1+Q^2$ method and paralleled-reactances technique. Transmission-line matching applications for complex source and load are described by less well-known methods. Levy's broadband-matching adaptation of Fano's original theory is reviewed, and programs are provided. Standard lowpass prototype filter notation, lowpass-to-bandpass transformation, and Norton transformers are used in practical examples. The last two topics in Chapter Six are new methods of broadband matching. Carlin's method for matching measured impedances is developed on the basis of several computing aids that include a Hilbert transform application with quite general significance. Cottee's pseudobandpass (lowpass transformer) matching method employs numerical integration of the Chebyshev squared-frequency function. This is accomplished with the Romberg integration program from Chapter Two.

Chapter Seven contributes uniquely to reader background. Amplifier designers experienced in scattering parameters, the equipment for their measurement, and the body of technique for their use are probably aware of the large and growing number of computer programs available for the methods involved. The better-known programs exist on timeshare computing services and provide stability, gain, impedance, selectivity, optimization, and device data base information for amplifiers and their matching networks. There are also numerous smaller programs of reduced scope available for desktop and hand-held computers. Furthermore, the trade journals and institutional literature are full of design articles about scattering parameter applications. Chapter Seven provides the perspective and computing tools that are not readily available and yet are the basis for the popular methods. Generalized reflection coefficients for power waves are defined and related to scattering parameters for two- and three-port linear networks. A convenient means for ladder network analysis with embedded circulators is noted. The bilinear function of complex variable theory is introduced and arranged to represent a Smith chart of all possible branch impedance values on a linear network response plane. Convenient methods and computer programs are given for determining the coefficients, the relationship of the Smith chart to two-port power, and geometric models of important network behavior. Concise unification is provided for earlier Linvill models, gain analysis, and impedance-mapping methods for linear networks. This insight also applies to oscillator, filter, and impedance-matching design. A new gain design method based on the older Linvill model is described.

Chapter Eight introduces a new method for direct-coupled filter design based on a loaded-Q parameter that is well known to early radio-manufacturing engineers. The great strength of the method is the wide range of element values that can be selected by the designer with guidance by its clear principles. Direct-coupled-filter principles are widely utilized in design of microwave filters based on the inverter principle. They have important applications at all frequencies down to vlf. This topic is developed by a practical relationship of resonators (tanks), inverters, and end-coupling methods, and the selectivity effects that result. Also, the resonator slope-equivalence technique is described to extend the method to adapted elements other than the ideal lumped elements usually considered. The full range of response shapes—from overcoupled (equal ripple), through maximum flatness, to undercoupled—is described; the important minimum-loss case is covered too. Tuning methods and sensitivity relationships are explained in terms of laboratory methods and the loaded-Q parameter.

The last chapter deals with other direct filter design methods, especially those that depend on recursive formulas for element values or means other than synthesis. This potpourri was chosen because the methods are useful, frequently applicable, and demonstrate worthwhile principles. Chapter Nine begins with Taub's equal-stub microwave filter design method. Then, a new elliptic filter design method is introduced by a general discussion of filter types and performance parameters. The entire family of related selectivity functions is reviewed, and a standard nomograph and program provide performance estimates. Next, the basis of two new and powerful programs for doubly terminated filter design by Amstutz is explained, and program operation is illustrated. Useful tables of lumped-element equivalence transformations are included.

Chapter Nine also contains theory and design tools to estimate load effects on passive networks, and maximum possible efficiency is shown to be the controlling parameter. This topic is important because filters designed to operate properly between carefully defined load resistances are likely to be operated between quite different terminations. The last topic in Chapter Nine extends the load effects concept to invulnerable (intrinsic) filters that absorb rather than reflect energy. These may be regarded as selective attenuators; they are quite valuable in mixer and low-frequency applications where circulators are not feasible. Equations and a design chart for a lowpass, invulnerable network are derived.

Another way to view the contents of this book is according to the mathematical subjects treated, even though the material was organized according to design applications. Matrix algebra topics include multiplication, exponentiation, inner products and norms, quadratic forms and conics, and partitioning. Polynomial algebra of real and complex variables touches on power series and product forms, as well as rational-polynomial continued fractions, partial fractions, and Chebyshev expansions. Calculus tools include multivariate Taylor series, partial derivatives and gradient vectors, the Cauchy–Riemann principle, numerical differentiation and integration, and infinite summations and products. Complex variables appear throughout, and special attention is given to bilinear transformations and the generalized Smith chart. Hilbert, Fourier, and Laplace transforms and the convolution integral are employed.

The material that follows has been tested in industry and can become an important part of your set of engineering tools.

Chapter 2 _____

Some Fundamental Numerical Methods

It is necessary to create several computing aids before addressing specific design tasks. Certainly the most elementary of these is a hand-held computer program to calculate the complex four functions. Also, the solution of linear systems of equations, both in real and complex variables, and numerical integration are useful in many electrical engineering applications. The former is required in the last part of this chapter to fit discrete, complex data by a rational polynomial in the frequency variable to the least-squared-error criterion. Before that, a piecewise linear function will be approximated in the minimum-of-maximum-errors (minimax) sense by a polynomial in a real variable. This is a useful tool that allows the introduction of the versatile Chebyshev polynomials, which will make several later appearances.

2.1. Complex Four Functions

The convenience of addition, subtraction, multiplication, and division of complex numbers on a hand-held calculator, both manually and within programs, cannot be overrated. Program A2-1 in Appendix A provides these subroutines on function keys B, C, and D for manual keying or for GSB (Go Subroutine) commands within programs. As explained in the program description, the more frequently required polar complex number format has been assumed.

Hopefully, the reverse Polish (RPN) stack concept is somewhat familiar to the reader, since it has been used by many calculator manufacturers in several countries. Owners of calculators with the algebraic operating system (AOS) are at no great disadvantage, because RPN programs are easily converted (see Murdock, 1979). In Program A2-1 and in similar programs to follow, the polar complex number(s) are entered into the calculator's XYZT "stack" as angle in degrees, then magnitude. For the operation $Z_1 + Z_2$, it is necessary to enter $\deg Z_2$, $\operatorname{mag} Z_2$, $\deg Z_1$, $\operatorname{mag} Z_1$ and press key B to see $\operatorname{mag}(Z_1 + Z_2)$ in the X register (and the angle in the Y register by pressing key A to swap X and Y registers). Complex subtraction depends on a feature of the HP calculator in which a negative-magnitude number adds 180 degrees to the angle during the next operation. Thus a separate key for complex subtraction is not required; just key in the sequence for $Z_1 + Z_2$, but press the CHS (change sign) key before pressing B (+) key. The answer is $Z_1 - Z_2$. A complex-division key is made unnecessary by providing the complex inverse function $1/Z$ on key C. Thus to compute Z_1/Z_2, the stack entries (in order) are: $\deg Z_1$, $\operatorname{mag} Z_1$, $\deg Z_2$, $\operatorname{mag} Z_2$. Then press key C to obtain $1/Z_2$ (the answer is placed properly in stack registers X and Y without disturbing registers Z and T), followed by pressing key D for the complex multiplication. Again, the answer appears in stack positions X and Y. Example 2.1 shows that manual or programmed steps with complex numbers are as e?·y as with real numbers.

Example 2.1.   Consider the bilinear function from Chapter Seven:

$$w = \frac{a_1 Z + a_2}{a_3 Z + 1}. \tag{2.1}$$

All variables may be complex; suppose that $a_1 = 0.6 \ \underline{/75°}$, $a_2 = 0.18 \ \underline{/-23°}$, and $a_3 = 1.4 \ \underline{/130°}$. Given $Z = 0.5 \ \underline{/60°}$, what is w? The manual or programmed steps are the same: enter Z in the stack and also store its angle and magnitude in two spare registers. Then enter a_3 and multiply, enter 0 degrees and unity magnitude and add, saving the two parts of the denominator value in two more spare registers. The numerator is computed in the same way, the denominator value is recalled into the stack and inverted, and the two complex numbers in the stack are multiplied. The correct answer is $w = 0.4473 \ \underline{/129.5°}$. Normally, a given set of coefficients (a_1, a_2, and a_3) are fixed, and a sequence of Z values are input into the program. A helpful hint for evaluating bilinear functions is to rewrite them by doing long division on (2.1):

$$w = \frac{a_1}{a_3} + \frac{a_2 - a_1/a_3}{a_3 Z + 1}. \tag{2.2}$$

Then store a_1/a_3, a_2, and a_3. Now the operations for evaluating (2.2) do not require storing Z, although a zero denominator value should be anticipated by always adding $1.E-9$ (0.000000001) to its magnitude before inverting. If there is a fourth complex coefficient in place of unity in the denominator of (2.1), the standard form of (2.1) should be obtained by first dividing the other coefficients by the fourth coefficient.

2.2.   Linear Systems of Equations

Every engineering discipline requires the solution of sets of linear equations with real coefficients; this will also be required in Section 2.5 of this chapter.

Although the Gauss–Jordan method considered here is well known, it is less well known that the real-coefficient case can easily be extended to solve systems having both complex coefficients and variables. BASIC language Program B2-1 for the Gauss–Jordan method is contained in Appendix B, and its preamble for coping with the complex system case is Program B2-2.

2.2.1. *The Gauss–Jordan Elimination Method.* The Gauss–Jordan elimination method is but one of several acceptable means to solve systems of real linear equations (see Hamming, 1973, for a commentary). The problem to be solved is to find **x** when

$$\mathbf{Ax} = \boldsymbol{\alpha}, \tag{2.3}$$

in matrix notation, or, written out,

$$a_{11}x_1 + a_{12}x_2 + a_{13}x_3 = a_{14},$$

$$a_{21}x_1 + a_{22}x_2 + a_{23}x_3 = a_{24}, \tag{2.4}$$

$$a_{31}x_1 + a_{32}x_2 + a_{33}x_3 = a_{34}.$$

The order $N = 3$ case will be discussed without loss of generality. Readers not familiar with the matrix notation in (2.3) are urged to refer to an introductory book on linear algebra, such as that of Noble (1969). There will be frequent need for this shorthand notation, although the subject will not be much more rigorous than understanding the equivalence of (2.3) and (2.4). It is also helpful to sketch the $N = 2$ case of two lines in the $x_1 - x_2$ plane and to recall that the solution is merely the intersection of these two lines. The concept extends to hyperplanes in N-dimensional space.

The Gauss–Jordan algorithm evolves (2.4) into the solution form

$$x_1 + 0 + 0 = b_{14},$$

$$0 + x_2 + 0 = b_{24}, \tag{2.5}$$

$$0 + 0 + x_3 = b_{34},$$

by scaling adjacent rows so that subtractions between rows produce the zeros in the columns in (2.5), working from left to right. Recall that scaling a given equation or adding one to another does not change a system of linear equations.

A specific example (Ley, 1970) begins with the "augmented" matrix formed from (2.4):

$$\begin{bmatrix} a_{11} & a_{12} & a_{13} & a_{14} \\ a_{21} & a_{22} & a_{23} & a_{24} \\ a_{31} & a_{32} & a_{33} & a_{34} \end{bmatrix}. \tag{2.6}$$

Consider the array

$$\begin{bmatrix} -2 & -1 & 1 & -1 \\ 1 & 1 & 1 & 6 \\ 3 & 1 & -1 & 2 \end{bmatrix}. \tag{2.7}$$

First normalize the first row with respect to "pivot coefficient" a_{11}:

$$\begin{bmatrix} 1 & 0.5 & -0.5 & 0.5 \\ 1 & 1 & 1 & 6 \\ 3 & 1 & -1 & 2 \end{bmatrix}. \tag{2.8}$$

This is done to avoid potential numerical overflow or underflow in the next multiplication. In order to make the a_{21} coefficient zero, form a new row 2 by multiplying row 1 by a_{21} and subtracting this from row 2. Also, form a new row 3 by multiplying row 1 by a_{31} and subtracting this from row 3. The result is

$$\begin{bmatrix} 1 & 0.5 & -0.5 & 0.5 \\ 0 & 0.5 & 1.5 & 5.5 \\ 0 & -0.5 & 0.5 & 0.5 \end{bmatrix}. \tag{2.9}$$

The next cycle is to normalize the coefficients of row 2 with respect to the new pivot coefficient a_{22}:

$$\begin{bmatrix} 1 & 0.5 & -0.5 & 0.5 \\ 0 & 1 & 3 & 11 \\ 0 & -0.5 & 0.5 & 0.5 \end{bmatrix}. \tag{2.10}$$

Note that after normalization the new coefficient is always

$$a_{IJ} - a_{IK} a_{KJ}, \tag{2.11}$$

where K is the pivot row, I is the new row being formed, and J is the coefficient (column) being formed. Continue by forming new rows 1 and 3 in (2.10):

$$\begin{bmatrix} 1 & 0 & -2 & -5 \\ 0 & 1 & 3 & 11 \\ 0 & 0 & 2 & 6 \end{bmatrix}. \tag{2.12}$$

In the final cycle, normalize the coefficients of row 3 with respect to coefficient a_{33}:

$$\begin{bmatrix} 1 & 0 & -2 & -5 \\ 0 & 1 & 3 & 11 \\ 0 & 0 & 1 & 3 \end{bmatrix}. \tag{2.13}$$

Finally, form new rows 1 and 2 using (2.11):

$$\begin{bmatrix} 1 & 0 & 0 & 1 \\ 0 & 1 & 0 & 2 \\ 0 & 0 & 1 & 3 \end{bmatrix}. \tag{2.14}$$

Because (2.14) now represents the system in (2.5), the solution is in column 4: $x_1 = 1$, $x_2 = 2$, and $x_3 = 3$.

The BASIC language Program B2-1 in Appendix B implements this algorithm. Note that new coefficients are generated according to (2.11) in program line number 9330. Lines 9140–9240 implement a feature not yet mentioned. If any pivot coefficient is too small (taken to be less than or equal to $1E - 6$ in line 9020), then the rows are interchanged. The reader is encouraged to first work this procedure on paper and then run Program B2-1 with the same data; it is easy to appreciate the advantages of readily available computers! Program B2-1 documentation also contains some preamble code to change the input from the user prompting mode to input by READ commands related to DATA statements. Statement numbers lower than 9110 in the original code may be replaced by the alternate code, so that extensive input data need not be flawlessly entered in real time.

Example 2.2. The DATA statements in Program B2-1 contain the element values of the matrix in Figure 2.1. This system of 10 Kirchhoff current and voltage equations is solved in 42 seconds on a Commodore PET computer. The currents thus calculated should sum to zero at each node.

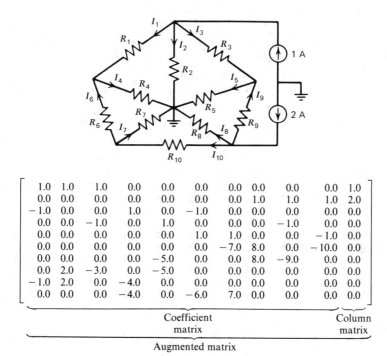

$$
\begin{bmatrix}
1.0 & 1.0 & 1.0 & 0.0 & 0.0 & 0.0 & 0.0 & 0.0 & 0.0 & 0.0 & 1.0 \\
0.0 & 0.0 & 0.0 & 0.0 & 0.0 & 0.0 & 0.0 & 1.0 & 1.0 & 1.0 & 2.0 \\
-1.0 & 0.0 & 0.0 & 1.0 & 0.0 & -1.0 & 0.0 & 0.0 & 0.0 & 0.0 & 0.0 \\
0.0 & 0.0 & -1.0 & 0.0 & 1.0 & 0.0 & 0.0 & 0.0 & -1.0 & 0.0 & 0.0 \\
0.0 & 0.0 & 0.0 & 0.0 & 0.0 & 1.0 & 1.0 & 0.0 & 0.0 & -1.0 & 0.0 \\
0.0 & 0.0 & 0.0 & 0.0 & 0.0 & 0.0 & -7.0 & 8.0 & 0.0 & -10.0 & 0.0 \\
0.0 & 0.0 & 0.0 & 0.0 & -5.0 & 0.0 & 0.0 & 8.0 & -9.0 & 0.0 & 0.0 \\
0.0 & 2.0 & -3.0 & 0.0 & -5.0 & 0.0 & 0.0 & 0.0 & 0.0 & 0.0 & 0.0 \\
-1.0 & 2.0 & 0.0 & -4.0 & 0.0 & 0.0 & 0.0 & 0.0 & 0.0 & 0.0 & 0.0 \\
0.0 & 0.0 & 0.0 & -4.0 & 0.0 & -6.0 & 7.0 & 0.0 & 0.0 & 0.0 & 0.0
\end{bmatrix}
$$

	Coefficient matrix		Column matrix

Augmented matrix

Figure 2.1. A six-node, 10-branch resistive network. [From Ley, 1970.]

2.2.2. Linear Equations With Complex Coefficients.
There is a simple way to apply any real-coefficient method, such as the preceding Gauss–Jordan method, to solve systems of linear equations with both complex coefficients and variables. Without loss of generality, consider the following two equations:

$$
\begin{bmatrix}
a_{11}+jb_{11} & a_{12}+jb_{12} \\
a_{21}+jb_{21} & a_{22}+jb_{22}
\end{bmatrix}
\begin{bmatrix}
x_1+jy_1 \\
x_2+jy_2
\end{bmatrix}
=
\begin{bmatrix}
a_{13}+jb_{13} \\
a_{23}+jb_{23}
\end{bmatrix}. \tag{2.15}
$$

Write (2.15) as two equations in two unknowns in the same way that (2.3) and (2.4) were related; then perform the algebraic multiplication of the complex products and collect real and imaginary parts on each side of the equality signs. Recall that in complex equations the real parts on the left of the equality sign must equal the real parts on the right side, and the same holds true for imaginary parts. Four equations result from these operations; in matrix notation, they are represented as follows:

$$\begin{bmatrix} a_{11} & -b_{11} & a_{12} & -b_{12} \\ b_{11} & a_{11} & b_{12} & a_{12} \\ a_{21} & -b_{21} & a_{22} & -b_{22} \\ b_{21} & a_{21} & b_{22} & a_{22} \end{bmatrix} \begin{bmatrix} x_1 \\ y_1 \\ x_2 \\ y_2 \end{bmatrix} = \begin{bmatrix} a_{13} \\ b_{13} \\ a_{23} \\ b_{23} \end{bmatrix}. \tag{2.16}$$

There is a general pattern in (2.15) and (2.16) for transforming a complex augmented matrix. All odd-row coefficients in the new matrix (2.16) alternate in sign, beginning with a positive sign and ending with a negative sign. All imaginary coefficients (b_{ij}) in the even rows of (2.16) are the same as the coefficients diagonally above, except for the sign. The solution of NC complex equations requires a 2NC by 2NC + 1 real, augmented matrix in the preceding Gauss–Jordan algorithm.

BASIC language Program B2-2 in Appendix B requests the complex coefficients in rectangular form, as in (2.15). The program then forms (2.16) and outputs the solution in the sequence x_1, y_1, x_2, y_2, etc. For example, evaluate each equation in the system given in problem 2.10 by using hand-held computer Program A2-1. Then enter the matrix and right-hand-side coefficients into Program B2-2 to find the solution elements $1+j3$ and $-3+j5$.

As in the real-coefficient program, a READ and DATA modification preamble has been added to Program B2-2 documentation to replace all statements numbered less than 100.

Example 2.3. The DATA statements in Program B2-2 relate to the six complex equations described in Figure 2.2. These are the mesh equations for the sinusoidal steady-state condition of the network, and they are equivalent to 12 real equations. The solution is obtained in 70 seconds on a Commodore PET computer. The six complex mesh equations can be checked by using the solved mesh currents and hand-held computer Program A2-1.

2.2.3. Linear Equations Summary.   The Gauss–Jordan algorithm is reasonably fast, accurate, brief, and solves real and complex equations. Systems of NC complex equations can be solved easily by a simple, programmable transformation to an equivalent system of 2NC real equations. Programs are commercially available to solve real systems of equations. However, adapting these for the user's purposes, especially incorporating them into other programs, is often quite difficult. Programs B2-1 and B2-2 should be suitable for

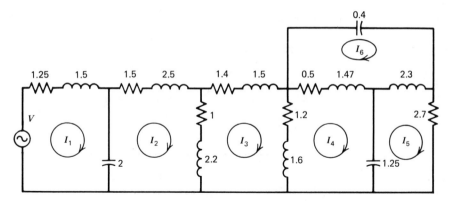

Mesh equations for the sinusoidal steady state:

$$(1.25+j1.0)I_1+j0.5I_2=50+j86.6$$
$$j0.5I_1+(2.5+j4.2)I_2-(1.0+j2.2)I_3=0$$
$$-(1.0+j2.2)I_2+(3.6+j5.3)I_3-(1.2+j1.6)I_4=0$$
$$-(1.2+j1.6)I_3+(1.7+j2.27)I_4+j0.8I_5-(0.5+j1.47)I_6=0$$
$$j0.8I_4+(2.7+j1.5)I_5-j2.3I_6=0$$
$$-(0.5+j1.47)I_4-j2.3I_5+(0.5+j1.27)I_6=0$$

The corresponding augmented matrix is therefore given by:

$$
\begin{bmatrix}
(1.25+j1.0) & j0.5 & 0 & 0 & 0 & 0 & (50+j86.6) \\
j0.5 & (2.5+j4.2) & -(1.0+j2.2) & 0 & 0 & 0 & 0 \\
0 & -(1.0+j2.2) & (3.6+j5.3) & -(1.2+j1.6) & 0 & 0 & 0 \\
0 & 0 & -(1.2+j1.6) & (1.7+j2.27) & j0.8 & -(0.5+j1.47) & 0 \\
0 & 0 & 0 & j0.8 & (2.7+j1.5) & -j2.3 & 0 \\
0 & 0 & 0 & -(0.5+j1.47) & -j2.3 & (0.5+j1.27) & 0
\end{bmatrix}
$$

Figure 2.2. A six-mesh network; $v=50+j86.6$, 1 radian/second. [From Ley, 1970.]

special user applications. Further application of Gauss–Jordan Program B2-1 will be made in Section 2.5.

2.3. Romberg Integration

Numerical integration, or quadrature, is usually accomplished by fitting the integrand with an approximating polynomial and then integrating this exactly. Many such algorithms exhibit numerical instability because increasing degrees of approximation can be shown to converge to a limit that is not the correct answer. However, the simple trapezoidal method assumes that the integrand is linear between evenly spaced points on the curve, so that the area is the sum of 2^i small trapezoids for a large enough i (see Figure 2.3). The trapezoidal method is numerically stable. There are other numerically stable integration

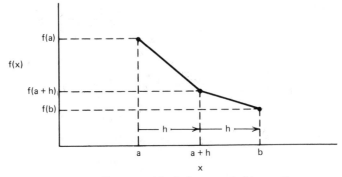

Figure 2.3.   The trapezoid rule for numerical integration.

methods, such as Gaussian quadrature, based on weighted-sample schemes, but calculation of the weights consumes time and memory. The latter methods pose difficulties in recursive calculation of estimates of increasing order, thus limiting their use as computer aids.

The Romberg integration method first approximates the integral as the area of just one trapezoid in the range of integration, then two, continuing for 2^i evenly spaced trapezoids until a larger i does not change the answer significantly. The other main feature of the Romberg method is deciding how many trapezoids are enough. The width of each trapezoidal area starts at $h = b - a$, then $h/2$. The areas found for these values are linearly extrapolated versus h^2 to $h = 0$; when the estimate using width $h/4$ is found, the extrapolation to $h = 0$ is quadratic, and this is tested against the linearly extrapolated answer for convergence. There is a sequence of estimates for decreasing trapezoid widths and increasing degrees of extrapolation until either convergence or a state of numerical noise is obtained. The Romberg method is very efficient, stable, and especially suitable for digital computing. However, the integrand must be computed from an equation, as opposed to using measured data.

In the next four sections it will be shown how the formulas for trapezoidal integration, repeated linear interpolation, and the Romberg recursion are obtained. A BASIC language program will then be described, and an example will be considered. Finally, a once-repeated trapezoid rule will be shown to yield Simpson's rule for integration; this will be used in Chapter Four.

2.3.1.   Trapezoidal Integration.   The integration problem is to find the value of the integral T given the integrand f(x) and the limits of integration a and b:

$$T(a, b) = \int_a^b f(x)\, dx. \tag{2.17}$$

Summing the two trapezoidal areas in Figure 2.3 yields

$$T = h\left(\frac{f_a + f_{a+h}}{2} + \frac{f_{a+h} + f_b}{2}\right), \tag{2.18}$$

and it is convenient to rearrange (2.18) as

$$T = h\left[(f_a + f_{a+h} + f_b) - \tfrac{1}{2}(f_a + f_b) \right]. \qquad (2.19)$$

Similar equations for four trapezoids can be written and then expanded to obtain the general rule for 2^i trapezoids:

$$T_{0,i} \doteq h_i \left\{ \left(\sum_{k=0}^{2^i} f_{i,k} \right) - \frac{1}{2}\left[f(a) + f(b) \right] \right\}, \qquad (2.20)$$

where the trapezoids have the width

$$h_i = \frac{(b-a)}{2^i}. \qquad (2.21)$$

The error in the trapezoid rule estimate is proportional to h_i^2. The interested reader is referred to McCalla (1967) for more details. The zero subscript on T in (2.20) indicates that the estimate was obtained without the extrapolation discussed next, i.e., a zero-order extrapolation.

2.3.2. Repeated Linear Interpolation and the Limit. A linear interpolation formula will be derived in terms of Figure 2.4. Equating the slopes between the two line segments in Figure 2.4 gives

$$\frac{q - q_1}{x - x_1} = \frac{q_2 - q_1}{x_2 - x_1}, \qquad (2.22)$$

which reduces to the standard interpolation (or extrapolation) formula:

$$q(x) = \frac{q_2(x_1 - x) - q_1(x_2 - x)}{x_1 - x_2}. \qquad (2.23)$$

Now suppose that the q(x) function is the integral function T, and two particular estimates, $T_{0,i}$ and $T_{0,i+1}$, have been obtained by (2.20). For the function of h, $T_{0,i}(h)$, use (2.23) with the interpolation variable h^2:

$$T_{1,i} = \frac{T_{0,i+1}\left(h_i^2 - h^2\right) - T_{0,i}\left(h_{i+1}^2 - h^2\right)}{h_i^2 - h_{i+1}^2}, \qquad (2.24)$$

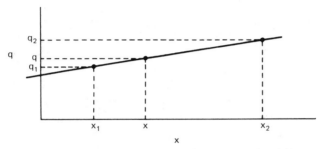

Figure 2.4. Interpolation of a real function of a real variable.

where $T_{1,i}$ indicates a degree-1 (linear) extrapolation. To extrapolate trapezoid widths to zero, set h equal to zero in (2.24) and simplify the result by using (2.21):

$$T_{1,i} = \frac{3T_{0,i+1} + T_{0,i+1} - T_{0,i}}{2^2 - 1}. \tag{2.25}$$

The linear extrapolation in (2.25) is rewritten in a form for later use:

$$T_{1,i} = T_{0,i+1} + \frac{T_{0,i+1} - T_{0,i}}{2^2 - 1}. \tag{2.26}$$

This has an error from the true integral value proportional to h_i^4, and is thus a more accurate estimate than the individual trapezoidal estimates.

Again note that the linear extrapolation is versus h^2 and not simply versus h; the reasons for this choice and the following general formula are explained by McCalla (1967) and, in more detail, by Bauer et al. (1963). Briefly, the trapezoid rule estimate may be expanded as a finite Taylor series with a remainder in the variable h, the true value being the constant term. Since the error is of the order h^2, the remainder term is proportional to $f''(\xi)h^2$, with ξ somewhere on the interval h. McCalla (1967, p. 289) argues that $f''(\xi)$ should be about equal over h_i and its subdivided intervals $h_{i+1} = h_i/2$. This leads directly to (2.25), thus justifying the linear extrapolation to zero of successive trapezoidal estimates in the variable h^2.

The scheme is simply this: one, two, and then four trapezoids in the range of integration enable two linear extrapolations, as described. The two extrapolated results can then be extrapolated again for a new estimate. McCalla (1967) shows that repeating linear extrapolations once is equivalent to quadratic (second-degree) extrapolation. The concept of estimating performance at a limit, here at $h=0$, is known as Richardson extrapolation; it will appear again in Chapter Five.

Using this rationale, a general expression for Romberg integration is obtained from (2.26):

$$T_{k,j} = T_{k-1,j+1} + \frac{T_{k-1,j+1} - T_{k-1,j}}{2^{2k} - 1}, \qquad \begin{cases} k+j = i, \\ j = i-1, i-2, \ldots, 1, 0. \end{cases} \tag{2.27}$$

Index k is the order of extrapolation, and there are i bisections of the integration interval $(b-a)$. The compactness of (2.27) makes it ideal for programming. The table in Figure 2.5 illustrates the Romberg extrapolation process. The step length, or trapezoid width, is shown in the left-hand column. The brackets indicate pairs of lower-order estimates that produce an estimate of next higher order by linear extrapolation to step length $h=0$. Then the better estimates are similarly paired for extrapolation. Accuracy is ultimately limited, because the estimates are the result of the subtraction of two numbers. Eventually, the significant digits will diminish on a finite word-length computer, and the process should be terminated.

Truncation Error		$O(h_i^2)$	$O(h_1^4)$	$O(h_i^6)$	$O(h_i^8)\cdots$
Step Length h_i	j \ i	0	1	2	3
$b-a=h_0$	0	$T_{0,0}$	$T_{1,0}$	$T_{2,0}$	$T_{3,0}\cdots$
$\dfrac{b-a}{2}=h_1$	1	$T_{0,1}$	$T_{1,1}$	$T_{2,1}$	$\cdots$
$\dfrac{b-a}{4}=h_2$	2	$T_{0,2}$	$T_{1,2}$	$\cdots$	
$\dfrac{b-a}{8}=h_3$	3	$T_{0,3}$	$\cdots$		
$\vdots$	$\vdots$				

$h=1$	0.750000000	0.694444445	0.693174603	0.693147478	0.693147182
$h=0.5$	0.708333333	0.693253968	0.693147902	0.693147183	
$h=0.25$	0.69702381	0.693154531	0.693147194		
$h=0.125$	0.69412185	0.693147653			

$$f(x)=\frac{1}{x} \quad \begin{array}{l} a=1 \\ b=2 \end{array}$$

Figure 2.5. Table of T values in the Romberg integration algorithm.

2.3.3. Romberg Integration Program. BASIC language Program B2-3 in Appendix B implements (2.27), as illustrated in Figure 2.5. The only storage required is in vector (single-subscript array) $AU(\cdot)$. The integrand function should be coded by the user beginning in line 10,000; the values returned by the user's subroutine are expected to be labeled by the variable name "FC." The table of values in the format of Figure 2.5 can be compared to the program's computing sequence by adding lines to Program B2-3, as shown in Table 2.1.

Run the example in the subroutine programmed in line 10,000 and after. The integrand is $1/x$, so that the integral is known in closed form, namely $\ln x$. Input limits $a=1$ to $b=2$, so that the answer should be $\ln 2$. The progress of the Romberg algorithm for this example is shown in Figure 2.5, and the answer at termination is underlined. Parameter $ND=11$ in Appendix-B Program B2-3 limits the algorithm to a maximum of 1025 evaluations of the integrand function. The accuracy parameter $EP=1.E-5$ usually produces at

Table 2.1. Statements to Output the Romberg Table

```
9062 PRINT"II AU(II)"
9064 PRINTØ; AU(1)
9282 PRINT
9284 PRINTØ; AU(I)
9352 PRINTII; AU(II)
```

least six decimal places of true value. The value for $\ln 2$ in Figure 2.5 is off in the ninth place.

2.3.4. Simpson's Integration Rule.

The order of truncation errors for repeated linear interpolation is shown in the top row of Figure 2.5. The $j=0$ column is the trapezoid rule, and the $j=1$ column happens to be the well-known Simpson rule. The other columns represent increasing orders of accuracy, but they do not coincide with other frequently used methods, such as Weddle's rule (see Ley, 1970, p. 246). Simpson's rule is to be applied in Section 4.6, where independently incremented function data will be integrated. Therefore, it will be convenient to obtain a closed formula for Simpson's rule, comparable to (2.19) for the trapezoid rule. Recall that the area in Figure 2.3 is an estimate of an integral of $f(x)$ from a to b; call it $T_{0,1}$. From Figures 2.3 and 2.5, $T_{0,0} = h(f_a + f_b)$. So, with $i=0$ in (2.26), $T_{1,0}$ becomes

$$T_{1,0} = \frac{h}{3}(f_a + 4f_{a+h} + f_b),\tag{2.28}$$

where

$$h \triangleq \frac{b-a}{2}.\tag{2.29}$$

This is Simpson's three-point rule.

The general formula for Simpson's rule can be recognized by first finding the five-point rule, namely, $T_{1,1}$, using $h_2 = h/2$. Extending the analysis evident in Figure 2.3, the five-point trapezoid rule is

$$T_{0,2} = \frac{h}{2}\left(\frac{1}{2}f_a + f_{a+h/2} + f_{a+h} + f_{a+3h/2} + \frac{1}{2}f_b\right).\tag{2.30}$$

Substituting (2.30) and (2.28) into (2.26), with $i=1$, yields

$$T_{1,1} = \frac{h/2}{3}(f_a + 4f_{a+h/2} + 2f_{a+h} + 4f_{a+3h/2} + f_b).\tag{2.31}$$

Deducing Simpson's rule from (2.29) and (2.31) and putting it into standard form, using variable t, we obtain

$$\int_{t_0}^{t_n} f(t)\,dt \simeq \frac{\Delta t}{3}(f_0 + 4f_1 + 2f_2 + \cdots + 4f_{n-1} + f_n),\tag{2.32}$$

where n is even and

$$\Delta t = \frac{t_n - t_0}{n}.\tag{2.33}$$

Recall that errors in the trapezoid rule were proportional to $(\Delta t)^2$. Simpson's rule errors are proportional to $(\Delta t)^4$.

2.3.5. Summary of Integration.

Romberg integration is based on numerically stable trapezoidal integration. The number of trapezoid sections necessary to produce an accurate estimate of the integral value is obtained by repeated linear extrapolations. The recursive algorithm is compact and requires only one small vector in memory. The repeated extrapolation to zero of the squared trapezoid width (h^2) is the classical method of Richardson. The first extrapolations of pairs of trapezoidal estimates produce Simpson's rule estimates. Subsequent higher-order estimates do not coincide with other well-known integration rules, but they are well behaved and can be calculated in an efficient manner.

2.4. Polynomial Minimax Approximation of Piecewise Linear Functions

There are many instances in engineering when a mathematical expression is required to represent a given graphic function. This is often required to be a real function of real variables. (Rational complex functions of a complex variable are the next topic). Many methods require the approximating function to pass through the given data points, perhaps matching slopes as well. Others require the function values at selected, independent-variable values to differ from given function values by a minimum aggregate error, e.g., least-squared errors (LSE). The minimax criterion specifies that the approximating polynomial minimize the maximum magnitude in the set of errors resulting from not passing through the given data points. In ideal cases, the minimax criterion results in "equal-ripple" behavior of a plotted error function (see Ralston, 1965, for more details).

This section describes a minimax approximation to a function that is given graphically by a series of connected line segments, i.e., a piecewise linear function. This function description is often convenient because involved integral relationships are simplified considerably. (This is also the case in Section 6.7). However, the approximation technique to be described in this section could easily be adapted to numerical integration of analytical functions by using the Romberg integration just described. Either way, this technique relates polynomials in x (power series) to weighted summations of classical Chebyshev polynomials of the first kind. These truly remarkable functions appear throughout mathematics, as well as in several places in this text. The basis of this method will be described, and a BASIC language program with two variations will be provided.

2.4.1. Chebyshev Functions of the First Kind. Chebyshev functions of the first kind can be expressed in polynomial or trigonometric forms. The former are given in Table 2.2, where $T_i(x)$ are Chebyshev functions of the first kind. Chebyshev functions of the first kind oscillate between amplitudes $+1$ and -1 in the interval $-1 \leqslant x \leqslant +1$, as shown in Figure 2.6. This equal ripple is crucial, but it is only one of many interesting characteristics.

The polynomials in Table 2.2 can be calculated from a single recursive

**Table 2.2.   Chebyshev Polynomials
of the First Kind**

$$T_1(x) = x$$
$$T_2(x) = 2x^2 - 1$$
$$T_3(x) = 4x^3 - 3x$$
$$T_4(x) = 8x^4 - 8x^2 + 1$$
$$T_5(x) = 16x^5 - 20x^3 + 5x$$
$$T_6(x) = 32x^6 - 48x^4 + 18x^2 - 1$$

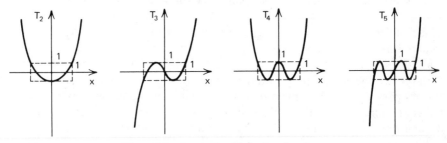

Figure 2.6.   Low-order Chebyshev functions.

formula, and they are equivalent to the following trigonometric expressions:

$$T_i(x) = \cos(i \cos^{-1} x), \tag{2.34}$$

where $-1 \leqslant x \leqslant +1$, or

$$T_i(x) = \cosh(i \cosh^{-1} x), \tag{2.35}$$

where $|x| > 1$. The interested reader can consult Guillemin (1957) for more details.

2.4.2.   Chebyshev Expansions.   Given some real function $g(x)$ over a range of real x values, it is desired to find some approximating polynomial in power series form:

$$f(x) = b_0 + b_1 x + b_2 x^2 + \cdots + b_n x^n. \tag{2.36}$$

If $g(x)$ is specified at a finite set of x_r values, then the objective is to minimize the error,

$$E = \max |g(x_r) - f(x_r)|, \tag{2.37}$$

for $r = 1, 2, \ldots, M$. The unknowns are the $n+1$ coefficients $b_0, b_1, \ldots, b_n$. For reasons of scale and formulation, it is necessary to work on the range $-1 \leqslant x \leqslant +1$. Suppose that the given function $g(x)$ is defined in the variable y

over the range $a \leqslant y \leqslant b$. Then the linear translation

$$x = \frac{2y - (b+a)}{b-a} \tag{2.38}$$

relates g(y) values to the chosen x range. Once an approximating function, f(x), is obtained, the inverse relation of (2.38) can be used to find f(y).

Example 2.4. Suppose that a given function, g(y), is defined by samples over the range from $a = 5$ to $b = 15$. Then (2.38) reads $x = (2y - 20)/10$, so that every value of g(y) can be considered as g(x). Once the approximating function f(x) in (2.36) is determined, then every value of f(x) can be considered as f(y) by the inverse relation $y = 5x + 10$.

Usually, the problem of finding the unknown coefficients in (2.36) is badly conditioned, i.e., the solution is difficult on a finite word-length computer. Therefore, another remarkable property of Chebyshev polynomials will be utilized by redefining the approximating function in terms of a weighted sum of Chebyshev polynomials:

$$f(x) = a_0 T_0 + a_1 T_1(x) + a_2 T_2(x) + \cdots + a_n T_n(x). \tag{2.39}$$

The concept of weighting is seen by referring to Table 2.2: there are some scaling coefficients (a_i) that multiply each Chebyshev polynomial, $T_i(x)$, so that their sum suitably approximates the given polynomial g(x) over the range $-1 \leqslant x \leqslant +1$. The set of unknowns that is chosen for solution contains all the a_i, $i = 0, 1, \ldots, n$, and this problem is almost always well conditioned. Once these are known, they can be used directly in the form (2.39), or the b_i coefficients in (2.36) can be found by collecting contributions to coefficients of like powers of x (see (2.36) and Table 2.2). There is a simple recursion to convert the a_i set to the b_i set (see Abramowitz and Stegun, 1972). The algorithm requires little coding. Determination of the a_i coefficients in (2.39) is classical (see Vlach, 1969, p. 176):

$$a_i = \frac{2}{\pi} \int_{-1}^{+1} \frac{g(x) T_i(x)}{\sqrt{1 - x^2}} dx, \qquad i = 0 \text{ to } n, \tag{2.40}$$

where $T_0 = \frac{1}{2}$ is defined for convenience. This integral can be evaluated numerically for any analytic function g(x), since the integrand is thus known (see Section 2.3). Even so, it may be suitable to approximate the given function or a given discrete data point set by connected line segments of arbitrary lengths. Then the integration in (2.40) is analytically simplified, as shown in the next section.

2.4.3. Expansion Coefficients for Piecewise Linear Functions. Integration of (2.40) can be avoided by assuming that the given function g(x) is composed of linear segments:

$$g(x) = kx + q, \tag{2.41}$$

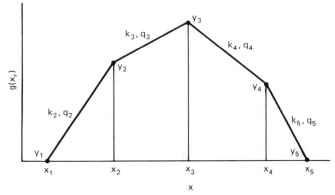

Figure 2.7.   Piecewise linear function to be approximated. [From Vlach, 1969.]

where $x_r \leqslant x \leqslant x_{r+1}$, as shown in Figure 2.7. It is also helpful to introduce a new variable, angle ϕ, that is clearly related to (2.34):

$$x = \cos\phi. \tag{2.42}$$

Then (2.41) and (2.40) yield

$$a_{i,r} = \frac{2}{\pi} \int_{x_r}^{x_{r+1}} \frac{(kx+q)T_i(x)}{\sqrt{1-x^2}} dx, \tag{2.43}$$

and (2.42) and (2.34) in (2.43) yield

$$a_{i,r} = \frac{-2}{\pi} \int_{\phi_r}^{\phi_{r+1}} (k\cos\phi\cos i\phi + q\cos i\phi)\, d\phi. \tag{2.44}$$

This integrates easily for $i=0$ and $i=1$. For $i=2,\ldots,n$ it is

$$a_{i,r} = \frac{-k}{\pi}\left[\frac{\sin(i+1)\phi}{i+1} + \frac{\sin(i-1)\phi}{i-1}\right]_{\phi_r}^{\phi_{r+1}} - \frac{2q}{\pi}\left[\frac{\sin i\phi}{i}\right]_{\phi_r}^{\phi_{r+1}}. \tag{2.45}$$

Each $a_{i,r}$ is the contribution to the (2.40) integral by each trapezoid between x_r and x_{r+1} in Figure 2.7. The final expression for the a_i coefficients in (2.39) applicable to piecewise linear $g(x)$ is

$$a_i = \sum_{r=1}^{M-1} a_{i,r}, \tag{2.46}$$

where M is the number of given $g(y)$ data pairs.

2.4.4.   A Minimax Approximation Program.   Program B2-4 in Appendix B performs the preceding calculations from given sets of data in the range $-1 \leqslant x \leqslant +1$. The end points at $x=-1$ and $x=+1$ must be included. The output first shows the a_i weighting coefficients for the Chebyshev functions in (2.39) and then the b_i power series coefficients of x in (2.36) for various values of degree n.

Table 2.3. The Data Pairs Defining a Piecewise Linear Function

x	−1	0	+1
y	0	0.6	0.6

Table 2.4. Chebyshev and Power Series Coefficients for $n = 4$

$a_0 =$	0.409014	$b_0 =$	0.561803
$a_1 =$	0.30	$b_1 =$	0.30
$a_2 =$	−0.127324	$b_2 =$	−0.458366
$a_3 =$	0.	$b_3 =$	0.
$a_4 =$	0.0254648	$b_4 =$	0.203718

From Vlach, 1967. ©1967 IEEE.

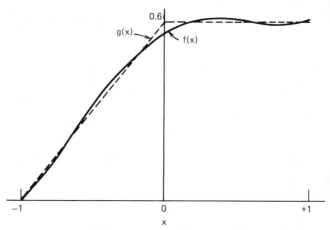

Figure 2.8. Degree-4 approximation to a three-point function. [From Vlach, 1967. © 1967 IEEE.]

Example 2.5.   Consider the data given in Table 2.3. Input these into Program B2-4 and ask for polynomial degrees 4 through 6. The program will display all seven coefficients for the Chebyshev expansion first, then the degree-4 power series coefficients. These are shown in Table 2.4, and the graph of either of these representations is plotted in Figure 2.8. Note that the program also lists the approximating function values and the errors at each sample point in x.

Program B2-4 documentation also indicates two variations on the coding, also given by Vlach (1969). These make the data input and calculations more efficient when strictly even or odd functions of x are approximated. The data samples must be on the closed (end points included) interval $0 \leqslant x \leqslant +1$. Run the modified programs with three data pairs to approximate a constant and a 45-degree line for the even and odd cases, respectively.

2.4.5. Piecewise Linear Function Approximation Summary. Arbitrary real functions of real variables can be expressed as a linear weighted sum of Chebyshev polynomials of the first kind. The coefficients are determined by an integral formula, but for piecewise linear functions the Chebyshev coefficients are found by an algebraic formula. Chebyshev polynomials have many amazing characteristics, one being the minimax error property (see Hamming, 1973, for a commentary).

A linear (weighted) summation of Chebyshev polynomials is easily restated as a power series polynomial in real, independent variable x. Since these approximations are found on the normalized interval $-1 \leqslant x \leqslant +1$, a simple linear mapping is required in the usual case where given functions are defined otherwise. It is not commonly observed, but the approximation described in this section has close connections with Fourier series approximations, which are more familiar to electrical engineers. Many other closed-form approximations are related to the method described (see Ralston, 1965, p. 286).

2.5. Rational Polynomial LSE Approximation of Complex Functions

There are many applications in electrical engineering for complex curve fitting, i.e., finding a complex function of a complex variable such as frequency. Examples include modeling an antenna impedance versus frequency for interpolation or for synthesis of an equivalent network; the latter might be used as a "dummy load" in place of the real antenna. Another example is approximation of a higher-order-system transfer function by a lower-order one over a limited frequency range.

A rational polynomial in complex (Laplace) frequency s has more approximating power than an ordinary polynomial in s and can be an intermediate step to synthesizing an equivalent network. Such rational polynomials take the form

$$Z(s) = \frac{a_0 + a_1 s + a_2 s^2 + \cdots + a_p s^p}{1 + b_1 s + b_2 s^2 + \cdots + b_n s^n}, \qquad (2.47)$$

where $s = j\omega$ will be used interchangeably. The relationship between functions of complex s and real ω is rooted in the concept of analytic continuation, which is described in most network synthesis textbooks (see Van Valkenburg, 1960). Although the method to be presented will generally assume that $s = j\omega$, it also applies to the less general real-variable case $s^2 = -\omega^2$ for approximating even functions.

The kind of problem to be solved is shown in Figure 2.9; it was given by E. C. Levy (1959), who published the algorithm to be described in this section.

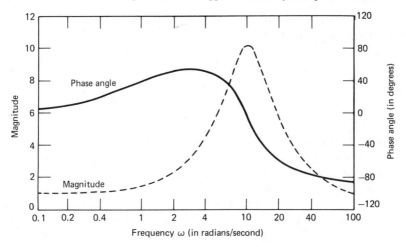

Frequency response characteristics of a dynamic system with
a transfer function given as

$$F(j\omega) = \frac{1+j\omega}{1+2(0.5)(j\omega/10)+(j\omega/10)^2} .$$

k	ω_k	Magni-tude	Phase Angle	R_k	I_k
0	0.0	1.00	0	1.00	0.000
1	0.1	1.00	5	1.00	0.090
2	0.2	1.02	10	1.00	0.177
3	0.5	1.12	24	1.02	0.450
4	0.7	1.24	31	1.05	0.630
5	1.0	1.44	39	1.10	0.900
6	2.0	2.27	51.5	1.41	1.78
7	4.0	4.44	50.5	2.82	3.42
8	7.0	8.17	28	7.23	3.82
9	10.0	10.05	− 6	10.00	− 1.00
10	20.0	5.56	−59	2.85	−4.77
11	40.0	2.55	−76	0.602	−2.51
12	70.0	1.45	−82	0.188	−1.43
13	100.0	1.00	−84	0.091	−1.01

$R_k = $ (Magnitude at ω_k) × cos(phase angle at ω_k)
$I_k = $ (Magnitude at ω_k) × sin(phase angle at ω_k).

Figure 2.9. Frequency response and discrete data for a second-degree system. [From Levy, etc., *IRE Trans. Auto. Control*, Vol. AC-4, No. 1, p. 41, May 1959. © 1959 IRE (now IEEE).]

The table of values is the given data. Although measured data are often inaccurate (noisy), this particular data set was computed from the $F(\omega)$ values shown in Figure 2.9 for purposes of illustration. The graph shows the magnitude and angle components of the function. The technique will be to approximate only the magnitude function by finding the unknown coefficients of (2.47).

The error criterion will be the weighted least-squared-error (LSE) function over the frequency samples $0, 1, \ldots, m$:

$$E = \sum_{k=0}^{m} \left[W_k |\varepsilon_k| \right]^2, \tag{2.48}$$

where

$$\varepsilon_k = F(\omega_k) - Z(\omega_k). \tag{2.49}$$

The complex numbers $F(\omega_k)$ are the given data to be fitted, i.e., the target function. The complex approximating function $Z(\omega_k)$ is given in (2.47). The W_k values in (2.48) are the weighting values at each frequency ω_k. The necessary condition for a minimum value of E (generally not zero) is that the partial derivatives of E with respect to the coefficients $a_0, a_1, \ldots, a_p, b_1, b_2, \ldots, b_n$ in (2.47) be equal to zero. A set of simultaneous nonlinear equations will result if the formulation in (2.48) and (2.49) is used with independent weights W_k. The equations are badly conditioned and extremely difficult to solve. Gradient optimizers (Chapter Five) usually are not successful in finding a solution (according to Jong and Shanmugam, 1977).

E. C. Levy's method will be described. It employs a weighted LSE objective function similar to (2.48), except that the weights are dependent functions. This produces a system of simultaneous linear equations that are readily solved by the Gauss–Jordan program described in Section 2.2. The derivation will be outlined, the matrix of linear equation coefficients will be tabulated, and a brief BASIC language program will be furnished to calculate the four kinds of matrix coefficients. An example will be provided here, and others will be given in Section 6.7.

2.5.1. The Basis of Levy's Complex Curve-Fitting Method.
The definition of $Z(s)$ in (2.47) is expanded, with $s = j\omega$, to produce a set of linear equations:

$$Z(s) = \frac{\left(a_0 - a_2\omega^2 + a_4\omega^4 + \cdots\right) + j\omega\left(a_1 - a_3\omega^2 + a_5\omega^4 + \cdots\right)}{\left(1 - b_2\omega^2 + b_4\omega^4 + \cdots\right) + j\omega\left(b_1 - b_3\omega^2 + b_5\omega^4 + \cdots\right)}, \tag{2.50}$$

which is further defined by

$$Z(s) \overset{\Delta}{=} \frac{\alpha + j\omega\beta}{\sigma + j\omega\tau} = \frac{N(\omega)}{D(\omega)}. \tag{2.51}$$

The real terms in the numerator and denominator of (2.50) are even functions of frequency, and the imaginary terms are odd. Quantities in parentheses are equated by relative position with the variables appearing in (2.51), where the numerator and denominator functions are also identified.

With these definitions, the unweighted error function in (2.49) becomes

$$\varepsilon(\omega) = F(\omega) - \frac{N(\omega)}{D(\omega)}. \tag{2.52}$$

When (2.52) is multiplied through by $D(\omega)$, the squared magnitude is

$$|D(\omega)\varepsilon(\omega)|^2 = |D(\omega)F(\omega) - N(\omega)|^2 \qquad (2.53)$$

It is important for the reader to understand the following point in order to apply this scheme to practical situations. Compare (2.53) to (2.48) and (2.49); (2.53) shows that <u>the "weighting" at any frequency is the magnitude of the approximating function's denominator</u>. Suppose that a rational approximating function has been found; ordinarily its denominator is large when its value is small. The large denominator means that the function was most heavily weighted in the frequency "stopband." This weighting can be offset by taking more samples at "passband" frequencies than elsewhere, which is the price paid for making the method tractable.

The equations to be solved are found by first extending the definition of the target function $F(\omega)$ appearing in (2.49) and (2.52):

$$F(\omega) = R(\omega) + jI(\omega). \qquad (2.54)$$

Then (2.48), (2.51), and (2.54) yield

$$E = \sum_{k=0}^{m} \left[(R_k\sigma_k - \omega_k\tau_kI_k - \alpha_k)^2 + (\omega_k\tau_kR_k + \sigma_kI_k - \omega_k\beta_k)^2 \right]. \qquad (2.55)$$

So the necessary conditions for minimum E,

$$\frac{\partial E}{\partial a_i} = 0 = \frac{\partial E}{\partial b_j} \qquad \text{for all i and j,} \qquad (2.56)$$

can be written directly from (2.55) using the relations defined by (2.50) and (2.51). A large amount of ordinary calculus and algebra is involved in reducing the resulting linear equations to the compact form given by Levy (1959). The resulting matrix equations are given in Figure 2.10 in terms of the coefficients defined by (2.57).

$$\lambda_h = \sum_{k=0}^{m} \omega_k^h, \qquad S_h = \sum_{k=0}^{m} \omega_k^h R_k,$$

$$\qquad (2.57)$$

$$T_h = \sum_{k=0}^{m} \omega_k^h I_k, \qquad U_h = \sum_{k=0}^{m} \omega_k^h (R_k^2 + I_k^2).$$

2.5.2. Complex Curve-Fitting Procedure.

The basis of the procedure appearing in Figure 2.10 and (2.57) may seem complicated at first glance. This is remedied by a brief explanation and an example for the problem shown in Figure 2.9. The equations in (2.57) have been placed in BASIC language Program B2-5 in Appendix B. The $F(\omega)$ real and imaginary components are defined as $R(\omega)$ and $I(\omega)$ in (2.54); they are given versus frequency ω in the data shown in Figure 2.9 and are used in (2.57). Program B2-5 reduces this calculation to entering the $m+1$ data triples ω_k, R_k, and I_k. The matrix equations in Figure 2.10 have as unknowns the set of a and b coefficients that

The upper-left and lower-right submatrices must be square.

$$
\begin{bmatrix}
S_0 & T_1 & S_2 & T_3 & S_4 & T_5 & \cdots & \big| & 0 & U_2 & 0 & U_4 & 0 & U_6 & 0 & \cdots
\end{bmatrix}
$$

$$=$$

$$
\begin{bmatrix}
a_0 & a_1 & a_2 & a_3 & a_4 & a_5 & \cdots & \big| & b_1 & b_2 & b_3 & b_4 & b_5 & b_6 & b_7 & \cdots
\end{bmatrix}
$$

$$
\begin{bmatrix}
T_1 & S_2 & -T_3 & -S_4 & T_5 & -S_6 & \cdots & \big| & U_2 & 0 & U_4 & 0 & U_6 & 0 & \cdots\\
S_2 & T_3 & S_4 & T_5 & S_6 & T_7 & \cdots & \big| & 0 & U_4 & 0 & U_6 & 0 & U_8 & \cdots\\
-T_3 & S_4 & -T_5 & -S_6 & -T_7 & S_8 & \cdots & \big| & U_4 & 0 & U_6 & 0 & U_8 & 0 & \cdots\\
-S_4 & -T_5 & -S_6 & -T_7 & -S_8 & -T_9 & \cdots & \big| & 0 & -U_6 & 0 & -U_8 & 0 & -U_{10} & \cdots\\
T_5 & -S_6 & T_7 & S_8 & T_9 & S_{10} & \cdots & \big| & U_6 & 0 & U_8 & 0 & U_{10} & 0 & \cdots\\
-S_6 & T_7 & -S_8 & -T_9 & -S_{10} & T_{11} & \cdots & \big| & 0 & -U_8 & 0 & -U_{10} & 0 & -U_{12} & \cdots\\
\vdots & \vdots & \vdots & \vdots & \vdots & \vdots & & \big| & \vdots & \vdots & \vdots & \vdots & \vdots & \vdots\\
\hline
\lambda_0 & 0 & \lambda_2 & 0 & \lambda_4 & 0 & \lambda_6 & \cdots & \big| & T_1 & S_2 & T_3 & S_4 & T_5 & S_6 & T_7 & \cdots\\
0 & \lambda_2 & 0 & \lambda_4 & 0 & \lambda_6 & & \cdots & \big| & -S_2 & T_3 & -S_4 & T_5 & -S_6 & T_7 & \cdots\\
-\lambda_2 & 0 & -\lambda_4 & 0 & -\lambda_6 & 0 & & \cdots & \big| & T_3 & -S_4 & T_5 & -S_6 & T_7 & -S_8 & \cdots\\
0 & -\lambda_4 & 0 & -\lambda_6 & 0 & -\lambda_8 & & \cdots & \big| & S_4 & T_5 & S_6 & T_7 & S_8 & -T_9 & \cdots\\
\lambda_4 & 0 & \lambda_6 & 0 & \lambda_8 & 0 & \lambda_{10} & \cdots & \big| & T_5 & S_6 & T_7 & S_8 & T_9 & S_{10} & \cdots\\
0 & -\lambda_4 & -\lambda_6 & -\lambda_8 & & & & \cdots & \big| & -S_6 & T_7 & -S_8 & T_9 & -S_{10} & T_{11} & -S_{12} & \cdots\\
\vdots & \vdots & \vdots & \vdots & \vdots & & & & \big| & \vdots & \vdots & \vdots & \vdots & \vdots & \vdots
\end{bmatrix}
$$

Note: The upper-left and lower-right submatrices must be square.

Figure 2.10. Levy's matrix of linear equations.

28

determine (2.47). How many of each are contemplated becomes the basis for partitioning and selecting the equations in Figure 2.10. This is best shown by an example.

Example 2.6. Suppose that the given sampled data are those in Figure 2.9 and the rational polynomial required to fit these data is

$$Z(s) = \frac{a_0 + a_1 s + a_2 s^2}{1 + b_1 s + b_2 s^2}. \tag{2.58}$$

There are five variables: a_0, a_1, a_2, b_1, and b_2. The vector (column) of variables in Figure 2.10 appears just to the left of the equality sign. The horizontal dashed partition line should occur just below a_2, and the bottom of the matrix just below b_2. The vertical dashed partition line is placed so that the upper-left submatrix is square. The set of linear equations appropriate for (2.58) is thus

$$
\begin{bmatrix}
\lambda_0 & 0 & -\lambda_2 & T_1 & S_2 \\
0 & \lambda_2 & 0 & -S_2 & T_3 \\
\lambda_2 & 0 & -\lambda_4 & T_3 & S_4 \\
\hline
T_1 & -S_2 & -T_3 & U_2 & 0 \\
S_2 & T_3 & -S_4 & 0 & U_4
\end{bmatrix}
\begin{bmatrix}
a_0 \\ a_1 \\ a_2 \\ b_1 \\ b_2
\end{bmatrix}
=
\begin{bmatrix}
S_0 \\ T_1 \\ S_2 \\ 0 \\ U_2
\end{bmatrix}. \tag{2.59}
$$

Now Program B2-5 is used with the 14 data triples from Figure 2.9; the h subscript will vary from 0 to 4, the limit being obtained by inspection of entries required in (2.59). In this case, the program output is shown in Table 2.5. The system in (2.59) is then solved by Gauss–Jordan Program B2-1, and the resulting a and b coefficients are also listed in Table 2.5 for use in (2.58). Of course, these rational-polynomial coefficients agree fairly well with those in Figure 2.9 because the problem was constructed for confirmation purposes.

Table 2.5. Example 2.6 Levy Coefficients

h	λ_h	S_h	T_h	U_h
0	14.	31.361	0.5470	241.1188
1	255.5	270.695	−361.3096	2703.4268
2	17070.79	5341.3495	−22880.6508	57201.6953
3	1416416.48	229485.25	−1698745.45	2540544.42
4	126742674	15729105	−142523023	175956492

$a_0 = 0.9993 \qquad a_1 = 1.0086 \qquad a_2 = -1.59E-5$
$b_0 = 1 \qquad b_1 = 0.10097 \qquad b_2 = 0.0100$

2.5.3.   Summary of Complex Curve Fitting by Rational Polynomials.   Levy's method for fitting complex data at sampled frequencies in the weighted least-squared-error sense is straightforward. The weighting versus frequency is inversely proportional to the value of the rational polynomial thus found. Since the polynomial should roughly correspond to data values to be of any use, nonuniform samples versus frequency should produce emphasis on the frequencies where the magnitude of the complex data is least. For example, if a lowpass function is to be fitted over several decades, then the fit to the passband (lower) frequency data may be poor unless samples are spaced more closely in this frequency range. Proposed iterative schemes have been based on a sequence of solutions similar to those presented here; they tend to converge to a situation equivalent to uniform weighting (see Jong and Shanmugam, 1977, and Sanathanan and Koerner, 1963). However, equal weighting may still require some experimenting. Thus the built-in inverse weighting does not seem too severe a limitation.

The method requires the user to input real and imaginary data parts, with the associated frequency, into Program B2-4 to obtain coefficients for a system of linear equations. The system's matrix elements are partitioned from a general matrix format (Figure 2.10) according to the approximating rational polynomial's numerator and denominator degrees. The system of linear equations is then solved by Gauss–Jordan Program B2-1 or by any other program that solves linear systems of real equations. This method will play an important role in Carlin's broadband impedance-matching technique in Section 6.7.

Problems

2.1.   If

$$\rho(Z) = \frac{Z - Z_c}{Z + Z_c^*},$$

where $Z_c = 2 + j3$, find ρ when $Z = 3 - j5$.

2.2.   Given that $V = V_r + jV_i$ and $I = I_r + jI_i$, show that

$$Re(VI^*) = Re(V^*I) = V_r I_r + V_i I_i.$$

2.3.   Show that $|Z|^2 = ZZ^*$.

2.4.   Show that $2 Re(Z) = Z + Z^*$.

2.5.   If

$$a = \frac{V + ZI}{2\sqrt{R}}, \quad b = \frac{V - Z^*I}{2\sqrt{R}}, \quad \text{and} \quad Z = R + jX,$$

show that $|a|^2 - |b|^2 = Re(IV^*)$.

2.6. If

$$w = \frac{a_1 Z + a_2}{a_3 Z + 1},$$

find the derivative $w' = dw/dZ$.

2.7. If $z = x + jy$ and $z^p = X_p + jY_p$,

(a) Calculate $z^2 = (x + jy)(x + jy)$.

(b) Find X_0, Y_0, X_1, and Y_1.

(c) Given

$$X_k = 2xX_{k-1} - (x^2 + y^2)X_{k-2},$$
$$Y_k = 2xY_{k-1} - (x^2 + y^2)Y_{k-2},$$

find X_2 and Y_2.

(d) Do (a) for $p = 3$ and (c) for $k = 3$.

2.8. Given

$$\begin{bmatrix} -2 & -1 & 1 \\ 1 & 1 & 1 \\ 3 & 1 & -1 \end{bmatrix} \begin{bmatrix} 1 \\ 2 \\ 3 \end{bmatrix} = \begin{bmatrix} b_{14} \\ b_{24} \\ b_{34} \end{bmatrix},$$

find the values of b_{14}, b_{24}, and b_{34}.

2.9. Solve the following system for x_1 and x_2 by the Gauss–Jordan method, showing the sequence of augmented matrices.

$$4x_1 + 7x_2 = 40,$$
$$6x_1 + 3x_2 = 30.$$

2.10. Given

$$\begin{bmatrix} 2 - j3 & -4 + j7 \\ 6 + j1 & 8 - j10 \end{bmatrix} \begin{bmatrix} 1 + j3 \\ -3 + j5 \end{bmatrix} = \begin{bmatrix} a_{13} + jb_{13} \\ a_{23} + jb_{23} \end{bmatrix},$$

find the values of a_{13}, b_{13}, a_{23}, and b_{23}.

2.11. For the matrix equation

$$\begin{bmatrix} 1 + j2 & 2 + j3 & 3 - j4 \\ -3 + j4 & -3 - j1 & 6 + j9 \\ -2 - j3 & -3 + j2 & 7 + j5 \end{bmatrix} \begin{bmatrix} 1 - j3 \\ 6 - j7 \\ 2 + j3 \end{bmatrix} = \begin{bmatrix} z_1 \\ z_2 \\ z_3 \end{bmatrix},$$

find $z = (z_1, z_2, z_3)^T$ numerically.

2.12. Given

$$\begin{bmatrix} 0.5 \angle 60 & 1.1 \angle 250 \\ 0.3 \angle 0 & 0.9 \angle -60 \end{bmatrix} \begin{bmatrix} 2 \angle 40 \\ 0.2 \angle -10 \end{bmatrix} = \begin{bmatrix} m_1 \angle \theta_1 \\ m_2 \angle \theta_2 \end{bmatrix},$$

find the values of m_1, θ_1, m_2, and θ_2.

2.13. Write the trapezoidal integration formulas for $T_{0,0}$, $T_{0,1}$, and $T_{0,2}$. Then use extrapolation formula (2.27) to find Simpson rules $T_{1,0}$ and $T_{1,1}$.

2.14. Calculate

$$\int_3^5 xe^x\,dx$$

numerically, using:
(a) Simpson's rule with five samples.
(b) The trapezoidal rule with three evenly spaced samples.
(c) The trapezoidal rule with five evenly spaced samples.
(d) The Romberg extrapolation to the limit, using the preceding results in (b) and (c) above.

2.15. Evaluate Chebyshev polynomials $T_4(0.8)$ and $T_4(3.1)$ by:
(a) Horner's nesting method

$$T(x)=a_0+x\{a_1+x[a_2+x(\dots)]\}.$$

(b) Trigonometric or hyperbolic identities.
(c) Numerical recursion

$$T_i(x)=2xT_{i-1}(x)-T_{i-2}(x).$$

2.16. Chebyshev polynomials of the *second* kind are defined by

$$P_k(y)=yP_{k-1}-P_{k-2}.$$

where $P_1=1$ and $P_2=y$. Find P_3, P_4, P_5, and P_6 numerically for $y=1.5$.

2.17. Find an expression for x at the n–1 extreme values of

$$T_n(x)=\cos(n\cos^{-1}x).$$

2.18. Write the power series equivalent to

$$P(x)=T_1(x)+4T_2(x)+2T_3(x)-T_4(x)$$
$$=a_0+a_1x+a_2x^2+a_3x^3+a_4x^4;$$

in other words, find the a_i coefficients. See Table 2.2 for T_i.

2.19. Suppose that q(y) is defined on $-7\leqslant y\leqslant 25$. If (q)x is defined on $-1\leqslant x\leqslant 1$, find the value of y corresponding to $x=0.5$ using the linear mapping in (2.38).

2.20. The three points

x	-1	-0.5	$+1$
g(x)	0	1	0

define a piecewise linear function that can be fitted in the minimax sense by the sum of first-kind Chebyshev polynomials $T_i(x)$:

$$f(x) = a_0 T_0 + a_1 T_1 + a_2 T_2 + a_3 T_3 + a_4 T_4.$$

Find the value of a_3.

2.21. Given the function of two variables

$$F(x_1, x_2) = 612 - 60x_1 - 132x_2 + 13x_1^2 - 10x_1 x_2 + 13x_2^2,$$

find the values of x_1 and x_2 at the extreme value by equating the first partial derivatives to zero.

2.22. Given the fitting-function form

$$Z(s) = \frac{a_0 + a_1 s + a_2 s^2}{1 + b_1 s + b_2 s^2 + b_3 s^3},$$

write the third equation from the appropriate set from Figure 2.10.

2.23. Discrete, complex numerical data can be fitted versus frequency by a rational polynomial. For the polynomial

$$Z(s) = \frac{a_0 + a_1 s + a_2 s^2 + a_3 s^3 + a_4 s^4}{1 + b_1 s + b_2 s^2 + b_3 s^3 + b_4 s^4 + b_5 s^5},$$

write the first and last linear equations that result from Levy's method in terms of constants λ_i, S_i, T_i, and U_i for $i = 0, 1, \ldots$.

Chapter Three _____

Some Tools and Examples of Filter Synthesis

This chapter provides the necessary computing aids for manipulating polynomials in Laplace complex frequency s. These programs are explained and then applied to a meaningful sequence of modern network synthesis steps by way of example. The result is a sense of confidence, ease, and insight that is difficult to obtain by a purely academic approach to either computing methods or synthesis.

A reliable root finder based on useful, important principles begins the chapter. The synthesis process involves assembly as well as disassembly (factoring) of polynomials; so, programs that form polynomials from factors and by the polynomial four functions (add, subtract, multiply, and divide) are considered next. Also, programs for continued and partial fraction expansion are presented with some applications.

By the end of Chapter Three, those who have used the programs, tried the examples, and followed the fairly routine mathematical steps should be able to appreciate more detailed explanations of synthesis methods, for example, those of Temes and Mitra (1973).

3.1. Complex Zeros of Complex Polynomials

Finding complex zeros of polynomials ranks, along with solution of linear systems of equations, as a fundamental tool in engineering analysis. Textbooks usually give examples that factor by the quadratic formula or inspection, leaving the serious student to do his own numerical root finding by some system routine on a large and perhaps inconvenient computer. Moore (1967) described a conceptually interesting root finder that works well and fits easily into small computers. This is the time to eliminate the frustration or missed opportunity that yesterday's student suffered upon encountering the instruction, "In general, this will have to be done numerically."

The problem is to find the n values of z that make the following polynomial equal to zero:

$$f(z) = \sum_{k=0}^{n} (a_k + jb_k)z^k = 0, \tag{3.1}$$

The coefficients of this summation, a power series in z, may be complex. Certainly the independent variable z and the roots in z may be complex, with rectangular components

$$z = x + jy. \tag{3.2}$$

Clearly, given a value of z, the polynomial may have a complex value with components

$$f(z) = u + jv. \tag{3.3}$$

To be explicit, the problem is to find the roots z_i so that the product form of the summation in (3.1) is

$$f(z) = (a_n + jb_n)(z - z_1)(z - z_2)\ldots(z - z_n). \tag{3.4}$$

Polynomials in modern network synthesis commonly have only real coefficients, a condition that results in roots being either real or in conjugate complex pairs. Moore's root finder was formulated for the more general case having complex coefficients, as in (3.1), which occurs, for example, in solving the characteristic equations associated with complex matrices. The real-coefficient polynomial will be solved more than twice as fast if the suggestions that follow are incorporated. However, the more general case is retained for instructional and practical reasons. Moore's method employs derivatives of the polynomial. This causes some multiple-root inaccuracy not found in nonderivative methods, such as the popular method of Muller (1956). There are also root finders that utilize synthetic division in special ways, so that convergence depends upon initial conditions (e.g., the Newton–Raphson, Lin, and Bairstow methods). Some other methods that guarantee convergence are not straightforward and are often slow, for example, the Lehmer–Schur and Graeffe methods. See Ralston (1965) for descriptions of these six other root-finding techniques.

There are two intriguing ideas central to Moore's method. The first is the Cauchy–Riemann principle that defines the derivative of an analytic (regular) complex function in terms of the partial derivatives of u and v (3.3) with respect to x and y (3.2). Any student of complex-variable theory or its application will find this worth knowing. The second idea is the Mitrovic method for evaluation of a polynomial and its derivatives. This is a much more efficient means than the better-known "nesting" programming technique, especially on computers where polar complex arithmetic is either slow or nonexistent.

Topics in this section include Moore's search algorithm, synthetic division for linear and quadratic factors, the Mitrovic evaluation method, BASIC program ROOTS, and polynomial scaling.

3.1.1. Moore's Root Finder.

Moore's root-finder method adjusts the components of $z = x + jy$ until the squared magnitude of $f = u + jv$ is zero at $z = z_i$. The root factor $(z - z_i)$ is then removed from the polynomial by synthetic division, and the process is repeated on the remainder polynomial. The adjustments on x and y are made by the Newton–Raphson method. The method now will be developed in detail.

The error function to be minimized over the (x, y) space is

$$F = |f|^2 = u^2 + v^2, \tag{3.5}$$

as illustrated in Figure 3.1. The positive, real function F in (3.5) must have exactly n zeros, as does the given complex function f in (3.1) or (3.4). It is well known that u and v are well-behaved functions of x and y; i.e., they are continuous, and their derivatives exist. In such cases, the Cauchy–Riemann condition defines $f'(z)$, the derivative of f with respect to z:

$$f'(z) \overset{\Delta}{=} \frac{df}{dz} = \frac{\partial u}{\partial x} + j\frac{\partial v}{\partial x} = \frac{\partial v}{\partial y} - j\frac{\partial u}{\partial y}. \tag{3.6}$$

Furthermore, (3.6) defines a relationship between real parts and between imaginary parts; consequently, knowledge of partial derivatives with respect to x will furnish partial derivatives with respect to y without further work. Proceeding, the partial derivative of F with respect to x is written by inspection of (3.5):

$$\frac{\partial F}{\partial x} = 2\left(u\frac{\partial u}{\partial x} + v\frac{\partial v}{\partial x}\right). \tag{3.7}$$

The partial derivative of F with respect to y is similarly written, but the

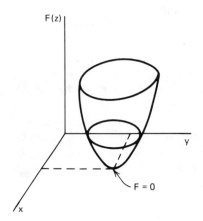

Figure 3.1. Polynomial error surface near a root.

equalities available from (3.6) enable an expression again using only partial derivatives with respect to x:

$$\frac{\partial F}{\partial y} = 2\left(-u\frac{\partial v}{\partial x} + v\frac{\partial u}{\partial x}\right). \tag{3.8}$$

The slopes of the error surface in the x and y directions are now available to guide the search for one of the zero-function values illustrated in Figure 3.1.

Suppose that the search is at some particular coordinate intersection in the $x-y$ plane. The adjustment of each of these values is

$$\Delta x = -0.5\frac{\partial F/\partial x}{|f'|^2}, \tag{3.9}$$

$$\Delta y = -0.5\frac{\partial F/\partial y}{|f'|^2}, \tag{3.10}$$

where the steps are damped by $\frac{1}{2}$ and scaled by the squared length (norm) of the gradient

$$|f'|^2 = \left(\frac{\partial u}{\partial x}\right)^2 + \left(\frac{\partial v}{\partial x}\right)^2. \tag{3.11}$$

The gradient is the vector that points in the uphill direction of the steepest slope, and its components are just the partial derivatives in (3.7) and (3.8). The square root of (3.11), the gradient's magnitude, expresses the steepness of the slope. These are matters that will be considered in more detail in Chapter Five. The Newton–Raphson search scheme for several variables also will be derived there. It happens that the Moore search steps defined in (3.9)–(3.11) are exactly the steps in the Newton–Raphson method, which converge very rapidly. If these steps are too large, so that the new value of F exceeds the last one, then the step sizes are reduced by a factor of 4 until a decrease in function value is obtained. The details will be considered in Section 3.1.4.

3.1.2. Synthetic Division.   Once a root is found by the search procedure just described, then that factor is removed by synthetic division. Without loss of generality, real coefficients will be used in a third-degree polynomial for illustration of the synthetic division process. Consider the polynomial

$$f(z) = a_0 + a_1 z + a_2 z^2 + a_3 z^3 \tag{3.12}$$

and its equivalent product form

$$f(z) = (z - z_i)(c_0 + c_1 z + c_2 z^2), \tag{3.13}$$

where z_i is the root. The unknowns are the coefficients c_k, where $k = 0$, 1, and 2 in (3.13), since the right-hand term is the next polynomial to be used in the root search algorithm of Section 3.1.1. Ralston (1965, p. 371) shows that the recursion is

$$c_k = a_{k+1} + z_i c_{k+1}, \quad \begin{cases} k = n-1, \ldots, 0 \\ c_n = 0. \end{cases} \tag{3.14}$$

Example 3.1. Consider the factors

$$f(z) = (z+2)(3+2z+z^2), \tag{3.15}$$

which are equal to the polynomial

$$f(z) = 6 + 7z + 4z^2 + z^3. \tag{3.16}$$

The algorithm in (3.14) will be used to find the quadratic factor in (3.15), which is the unknown in real problems. Proceeding with (3.14):

$$
\begin{aligned}
k=2: & \quad c_2 = 1 + (-2) \times 0 = 1, \\
k=1: & \quad c_1 = 4 + (-2) \times 1 = 2, \\
k=0: & \quad c_0 = 7 + (-2) \times 2 = 3.
\end{aligned} \tag{3.17}
$$

There is no change in the algebra when coefficients a_k in (3.12) and c_k in (3.13) are complex; complex arithmetic is employed in (3.14) instead of the real arithmetic previously indicated. However, when all b_k in (3.1) are zero, so that coefficients a_i in (3.12) are known to be real, then there may be one or more real roots and any complex roots will occur in conjugate pairs. This will be the case in ordinary filter synthesis, so that computing effort can be reduced substantially in both synthetic division and evaluation of the polynomial and its derivatives. Assuming real coefficients, real roots are removed, as in (3.14), using only real arithmetic. When a root's imaginary part is not essentially zero, then the quadratic factor containing the root and its conjugate is removed.

Consider the identity

$$(z - z_i)(z - z_i^*) = z^2 + p_i z + q_i, \tag{3.18}$$

where $p_i = -2x_i$, $q_i = x_i^2 + y_i^2$, and $z^* = x - jy$ (see (3.2)). Ralston (1965, p. 372) described removal of quadratic factors; no complex arithmetic is involved. Without loss of generality, consider the polynomial

$$f(z) = a_0 + a_1 z + a_2 z^2 + a_3 z^3 + a_4 z^4 + a_5 z^5 \tag{3.19}$$

and its equivalent product form

$$f(z) = (z^2 + p_i z + q_i)(c_0 + c_1 z + c_2 z^2 + c_3 z^3), \tag{3.20}$$

where the quadratic term corresponds to (3.18) with the one discovered root z_i. The recursion is

$$c_k = a_{k+2} - p_i c_{k+1} - q_i c_{k+2}, \qquad \begin{cases} k = n-2, \dots, 0, \\ c_n = c_{n-1} = 0. \end{cases} \tag{3.21}$$

Example 3.2. Consider the factors

$$f(z) = (z^2 + 3z + 2)(60 + 47z + 12z^2 + z^3), \tag{3.22}$$

which are equal to the polynomial

$$f(z) = 120 + 274z + 225z^2 + 85z^3 + 15z^4 + z^5. \tag{3.23}$$

The algorithm in (3.21) will be used to find the cubic factor in (3.22), which is the unknown in actual problems. Proceeding with (3.21):

$$
\begin{aligned}
k=3: \quad & c_3 = 1 - (3) \times 0 - (2) \times 0 = 1, \\
k=2: \quad & c_2 = 15 - (3) \times 1 - (2) \times 0 = 12, \\
k=1: \quad & c_1 = 85 - (3) \times 12 - (2) \times 1 = 47, \\
k=0: \quad & c_0 = 225 - (3) \times 47 - (2) \times 12 = 60.
\end{aligned}
\tag{3.24}
$$

So far, a means to find and remove roots of a defined polynomial has been described. It has been assumed that, given a trial value of the independent variable $z = x + jy$, the polynomial's real and imaginary parts (u and v) and their partial derivatives with respect to x can be evaluated.

3.1.3. Efficient Evaluation of a Polynomial and Its Derivatives.

Given a value of z, many programmers are aware that evaluation of (3.16) is better accomplished by the nesting

$$f(z) = 6 + z[7 + z(4 + z)]. \tag{3.25}$$

However, the indicated multiplications are neither convenient nor fast on most small computers, which either lack polar complex arithmetic or execute slowly in that mode. Kokotovic and Siljak (1964) have described the Mitrovic method, which uses only rectangular components (real numbers), in an efficient scheme for evaluating both the polynomial and its derivative, as in (3.1) and (3.6).

Consider a defined expression for the independent variable raised to some power p:

$$z^p = (x + jy)^p \triangleq X_p + jY_p, \tag{3.26}$$

where the upper- and lower-case x and y variables are different; for example,

$$z^2 = (x + jy)(x + jy) = (x^2 - y^2) + j(2xy), \tag{3.27}$$

where it is seen that $X_2 = x^2 - y^2$ and $Y_2 = 2xy$. It can be shown in general that

$$
\begin{aligned}
X_k &= 2xX_{k-1} - (x^2 + y^2)X_{k-2}, \\
Y_k &= 2xY_{k-1} - (x^2 + y^2)Y_{k-2},
\end{aligned}
\tag{3.28}
$$

where $k = 2, 3, \ldots, p$; $X_0 = 1$; $Y_0 = 0$; $X_1 = x$; and $Y_1 = y$. Although (3.28) will be used numerically, the reader is urged to verify (3.26) by using (3.28) algebraically for $p = 2$ and $p = 3$; this will agree with (3.27) for $p = 2$ and similarly for $p = 3$.

The desired results are obtained from (3.28) and the following equations, which are derived by substituting (3.26) into (3.1) and associating real and

$$f(z) = \sum_{k=0}^{n} (a_k + jb_k)(X_k + jY_k)$$

$$= \sum_{0}^{n} (a_k X_k - b_k Y_k + a_k j Y_k + j b_k X_k)$$

imaginary parts with (3.3); this straightforward process yields

$$u = \sum_0^n (a_k X_k - b_k Y_k), \tag{3.29}$$

$$v = \sum_0^n (a_k Y_k + b_k X_k). \tag{3.30}$$

Furthermore, differentiating (3.1) with respect to z and following the same procedure yields

$$\frac{\partial u}{\partial x} = \sum_1^n k(a_k X_{k-1} - b_k Y_{k-1}), \tag{3.31}$$

$$\frac{\partial v}{\partial x} = \sum_1^n k(a_k Y_{k-1} + b_k X_{k-1}). \tag{3.32}$$

Clearly, (3.29)–(3.32) can be programmed easily in the BASIC language, especially since complex-variable calculations have been avoided. For the common situation where the given polynomials have only real coefficients, half the work in (3.29)–(3.32) can be eliminated, because all b_k are zero. This and savings in synthetic division by quadratic factors make it worthwhile to have a separate real-coefficient, root-finding program.

3.1.4. Root-Finder Program.

BASIC language Program B3-1 is documented in Appendix B, including a flowchart and listing. This is similar to the Hewlett–Packard Co. (1976a) program in the RPN language. Given a polynomial as in (3.1), the program always starts at the point $z = 0.1 + j1$ (see Figure 3.1). Subroutine 3000 calculates (3.28), (3.29), and (3.30). Only program lines 2040–2070 are required to obtain the derivatives in (3.31) and (3.32), so that the adjustments in x and y can be calculated for the Newton–Raphson step in (3.9) and (3.10). If taking that step increases the objective function (goes too far up an opposite hill in Figure 3.1), then the steps are reduced by a factor of 4 in the flowchart loop to reentry point 2190 until a lower objective value is obtained. Note that while in that cutback loop, new derivatives are not required, because the search direction is unchanged. It is interesting to observe how seldom cutback is required by temporarily adding the lines in Table 3.1.

Table 3.1.   Temporary Code to Print Search Cutback in Program B3-1

```
4005 PRINT "***CUTBACK***ON ITER#"; L
5035 PRINT" ITERS="; L
```

Example 3.3. Input the coefficient real and imaginary parts for the polynomial $f(z) = 1 - z^8$. The roots are on the unit circle; they are located at the four axis intersections and spaced between these at 45 degrees. Running the program shows that these roots have coordinates equal to either unity or $1/\sqrt{2}$, with agreement through eight significant figures on most computers. Adding the temporary statements in Table 3.1 and running the program again show how few times the algorithm needs to reduce the step length in a chosen search direction. Such reductions usually occur early in the search at some distance from the root (minimum) location.

The roots are printed whenever changes in x and y are less than $1.E-5$ or, following 10 step-size reductions, when F is no greater than $1.E-8$. The algorithm is aborted if the latter condition fails or when there have been more than 50 iterations (search directions). Little memory is required; there are two vectors (single-subscript arrays) for the coefficient's rectangular components a_k and b_k and two more vectors for X_k and Y_k in (3.26). These are dimensioned to hold N elements, where N is the maximum polynomial degree. However, on computers with exponent ranges of about $10\exp(+/-37)$, numerical overflow occurs for polynomials of degree greater than 20. Exponent ranges to $10\exp(+/-99)$ usually solve polynomials up to degree 35. The difficulty occurs in the large polynomial value because of the poor initial root guess of $z=0.1+j1$.

Gradient root finders such as Moore's suffer from a chronic problem with multiple roots. Consideration of a function such as $y=(x-1)^2$ and its derivative shows that repeated (multiple) roots cause gradients (coordinate derivatives) that tend to zero in the neighborhood of the root. This causes some inaccuracy in repeated root values, because Moore's method depends on gradient scaling in the step length formulas (3.9)–(3.11). The code in Table 3.2 can be added to print the value of (3.11).

Table 3.2. Temporary Code to Print the Squared Length of a Gradient

2085 PRINT"GRAD MAG SQD=";PM

Example 3.4. Add the program code in Table 3.2 to root-finder Program B3-1 and solve the polynomial

$$1080+2466z+2025z^2+765z^3+135z^4+9z^5=9(z+1)(z+2)(z+3)(z+4)(z+5).$$

Note that the "GRAD MAG SQD" value (3.11) is well scaled. Then solve the polynomial

$$54+135z+126z^2+56z^3+12z^4+z^5=(z+1)(z+2)(z+3)^3.$$

Note that the squared gradient length used as a divisor in the search step adjustment is well behaved until the $z=-3$ repeated root is encountered.

Repeated roots are usually determined to within two or three significant figures; this may be adequate for most but certainly not all engineering work.

3.1.5. Polynomial Scaling.   By the initial guess $z = 0.1 + j1$ for the root location, there is an assumption that the roots are not too far from the origin. Some polynomials may require scaling of coefficients to obtain the assumed condition, and the roots will require subsequent rescaling to correspond to the original problem. Two methods will be described, as given by Turnbull (1952): (1) decreasing all roots by the factor 10; (2) decreasing all roots by subtracting some fixed amount. The choice of method and amount depends on the problem being solved; there is usually adequate information to make those choices.

To reduce all root real and imaginary components by a factor of 10, reduce all polynomial coefficients of the kth-power terms by $10 \exp(n - k)$, where the polynomial degree is n. The following example clarifies the procedure.

Example 3.5.   Consider the polynomial

$$f(z) = 19404 - 394z + 2z^2, \tag{3.33}$$

which has roots $98 + j0$ and $99 + j0$ (available from the root-finder program). Rewrite the polynomial with revised coefficients using the rule given above:

$$f_1(z) = 194.04 - 39.4z + 2z^2. \tag{3.34}$$

The root-finder program will show that the roots of (3.34) are $9.8 + j0$ and $9.9 + j0$. Similarly, the roots of

$$f_2(z) = 1.9404 - 3.94z + 2z^2 \tag{3.35}$$

are $0.98 + j0$ and $0.99 + j0$.

The method for shifting the roots by a given amount is somewhat more involved but uses synthetic division in an interesting way. Again, consider the degree-3 polynomial in (3.12) without loss of generality. Suppose that variable z is decreased by amount h:

$$z = s + h \quad \text{or} \quad s = z - h. \tag{3.36}$$

Making that substitution in (3.12), there must be an equivalent polynomial, $F(s)$, in the new variable s:

$$f(z) = F(s) = b_0 + b_1 s + b_2 s^2 + b_3 s^3. \tag{3.37}$$

This is rewritten two more ways:

$$F(s) = b_0 + s(c_0 + c_1 z + c_2 z^2), \tag{3.38}$$

$$F(s) = b_0 + s[b_1 + s(c_0' + zc_1')]. \tag{3.39}$$

Note that (3.39) is nested in the same fashion as (3.25). Now (3.13) is written

Table 3.3.   Procedure for Decreasing Roots by Amount h

1. Set $z_i = h$ in (3.14) and find c_{-1} using $k = n - 1, \ldots, 0$, -1; note the extra subscript added to (3.14).
2. Set $b_0 = c_{-1}$; replace a_i with c_i, $i = n - 1, \ldots, 0$; and replace n with $n - 1$.
3. Do steps 1 and 2 again, but equate $b_1 = c_{-1}$.
4. Continue finding $b_2, b_3, \ldots, b_n$ through the $n = 0$ cycle.
5. Find the roots of (3.37); then the roots of (3.12) are $z_i = s_i + h$.

in a more general application of synthetic division:

$$f(z) = f(z_i) + (z - z_i)(c_0 + c_1 z + c_2 z^2). \tag{3.40}$$

The first term on the right side of (3.40) is zero by definition if z_i is a root of $f(z)$; but (3.40) is valid for evaluating $f(z)$ for any z, not necessarily a root. That first term is found as the value of c_{-1} when (3.14) is calculated through $k = -1$ instead of just through $k = 0$ as previously applied. Suppose that $z_i = h$, and (3.36) is substituted for the linear term in (3.40). Then (3.38) is the result of synthetic division cycle (3.14) on (3.12), and b_0 is obtained as c_{-1} when that cycle is carried on through $k = 0$. Now note that b_1 in (3.39) relates to (3.38) as b_0 in (3.38) is related to (3.12). So synthetic division starting with c_0, c_1, and c_2 (found by the last synthetic division cycle) will yield b_1, c_0', and c_1' in (3.39).

The procedure in Table 3.3 finds a new polynomial, $F(s)$, as in (3.37), given polynomial $f(z)$, as in (3.12), so that $s = z - h$.

Example 3.6.   Given polynomial $f(z)$ in (3.33) with roots $98 + j0$ and $99 + j0$, find the corresponding polynomial $F(s)$ having roots that are 100 less.

$$n = 2, \quad h = 100, \quad a_2 = 2, \quad a_1 = -394, \quad a_0 = 19404;$$

$$k = 1: \quad c_1 = 2 + (100) \times 0 = 2$$

$$k = 0: \quad c_0 = -394 + (100) \times 2 = -194$$

$$k = -1: \quad c_{-1} = 19404 + (100) \times (-194) = 4 = b_0$$

$$n = 1, \quad h = 100, \quad a_1 = 2, \quad a_0 = -194; \tag{3.41}$$

$$k = 0: \quad c_0 = 2 + (100) \times 0 = 2$$

$$k = -1: \quad c_{-1} = -194 + (100) \times 2 = 6 = b_1$$

$$n = 0, \quad h = 100, \quad a_0 = 2;$$

$$k = -1: \quad c_{-1} = 2 + (100) \times 0 = 2 = b_2$$

The polynomial with roots $-2 + j0$ and $-1 + j0$ is thus found to be

$$F(s) = 4 + 6s + 2s^2. \tag{3.42}$$

3.1.6. Root-Finder Summary.

Moore's root finder is a practical tool that is accurate and robust except for repeated roots, when accuracy is reduced. It is based on the Cauchy–Riemann condition and the Mitrovic method for evaluation of the polynomial and its derivative. Computation is reduced by more than half when all polynomial coefficients are real, which is usually the case in modern network synthesis. There are many other applications for this fast root finder, such as in root locus plotting versus gain factors and in z-transform calculations in sampled data system design. The structure of the particular problem may result in roots being far from the origin of the complex plane; in these cases, where the root finder may be slow or may fail to converge, scaling of polynomial coefficients can reduce each root by either a factor or a fixed amount. Roots thus found closer to the origin can then be moved back to their original location by shifting in the opposite fashion.

The following sections will employ this root finder for network synthesis steps and partial fraction expansions.

3.2. Polynomials From Complex Zeros and Products

The next two sections describe the composition of polynomials by multiplication and addition, respectively. The computer programs provided will continue to be in BASIC language, although these calculations are just as feasible in hand-held computers. This section begins with composition of polynomials from known root factors as available in the preceding root-finder section. Complex factors will be multiplied to find the generally complex coefficients of the resulting polynomial. Then a program will be given that multiplies a sequence of polynomials having real coefficients.

The last half of this section includes the beginning steps in doubly terminated network synthesis; both the ideas and the use of the computing aids are important in what follows. Power transfer from a complex source to a complex load will be introduced and then specialized to the real-source impedance case. The generalized reflection coefficient will be defined, and the Feldtkeller energy equation will be discussed for a given steady-state frequency of excitation. Finally, polynomials used in network synthesis will be described, and the fundamental polynomial relationship will be derived from power transfer considerations of a lossless two-port network.

3.2.1. Polynomials From Complex Zeros.

Only polynomials with real coefficients are considered, so that their roots must be real or occur in conjugate pairs. A conjugate pair of complex numbers can always be expressed as a quadratic factor, as previously described by (3.18). Program B3-2 in appendix B asks for a set of complex zeros in rectangular components, then outputs the resulting polynomial coefficients, also in their rectangular components. It is interesting to confirm some of the previously described characteristics of polynomials by use of this program.

Example 3.7.   Use Program B3-2 to multiply the root factors

$$F(s) = (s - s_1)(s - s_2), \tag{3.43}$$

choosing pairs of roots from $\pm 4 \pm j5$. Note that conjugate pairs produce the quadratic factor described by (3.18) and that conjugate pairs from the left-half plane ($\sigma < 0$ in Figure 3.2) yield all positive coefficients. The zeros of transmission (loss poles) of the two major synthesis polynomials, to be described in Section 3.2.4, must be accompanied by both their conjugate and negative roots, as shown by the "quad" in Figure 3.2. The special cases of real, imaginary, or zero roots are also indicated. Also multiply all four possible roots from the data above to obtain the quadratic polynomial with real coefficients; further multiplication by factors with real roots does not change this condition, of course.

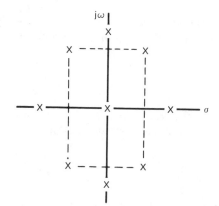

Figure 3.2. Possible locations for transmission zeros in p(s).

For further exercise of Program B3-2, multiply the root factors given in Example 3.4. Note that the coefficient of the highest degree term is always unity. The interested reader might wish to add a scaling feature to multiply all coefficients by any desired factor; this is often a requirement in network synthesis. The actual computation in Program B3-2 occurs in lines 170–320; those interested in details of the scheme are referred to Vlach (1969).

3.2.2. Polynomials From Products of Polynomials.   The need to multiply two polynomials having real coefficients will be encountered throughout network synthesis. The appropriate algorithm is not complicated; Program B3-3 in Appendix B is adapted from Vlach (1969), where it is explained in detail. A chaining feature has been added, so that the last product computed exists as the first of the next polynomial pair to be multiplied. Note that the main calculation in Program B3-3 requires only lines 210–300.

Example 3.8. Use Program B3-3 to multiply

$$(s^2+3s+2)(5s^2+4s-10)(3s^2+1) = 15s^6+57s^5+41s^4-47s^3-48s^2-22s-20,$$

(3.44)

using the program's chaining feature. Also, multiply the left-half-plane and right-half-plane quadratic factors found in Example 3.7 to confirm the earlier results.

3.2.3. *Power Transfer.* Power delivered from a complex source to a complex load will be encountered repeatedly in the following sections. It will be specialized to the real-source case for classical network synthesis in this chapter. Consider the source and load connection shown in Figure 3.3. It is well known that the maximum available source power is

$$P_{as} = \frac{|E_s|^2}{4R_s},$$

(3.45)

which occurs when $Z=Z_s^*$. Kurokawa (1965) developed relationships for less power transferred into other load impedance values. An important parameter is the generalized reflection coefficient

$$\alpha = \frac{Z-Z_s^*}{Z+Z_s}.$$

(3.46)

It defines a Smith chart with the center corresponding to Z_s^*; this will be explained in detail in Section 7.2. The power delivered to the load relative to the maximum available turns out to be

$$\frac{P}{P_{as}} = 1-|\alpha|^2.$$

(3.47)

The numerator of the reflection coefficient indicates that its magnitude is zero when $Z=Z_s^*$, so that $P=P_{as}$, as mentioned. Program A2-1, introduced in Section 2.1, makes the evaluation of the preceding two equations quite elementary for any range of load impedances, given a fixed source impedance.

In this chapter the source impedance is considered to be resistor R_1, and Z

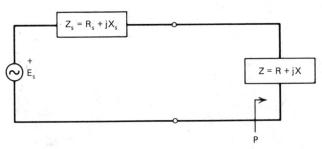

Figure 3.3. Power transfer from a fixed complex source to a variable complex load impedance.

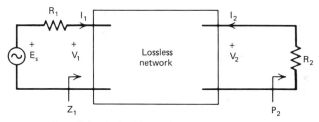

Figure 3.4. A doubly terminated lossless network.

is the input impedance Z_1 for the lossless network in Figure 3.4. The network's load termination is R_2; the resistive terminations at both ends make this a doubly terminated network. Any power that enters the network must exit into R_2, so that the reflection coefficient of interest is

$$\rho = \frac{Z_1 - R_1}{Z_1 + R_1}. \tag{3.48}$$

Consequently, power delivered to Z_1 and R_2 in Figure 3.4 is described by the famous Feldtkeller energy equation:

$$|H(j\omega)|^2 = \frac{P_{as}}{P_2} = \frac{1}{1 - |\rho|^2}, \tag{3.49}$$

where H is the transducer function.

3.2.4. Network Synthesis Polynomials. The network in Figure 3.4 is now assumed to be composed of lumped inductors and capacitors, so that Z_1 and other impedances will be well-behaved functions of complex Laplace frequency s. If $Z(\omega) = R(\omega) + jX(\omega)$, then one should know that $R(\omega)$ is always an even function of ω and that $X(\omega)$ is always an odd function of ω. Thus, brief consideration will lead to the conclusion that $Z^*(j\omega) = Z(-j\omega)$. The imaginary axis in the s variable is $j\omega$. According to the analytic continuation principle, $j\omega$ may be replaced by s in expressions where it occurs. Furthermore, the resulting functions of s have significance over the entire s plane. This concept leads to an identity with considerably greater importance than is at first apparent:

$$|f(j\omega)|^2 = f(s)f(-s), \qquad s = j\omega. \tag{3.50}$$

This is the squared-magnitude function, and it is also an even function of ω.

Example 3.9. Suppose that a given function is

$$f(s) = -76s^4 + 11s^3 - 33s^2 + 12s - 4. \tag{3.51}$$

Compute $f(s)f(-s)$ using Program B3-3, and save the result. Note that the resulting magnitude function is even in s.

The transducer magnitude function in (3.49) implies the existence of H(s), and it will be apparent later as to the convenience of defining a companion function,

$$K(s) = \rho H(s), \tag{3.52}$$

called the characteristic function. Using (3.49) and (3.50), an important energy relationship between the transducer and characteristic functions is obtained:

$$H(s)H(-s) = 1 + K(s)K(-s). \tag{3.53}$$

This shows that $|H(j\omega)| \geqslant = 1$, as required. Both H(s) and K(s) are rational functions with numerators and denominators identified as

$$H(s) = \frac{e(s)}{p(s)}, \tag{3.54}$$

$$K(s) = \frac{f(s)}{p(s)}. \tag{3.55}$$

A concise statement can be made about the nature of the individual polynomials e, f, and p. The roots of e(s) and f(s) are real or in conjugate pairs. The roots of e(s) lie in the open (not on $j\omega$ axis) left-half plane and are the natural modes of the LC network; the roots of f(s) are called reflection zeros or zero-loss frequencies. Polynomial f(s) is either even or odd, with degree no greater than that of e(s). As in Figure 3.2, the roots of p(s) are conjugate by pairs, are purely imaginary (on the $j\omega$ axis) for ladder networks, and are called the loss poles (peaks) or transmission zeros. Polynomial p(s) is either even or odd.

Using (3.53)–(3.55), the fundamental polynomial relationship in doubly terminated network synthesis is

$$e(s)e(-s) = p(s)p(-s) + f(s)f(-s). \tag{3.56}$$

Either H or K is given, so that either f or e must be found from (3.56), respectively. The latter is illustrated in the example from Temes and Mitra (1973).

Example 3.10.   Find H(s) given

$$K(s) = \frac{-76s^4 + 11s^3 - 33s^2 + 12s - 4}{4\sqrt{3}\,(s^2 + 4)}. \tag{3.57}$$

Compare (3.57) with (3.55) to identify f(s) and p(s); e.g., f(s) is shown in (3.51). Use Program B3-3 to calculate $p(s)p(-s)$ and $f(s)f(-s)$. Adding these manually (a program to do this will be described in Section 3.3.1), (3.56) yields

$$e(s)e(-s) = 5776s^8 + 4895s^6 + 1481s^4 + 504s^2 + 784. \tag{3.58}$$

The eight roots of (3.58) are found easily using Program B3-1:

$$\pm 0.226127 \pm j0.828392; \quad \pm 0.596242 \pm j0.379658. \tag{3.59}$$

The last step is to associate the left-half-plane roots with e(s) as required above. Using the four left-half-plane roots from (3.59) and Program B3-2, the e(s) polynomial is obtained, except for a constant. By (3.58), that constant must be $\sqrt{5776} = 76$, so that

$$e(s) = 76s^4 + 125s^3 + 135s^2 + 84s + 28. \qquad (3.60)$$

Using the denominator of (3.57) as p(s) and (3.60), the rational polynomial H(s) is thus found according to (3.54).

3.2.5. Summary of Polynomials From Zeros and Products. Programs that calculate polynomial coefficients given complex zeros or given a sequence of polynomials to multiply are easy to program and require very little computer memory. Quadratic factors, magnitude functions, and polynomial factors having roots in the left-half plane are important parts of the mathematics of network synthesis.

The basis of doubly terminated network selectivity behavior is the Feldtkeller energy equation (3.49), which describes the power transfer from a source, relative to maximum available power, in terms of the reflection coefficient at that interface. This leads to the transducer and characteristic functions that are polynomials in complex frequency $s = \sigma + j\omega$. There is a free exchange of s and $j\omega$ in the magnitude–function relationships (the interested reader is referred to Van Valkenburg, 1960, for details of the underlying analytic continuation principle). There is a straightforward procedure for finding the transducer numerator polynomial given the characteristic function numerator and denominator, and vice versa. The programs in this chapter make these computations relatively easy.

3.3. Polynomial Addition and Subtraction of Parts

The transducer and characteristic functions H and K have been introduced by way of the Feldtkeller energy equation. The chain (or ABCD) parameters for two-port networks are commonly encountered as complex numbers at a frequency, but also may be rational functions of complex frequency s. This section will introduce a simple program for adding and subtracting polynomials, the main step required to use H(s) and K(s) to find the polynomials A, B, C, and D prior to finding an LC network that corresponds to the given data. The program and the synthesis steps will be described.

3.3.1. Program for Addition and Subtraction of Parts. Program B3-4 in Appendix B adds or subtracts coefficients of like powers of s in two given polynomials, or just those coefficients of even powers or of odd powers. It is written in BASIC, but the single-subscript array R(·) is the basis of the memory assignment; this makes its translation to hand-held calculators especially elementary. The computation occurs in lines 200–370.

Example 3.11.   Consider the polynomials

$$P_1(s) = 9s^2 + 3s + 4, \qquad P_2(s) = 10s^3 + 2s + 1, \qquad (3.61)$$

which are neither even nor odd. Try these in Program B3-4; note that the two polynomials stay intact for subsequent operations (add or subtract; all, even, or odd parts). The answers can be checked by inspection; real problems are seldom this simple.

3.3.2.   The ABCD Matrix of Rational Polynomials.

The ABCD two-port parameters are defined in terms of the standard voltages and currents shown in Figure 3.4:

$$V_1 = AV_2 - BI_2, \qquad (3.62)$$

$$I_1 = CV_2 - DI_2. \qquad (3.63)$$

This form of expressing two-port behavior has a number of important properties that will be useful in many later sections. An input impedance expression will be of use here:

$$Z_1 = \frac{V_1}{I_1}. \qquad (3.64)$$

Similarly, the load resistance at port 2 is related to its voltage and current by

$$R_2 = \frac{-V_2}{I_2}. \qquad (3.65)$$

Solve (3.65) for V_2 and substitute in (3.62) and (3.63); then the resulting equations reduce (3.64) to

$$Z_1 = \frac{AR_2 + B}{CR_2 + D}. \qquad (3.66)$$

The goal is to find the ABCD polynomials in terms of H and K. It can be seen from (3.45) and (3.49) that

$$|H| = \frac{1}{2\sqrt{R_1 R_2}} \left| \frac{E_s}{I_2} \right|. \qquad (3.67)$$

But Figure 3.4 shows that $E_s = I_1 \times R_1 + V_1$; substituting this relationship and (3.62) and (3.63) into the numerator of (3.67) yields

$$H(s) = \frac{(AR_2 + DR_1) + (B + CR_1 R_2)}{2\sqrt{R_1 R_2}}. \qquad (3.68)$$

To find a similar expression for K, substitute (3.66) into (3.48), and substitute the result obtained, along with (3.68), into the definition of K in (3.52). The result is

$$K(s) = \frac{(AR_2 - DR_1) + (B - CR_1 R_2)}{2\sqrt{R_1 R_2}}. \qquad (3.69)$$

Note that the magnitude symbols have been omitted in the last two equations and that the substitution $s = j\omega$ has been made on the assumption that magnitude functions such as (3.50) are involved. Also, the grouping of parameters is strategic, because it can be shown that, for lossless networks, A and D are even functions of s, while B and C are odd. Further, reciprocity requires that $AD - BC = 1$. Beyond that, the grouping is convenient because adding or subtracting H and K cause major cancellations. One good reason for defining K at all is the following important result:

$$\begin{bmatrix} A & B \\ C & D \end{bmatrix} = \frac{1}{\sqrt{R_1 R_2}} \begin{bmatrix} (H_e + K_e)R_1 & (H_o + K_o)R_1 R_2 \\ (H_o - K_o) & (H_e - K_e)R_2 \end{bmatrix}, \qquad (3.70)$$

where the e subscript denotes an even polynomial and o an odd polynomial.

Example 3.12. In Example 3.10, $K(s)$ was given in (3.57) and the numerator of $H(s)$ was found as (3.60). Note that the denominators of H and K are the same. Enter the numerators of H and K into Program B3-4 in that order; then (3.70) yields the ABCD matrix numerators without difficulty. The result is:

$$\begin{bmatrix} A & B \\ C & D \end{bmatrix} = \begin{bmatrix} \dfrac{102s^2 + 24}{4\sqrt{3}\,(s^2 + 4)} & \dfrac{136s^3 + 96s}{4\sqrt{3}\,(s^2 + 4)} \\[2ex] \dfrac{114s^3 + 72s}{4\sqrt{3}\,(s^2 + 4)} & \dfrac{152s^4 + 168s^2 + 32}{4\sqrt{3}\,(s^2 + 4)} \end{bmatrix}, \qquad (3.71)$$

where $R_1 = R_2 = 1$ is assumed, as explained in Section 3.4.4.

3.3.3. Summary of Polynomial Addition and Subtraction of Parts. This section began with a simple BASIC language program to add and subtract even, odd, or all parts of polynomials. It continued with a look at the well-known ABCD (chain) parameters for two-port networks. The H(s) and K(s) functions were related to the ABCD parameters by considering input power transfer and input impedance and then assuming $s = j\omega$ for implied magnitude functions. The right tools make the task quite simple along theoretical lines that are easy to remember after a little practice.

The strategy behind the convenient ABCD development is to obtain simple expressions for LC impedance and admittance parameters in terms of the ABCD polynomials already found. A continued fraction expansion of these produces the corresponding network element values, as shown next.

3.4. Continued Fraction Expansion

Continued fraction expansion of reactance functions (Z_{LC}) will be described and used to realize a lowpass network as the last step in the LC network synthesis procedure. These functions are the port impedance or admittance of lossless LC networks when open or short-circuited at the opposite port (see Figure 3.4). They are rational polynomials that are always an even polynomial over an odd polynomial or vice versa. Such expansions provide lowpass or highpass network element values and can also be used to determine if a given polynomial is a Hurwitz polynomial (all roots in the left-half plane).

3.4.1. Lowpass and Highpass Expansions. Continued fraction expansions may be finite or infinite. Two finite examples and their equivalent rational polynomials are:

$$Z_1(s) = 2s + \cfrac{1}{3s + \cfrac{1}{4s + \cfrac{1}{5s}}} = \frac{120s^4 + 36s^2 + 1}{60s^3 + 8s}, \tag{3.72}$$

$$Z_2(s) = \frac{1}{2s} + \cfrac{1}{\cfrac{1}{3s} + \cfrac{1}{\cfrac{1}{4s} + \cfrac{1}{1/5s}}} = \frac{1 + 38s^2 + 120s^4}{2s + 64s^3}. \tag{3.73}$$

A convenient shorthand for representing continued fraction expansions is described by Vlach (1969); applied to (3.72) it is

$$Z_1(s) = 2s + \frac{1}{3s} + \frac{1}{4s} + \frac{1}{5s}. \tag{3.74}$$

3.4.2. A Continued Fraction Expansion Program. Consider the rational polynomial to be in one of the following forms or their reciprocals:

$$\frac{a_0 + a_2 s^2 + a_4 s^4 + \cdots + a_n s^n}{a_1 s + a_3 s^3 + a_5 s^5 + \cdots + a_{n-1} s^{n-1}}, \qquad \text{n is even;} \tag{3.75}$$

$$\frac{a_0 + a_2 s^2 + a_4 s^4 + \cdots + a_{n-1} s^{n-1}}{a_1 s + a_3 s^3 + a_5 s^5 + \cdots + a_n s^n}, \qquad \text{n is odd.} \tag{3.76}$$

Program B3-5 in Appendix B is adapted from Vlach (1969); it requires only lines 210–340 for computation.

Example 3.13. Program B3-5 will be run using (3.72) for the cases where the rational polynomial represents an LC, two-port input impedance with an open-circuit load or an input admittance with a short-circuit load. Consider Case 1 for maximum degree N=4 in Figure 3.5. Certainly, the input impedance of the network shown must be Z=sL+remainder, according to the form of (3.72). Therefore, the first element must be an inductor with a value of 2 henrys. If the remainder polynomial Z_r is inverted to provide $Y_r = 1/Z_r$, then the next term removed must be Y=sC, where C=3 farads. Comparison of this case with (3.72) shows how each element value was obtained for the lowpass network. Note that a short circuit across the 5-farad capacitor would

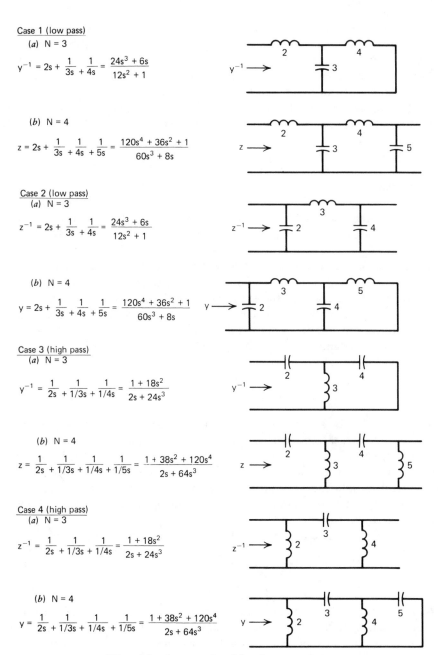

Case 1 (low pass)

(a) N = 3

$$y^{-1} = 2s + \cfrac{1}{3s} + \cfrac{1}{4s} = \frac{24s^3 + 6s}{12s^2 + 1}$$

(b) N = 4

$$z = 2s + \cfrac{1}{3s} + \cfrac{1}{4s} + \cfrac{1}{5s} = \frac{120s^4 + 36s^2 + 1}{60s^3 + 8s}$$

Case 2 (low pass)

(a) N = 3

$$z^{-1} = 2s + \cfrac{1}{3s} + \cfrac{1}{4s} = \frac{24s^3 + 6s}{12s^2 + 1}$$

(b) N = 4

$$y = 2s + \cfrac{1}{3s} + \cfrac{1}{4s} + \cfrac{1}{5s} = \frac{120s^4 + 36s^2 + 1}{60s^3 + 8s}$$

Case 3 (high pass)

(a) N = 3

$$y^{-1} = \cfrac{1}{2s} + \cfrac{1}{1/3s} + \cfrac{1}{1/4s} = \frac{1 + 18s^2}{2s + 24s^3}$$

(b) N = 4

$$z = \cfrac{1}{2s} + \cfrac{1}{1/3s} + \cfrac{1}{1/4s} + \cfrac{1}{1/5s} = \frac{1 + 38s^2 + 120s^4}{2s + 64s^3}$$

Case 4 (high pass)

(a) N = 3

$$z^{-1} = \cfrac{1}{2s} + \cfrac{1}{1/3s} + \cfrac{1}{1/4s} = \frac{1 + 18s^2}{2s + 24s^3}$$

(b) N = 4

$$y = \cfrac{1}{2s} + \cfrac{1}{1/3s} + \cfrac{1}{1/4s} + \cfrac{1}{1/5s} = \frac{1 + 38s^2 + 120s^4}{2s + 64s^3}$$

Figure 3.5. Some continued fraction expansions.

be senseless. Run Program B3-5, answering "YES" to indicate that the first element is a series L, because (3.72) is considered an impedance function. However, note that the same programmed solution applies to the network in Figure 3.5 for Case 2, $N = 4$, if (3.72) is considered a two-port input admittance function (with a short-circuit load). In that case, an open circuit after the 5-farad inductor would be senseless. The reader is urged to run Program B3-5 for all the possible combinations shown in Figure 3.5.

Program B3-5 can also be used to determine whether all roots of a polynomial $(a_0 + a_1 s + \cdots + a_n s^n)$ are in the left-half s plane, i.e., whether the polynomial is "Hurwitz." If any of the continued fraction expansion coefficients are negative or zero, or if the program fails with a "divide by zero" error, then the polynomial was not "Hurwitz." The polynomials being tested in this way are not rational, but are just the sum of all terms in (3.75) or (3.76).

3.4.3.   Finding LC Values From ABCD Polynomials.

It should be clear from the last section that two-port networks subjected to open- or short-circuit port conditions are relevant to the synthesis procedure. Equations using ABCD parameters to describe two-port networks were introduced in Section 3.3.2. Two more of the infinite set of such descriptions are now introduced, based on the port voltages and currents and terminal conditions. Consider the equations based on Figure 3.4:

$$V_1 = I_1 z_{11} + I_2 z_{12}, \tag{3.77}$$

$$V_2 = I_1 z_{21} + I_2 z_{22}. \tag{3.78}$$

These characterize any two-port network, lossless or not. It is important to understand what the coefficients mean. For example, z_{21} is V_2/I_1 when $I_2 = 0$, as seen from (3.78). $I_2 = 0$ says that the output port is terminated by an open circuit. These two equations are known as the open-circuit impedance parameters because both independent variables are the port currents. Look at z_{21} another way: it is the output voltage into an open circuit when the input current is 1 ampere.

A similar characterization is based on short-circuit terminal conditions where

$$I_1 = V_1 y_{11} + V_2 y_{12}, \tag{3.79}$$

$$I_2 = V_1 y_{21} + V_2 y_{22}. \tag{3.80}$$

Now, for example, y_{21} is the current entering port 2 in Figure 3.4, carried by a short-circuit load, for 1 volt applied across the input port. It is convenient to write the open- and short-circuit equation systems in matrix notation:

$$\mathbf{V} = \mathbf{Z} \mathbf{I} \tag{3.81}$$

$$\mathbf{I} = \mathbf{Y} \mathbf{V} \tag{3.82}$$

It is well known that matrix $\mathbf{Y}$ is the inverse of matrix $\mathbf{Z}$ and vice versa; doing this algebra provides relationships between z and y parameters.

It was shown in Section 3.3.2 how to find the rational polynomials for A, B, C, and D. The ABCD linear equations were given in (3.62) and (3.63), comparable to the z- and y-parameter equations above. In order to find a port immittance (impedance or admittance) as functions of ABCD when the opposite port is terminated by either a short or open circuit, it is necessary to find the z and y parameters in terms of the ABCD parameters. For example, solve for V_2 in (3.63):

$$V_2 = \frac{I_1 + DI_2}{C}.$$

(3.83)

Using this in (3.62) yields

$$V_1 = I_1\left(\frac{A}{C}\right) + I_2\left(D\frac{A}{C} - B\right).$$

(3.84)

Comparison of (3.84) with (3.77) provides z_{11} and z_{12} in terms of the ABCD parameters. The coefficient of I_2 in (3.84) is further simplified for lossless two ports because $AD - BC = 1$ in that case. Therefore, the following identities apply for two-port networks:

$$Z = \frac{1}{C}\begin{bmatrix} A & 1 \\ 1 & D \end{bmatrix},$$

(3.85)

$$Y = \frac{1}{B}\begin{bmatrix} D & -1 \\ -1 & A \end{bmatrix}.$$

(3.86)

These are valid for complex numbers or for rational functions; the latter will illustrate the last step in network synthesis. For example, (3.85) says that the open-circuit impedance parameter $z_{11} = A/C$ and both A(s) and C(s) were described in terms of H(s) and K(s) in (3.70). The numerator and denominator polynomials of H and K were defined in (3.54) and (3.55). For z_{11}, the result is

$$z_{11}(s) = R_1\frac{e_e(s) + f_e(s)}{e_o(s) - f_o(s)}.$$

(3.87)

This is the impedance for Case 1, $N = 4$, in Figure 3.5. Note that $1/y_{11}$ was relevant to $N = 3$ but not to $N = 4$. An example from Temes and Mitra (1973) is given below.

Example 3.14. Given the characteristic function $K = s^3$, find z_{11} and $1/y_{11}$ and the related networks. It is seen from (3.55) that $f(s) = s^3$ and $p(s) = 1$. Then (3.56) is

$$e(s)e(-s) = 1 - s^6 = (1+s)(1-s)(1+s+s^2)(1-s+s^2).$$

(3.88)

As noted in Section 3.2.4, the roots of e(s) are the natural modes, which must be in the left-half plane. Therefore, the transducer function according to (3.54) is

$$H(s) = \frac{e(s)}{p(s)} = (1+s)(1+s+s^2) = 1 + 2s + 2s^2 + s^3.$$

(3.89)

Figure 3.6. Network realizations for Example 3-14 using (a) z_{11} (or z_{22}) and (b) the reciprocal of y_{11}.

Using the even and odd parts of H and K in (3.70) and assuming that $R_1 = R_2 = 1$, the chain matrix is found to be:

$$\begin{bmatrix} A & B \\ C & D \end{bmatrix} = \begin{bmatrix} 1 + 2s^2 & 2s + 2s^3 \\ 2s & 1 + 2s^2 \end{bmatrix}. \tag{3.90}$$

Therefore, (3.85) yields

$$z_{11} = \frac{A}{C} = \frac{2s^2 + 1}{2s}, \tag{3.91}$$

and (3.86) yields

$$\frac{1}{y_{11}} = \frac{B}{D} = \frac{2s^3 + 2s}{2s^2 + 1}. \tag{3.92}$$

A network for this example is shown in Figure 3.6, as found by continued fraction Program B3-5. There must be three elements according to the degree of (3.89). Figure 3.6a uses z_{11} to find only the first two elements. (Why?) Figure 3.6b uses $1/y_{11}$ to find all three elements, because y is a short-circuit parameter, and the last element is in series. Note that both z_{11} and z_{22} could have been used to find all three elements, two at a time, including the shunt C in the middle twice. That would have shown whether or not $R_1 = R_2$ (Why?) and could provide greater numerical accuracy. Mellor (1975) has estimated that computer decimal-digit word length (N_d) and filter synthesis degree (N) are compatible if $N \leqslant N_d/2$. However, Lind (1978) gives a simple method for increasing accuracy.

3.4.4.   Comments on Continued Fraction Expansion.   Continued fraction expansions are an important mathematical tool with many applications, e.g., for LC ladder network realization and the polynomial Hurwitz test. The synthesis procedure described above is based on reactance functions (Z_{LC}), not the input impedance of a resistively terminated two-port network (Z_{RLC}). However, Z_{RLC} can be reduced to the corresponding Z_{LC}, as described in Section 3.5.3.

Example 3.14 gave K and found H; conversely, (3.56) can be rewritten to be explicit in $f(s)f(-s)$ when given H to find K. In the latter case, allocation of roots (reflection zeros) to $f(s)$ and $f(-s)$ is more arbitrary; it is necessary only to keep roots in conjugate pairs and to place in $f(-s)$ the negative of each root in $f(s)$. Each arrangement of root allocation in $f(s)f(-s)$ will result in a different chain matrix and therefore a different network. They will all have the same transducer magnitude function versus frequency, but their input impedance functions will differ (see Temes and Mitra, 1973).

Example 3.14 gave $K = f/p$ so that $p = 1$. As noted in Section 3.2.4, any roots of $p(s)$ occur on the $j\omega$ axis for lossless ladder networks. A more general case would be the $K(s)$ given in (3.57), where the roots of p are at $\omega = \pm 2$. A lowpass function would then produce a network with "traps" to produce zero transmission (loss peaks) at these root frequencies of $p(s)$. A very effective method for designing networks of this sort without resorting to synthesis will be described in Sections 9.2 and 9.3. The continued fraction expansion described here will not suffice for the synthesis of these more general networks. However, Temes and Mitra (1973) provide a compact summary of Orchard's elegant method for networks containing the four possible arrangements of traps; the method is well suited for small computers.

Finally, as noted in Example 3.14, $R_1 = R_2$ is not the general case. However, it is a fairly standard procedure to make this assumption, then derive one or more elements by synthesis from opposing ends of the network, and then decide (by any difference in answers) what the impedance scaling must be, i.e., how R_1 is related to R_2.

3.5. Input Impedance Synthesis From Its Real Part

Sections 3.2 through 3.4 developed a method of doubly terminated network synthesis, along with the introduction of various computer aids for a variety of engineering applications. The specification related to power transferred from a source to a resistively terminated lossless network, and the power was relative to the maximum available from the source. There are many situations where the source impedance has no real part, so that the maximum power available is infinite in theory. An equivalent case is the situation where the complex source is connected to an unterminated lossless network, so that no power can be transferred to the network. In either case, there is often an interest in the output voltage function versus frequency. These cases arise from singly terminated networks.

It is important to understand that the discussion of singly terminated networks and the synthesis of input impedance from its real part are the same thing. The need to realize an input impedance function might occur, for example, in building a lumped-element dummy antenna to approximate the real antenna behavior over a band of frequencies. Suppose that a constant current source is connected to the singly terminated lossless network, as shown in Figure 3.7. The input power must be $P_1 = |I_1|^2 R_{in}$, and the power in the output resistor must be $P_2 = |V_2|^2 / R_2$. Since the network is lossless, $P_1 = P_2$, and the impedance transfer function is thus

$$|Z_{21}(\omega)|^2 = R_2 \, Re \, Z_{in}(\omega). \tag{3.93}$$

Networks with only one possible signal path are called minimum-phase

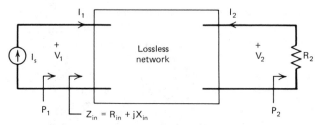

Figure 3.7. A singly terminated, lossless network.

networks. A ladder network is a minimum-phase network, but a bridge circuit is not. If the real part of the input impedance of a minimum-phase network is known for all frequencies, then its imaginary part (reactance) is dependent and can be found. A desktop computer program for finding the reactance at any frequency, given the piecewise linear resistance function versus frequency, will be furnished in Section 6.7. Here, a regular resistance function of frequency will be given in polynomial form, and the entire $Z(s)$ rational function will be found. This will be the Z_{RLC} shown in Figure 3.7. Then, a method will be described for finding the corresponding reactance function Z_{LC}, so that the continued fraction realization previously given may be employed to find the network element values.

3.5.1. Synthesis Problem Statement. Suppose that a resistance function is given as the rational polynomial

$$R(\omega) = \frac{A_0 + A_1\omega^2 + \cdots + A_m\omega^{2m}}{B_0 + B_1\omega^2 + \cdots + 1\omega^{2n}}. \qquad (3.94)$$

Such a function may result from the fitting procedure of Section 2.5. However, note that the denominator in (3.94) has a nonunity coefficient (B_0), and the coefficient of highest degree is unity. As mentioned earlier, resistance functions are even, so that all powers of ω are even. The goal is to find the corresponding impedance function:

$$Z(s) = \frac{a_0 + a_1 s + a_2 s^2 + \cdots + a_m s^m}{b_0 + b_1 s + b_2 s^2 + \cdots + 1 s^n}. \qquad (3.95)$$

Remarks similar to those regarding the denominator coefficients in (3.94) apply to the denominator of (3.95). There are at least two ways to solve this problem: Bode's method and Gewertz's method, as described by Guillemin (1957). The latter, which follows, is more compact.

3.5.2. Gewertz Procedure to Find RLC Input Impedance. Form an even function of complex frequency by substituting $\omega^2 = -s^2$ in the given resistance function (3.94):

$$Z_e(s) = \frac{A_0 - A_1 s^2 + A_2 s^4 - \cdots + (-1)^m A_m s^{2m}}{B_0 - B_1 s^2 + B_2 s^4 - \cdots + (-1)^n s^{2n}}. \qquad (3.96)$$

Find the left-half-plane roots of the denominator in (3.96). The product of these left-half-plane root factors is the denominator polynomial in (3.95). It remains to find the numerator of (3.95).

The remaining unknowns, $a_0, a_1, \ldots, a_m$ in (3.95) can be found by solving a linear system of equations using the known terms $b_0, b_1, \ldots$ just found and the easily derived $A_0, A_1, \ldots, A_m$ in (3.96). Gewertz solves the linear system

$$
\begin{bmatrix}
b_0 & 0 & 0 & 0 \\
-b_2 & b_1 & -b_0 & 0 \\
b_4 & -b_3 & b_2 & -b_1 \\
-b_6 & b_5 & -b_4 & b_3
\end{bmatrix}
\begin{bmatrix}
a_0 \\
a_1 \\
a_2 \\
a_3
\end{bmatrix}
=
\begin{bmatrix}
A_0 \\
A_1 \\
A_2 \\
A_3
\end{bmatrix}. \tag{3.97}
$$

Gauss–Jordan Program B2-1 solves linear systems of this sort with ease. The solution yields the numerator of (3.95), which is the desired input impedance Z_{RLC} of the terminated lossless network. An example from Carlin (1977) follows.

Example 3.15. Suppose that a given resistance function is

$$
R(\omega) = \frac{2.2}{1 + 2.56\omega^2 - 4.44\omega^4 + 4.29\omega^6}. \tag{3.98}
$$

Substituting $\omega^2 = -s^2$ in (3.98) yields

$$
f(s) = \frac{2.2}{1 - 2.56s^2 - 4.44s^4 - 4.29s^6}, \tag{3.99}
$$

which must be divided in both numerator and denominator by 4.29 in order to be in the form of (3.96). The roots of the denominator are found using Program B3-1. Roots s_1 and s_2 are $\pm 0.502752 + j0$, and roots s_3 through s_6 are $\pm 0.397782 \pm j0.895596$. The left-half-plane roots define a polynomial obtained by Program B3-2; the resulting b_i coefficients in the denominator of (3.95) are shown in Table 3.4. Using these, $b_4 = b_5 = b_6 = 0$, and $A_0 = 2.2/4.29$ in (3.97), the Gauss–Jordan Program B2-1 yields the a_i coefficients in the numerator of (3.95); these are also shown in Table 3.4. The rational input impedance polynomial is

$$
Z(s) = \frac{1.074610s^2 + 1.395184s + 1.062172}{s^3 + 1.298316s^2 + 1.360294s + 0.482804} \triangleq \frac{P(s)}{Q(s)}. \tag{3.100}
$$

Table 3.4. Input Impedance Coefficients for a Gewertz Example

$b_0 = 0.482804$	$a_0 = 1.06270$
$b_1 = 1.360294$	$a_1 = 1.395182$
$b_2 = 1.298316$	$a_2 = 1.074609$
$b_3 = 1.00$	$a_3 = 0$

The real part of (3.100) evaluated at any $s = j\omega$ gives the same answer as (3.98) for that value of ω.

What is obtained by Gewertz's procedure is the input impedance of a terminated lossless network; the corresponding reactance function is required for the continued fraction expansion of Section 3.4 to apply. The conversion of Z_{RLC} to Z_{LC} is discussed next.

3.5.3. Reactance Functions From Impedance Functions. A particular expression for the input impedance of a two-port network will be required in order to find Z_{LC} given the corresponding Z_{RLC}, as found in the preceding section. In Section 3.3.2 the input impedance of a two-port network was found using the ABCD equations and the load impedance. We proceed similarly with the open-circuit parameter equations by substituting $V_2 = -I_2 Z_L$ in (3.78) and solving that for I_2. But $Z_{in} = V_1/I_1$; so (3.77) readily yields

$$Z_{in} = z_{11} - \frac{z_{12}z_{21}}{z_{22} + Z_L}. \tag{3.101}$$

Using $Z_L = 1$, this can be written

$$Z_{in} = z_{11}\frac{1 + \Delta z/z_{11}}{1 + z_{22}}, \tag{3.102}$$

where the open-circuit-parameter determinant is

$$\Delta z = z_{11}z_{22} - z_{12}z_{21}. \tag{3.103}$$

A means for finding y parameters in terms of z parameters was suggested in Section 3.4.3. An equivalent expression for (3.102) turns out to be

$$Z_{in} = z_{11}\frac{1 + 1/y_{22}}{1 + z_{22}}, \tag{3.104}$$

where z_{11}, z_{22}, and y_{22} are ratios of even and odd polynomials.

Now consider a Z_{RLC} expression such as (3.100):

$$Z_{in}(s) = \frac{P(s)}{Q(s)} = \frac{P_e(s) + P_o(s)}{Q_e(s) + Q_o(s)}, \tag{3.105}$$

where the e and o subscripts denote the even and odd parts, respectively, of polynomials P(s) and Q(s). Two ways of writing (3.105) are

$$Z_{in} = \frac{P_e}{Q_o}\frac{1 + P_o/P_e}{1 + Q_e/Q_o}, \tag{3.106}$$

$$Z_{in} = \frac{P_o}{Q_e}\frac{1 + P_e/P_o}{1 + Q_o/Q_e}. \tag{3.107}$$

Comparison of the last two equations with (3.104) enables the construction of Table 3.5. The left-hand column represents cases where the Z_{RLC} numerator is even and the denominator is odd, so that there is a pole at the origin. The following example illustrates the use of Table 3.5 and a continued fraction expansion.

**Table 3.5.   Open- or Short-Circuit Z_{LC}
Impedance Functions**

Pole at Origin	No Pole at Origin
$z_{11} = P_e/Q_o$	$z_{11} = P_o/Q_e$
$z_{22} = Q_e/Q_o$	$z_{22} = Q_o/Q_e$
$y_{22} = P_e/P_o$	$y_{22} = P_o/P_e$

Example 3.16.   Consider the pi network in Figure 3.8. Suppose that the impedance Z_{RLC} looking back into the terminated network at port 2 is the same as (3.100). Open-circuit impedance function z_{22} is selected because y_{22} implies a short circuit that would prevent determination of C_1. But the presence of C_3 means that a port admittance function is required, so that the selection from Table 3.5 is

$$z_{22}^{-1} = \frac{Q_e}{Q_o} = \frac{1.298316s^2 + 0.482804}{s^3 + 1.360294s}. \tag{3.108}$$

Continued fraction expansion Program B3-5 applied to (3.108) yields $C_1 = 0.350$ farad, $L_2 = 2.890$ henrys, and $C_3 = 0.931$ farad after scaling from the 1-ohm source to the 2.2-ohm source shown in Figure 3.8.

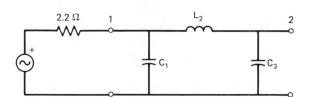

Figure 3.8.   A three-pole normalized lowpass network.

3.5.4.   Impedance Real-Part Synthesis Summary.   It has been shown that lossless networks terminated on only one end can be synthesized according to input impedance behavior. This is based on the fact that, for constant input current, the input power (and consequent output power) is proportional to input resistance. A similar statement can be made concerning input conductance in the case of constant voltage sources. Singly terminated instances of resistive sources connected to unterminated lossless networks are equivalent by proper consideration of the reciprocity theorem.

The Gewertz procedure was described for problems beginning with the even resistance function of frequency. Substitution of $\omega^2 = -s^2$ produces a polynomial whose denominator left-half-plane roots produce the input impedance denominator. The input impedance numerator coefficients are obtained by solving a system of real, linear equations involving these roots and known coefficients. The result is the input impedance Z_{RLC} of a terminated network.

Utilization of the continued fraction expansion of reactance functions from Section 3.4 requires the conversion of Z_{RLC} to its corresponding Z_{LC}. This is obtained by inspection of Z_{RLC} behavior at $s=0$ and reference to a standard table, which was derived. The worked example relates to the looking-back impedance at the output of a resistively driven LC two-port network. Starting from a given resistance function polynomial, which could have been obtained by the fitting procedure of Section 2.5, pi-network element values that realized this behavior versus frequency were obtained. This example will be of central importance as the final operation in a relatively new broadband impedance–matching procedure considered in Section 6.7.

3.6. Long Division and Partial Fraction Expansion

The last section of Chapter Three describes an important design tool that is useful for network synthesis in the frequency domain as well as for Laplace analysis in the time domain. The former is illustrated by Bode's alternative to the Gewertz procedure (see Guillemin, 1957). The time domain application of partial fraction expansions will be illustrated next (from Blinchikoff and Zverev, 1976).

Suppose that a given system transfer function is

$$H_1(s) = \frac{s^4 + 6s^3 + 22s^2 + 30s + 14}{s^4 + 6s^3 + 22s^2 + 30s + 13}. \qquad (3.109)$$

As will be demonstrated, it can also be expressed in the form

$$H_1(s) = 1 + \frac{0.1}{(s+1)^2} - \frac{0.02}{(s+1)} + \frac{0.02s + 0.04}{(s-s_1)(s-s_1^*)}, \qquad (3.110)$$

where s is the Laplace complex frequency variable, and root s_1 is $s_1 = -2 + j3$. Using a standard table of Laplace transforms for time and frequency functions, it is easy to show that the time response corresponding to (3.110) is

$$h_1(t) = \delta(t) + \left[0.1te^{-t} - 0.02e^{-t} + 2e^{-t}\left(0.01\cos 3t - \frac{\sin 3t}{75} \right) \right] u(t), \qquad (3.111)$$

where $\delta(t)$ is an impulse function, and u(t) is a unit-step function.

The algorithm to be described operates on proper rational functions, i.e., those whose numerator degree is lower than the denominator degree. Clearly, (3.109) is not proper, but would be if one long-division step were accomplished. The first subject treated in this section will be a compact long-division algorithm, both for obtaining proper fractions and to convince the reader that it is not complicated to program. This is important, because long division is one of two main features of the partial fraction expansion algorithm to follow.

3.6.1. Long Division.   Vlach (1969) gives a brief FORTRAN program for long division; it is adapted to BASIC language in Appendix-B Program B3-6. The calculation occurs in the last 10 lines of the program.

Example 3.17. Using Program B3-6 and by longhand, show that

$$H_2(s) = \frac{4s^2 + 9s + 3}{s + 2} = 4s + 1 + \frac{1}{s + 2}. \qquad (3.112)$$

Also, perform one division on (3.109) to show the constant-plus-proper-fraction form

$$H_1(s) = 1 + \frac{1}{s^4 + 6s^3 + 22s^2 + 30s + 13}. \qquad (3.113)$$

Note that the program is not dependent on having coefficients input in ascending or descending powers, because the algorithm proceeds the same in either case if the user is consistent.

3.6.2. A Partial Fraction Expansion Program.

Chin and Steiglitz (1977) have presented a partial fraction expansion algorithm that is claimed to reduce the number of computer operations by a factor of about 2. They correctly explain that this is important in spite of existing brief algorithms, because the calculations may occur many times in an iterative process, and they may be programmed on small computers, where program and storage size and speed are important.

The algorithm is based on two operations, the first being long division with a remainder (see Figure 3.9). Note that the given problem must be posed as a proper fraction and that the numerator is in polynomial form and the denominator is in factored form, i.e., the denominator roots must be known.

$$P(x) = \frac{2x^5 + 9x^4 - x^3 - 26x^2 + 5x - 1}{(x+1)^2(x-1)^3(x+2)}$$

$$= \frac{1}{(x+1)(x-1)^3(x+2)}\left[2x^4 + 7x^3 - 8x^2 - 18x + 23 - \frac{24}{x+1}\right]$$

$$= \frac{1}{(x-1)^3(x+2)}\left[2x^3 + 5x^2 - 13x - 5 + \frac{1}{x+1}\left(28 - \frac{24}{x+1}\right)\right]$$

$$= \frac{1}{(x-1)^2(x+2)}\left[2x^2 + 7x - 6 + \frac{1}{x-1}\left(-11 + \frac{28}{x+1} - \frac{24}{(x+1)^2}\right)\right]$$

$$= \frac{1}{(x-1)(x+2)}\left[2x + 9 + \frac{1}{x-1}\left(3 + \frac{-11+8}{x-1} + \frac{-8}{x+1} + \frac{12}{(x+1)^2}\right)\right]$$

$$= \frac{1}{x+2}\left[2 + \frac{1}{x-1}\left(11 + \frac{3-1}{x-1} + \frac{-3}{(x-1)^2} + \frac{1}{x+1} + \frac{-6}{(x+1)^2}\right)\right]$$

$$= \frac{1}{x+2}\left[2 + \frac{11-1}{x-1} + \frac{2}{(x-1)^2} + \frac{-3}{(x-1)^3} + \frac{1}{x+1} + \frac{3}{(x+1)^2}\right]$$

$$= \frac{1}{x+2} + \frac{3}{x-1} + \frac{1}{(x-1)^2} + \frac{-1}{(x-1)^3} + \frac{-2}{x+1} + \frac{3}{(x+1)^2}.$$

Figure 3.9. Algebraic flow of a particular example. [From Chin, F. Y., and Steiglitz, K. *IEEE Trans. Circuits Syst.*, Vol. CAS-24, No. 1, p. 44, January 1977. ©1977 IEEE.]

Figure 3.9 shows that two successive long divisions by the factor $x+1$ were accomplished with remainder numerators -24 and $23-24/(x+1)$, respectively. It is helpful to follow this process by doing the division either manually or with Program B3-6 and writing the results of each separate division. Then a division step by the factor $x-1$ occurs, leaving the constant -11 plus the prior rational remainder. The process is fairly clear up to the point where the second main operation occurs. However, the next (second) division by factor $x-1$ leaves the constant 3 plus the expression

$$\frac{1}{x-1}\left(-11+\frac{28}{x+1}-\frac{24}{(x+1)^2}\right)=\frac{-11+8}{x-1}+\frac{-8}{x+1}+\frac{12}{(x+1)^2}. \quad (3.114)$$

In this identity the right side preserves the form of the preceding collection of terms, and thus preserves the algorithm as different root factors are encountered. This illustrates the general scheme; interested readers are referred to Chin and Steiglitz (1977) for further detail.

Two more comments are appropriate. Some ill-conditioned roots may cause rounding errors to accumulate unless the roots are processed in order of ascending magnitude. Note that the example in Figure 3.9 employs real roots; the roots may be complex and therefore in conjugate complex pairs. They are processed separately in Program B3-7 using complex arithmetic. As in the root-finder Program B3-1, this can be avoided by dealing only with quadratic factors, as mentioned by Chin and Steiglitz (1977).

Example 3.18. First run the example in Figure 3.9 to be sure that the output sequence of residues is understood. Then perform a partial fraction expansion of (3.109) by first obtaining the proper fraction in (3.113) by one long-division step (Program B3-6). Use root-finder Program B3-1 to find denominator roots $-1+j0$, $-1+j0$, $-2+j3$, and $-2-j3$. Enter these roots, in that order, into partial fraction expansion Program B3-7 to find the residues of each term. These are shown in (3.110), except for the combined conjugate roots term. This is obtained with the useful identity

$$\frac{K_i}{s-s_i}+\frac{K_i^*}{s-s_i^*}=\frac{(K_i+K_i^*)s-(K_is_i^*+K_i^*s_i)}{(s-s_i)(s-s_i^*)}, \quad (3.115)$$

where K_i is a residue. Note that residues of complex conjugate roots also occur as complex conjugates.

3.6.3. Summary of Partial Fraction Expansion. A long-division algorithm that is simple enough for even hand-held computers is furnished in BASIC language. It is useful in reducing rational polynomials to proper form, i.e., numerator degree less than denominator degree. Long division is also one of the two main features of an efficient algorithm that is also especially suitable for small computers.

The input to the partial fraction expansion algorithm consists of the numerator real coefficients and the denominator roots in order of ascending magnitude. The program provides the residues corresponding to the order in which the roots were furnished and in descending root multiplicity. The residue output order can be understood best by running the example appearing in Figure 3.9 and comparing the results. A Laplace transformation to the time domain was illustrated as one of many important applications of the partial fraction expansion program.

Problems

3.1. Differentiate

$$f(z) = (z-1)(z-2)^2$$

using the calculus formula

$$d(uvw) = vw\,du + uw\,dv + uv\,dw.$$

Differentiate

$$f(z) = z^3 - 5z^2 + 8z - 4$$

and evaluate $f'(2)$. Note why the derivative of polynomials with multiple roots is zero at the root.

3.2. Given the polynomial

$$f(z) = z^3 = (x+jy)^3 = (x^3 - 3xy^2) + j(3x^2y - y^3) = u + jv.$$

(a) Find derivative df/dz by differentiation.
(b) Find derivative df/dz using the Cauchy–Riemann identity.
(c) Show, using $f(z)$, that

$$\frac{\partial u}{\partial x} = \frac{\partial v}{\partial y} \quad \text{and} \quad \frac{\partial v}{\partial x} = -\frac{\partial u}{\partial y}.$$

3.3. Given the complex polynomial

$$f(z) = 5 + 3z + 2z^2 + 4z^3 - 2z^4 = u + jv$$

for $z = x + jy$, use the Mitrovic method to find numerically the values of u, v, and the following derivatives when $z = 1 + j3$:

$$\frac{\partial u}{\partial x}, \quad \frac{\partial v}{\partial x}, \quad \frac{\partial u}{\partial y}, \quad \text{and} \quad \frac{\partial v}{\partial y}.$$

3.4. A root finder has located root $z_i = -\frac{3}{2} - j\sqrt{7}/2$ of the polynomial equation

$$f(z) = 2z^5 + 9z^4 + 13z^3 + z^2 - 13z + 4 = 0.$$

Find the polynomial that remains when the quadratic factor related to this root is removed using synthetic division.

3.5.   Derive p_i and q_i, defined by

$$(z-z_i)(z-z_i^*)=z^2+p_iz+q_i.$$

3.6.   Linear synthetic division in Example 3.1 on

$$f(z)=6+7z+4z^2+z^3,$$

using root $z_i=-2+j0$, gave the remainder coefficients when the $z+2$ factor was divided out. Do procedure (3.14) on this polynomial, but with $k=2$, 1, 0, *and* -1, with $z_i=-1+j0$. Compare coefficient c_{-1} with $f(-1)$.

3.7.   Show why quadratic factors of conjugate-pair left-half-plane roots will have all positive coefficients.

3.8.   Given the lowpass network

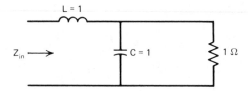

(a)   Write the $Z_{in}(s)$ expression using immittances

$$Z_L=sL, \qquad Y_C=sC.$$

(b)   Let $s=j\omega$ and express $Z_{in}(\omega)=R(\omega)+jX(\omega)$. Show that

$$Z_{in}^*(\omega)=Z_{in}(-\omega).$$

(c)   Evaluate Z_{in} at $\omega=0.1$ and $\omega=7.91$, using the expression obtained in (a).

3.9.   A 1-volt rms source with $Z_s=3-j2$ is connected to the network shown in Problem 3.8. At $\omega=2$, $Z_{in}=0.20+j1.60$. Find the input reflection coefficient α and the power delivered to the 1-ohm load resistor.

3.10.   A fixed sinusoidal voltage source with impedance $Z_s=R_s+jX_s$ is connected to a variable load impedance $Z=R+jX$, as in Figure 3.3. Given the definitions in (3.45) and (3.46), verify algebraically that (3.47) is true, i.e., the power P delivered to Z is $P=P_{as}(1-|\alpha|^2)$.

3.11.   Find the inverse of the two-dimensional chain matrix

$$\mathbf{T}=\begin{bmatrix} A & B \\ C & D \end{bmatrix}.$$

3.12. Given the resistive network

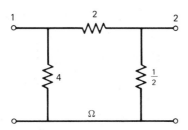

find chain parameters A, B, C, and D.

3.13. Find short-circuit parameter y_{22} in terms of open-circuit parameters z_{ij}.

3.14. Derive algebraically the ABCD chain parameters in terms of the short-circuit admittance parameters; i.e., derive (4.34).

3.15. Suppose that the characteristic polynomial $K(s) = s^3$. Find the associated transducer function $H(s)$.

3.16. Given the transducer function

$$|H(\omega)|^2 = 1 + \epsilon^2 T_5^2(\omega), \qquad \epsilon = \tfrac{1}{2},$$

find $H(s)$, $K(s)$, the ABCD polynomials, and the lowpass network L- and C-element values.

3.17. Synthesize the four-element, lowpass, doubly terminated LC network providing the Chebyshev response

$$|H(\omega)|^2 = \frac{P_{as}}{P_2} = 1 + 0.25 T_4^2(\omega),$$

where $T_4(\omega)$ is the degree-4 Chebyshev polynomial of the first kind. Show the $K(s)$, $H(s)$, $A(s)$, $B(s)$, $C(s)$, and $D(s)$ polynomials. Find the four element values and the termination resistances at the input and output ends of the LC two-port network.

3.18. Suppose that an LC network terminated in 1-ohm resistance has the following Chebyshev input resistance function:

$$R(\omega) = \frac{1.25}{1 + 0.25 T_4^2(\omega)}.$$

Use the Gewertz method to find the coefficients of the network's input impedance function:

$$Z_{RLC}(s) = \frac{a_0 + a_1 s + a_2 s^2 + a_3 s^3}{b_0 + b_1 s + b_2 s^2 + b_3 s^3 + s^4}.$$

Check the result by evaluating the two equations at $\omega = 0$ and $\omega = 1$ radian.

3.19. The Butterworth input resistance function for a three-element lowpass network is

$$R(\omega) = \frac{1}{1 + \omega^6}.$$

The corresponding $Z_e(s)$ function would be

$$Z_e(s) = \frac{1}{1 - s^6}.$$

The Gewertz method yields the corresponding input impedance function:

$$Z_{RLC}(s) = \frac{\frac{2}{3}s^2 + \frac{4}{3}s + 1}{s^3 + 2s^2 + 2s + 1}.$$

Obtain the partial fraction expansions of $Z_e(s)$ and of $Z_{RLC}(s)$. Compare these results and describe the similarities briefly.

3.20. Find the component values of a five-element lowpass filter having the response shape determined by the Legendre polynomial. The polynomial recursion expression is

$$P_{n+1}(x) = \frac{(2n+1)xP_n(x) - nP_{(n-1)}(x)}{(n+1)}.$$

The polynomial starting values are $P_0(x) = 1$, and $P_1(x) = x$. Scale the function for a $1 - dB$ response at an $\omega = 1$ radian passband edge.

Chapter Four

Ladder Network Analysis

Nearly all the design procedures in this book lead to ladder networks; these occur commonly in engineering practice. Ladder network analysis is quite practical for hand-held calculators. Computers having only 224 program steps can accommodate ladder analysis routines for networks with nine dissipative lumped elements, and the newer hand-held computers can do much better than that. More general network analysis, e.g., the nodal admittance matrix with LU factorization, is largely wasted on ladder networks, where most nodal matrix entries are zero. There is a great need for efficiency in ladder network analysis beyond fitting routines into small computers. Iterative (repeated) analysis at many frequencies and for many combinations of network component values occurs in optimization—the computer adjustment of components to obtain improved performance (Chapter Five). So ladder network analysis is extremely important for design confirmation, automatic design adjustment, and insight into certain impedance-matching and selectivity functions.

Chapter Four is based on a well-known method. An output current is assumed to exist. It is then traced back to the input by successive application of Kirchhoff's current and voltage laws to find the input current and voltage that would produce the assumed output. Since the ladder network is assumed to be linear, all voltages and currents thus found can be scaled by any factor representing steady-state changes in the input excitation. The discussion will almost always concern steady-state sinusoidal excitation. However, a convenient method of frequency sampling for a band-limited function will be shown to provide the impulse and other time responses of that network. This amount of calculation requires the speed and memory of at least a desktop computer.

The ladder networks considered here are quite general. The "menu" of element types can include nearly any one- or two-port subnetwork that can be programmed in a describing subroutine. Dissipative lumped elements (R, L, and C); dissipative uniform transmission lines in cascade or as terminated stubs; bridged-T networks; embedded two-port networks, including those described by data sets at each frequency; and two-terminal elements bridging nonadjacent nodes can be accommodated. Each branch of the basic ladder network may contain a large variety of series-connected-element subsets connected in parallel, and vice versa. A compact means for describing the network topology is an important part of Chapter Four.

All branch voltages and currents are available by the ladder network analysis method employed. Beyond direct applications, these provide exact sensitivity (partial derivative) information about how network performance changes with respect to each component value. This is an important part of gradient optimization methods and plays a significant role in manual and automatic network-tuning considerations.

A convenient, accurate, and familiar ladder network analysis program is one of the most important tools an individual can have in the world of radio frequency (rf) engineering.

4.1. Recursive Ladder Method

A definite form, nomenclature, and convention will be employed throughout Chapter Four. Various parts have been discussed in numerous references. The ladder network structure is shown in Figure 4.1.

4.1.1. Ladder Nomenclature.
Series (even-numbered) currents and shunt (odd-numbered) voltages are shown in Figure 4.1, with numbering beginning at the branch across the load impedance Z_L and proceeding back to the input, which may be either a series or a shunt branch. All voltages and currents will be rms (root mean square) values. Voltages between nodes (across series branches) may be obtained as the differences between the node voltages, and currents in shunt branches may be obtained in a similar way. If a branch does not exist physically, its immittance (impedance or admittance, as appropriate) is set equal to zero.

Each branch might contain only a single lumped element; e.g., $Y_1 = j\omega C$ and $Z_2 = j\omega L$. If these occurred in reverse order, the immittances would be $Y_1 = 1/j\omega L$ and $Z_2 = 1/j\omega C$ for nonzero elements. The load branch $Z_L = R_L + jX_L$ might be set to a very large real part (1E10) and a zero imaginary part if an open-circuit load is to be simulated. The load real part, R_L, must never be zero, as explained below.

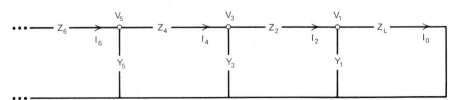

Figure 4.1. The ladder structure with alternating shunt admittances and series impedances.

4.1.2. Complex Linear Update. Load current I_0 in Figure 4.1 can be selected arbitrarily, but for several reasons (to appear later in the chapter) it is much more useful to specify load power P_L and load impedance Z_L, and thus determine the load current:

$$I_0 = \sqrt{\frac{P_L}{R_L}} \; . \tag{4.1}$$

All other branch voltages and currents correspond to this condition; this choice in no way precludes the later rescaling of all voltages and currents by some meaningful factor. Again, this decision means that R_L must never be zero, although $R_L = 1E-10$ is perfectly satisfactory.

The recursive calculation of node voltages and series currents is shown in Table 4.1. Load current I_0 is found from (4.1) and multiplied by Z_L to produce the complex number V_1. The current in the Y_1 branch is V_1Y_1. Admittance Y_1 is calculated at this time, and the branch-1 current is computed and added to the load current. Kirchhoff's current law states that this sum is equal to branch current I_2. These operations are easily accomplished with Program A2-1, for example.

Each line in Table 4.1 has the general form

$$\mathcal{A} = \mathcal{B} \mathcal{C} + \mathcal{D}, \tag{4.2}$$

where the variables are not the ABCD parameters. The variable $\mathcal{C}$ is either an impedance Z or an admittance Y, as they appear in Table 4.1. There are two good reasons for performing the operations in (4.2) in the rectangular format shown in (4.3) rather than in a polar format such as Program A2-1.

$$a_r = b_r c_r - b_i c_i + d_r, $$
$$a_i = b_i c_r + b_r c_i + d_i. \tag{4.3}$$

Table 4.1. Typical Ladder Network Recursion Scheme

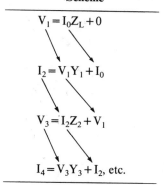

$$V_1 = I_0 Z_L + 0$$

$$I_2 = V_1 Y_1 + I_0$$

$$V_3 = I_2 Z_2 + V_1$$

$$I_4 = V_3 Y_3 + I_2, \text{ etc.}$$

The r and i subscripts indicate real and imaginary parts, respectively; e.g., $\ell = a_r + ja_i$. First, polar-to-rectangular conversions require cosine and sine functions that execute slowly in nearly all computers. Second, few if any versions of BASIC language allow variables to be declared complex, nor is the polar–rectangular conversion provided as a single operation. Providing a subroutine to remedy this deficiency is not convenient because few BASIC language sets have subroutine argument lists to transfer the several independent and dependent variables to and from the subroutine.

4.1.3. An Elementary Topology Code.   The means for specifying the arrangement of two-terminal elements for the ladder structure in Figure 4.1 is now described. The concepts will be extended for paralleled combinations of elements in series, and vice versa, in Section 4.1.5. Inclusion of arbitrary two-port networks will be described in subsequent sections of this chapter. The "menu" at this point will consist of just three kinds of components: resistors, dissipative inductors, and dissipative capacitors, assigned by integers 1, 2, and 3, respectively. Provisions for as many as nine different component types will be assumed, the arbitrary limitation (see Section 4.1.5) being that the descriptor must be a single, nonzero integer. The scheme employs a triple of component type number, a value for each of the element kinds, and its quality factor Q. A program using this scheme requires two integer pointers to keep track of its progress in the recursion shown in Table 4.1. Figure 4.2 shows a typical network and these parameters. Only the right-hand three columns are input by the user. The program will know when it has worked back to the network's input, because it will encounter a type-number zero in the next memory location, signalling that the input element has been processed.

The component types (1, 2, or 3) are shown in a column and correspond to the appearance of components in order from the load end back to the input end. Note the important use of a minus sign on some element-type numbers.

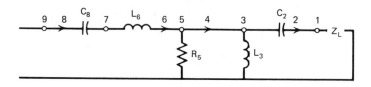

N	K	Type	Value	Q
1	2	−3	275	500
2	3	2	58	100
3	5	−1	94	0
4	6	2	43	250
5	8	−3	325	2000

Figure 4.2.   A typical lumped-element network with pointers and component triples.

This always indicates to the program that the preceding ladder branch immittance was zero, i.e., a null prior branch. If the first -3 in the type column had been a positive 3, then the program would think that capacitor C_3 would have been in branch 1 (Figure 4.1). The value is in the units specified separately, e.g., in microhenrys or picofarads. The Q column is the component quality factor; it is meaningless for resistors, of course. A program feature allows $Q = 0$ to indicate a lossless element (see Section 4.1.4). The two integer pointers N and K are used in the program to keep track of component number and branch number, respectively. It is very important that the reader understand this simple scheme. It puts some of the work on the user, but programs employing this scheme are very efficient in both memory and speed. The scheme can also be extended in many ways; for example, the column of component values is often the set of variables that an optimizer can adjust for improved performance. Subsequent sections in this chapter extend the topology capability in many ways.

4.1.4. Ladder Analysis Program. Program B4-1 in Appendix B is written in BASIC language. Some adaptation to make it more appropriate for hand-held computers is discussed in Section 4.1.5. The program will be explained and illustrated using the concepts previously discussed.

Program input begins with a request for the frequency, inductance, and capacitance units; typically, this might be 1E6, 1E$-$6, and 1E$-$12 for megahertz, microhenry, and picofarad, respectively. Then the load resistance and reactance values, in ohms, are requested. They are assumed to be frequency independent in this program, but that can be changed without great difficulty. The power delivered to the load is requested next; this enables calculation of the load current according to (4.1). Referring to Program B4-1 in Appendix B, the main analysis loop at each frequency begins at line 1200, where the frequency is input in the units previously specified. Radian frequency is then calculated for subsequent use.

The recursion in Table 4.1 is implemented in the loop from lines 1300 through 1390. It is first initialized with load current magnitude (4.1) and phase angle zero in code line 1220. Variable F1 is a flag to indicate that a null branch was processed in the previous complex linear update cycle. This is set up in subroutine 9000, where a zero value is assigned to the null-branch immittance. Otherwise, branch immittance is assigned by the calculated subroutine call in line 1385. The variable MK had previously been assigned from the component type array $M(\cdot)$; in line 1385, type MK $= 1$ would send the program to subroutine 9100, MK $= 2$ to subroutine 9200, etc. The actual complex linear update (4.3) occurs in subroutine 9900, called at line 1370. A little thought will show how elementary yet effective this ladder network analysis scheme can be.

It is important to understand the operation of the element-type subroutines 9100, 9200, and 9300 in Program B4-1. There is a small amount of standard overhead. If the branch number is odd (an admittance is anticipated), then the impedance, which is always calculated for a resistance or an inductance, must be inverted. A similar test of opposite properties is made for the capacitance subroutine (line 9300), where it is most convenient to calculate the admittance and invert it if the branch number is even. The Q parameter has been applied to add a series resistance to inductors and a parallel conductance to capacitors according to

$$Z = R + jX = X(d + j1); \tag{4.4}$$

$$Y = G + jB = B(d + j1), \tag{4.5}$$

where the decrement d is equal to $1/Q$ (line 1150). Also, $X = \omega L$ and $B = \omega C$. Note that a lossless element may be described by $Q = 0$ and yet avoid a "divide-by-zero" in the decrement calculation because of the test and replacement in line 1140, which sets $Q = 1E10$. There is some question as to whether Q is frequency independent. It is always possible to calculate Q in an arbitrary way in subroutines 9200 and 9300, where the frequency information is available. However the decrement is determined, inversion of (4.4) or (4.5) requires the identity

$$\frac{1}{d+j1} = \frac{d-j1}{d^2+1}. \tag{4.6}$$

This is coded in lines 9240–9260, which are potentially in common between subroutines 9200 and 9300.

Example 4.1. Run Program B4-1 for the three examples specified in Figure 4.3. The topological input is terminated by entering $0,0,0$. Note that any number of frequencies may be analyzed sequentially once the basic information has been input.

The input impedance is calculated last by lines 9955–9985; these will be discussed in Section 4.5.1.

4.1.5. Branch Topology Levels and Packing. The flexibility of the topological description may be extended considerably by defining branch levels, as illustrated in Figure 4.4. The analysis program keeps track of which branch number is being processed, and even-numbered branches are processed using a branch impedance value. If the branch were to contain several paralleled elements, their admittance should be calculated, added, and then inverted to give the branch impedance. This state of paralleling admittances in an even-numbered branch will be called level 1. Suppose that the branches to be paralleled are composed of elements in series; then these impedances should first be added, and the separate results should be inverted, so that the level-1 operation can proceed. The state of adding series impedances to obtain subsets to be paralleled in an even-numbered branch is called level 2. Branch 2 in Figure 4.4 contains two level-2 subset branches and one level-1 branch. The dual case is shown in branch 5 of Figure 4.4.

Example a:
3, 325, 200
2, 400, 100

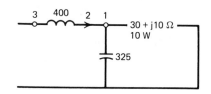

50 MHz; nH, pF.

$V_1 = 17.3205 + j5.7735$ $= 18.2574 \underline{/18.4349°}$

$I_2 = -3.2922E - 3 + j1.7714 = 1.7714 \underline{/90.1065°}$

$V_3 = -205.2845 + j7.5860$ $= 205.4246 \underline{/177.8837°}$

Example b:
$-2, 400, 100$
3, 325, 200

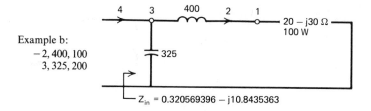

$Z_{in} = 0.320569396 - j10.8435363$

50 MHz; nH, pF

$V_1 = 44.7216 - j67.0820$ $= 80.6227 \underline{/-56.3098°}$

$I_2 = 2.2361 + j0$ $= 2.2361 \underline{/0°}$

$V_3 = 47.5313 + j213.9106 = 219.1277 \underline{/77.4723°}$

$I_4 = -19.5803 + j4.9622 = 20.1993 \underline{/165.7790°}$

Example c:
$-3, 325, 0$
2, 400, 100

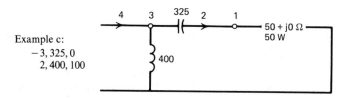

50 MHz; nH, pF

$V_1 = 50 + j0$ $= 50 \underline{/0°}$

$I_2 = 1 + j0$ $= 1 \underline{/0°}$ (because $P_L = R_L$)

$V_3 = 50 - j9.7942$ $= 50.9502 \underline{/-11.0830°}$

$I_4 = 0.9260 - j0.3986 = 1.0081 \underline{/-23.2896°}$

Figure 4.3. Three ladder network examples with answers.

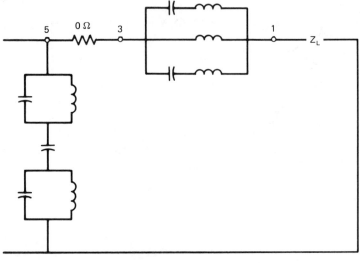

N	TYPE.d	VALUE
1	− 22.02	5.E − 3
2	23.001	1.75E − 6
3	12.02	4.5E − 3
4	22.02	5.E − 3
5	23.001	1.75E − 6
6	− 1.0	0.
7	23.001	11.5E − 6
8	22.02	2.05E − 3
9	13.001	6.25E − 6
10	22.02	2.05E − 3
11	23.001	11.5E − 6

Figure 4.4. Branch topological extensions to two levels.

Study of the component-type array (integer part of middle column) in Figure 4.4 shows that level 2 is described by adding 20 to the element-type code; i.e., an even-numbered branch having a paralleled subbranch consisting of L and C in series would be designated by 22 and 23. Similarly, an even-numbered branch consisting of just L and C in parallel would be described by adding only 10 to the level-1 designation (i.e., 12 and 13).

Consider Figure 4.4 in detail. The first component, − 22, is an inductor (type 2), and the minus sign indicates that the prior branch, namely, branch 1, is null. The second component, 23, is a capacitor; it is in series because branch 2 is an even (impedance) branch and level 2 is specified. The program should sum the L and C impedances. The third component, 12, indicates a change of level. The program code should recognize this, invert the impedance sum, and start an admittance sum. Then the 12 is processed as a capacitive admittance, which is added to the admittance sum. The next 22 and 23 begin a new impedance sum, which is terminated by the change of level indicated by − 1.

This last impedance sum is inverted, added to the admittance sum, and this is then inverted to become the final branch-2 impedance. Only then is the resistor processed; but it is branch 4, since the minus sign indicates a null-branch 3.

Note that the resistor is introduced with a zero value; this is one of two degenerate situations that may exist. The null resistor is needed to separate the branch-2 description from the branch-5 description. The second degenerate condition is covered by the following two rules: (1) level 1 can come before or after level 2, or not at all; (2) multiple level-2 entries must be separated by level 1, even if by a null element. These dummy elements might be null C in parallel or null R in series.

Depending on the mass storage capability of the small computer, it may be possible for the program owner to have another program to prompt him for input and arrange it in the proper form. However, the topological scheme just described is easily mastered by sole users.

There are several ways to save memory in hand-held computers that are register oriented. Referring to the topology data shown in Figure 4.4, one way to save registers is to store each component type and its decrement ($d = 1/Q$) in one register to the left and right of the decimal point, respectively. For example, the data in Figure 4.2 would be stored as -3.002, 2.01, -1.0, 2.004, and -3.0005. Unpacking is simplified by use of the integer and fractional operators. Calls to the component-type subroutines are still easy, because most calculators ignore the sign and the fractional part of the numbers. However, any level-1 and level-2 increases to the mantissa magnitude would need to be removed; this usually occurs anyway in the test to see if levels 1 and 2 are indicated. The unscaled component values would be paired with the registers; this occurs naturally in the HP-67/97 calculators, where primary/secondary register pairing is featured. In other programmable calculators, the pairing is by a fixed register number difference, e.g., registers 1 and 21, 2 and 22, etc. Packing the N and K components and branch integers into one N.K format also saves one register.

4.1.6. Recursive Ladder Analysis Summary.

The concept of working backward in a ladder network, from what is arbitrarily assumed to have occurred at the output end to what caused it at the input end, is well known. It is useful because the network is assumed to be linear. The method is valuable for both computations and algebraic formulations, as will be demonstrated in Section 8.3. There is only one complex functional form, which is solved repeatedly; it requires just one multiplication and one addition. This operation is best programmed in assembly language for fast evaluation on machines providing that opportunity along with a higher-level language, e.g., BASIC. The voltages and currents obtained for shunt and series branches, respectively, are often of direct interest; how they enable the exact calculation of component sensitivities will be shown in Section 4.7.

A detailed scheme for describing the network to the computer has been discussed. It requires minimal memory and controls program branching during numerical evaluation of the network at each frequency. An arbitrary component "type" number may be assigned to single- or even multiple-element component "types". The subroutines for each single element can be called separately and by the multielement types for maximum programming efficiency. A level-0 BASIC language program has been described, and several examples were run to illustrate its speed and simplicity.

Several enhanced features for the ladder network analysis approach were described, and other additions will be presented in the following sections. These include embedded two-port networks and bridging between nonadjacent nodes of the ladder network.

4.2. Embedded Two-Port Networks

An essential feature of any ladder network analysis is the ability to include two-port networks connected in cascade in the ladder network. The uniform, dissipative transmission line connected in cascade may be treated as such a two-port subnetwork. Transistors with feedback, three-port circulators with one port terminated, and bridged-T equalizers are other common examples that will be discussed later. The two-port networks may be described by various parameter sets, but this analysis will be accomplished using a unique application of the ABCD (chain) parameters. Conversion among various parameter sets has been described by Beatty and Kerns (1964).

The approach will be to reduce the ABCD characterization to an L section of two adjacent ladder branches. Then the standard complex linear update will apply with negligible modification. This topic is developed by first discussing additional properties of the ABCD parameters. This will expose the inefficiency in the common use of the ABCD parameters for ladder network analysis, such as in Hewlett–Packard (1976b).

4.2.1. Some Chain Parameter Properties.   It is convenient to redefine the output current direction and port subscripts for cascaded, two-port subnetworks (see Figure 4.5). Then (3.62) and (3.63) are equivalent to

$$V_a = AV_b + BI_b ; \qquad (4.7)$$

$$I_a = CV_b + DI_b . \qquad (4.8)$$

The purpose is to cascade a number of subnetworks, where Figure 4.5 might be the jth one. Then the input current and voltage are also the output current and voltage from the subnetwork immediately to its left. Denote the ABCD matrix of the jth subnetwork at T_j. Then it follows that the total network's ABCD matrix T is simply the product of all n of the subnetwork chain

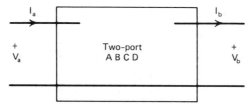

Figure 4.5. A typical cascaded, two-port subnetwork.

matrices:

$$T = T_1 \times T_2 \times T_3 \times \cdots \times T_j \times \cdots \times T_n. \qquad (4.9)$$

This is a popular means for analysis, but it will be shown to incorporate some gross inefficiencies.

The two-port networks for series Z and shunt Y are shown both separately and combined in Figure 4.6. The definition of the ABCD parameters is clear from (4.7) and (4.8); e.g., $A = V_a/V_b$ when $I_b = 0$. Applying this approach to the Z and Y two-port networks in Figure 4.6, their product is

$$\begin{pmatrix} 1 & Z \\ 0 & 1 \end{pmatrix} \begin{pmatrix} 1 & 0 \\ Y & 1 \end{pmatrix} = \begin{pmatrix} 1+ZY & Z \\ Y & 1 \end{pmatrix}. \qquad (4.10)$$

For emphasis and review, the L-section ABCD matrix on the right-hand side of (4.10) was obtained using complex arithmetic as follows. The upper left-hand corner element was obtained as $1 \cdot 1 + Z \cdot Y$. The upper right-hand element has even more trivial operations, namely $1 \cdot 0 + Z \cdot 1$. Similarly, the lower left-hand element was found from $0 \cdot 1 + 1 \cdot Y$. Finally, the lower right-hand element resulted from $0 \cdot 0 + 1 \cdot 1$. Further cascading will cause multiplication by another branch matrix having two 1's and a 0. Clearly, this is an ineffective technique for cascaded two-terminal elements such as the Z and Y branches most commonly encountered. It is reasonable if most of the subnetwork ABCD matrices are nontrivially full. It will obtain the complete ABCD matrix for the entire network. However, the reader should confirm what the fundamental ABCD definitions from (4.7) and (4.8) show: two network analyses with $Z_L = 1E - 10$ and $Z_L = 1E + 10$ will provide the overall ABCD matrix values. This matrix is sometimes required. Kajfez (1980) described its use in noise figure calculations, and (3.66) is another application that will be considered in much more detail. A more efficient ABCD calculation is now

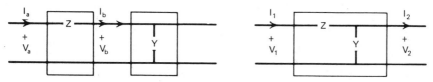

Figure 4.6. An L section by cascading Z and Y two-port networks.

considered for two-port subnetworks occasionally embedded in a Z and Y ladder network.

4.2.2. Chain Parameters in Complex Linear Updates.

Equations (4.7) and (4.8) are normalized to A and D, respectively, and (4.7) is reordered:

$$\frac{V_a}{A} = I_b \frac{B}{A} + V_b;$$ (4.11)

$$\frac{I_a}{D} = V_b \frac{C}{D} + I_b.$$ (4.12)

These two equations should be compared to those in Table 4.1. Consider the L section shown in Figure 4.7. The two-port subnetwork is given a component-type integer assignment, say 6. In the situation in Figure 4.7, the branch pointer K indicates that the next branch is to be an impedance, because $N = 4$ (an even number). Equation (4.11) is then relevant, since current $I_b = I_4$ (already known) is to multiply an impedance. Also, node voltage $V_b = V_3$ is already known from the back-to-front recursion in progress. These fit into the first equation shown in Figure 4.7, using $Z_4 = B/A$. To make (4.12) fit the next complex linear update equation, V_5 must be temporarily stored and not allowed to migrate to the second equation in Figure 4.7; $V_b = V_3$ is slipped into that place. I_4 migrates normally, $Y_5 = C/D$, and the solution I_6 is obtained, as shown in Figure 4.7. Finally, V_5 is recalled from storage, denormalized by multiplication by A, and placed as shown in the third equation in Figure 4.7. Current I_6 migrates normally, but is denormalized by multiplication by D. Then the ladder recursion continues normally. In fact, it is not really disturbed; when component-type 6 is encountered in the topological list, the program should go to the subroutine for ABCD two-port subnetwork type 6, where its ABCD parameters are computed and the several modifications are controlled. The reader should inspect the "next-branch-is-odd" case shown in Figure 4.8.

The technique just described is a little intricate, but it need be programmed only once in its own special subroutine. Other component-type subroutines can call it any number of times. The technique is independent of how the ABCD parameters involved were obtained. Also, much of the algorithm is

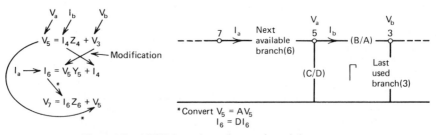

Figure 4.7. ABCD L sections: the next-branch-is-even case.

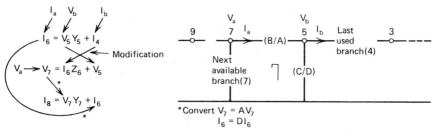

Figure 4.8. ABCD L sections: the next-branch-is-odd case.

especially suited to programmable calculators, e.g., register arithmetic for denormalizing variables.

4.2.3. Summary of Embedded Two-Port Networks.   The conventional use of ABCD parameters for cascaded subnetworks has been reviewed. It was shown that where only a few nontrivial ABCD subnetwork matrices are involved, the usual procedure, which multiplies all branch and other ABCD matrices to obtain the overall ABCD matrix, is quite wasteful. It usually amounts to complex multiplications by $0 + j0$ and $1 + j0$ many times. The review illustrated the mechanics of ABCD matrix multiplication, which will be applied algebraically in Section 8.1. It was also mentioned that two recursive analyses with extreme load impedance value would calculate the overall ladder network's ABCD parameters should they be required for special applications.

The normalization of the two ABCD-parameter equations resulted in their matching the form of the standard complex linear update formula. Then it was shown that the impedance and admittance quantities appropriate to the ladder branch were simply B/A and C/D, respectively. Thus an L section, turned in the direction to match the next two ladder branches to be considered, allowed the recursion to proceed. One complex linear update variable had to be denormalized, stored, and then recalled; another had to be swapped into a nonstandard position. The process is somewhat intricate, but is independent of how or where the ABCD parameters were obtained. Thus the ABCD L-section method requires programming only in one subroutine, and this can be done with little programming code.

4.3.   Uniform Transmission Lines

Dissipative, uniform transmission lines with real characteristic impedances will be considered (see Murdock, 1979, for the even more general case, if required). These will be treated as embedded, cascaded two-port subnetworks in the ladder network environment, as just described, or as short- or open-circuited stubs having only two terminals. The latter will reduce to the same Z or Y case as the dissipative lumped elements treated in Section 4.1. There are many occurrences of these components in filter models, especially in micro-wave filters; one of the latter will be designed and analyzed in Section 9.1.

Because of prior formulation, this topic reduces to a consideration of the ABCD-parameter calculations, some useful approximations, functional-form shortcuts in the programming associated with the calculations, and additions to the topological list technique.

4.3.1. Transmission Line ABCD Parameters.

4.3.1.   Transmission Line ABCD Parameters.   The ABCD parameters for dissipative, uniform transmission lines have been given in many places (for instance, Matthaei et al., 1964, p. 28):

$$A = D = \cosh(NP + j\theta), \tag{4.13}$$

$$B = Z_0 \sinh(NP + j\theta), \tag{4.14}$$

$$C = Y_0 \sinh(NP + j\theta). \tag{4.15}$$

Real characteristic impedance Z_0 is the reciprocal of admittance Y_0, angle θ is the line electrical length at some frequency ω, and NP is the frequency-independent loss, in nepers, for that length of transmission line. Note that 1 neper = 8.686 dB. The hyperbolic functions above have complex arguments. They may be evaluated by the following indentities from Dwight (1961, pp. 153, 4):

$$\sinh(NP + j\theta) = \frac{e^{NP}(\cos\theta + j\sin\theta) - e^{-NP}(\cos\theta - j\sin\theta)}{2}, \tag{4.16}$$

$$\cosh(NP + j\theta) = \frac{e^{NP}(\cos\theta + j\sin\theta) + e^{-NP}(\cos\theta - j\sin\theta)}{2}. \tag{4.17}$$

Note the functional similarity; only one interior sign is different, so that one program segment with a flag variable should suffice for evaluation. Programmers of hand-held calculators should also note the efficiency of the polar-to-rectangular conversion of unity at angle θ to obtain $\cos(\theta) + j\sin(\theta)$ in one operation.

4.3.2.   Lossy Transmission Line Stubs.   A compact means for calculating the input impedance of a short- or open-circuited dissipative transmission line will be described. The analysis of Section 3.3.2, leading to (3.66), is directly applicable for an arbitrary load impedance at port 2 of a two-port network:

$$Z_1 = \frac{AZ_L + B}{CZ_L + D}, \tag{4.18}$$

where Z_1 is the stub input impedance, and the ABCD parameters are given in (4.13)–(4.15). For Z_L approaching infinity and zero, it follows that

$$Z_{1,OC} = \frac{Z_0}{\tanh(NP + j\theta)}, \tag{4.19}$$

$$Z_{1,SC} = Z_0 \tanh(NP + j\theta), \tag{4.20}$$

for the open- and short-circuited-stub cases, respectively. It is possible to avoid (4.16) and (4.17) entirely by the identity

$$\tanh(NP+j\theta) = \frac{\tanh(NP)+jy}{1+jy\tanh(NP)},$$ (4.21)

where the definition for y will occur in many places throughout this book:

$$y = \tan\theta.$$ (4.22)

There is no point in stopping with this exact result for two reasons. First, short- or open-circuit terminations in the real world are only approximate. Second, this fact provides an opportunity to drastically shorten the calculations in both programming steps and execution time. Consider the series approximation

$$\tanh NP = NP - \frac{NP^3}{3} + \frac{2}{15}NP^5 - \cdots.$$ (4.23)

For dissipation less than 1 dB, the second term is less than $1/226$ of the first term. An approximate stub input impedance expression is obtained for these assumptions:

$$Z_{in} = Z_0\frac{K+jy}{1+jKy}; \quad y = \tan\theta; \quad K = \begin{cases} NP & \text{for SC,} \\ NP^{-1} & \text{for OC.} \end{cases}$$ (4.24)

Clearly, only one subroutine with a flag variable will suffice for both kinds of stubs in ladder network analysis, using either exact Equations (4.19)–(4.21) or approximate Equation (4.24) with (4.22). Component menu descriptions with sample data will be given in Section 4.3.4.

4.3.3. Lossy Transmission Lines in Cascade.
Section 4.2.2 shows that parameters A and D are required for normalization purposes in the L-section formulation for embedded ABCD two-port subnetworks. They are equal, according to (4.13) and (4.17). Furthermore, the L-section branches are

$$\frac{B}{A} = Z_0\tanh(NP+j\theta),$$ (4.25)

$$\frac{C}{D} = Y_0\tanh(NP+j\theta),$$ (4.26)

where the considerations above for the tanh function apply. The approximate form is recommended only for the more limited hand-held computers. The rest of the calculation for the cascaded transmission line is accomplished as for any cascaded two-port subnetwork (see Section 4.2.2). The transmission line component topology code will be considered next.

4.3.4. Transmission Line Topology Codes.
Each cascade and stub transmission line component will require five items to describe it: type, nepers loss,

Table 4.2. Transmission Line Component Topology and Numerical Data

Type	Name	Z_0	NP	$\theta°$	Z_L	Z_{IN}
4	SC stub	70	0.0567	40.0	0	6.8506 + j58.4061
4	OC stub	−70	0.0567	40.0	1E10	9.7068 − j82.7569
5	Cascade	70	0.0691	171.54	20 − j30	29.0996 − j39.9780

characteristic impedance (Z_0), radian frequency reference, and the electrical length, in degrees, at that frequency. Optimization is anticipated, and the two variables for adjustment could be Z_0 and electrical length, in degrees, at a reference frequency. Both of these quantities will be in a reasonable numerical range (well scaled). Electrical length at any arbitrary frequency will be

$$\theta = \theta_0 \frac{\omega}{\omega_0} . \tag{4.27}$$

Table 4.2 shows some typical data for illustration and numerical testing. Stubs are type-4 components with the sign of Z_0 indicating the termination. The dissipative, uniform transmission line subnetwork in cascade is type 5. The stub input impedance was calculated by the approximate relationship in (4.24); the exact equations were used for the cascaded transmission line.

Register packing for hand-held calculators is easy. Two pairs of registers are employed for each topological entry (see Table 4.3). The first register in the first pair contains the type integer in the integer part and the nepers in the fractional part (the loss thus being limited to the realistic maximum of less than 1 neper, or 8.686 dB). The second register in the first pair contains Z_0, a potential optimization variable. The first register in the second pair contains the reference radian frequency, and the second register contains the electrical length at that frequency, also a potential optimization variable. Each transmission line component thus requires two topological pairs instead of just one, as previously encountered. Actually, a component could occupy as many register

Table 4.3. Topological Input Data for the Filter in Figure 4.9

4.05	1.0
2.9531E10	90.0
5.05	1.0
2.9531E10	90.0
−4.05	1.0
2.9531E10	90.0
5.05	1.0
2.9531E10	90.0
−4.05	1.0
2.9531E10	90.0

pairs as required, since the topology pointer (N in Section 4.1.3) is incremented in the subroutine for that particular component.

Example 4.2. A three-stub transmission line filter, to be designed in Section 9.1, is shown in Figure 4.9. It is composed of three shorted stubs that are separated by cascaded transmission lines. All elements are 90 degrees long at a bandpass center frequency of 4700 MHz (2.9531E10 radians). The characteristic impedance is normalized, so that $Z_0 = 1$ is assumed. Loss for each of the line lengths is 0.05 nepers, or 0.4343 dB. The hand-held calculator, paired-register topological description is shown in Table 4.3.

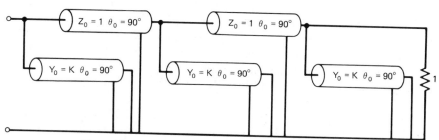

Figure 4.9. A microwave equal-stub filter from section 9.1.

4.3.5. *Transmission Line Summary.*

Dissipative, uniform transmission lines with real characteristic impedances (Z_0) have been analyzed using both exact and approximate methods. For two-terminal stubs with either open- or short-circuit terminations, the approximate analysis is suitable, because the terminations are realized with relative inaccuracy in practice. The faster execution and more easily programmed approximate method is especially attractive for hand-held calculators. There is less justification for an approximate calculation for cascaded transmission line subnetworks, but it is still an option for hand-held computers as opposed to desktop or larger machines. The approximation involves the computation of the hyperbolic tangent function with complex argument. Identity (4.21) utilizes only real arguments and enables the use of series approximation (4.23) for the tanh function. The first term of the loss-related series is usually satisfactory.

The menu of ladder network components was extended by type-4 stubs and type-5 cascaded transmission lines. The stub termination was indicated by the sign of the characteristic impedance, a negative number chosen to select the open-circuit termination. Optimization and other reasons lead to the choice of nepers, Z_0, radian reference frequency, and electrical degrees at that frequency as the four transmission line parameters. It was shown how these may be placed in calculator register pairs and remain consistent with the topological data previously defined. A microwave filter example indicated how elementary such a circuit description can be.

4.4. Nonadjacent Node Bridging

There are many applications where single components (e.g., R, L and C) are connected between nonadjacent nodes of a ladder network. The bridged-T network is often employed for servo lead/lag phase compensation and for group time delay equalization in radio circuits. There are also a number of narrow-band filter design methods that realize transmission loss poles by bridging nonadjacent nodes with L or C. This section considers convenient means for analyzing such networks without having to resort to the less efficient nodal analysis.

4.4.1. Derivation of Bridged-T Chain Parameters.

The approach for analyzing the bridged-T structure in Figure 4.10 is to find its ABCD parameters and then treat it as another cascaded two-port subnetwork, as described in Section 4.2. The four branch admittances may be composed of any number and kind of components. A specific delay equalizer, bridged-T arrangement will be considered in Section 4.4.2, and its ABCD parameters will be obtained using the following development.

Consider the separate two-port networks in Figure 4.11. The left one is the top branch of the bridged-T, and the right one is the remaining T structure. Paralleling these two structures produces the complete network in Figure 4.10. It will be shown that addition of the two separate short-circuit admittance matrices provides the short-circuit matrix of the entire bridged-T network.

To obtain the short-circuit parameters for the subnetworks, defining Equations (3.79) and (3.80) are recalled. It is easy to see for the subnetwork in Figure 4.11a that $y_{11a} = y_{22a} = Y_4$. Generally, $y_{21} = I_2/V_1$ when $V_2 = 0$. If $V_1 = 1$,

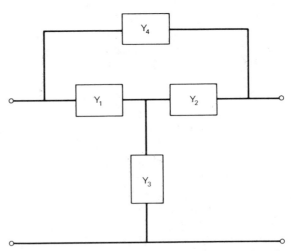

Figure 4.10. The bridged-T structure with admittance branches.

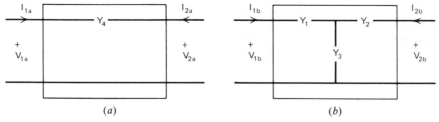

Figure 4.11. Two subnetworks that form a bridged-T network when paralleled. (*a*) Matrix $\mathbf{Y}_a$; (*b*) matrix $\mathbf{Y}_b$

then it is seen that $y_{21} = -Y_4$; the sign is due to the I_2 current convention. Furthermore, y_{12} has the same value (reciprocity).

The two-port subnetwork in Figure 4.11b has y_{11} equal to the admittance looking in when the output is shorted. It follows that

$$y_{11b} = \frac{Y_1(Y_2 + Y_3)}{Y_1 + Y_2 + Y_3}. \tag{4.28}$$

The output admittance y_{22b} is obtained from (4.28) by swapping the Y_1 and Y_2 variables. To find y_{21b}, the output must be shorted and the current-division rule applied to find the part of the input current that flows through Y_2 in Figure 4.11b. The input current I_1 is defined by $I_1 = y_{11b}$ when $V_1 = 1$. The current-division rule says that

$$-I_2 = I_1 \frac{Z_3}{Z_2 + Z_3}, \tag{4.29}$$

where $Z = 1/Y$ for each variable involved. This leads directly to

$$y_{21b} = \frac{-Y_1 Y_2}{Y_1 + Y_2 + Y_3}. \tag{4.30}$$

The two subnetworks in Figure 4.11 have been described by their y parameters. The independent variables are the terminal voltages; these coincide when the network terminals coincide. Also, $I_1 = I_{1a} + I_{1b}$ and $I_2 = I_{2a} + I_{2b}$, where I_1 and I_2 are the port currents for the complete bridged-T network in Figure 4.10. It follows from (3.79) and (3.80) that $y_{11} = y_{11a} + y_{11b}$, and an analogous argument applies for each of the other y parameters. This is simply an addition of the two subnetwork matrices. Thus the y parameters of the bridged-T network in Figure 4.10 are

$$y_{11} = \frac{Y_1(Y_2 + Y_3)}{Y_1 + Y_2 + Y_3} + Y_4; \tag{4.31}$$

$$y_{22} = \frac{Y_2(Y_1 + Y_3)}{Y_1 + Y_2 + Y_3} + Y_4; \tag{4.32}$$

$$y_{12} = y_{21} = -\frac{Y_1 Y_2}{Y_1 + Y_2 + Y_3} - Y_4. \tag{4.33}$$

A similar development could be accomplished by addition of open-circuit (z) matrices, as explained by Seshu and Balabanian (1959).

Section 3.4.3 showed how to find z and y parameters in terms of the ABCD parameters. The ABCD parameters in terms of the y parameters are found in an entirely analogous way. They are provided in the matrix identity

$$\begin{pmatrix} A & B \\ C & D \end{pmatrix} = \frac{-1}{y_{21}} \begin{pmatrix} y_{22} & 1 \\ y_{11}y_{22} - y_{21}y_{12} & y_{11} \end{pmatrix}. \tag{4.34}$$

It is tempting to substitute (4.31)–(4.33) into each element of (4.34). It is always worth checking algebraically to see if gross simplification or cancellation of mutual terms may be accomplished. But, for analysis purposes, it usually turns out that carrying forward the numerical results of important stages of the computation is by far the most effective procedure.

4.4.2. A Group Delay Equalizer.

Geffe (1963) has described the lowpass group delay equalizer shown in Figure 4.12. Suppose that the five component values have been determined, and what remains is the analysis task for this subnetwork. Using decrement $d = 1/Q$ for each component, a comparison of Figures 4.12 and 4.10 shows that

$$Y_1 = Y_2 = \omega C_1(d_{C1} + j1), \tag{4.35}$$

$$Y_4 = \left[\omega L_1(d_{L1} + j1) \right]^{-1}, \tag{4.36}$$

$$Y_3 = \left\{ \left[\omega C_2(d_{C2} + j1) \right]^{-1} + \omega L_2(d_{L2} + j1) \right\}^{-1}. \tag{4.37}$$

It is important to recognize that all parts of these calculations reside in the component type-2 and type-3 subroutines described in Section 4.1.4. So the bridged-T subroutine calls for the type-2 and type-3 subroutines to evaluate the main parts of (4.35)–(4.37). This means that the bridged-T topological input list must consist of four register pairs in the proper order, and it must control the topological pointer N appropriately before type-2 and type-3

Note: These sections may be used only if

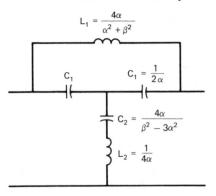

$$L_1 = \frac{4\alpha}{\alpha^2 + \beta^2}$$

$$C_1 = \frac{1}{2\alpha}$$

$$C_2 = \frac{4\alpha}{\beta^2 - 3\alpha^2}$$

$$L_2 = \frac{1}{4\alpha}$$

$$d_c = \frac{\alpha}{\beta} \leqslant \frac{1}{\sqrt{3}} = 0.577$$

Resonance Tests

Type IIIa:
1. L_1 resonates with $\frac{1}{2}C_1$ at
$\omega_1 = \sqrt{\alpha^2 + \beta^2}$

2. L_2 and C_2 resonate at
$\omega_2 = \sqrt{\beta^2 - 3\alpha^2}$

Figure 4.12. Geffe's type-IIIa lowpass group delay equalizer. [From Geffe, 1963.]

**Table 4.4. Topological Input
Data for the Geffe
Bridged-T Equalizer**

$6.d_{L_1}$	L_1 value
$6.d_{C_1}$	C_1 value
$6.d_{L_2}$	L_2 value
$6.d_{C_2}$	C_2 value

subroutines are called. The registers for storage of the ABCD parameters are available for all preceding computations. If the order of the calculations is well planned, it is not difficult to calculate the four complex numbers according to (4.35)–(4.37), then the short-circuit parameters in (4.31)–(4.33), and finally use the ABCD parameters in (4.34) to compute A, D, B/A, and C/D, as required in Section 4.2.2. The paired-register topological code for a type-6 bridged-T network appears as in Table 4.4.

4.4.3. Interpolation of Nonadjacent Node-Bridging Current. Figure 4.10 showed a T network bridged by a branch with admittance Y_4. It is often necessary to know what voltages and currents exist internal to such networks, e.g., the voltage across Y_3. Cases also occur where the bridging component bypasses more than one node. Moad (1970) described an approach that solves this problem efficiently. The following similar development is based on computing two of the bridged subnetwork's ABCD parameters and thus finding the bridging current. This establishes the correct output current for the bridged subnetwork, so that the ladder recursion may continue normally. In fact, the recursion also is used twice to find the two required ABCD parameters.

Consider the subnetwork in Figure 4.13, which is bridged by impedance Z_c. The ladder recursion method will arrive at port b with values determined for both node voltage, V_b, and I_K, the current in the K*th* (even) branch. If I_b (thus I_c) were known, the recursion could continue to the subnetwork's input

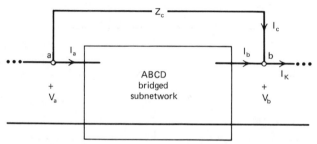

Figure 4.13. A subnetwork bridged by branch impedance Z_c.

port (node a in Figure 4.13), at which time I_c could be added to I_a, and the recursion could again proceed normally. Equation (4.7) applies to the bridged subnetwork. Kirchhoff's law for node b in Figure 4.13 yields

$$I_c = I_K - I_b. \tag{4.38}$$

Also, Figure 4.13 shows that

$$V_a - V_b = I_c Z_c. \tag{4.39}$$

Substitution of these two equations into (4.7) yields

$$I_b = \frac{I_K Z_c - V_b(A-1)}{B+Z_c}. \tag{4.40}$$

Only A and B need to be determined. But (4.7) shows that $A = V_a$ when $V_b = 1$ and $I_b = 0$; also, $B = V_a$ when $I_b = 1$ and $V_b = 0$. Two analyses of the ladder subnetwork using these output terminal conditions will provide A and B. Then (4.40) yields I_b, and (4.38) yields bridging current I_c.

This procedure begins when the recursive ladder method encounters a component-type code that indicates a node-b condition, as in Figure 4.13. V_b and I_K are saved and replaced by 1 and 0, respectively. The complex linear update is allowed to find $V_a = A$, and this is saved. Then the complex linear update is restarted at node b in Figure 4.13, with $V_b = 0$ and $I_b = 1$; it is allowed to find $V_a = B$. Then I_b is calculated according to (4.40); that and the saved V_b value are used to restart the complex linear update from node b for the third and last time. The subnetwork voltages and currents will then be correct. Upon arrival at node a in Figure 4.13, I_a is increased by I_c according to (4.38) and the ladder recursion method continued toward the ladder input terminals.

Example 4.3. Suppose that the embedded subnetwork is the bridged-T shown in Figure 4.14, with recursion variables $I_K = 2$ and $V_b = 3$. To find A according to its definition from (4.7), set $I_b = 0$, $V_b = 1$, and find V_a. Since there is no current through the 40-ohm branch, node-d voltage to ground must be 1, and the branch current must be 1/30. Therefore, $A = V_a = 1 + 50/30 = 8/3$. Setting $V_b = 0$ and $I_b = 1$, node-d voltage to ground must be 40 and

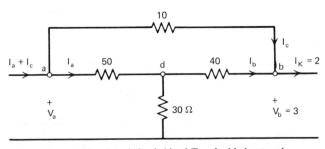

Figure 4.14. A resistive bridged-T embedded network.

$I_a = 1 + 40/30 = 7/3$. Thus $B = V_a = 40 + 50 \times 7/3 = 470/3$. Then (4.40) shows that $I_b = 3/100$ and (4.38) shows that $I_c = 1.970$.

4.4.4. Summary of Nonadjacent Node Bridging. A general method for finding the ABCD parameters of any bridged-T network was presented. Equations for the short-circuit (y) parameters were derived in terms of the four branch admittances. These can be converted to ABCD parameters for use in the embedded subnetwork technique described in Section 4.2. A time delay equalizer bridged-T network was discussed as an example. It was emphasized that existing subroutines for RLC impedance computation could be called by the bridged-T-component subroutine to minimize computer coding.

For T networks and more extensive subnetworks that are bridged, there often is a need to find the internal voltages and currents. The approach above does not provide this information and will not solve the larger problem in any case. An efficient technique, which uses the ladder recursion scheme two extra times to find the A and B chain parameters of the bridged subnetwork, was described. These values, the bridging branch impedance, and the bridged subnetwork's known output voltage and current enable the simple calculation of the bridging current. Thus the ladder recursion scheme may proceed through the bridged subnetwork, calculating correct voltages and currents as in the unbridged situation.

Network analysis that includes bridge subnetworks must be conducted with the possibility that a null condition might occur. Therefore, division such as in (4.40) should be protected by the addition of $1E-9+j0$ to the denominator.

4.5. Input and Transfer Network Responses

The methods described so far make it easy to obtain the input voltage and current of a ladder network given the topological data and load power. Several input and transfer response functions often required in practice will be described. Quantities related to impedance and power will be defined first. Then a definition of scattering parameters will be given as a basis for certain wave response functions and for important applications later in this book. Logarithms (log) in the following equations are with respect to base 10.

4.5.1. Impedance and Power Response Functions. Assume that ladder network input voltage and current are available (see V_1 and I_1 in Figures 3.4 or 3.7, for example). Then the input impedance is

$$Z_1 = \frac{V_1}{I_1} = R_1 + jX_1. \qquad (4.41)$$

This calculation is made in Program B4-1 by lines 9955–9985. Line 1381 detects that the input has been reached, because the component-type integer is zero. If the next branch number is even, then the last processed branch is in

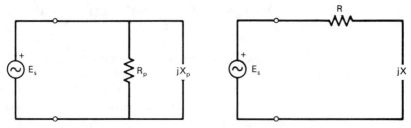

Figure 4.15.   Series and parallel impedance forms.

shunt (see Figures 4.3b and 4.3c). Then the voltage and current are in the complex linear update (Table 4.1), so that lines 9970–9985 are correct for impedance calculation. If the next branch is odd, then line 9965 in Program B4-1 swaps the respective real and imaginary parts of the voltage and current; otherwise, the calculation would have produced the admittance instead of the impedance.

The parallel impedance form is often required when a lossless voltage source exists and for various other reasons. The parallel-ohms form is much more widely accepted than admittance mhos; the latter, when it is used (primarily by microwave engineers), is often given in millimhos. The parallel input impedance form is

$$R_p = \frac{R^2 + X^2}{R} ; \qquad X_p = -\frac{(R^2 + X^2)}{X} , \qquad (4.42)$$

where R_p is the reciprocal conductance, and X_p is the negative of the reciprocal susceptance (see Figure 4.15).

It is well known that power P may be computed as the real part of the product of sinusoidal voltage and conjugated current:

$$P = Re(VI^*). \qquad (4.43)$$

Therefore, the power input to the network is

$$P_1 = V_{1r}I_{1r} + V_{1i}I_{1i} , \qquad (4.44)$$

where subscripts r and i denote real and imaginary parts, respectively.

It is assumed that power delivered to the load impedance was an arbitrary independent variable the user specified. Therefore, efficiency in dB loss is

$$\eta = 10 \log_{10} \frac{P_1}{P_L} dB. \qquad (4.45)$$

Negative values of (4.45) imply an active network with power gain, i.e., $P_L > P_1$.

4.5.2.   Scattering Parameters.   Two-port network equations have been written in terms of ABCD (chain) parameters in Section 3.3.2 and in terms of both

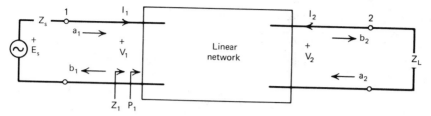

Figure 4.16. Linear network port scattering waves.

z and y open- and short-circuit parameters, respectively, in Section 3.4.3. In exactly the same way, the scattering parameter equations are

$$b_1 = S_{11}a_1 + S_{12}a_2,$$ (4.46)

$$b_2 = S_{21}a_1 + S_{22}a_2.$$ (4.47)

The variables labeled a_p (p=1 or 2) in Figure 4.16 are called the incident waves, and those labeled b_p are the emerging waves. Kurokawa (1965) defined the scattering variables in terms of the port voltages and currents:

$$a_p = \frac{V_p + Z_p^n I_p}{2\sqrt{R_p^n}},$$ (4.48)

$$b_p = \frac{V_p - Z_p^{n*} I_p}{2\sqrt{R_p^n}},$$ (4.49)

where p is 1 or 2, corresponding to the ports shown in Figure 4.16. The waves may also be interpreted in terms of signal flow graphs (see Hewlett–Packard, 1972).

Note the possibly complex, port-normalizing (reference) impedances Z_p^n in (4.48) and (4.49); they may or may not be equal to the actual source and load impedances. This subject will be treated in more detail in Section 7.1. It is seen from (4.48), with p=1 and $Z_s = Z_1^n$, that the numerator is equal to E_s. Then, the port-1 incident wave is

$$a_1 = \frac{V_1 + Z_1^n I_1}{2\sqrt{R_1^n}} = \frac{E_s}{2\sqrt{R_1^n}}.$$ (4.50)

The squared magnitude of (4.50) is recognized as the maximum power available from the Z_1 source, as defined by (3.45). This fact and the following development show that the coefficients in (4.46) and (4.47) have units that are the square root of power.

The net real power incident on a port turns out to be

$$P_p = |a_p|^2 - |b_p|^2.$$ (4.51)

This can be confirmed by substituting (4.48) and (4.49) into (4.51). The power identity in (4.43) results after some algebra, using the identities

$$|Z|^2 = ZZ^*, \qquad 2\,Re\,Z = Z + Z^*. \tag{4.52}$$

The reflection coefficients looking into port p are defined by

$$\rho_p = \frac{b_p}{a_p}. \tag{4.53}$$

But definitions (4.48) and (4.49) in (4.53) yield

$$\rho_p = \frac{Z_p - Z_p^{n*}}{Z_p + Z_p^n}. \tag{4.54}$$

It is important to note that this is essentially (3.46), which enabled a simple calculation of power transfer from a complex source to a complex load. Kurokawa (1965) discusses the differences in reflection of power waves and traveling waves on transmission lines with real or complex Z_0. In general, traveling waves are not closely related to power.

Finally, the transducer function, S_{21}, is defined by (4.47):

$$S_{21} = \frac{b_2}{a_1}\bigg|_{a_2=0} \tag{4.55}$$

The side condition that there be no reflection from the load is important in itself; it requires that $Z_L = Z_2^n$. (Why?) Using (4.48) and (4.49) in (4.55) and equating a_2 from (4.48) to zero yields the general transducer function

$$S_{21} = \frac{1 + Z_2^{n*}}{Z_2^n} \sqrt{\frac{R_1^n}{R_2^n}} \frac{V_2}{E_s}; \qquad Z_s = Z_1^n. \tag{4.56}$$

4.5.3. Wave Response Functions.   Scattering parameters normalized to complex port impedances will be used throughout Chapter Seven; the more familiar case of real port-normalizing impedances will be assumed. Also, the source impedance will be assumed to be equal to the port-1 normalizing resistance R_1. Then, the input reflection coefficient from (4.54) is

$$\rho_1 = \frac{Z_1 - R_1}{Z_1 + R_1}, \tag{4.57}$$

which is the same as (3.48). When $Z_L = Z_2 = R_2 + j0$, (4.57) is equal to coefficient S_{11} in (4.46). The reflection coefficient looking into port 2 is defined in a similar way, and will be used in Section 6.7. A low reflection coefficient magnitude indicates a high-quality impedance match as Z_1 approaches R_1. Three ways to express this condition are return loss, standing-wave ratio (SWR), and mismatch loss.

Return loss is commonly used in microwave design; it is defined to be:

$$RL = -20\log_{10}|\rho| \text{ dB.} \tag{4.58}$$

The standing-wave ratio is the ratio of voltage or current maxima to minima on uniform transmission lines, where a standing wave may exist as a result of load reflection. It is defined as

$$SWR = \frac{1 + |\rho|}{1 - |\rho|}, \tag{4.59}$$

where the reflection magnitude is usually obtained from (4.57). However, if (4.54) defines a general reflection coefficient, then (4.59) may be interpreted as an arbitrary, real function of a complex variable. It is also a well-behaved function that works well in network optimization. This will be discussed further in Section 5.5.

Mismatch loss is the ratio of power delivered to power available at an interface; it is simply (3.47) from Section 3.2.3, expressed in dB:

$$MIS = -10 \log_{10}(1 - |\rho|^2) \, dB. \tag{4.60}$$

The basis of network synthesis in Chapter Three was a lossless, two-port network. The issue was thus the power transferred from the source, since it had nowhere else to go but to the load impedance. The transducer function for general linear two-port networks considers both mismatch loss and dissipative loss (efficiency) or network activity (gain).

The transducer loss is the sum of mismatch loss (4.60) and efficiency loss (4.45). It also may be computed in terms of the forward scattering transfer parameter S_{21}:

$$TL = -20 \log_{10}|S_{21}|. \tag{4.61}$$

The expression for S_{21} in (4.56) allows a complex source and load if they are equal to their respective port-normalizing impedances. The important situation when they are not so related is discussed in Section 7.1. As mentioned, the more familiar case occurs with real terminations equal to their respective port-normalizing resistances. Then (4.56) simplifies to

$$S_{21} = 2\sqrt{\frac{R_1^n}{R_2^n}} \frac{V_2}{E_s} \quad \text{if } Z_2^n \text{ is real and} \quad Z_s = Z_1^n, \tag{4.62}$$

which is simply a scaled ratio of the voltages shown in Figure 4.16. The reader should remember the conditions that are attached to (4.62).

Finally, there is an extremely simple way to calculate (4.62) when using the ladder recursion scheme from Section 4.1: just add on a series branch, namely source resistance R_1, at the network's input terminal. Usually, the program is made to pause at the input terminals so that some of the other responses described above can be computed. When the recursion completes one more cycle, the source voltage E_s is obtained. The angle of S_{21} is available immediately:

$$\theta_{21} = -\arg E_s. \tag{4.63}$$

This is valid because the load current phase is the zero-degree reference and the load impedance is a resistance. Equation (4.62) shows that when the load power is set equal to

$$P_L = \frac{1}{4R_1^n},\tag{4.64}$$

then

$$TL = 20\log|E_s|.\tag{4.65}$$

4.5.4. Conclusion to Network Responses. Some impedance and power response functions were described. Scattering parameters were defined by two linear, complex equations in the same style as previously used for ABCD, z, and y parameters. The mathematics of scattering parameters is straightforward, but lack of familiarity with the general case makes its development worthwhile. For instance, scattering parameters are often described for a 50-ohm port reference impedance. This is the circumstance that has revolutionized accurate, automatic measuring equipment for all kinds of networks over extremely wide frequency ranges. However, there are some network responses that are explained better in terms of scattering parameters with some arbitrariness of port normalization, and this will be a necessity in Chapter Seven. Various wave response functions were then defined, and an efficient means for extending the ladder recursion analysis method for S_{21} calculation was explained.

Singly terminated (lossless source) responses have not been mentioned explicitly (e.g., V_2/V_1 in Figure 4.16). It is easy to extract these numbers from a ladder network analysis algorithm and calculate the logarithm of that magnitude. Unfortunately, the selectivity expression of interest typically is

$$SEL = 20\log_{10}\left|\frac{V_{ref}}{V_L}\right| dB,\tag{4.66}$$

where V_{ref} may be the input voltage at a midband frequency. This reference voltage may be contingent upon a certain input current or power, or similar load conditions. Then the excitation will have to be maintained at that level at all response frequencies. This can be confusing when the analysis scheme requires load power to be specified at every frequency. Experience has shown that one should not approach these definitions carelessly. Renormalizing at each frequency, by making the source excitation variable equal to unity, helps to eliminate confusion.

Finally, most response functions have an associated angle that makes the calculation of group time delay possible. Group delay, in seconds, is defined to be

$$T_G = -\frac{d\phi}{d\omega},\tag{4.67}$$

where angle ϕ is in radians. Time delay may be converted to degrees per megahertz by multiplying (4.67) by 360E6; this is especially useful for oscillator frequency stability calculations. Numerical differentiation will be discussed in Section 4.7.2. It is the easiest way to compute time delay, but it trades execution time for computer coding. Tellegen's theorem (Section 4.7.3) provides a basis for computing exact time delay and its exact sensitivities, but this particular application requires much more memory and coding. Therefore, it is recommended that the angle be computed at a frequency perturbed (increased) by 0.01% (a 1.0001 factor) and again at the desired frequency. The difference between these two angles, in degrees, is used in the numerator of the formula

$$T_G = \frac{-\Delta\phi}{0.036\,f_0}, \qquad (4.68)$$

which gives the delay in reciprocal frequency units. Suppose that the frequency of interest—f_0 in (4.68)—is 50 MHz. According to Program B4-1 in Section 4.1.4, the frequency units would be input as 1E6; therefore, the time delay would be in microseconds when calculated by (4.68). The only remaining problem is the occasional 360-degree jump in the calculated angle that might occur between the perturbed and desired frequencies. A simple program test can prevent this.

4.6. Time Response From Frequency Response

For most industrial engineers, there has been a gap between academic concepts and applied design and analysis. This section uses a desktop computer to close that gap for the Fourier and convolution integrals. A means for rapid steady-state frequency analysis of ladder networks has been developed that requires very little code and avoids most trivial calculations, such as complex multiplication by zeros and ones. This makes practical a method of numerical evaluation of the Fourier integral and, subsequently, numerical evaluation of the convolution integral. This enables the conversion of a system's band-limited frequency response to its impulse response in the time domain. Then the convolution integral enables the response to any arbitrary time excitation to be calculated in a reasonable amount of time, using a desktop microcomputer. Complicated networks and dense time samples could weaken this claim; the understanding of this process and its fundamental simplicity may be reward enough for design engineers. After all, bigger and faster computers are always available at some additional expense of time and convenience.

   This section begins with a review of the Fourier integral under the conditions that the system impulse time response is a real function and causal, i.e., cannot anticipate the excitation. Then Simpson's rule for numerical integration will be applied, as previously discussed. Finally, the convolution integral will also be evaluated by Simpson's rule according to a related general formula. This material follows Ley (1970), and the program has been adapted to BASIC language from the original FORTRAN.

4.6.1. Real, Causal Fourier Integrals.   The familiar Fourier integral

$$h(t) = \frac{1}{2\pi} \int_{-\infty}^{\infty} H(\omega) e^{j\omega t} \, d\omega \qquad (4.69)$$

may produce values of h(t) that are complex and nonzero in negative time. The system frequency response is $H(\omega)$ and the corresponding time impulse response is h(t). Those wishing to review its applications should see Blinchikoff and Zverev (1976). When h(t) must be real and also causal (zero in negative time), (4.69) may be written as

$$h(t) = \frac{2}{\pi} \int_{0}^{\infty} R(\omega) \cos \omega t \, d\omega, \qquad (4.70)$$

where $R(\omega)$ is the real part of the system frequency response function H. In practice, the integration is completed only to some finite frequency on the assumption that H is band limited, e.g., is zero above some limiting frequency.

Example 4.4.   Suppose that $H(\omega)$ is the impedance of a parallel RC network, i.e., the response function V/I. This could be calculated by an analysis program in the general case. Here, use the equation

$$H(s) = \frac{1}{s+1}. \qquad (4.71)$$

Then the real part is

$$R(\omega) = \frac{1}{1+\omega^2}. \qquad (4.72)$$

Appendix B BASIC Program B4-2 calculates 25 values of $R(\omega)$ from 0 to 12 radians in program lines 130–190. Clearly, (4.72) is a band-limited function. Running Program B4-2 shows that the real part is only 0.0069 (21.6 dB loss) at 12 radians. Program lines 200–350 evaluate (4.70) for the impulse response; particularly, lines 300–329 implement Simpson's rule (2.32) for the numerical integration required. Running Program B4-2 from the beginning shows the frequency samples and then the impulse response samples. They correspond reasonably well with

$$h(t) = e^{-t}, \qquad (4.73)$$

which is the exact impulse function corresponding to (4.71), the Laplace transform pair. The BASIC function on line 115 is there simply to slow the program output rate.

4.6.2. Numerical Convolution of Time Functions.   The convolution integral is defined by

$$F_o(t) = \int_{0}^{t} h(t - \tau) F_i(\tau) \, d\tau, \qquad (4.74)$$

where τ is the dummy variable of integration, F_i is the excitation function, and F_o is the system output function. The system impulse response is h(t). The case of (4.73) is shown in Figure 4.17. Convolution involves folding, shifting,

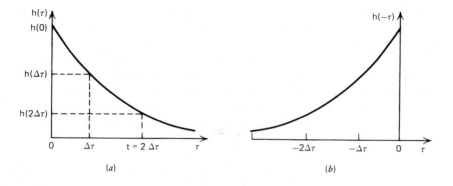

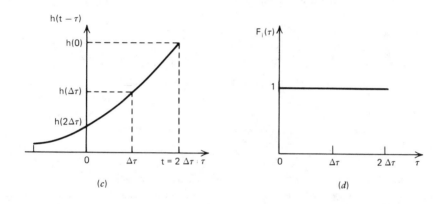

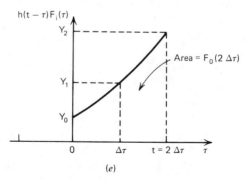

Figure 4.17. Graphical interpretation of the convolution integral for an exponential impulse response. (a) Impulse response; (b) folding; (c) shifting; (d) driving function; (e) product to be integrated. [From Ley, 1970.]

multiplying, and integrating the proper functions. Figure 4.17a is the impulse response function shown with values for τ at 0, $\Delta\tau$, and $2\Delta\tau$. It is folded by its negative argument in Figure 4.17b, then shifted by the amount $2\Delta\tau$ in Figure 4.17c. The unit-step excitation shown in Figure 4.17d (an arbitrary choice) multiplies the shifted function according to (4.74), with the resulting integrand in Figure 4.17e. This area is the output function F_o at time $t = 2\Delta\tau$, the shift interval.

The second application of Simpson's rule is the convolution process illustrated in Figure 4.17. Integration of Figure 4.17e uses the three samples

$$f_0 = h(2\Delta\tau)F_i(0),$$
$$f_1 = h(\Delta\tau)F_i(\Delta\tau), \tag{4.75}$$
$$f_2 = h(0)F_i(2\Delta\tau),$$

according to the integrand in (4.74). Then the integral estimate by (2.32) is

$$F_o(2\Delta\tau) \simeq \frac{\Delta\tau}{3}(f_0 + 4f_1 + f_2). \tag{4.76}$$

This result can be compared to a general expression by Ley (1970):

$$F_o(k\Delta\tau) = \frac{\Delta\tau}{3}(f_0 + 4f_1 + 2f_2 + \cdots + 2f_{n-2} + 4f_{n-1} + f_n), \tag{4.77}$$

where

$$f_n = h(t-\tau)F_i(\tau),$$
$$t = k\Delta\tau; \qquad k = 2, 4, 6, \ldots, \tag{4.78}$$
$$\tau = n\Delta\tau; \qquad n = 0, 1, 2, \ldots, k.$$

The algorithm calls for a choice of k, the even number of integration intervals, and letting n vary from 0 to k to obtain the output time response $F_o(k\Delta\tau)$. The reader is urged to write the algebraic expressions in (4.77) and (4.78) for $k = 2$ to confirm the (4.76) case shown in Figure 4.17e.

Lines 480–620 in Program B4-2 accomplish this numerical convolution. The unit-step excitation is computed in lines 410–430 and provides an estimate of the exact step response

$$g(t) = 1 - e^{-t}. \tag{4.79}$$

For engineering accuracy, about 100 frequency samples and 40-dB band limiting are required.

4.6.3. *Time Response Summary.*

Simpson's rule for numerical integration has been employed for both the Fourier and convolution integrals. The Fourier integral can be evaluated over a finite range for band-limited response functions. Furthermore, its integrand is the product of the system transfer function's real part and the cosine function when the system has a real impulse response that is zero in negative time.

It does not take long to compute and save 100 frequency response samples for fairly complicated ladder networks. These are used to compute and save the corresponding impulse response time samples. Finally, the numerical convolution integral can be evaluated for arbitrary time functions specified at matching time samples. The memory requirements for practical problems usually fit easily into desktop computers having 8–32 kilobytes of random-access memory.

The fast Fourier transform (FFT) must be mentioned before leaving this topic. It is clearly superior to the above and is available as standard software from several desktop computer manufacturers. Serious users of the frequency-to-time-domain transform should consider special programs built around this technique, which are more efficient by at least an order of magnitude. It was not described because of its computational complexity.

4.7. Sensitivities

Sensitivity quantifies the relative change in a response function (Z) with respect to a relative change in any one of several independent variables; i.e.,

$$S_{x_k}^Z = \frac{\Delta Z / Z}{\Delta x_k / x_k} \tag{4.80}$$

for small changes in the kth variable x_k. Often, Z is complex, and is evaluated at some given frequency. In this case, the sensitivity is also a complex number. For example, Z might be a ladder network input impedance, and x_k might be an inductance value in microhenrys. Alternatively, Z might be a time function evaluated at some given time. Hopefully, a system being built will have sensitivities with magnitudes less than unity, otherwise it might react badly to component tolerances and to its environment.

For each response, there are as many sensitivity numbers at a frequency or time value as there are variables in the problem. Applications include component tolerances, optimization (Chapter Five), and large-change calculations, e.g., network tuning. This section further defines real and complex sensitivities, relates them to partial derivatives, shows ways to obtain partial derivatives approximately by finite differences and exactly by Tellegen's theorem, and provides several examples. Most of the discussion is limited to the frequency domain, as justified in Section 4.7.4. Programs A2-1 and B4-1 will be used for calculations.

4.7.1. Sensitivity Relationships. The partial derivative operator abbreviation

$$\Lambda_k = \frac{\partial}{\partial x_k} \tag{4.81}$$

will be used throughout. As the change in the variable, Δx_k, approaches zero, (4.80) approaches the common sensitivity definition

$$S_{x_k}^Z = \frac{x_k}{Z} \frac{\partial Z}{\partial x_k}, \tag{4.82}$$

which can be viewed as a normalized partial derivative. Recalling the derivative formula for natural logarithms (ln), (4.82) can also be written as

$$S_{x_k}^Z = \frac{\Lambda_k \ln Z}{\Lambda_k \ln x_k}, \tag{4.83}$$

which shows that the normalized derivatives are describing relative changes in logarithmic space. Optimization (automatic component adjustment) in logarithmic space often is better behaved because of the normalization of partial derivatives that would otherwise be badly scaled (grossly different magnitudes).

Suppose that $Z = |Z|e^{j\theta}$. Then differentiation of Z in the right-hand term in (4.82) follows the rule for differentiation of a product, namely $d(uv) = v\,du + u\,dv$. It follows that

$$S_{x_k}^Z = \frac{x_k}{|Z|e^{j\theta}}\left[(\Lambda_k|Z|)e^{j\theta} + j|Z|(\Lambda_k\theta)e^{j\theta}\right], \tag{4.84}$$

which reduces to a useful identity:

$$S_{x_k}^Z = S_{x_k}^{|Z|} + j\theta S_{x_k}^\theta. \tag{4.85}$$

This says that when the complex sensitivity of a complex response function is obtained, the real sensitivities of both the magnitude and angle (phase) are immediately available.

First-order prediction of response behavior for small changes in several independent variables may be derived by recalling the total differential

Table 4.5.   Useful Identities for Partial Derivative Applications

1.	$Z = U + jW;\quad Z^* = U - jW.$
2.	$\Lambda Z = \dfrac{\partial Z}{\partial x_k} = |\Lambda Z|e^{j\theta}.$
3.	$\Lambda Z = \Lambda U + j\Lambda W.$
4.	$\Lambda(\alpha V + \beta I) = \alpha(\Lambda V) + \beta(\Lambda I);\qquad \alpha$ and β are scalars.
5.	For $Z = |Z|e^{j\phi}$: $\Lambda|Z| = Re\,\dfrac{Z^*(\Lambda Z)}{|Z|} = |\Lambda Z|\cos(\theta - \phi),$ $\Lambda\phi = Im\,\dfrac{(\Lambda Z)}{Z}$ seconds. Multiply by 360E6 to get degrees/MHz.
6.	$\Lambda|Z|^2 = 2[U(\Lambda U) + W(\Lambda W)] = 2\,Re[Z^*(\Lambda Z)].$
7.	$\dfrac{\partial Z_2}{\partial x_k} = \dfrac{\partial Z_2}{\partial Z_1}\dfrac{\partial Z_1}{\partial x_k}$ if x_k is only in domain of Z_1.
8.	$\Lambda \log_{10} U = \dfrac{\log_{10} e}{U}(\Lambda U).$
9.	$\log_{10} e = 0.434\ 294\ 482.$
10.	Insertion-loss ratio $= |s_{21}|^2\dfrac{(R_1 + R_2)^2}{4R_1 R_2}.$

formula:

$$dZ \simeq \Lambda_1 Z \Delta x_1 + \Lambda_2 Z \Delta x_2 + \cdots + \Lambda_n Z \Delta x_n. \qquad (4.86)$$

Dividing both sides by Z and placing x_k/x_k in each term on the right-hand side yields

$$\frac{dZ}{Z} \simeq S_{x_1}^Z \frac{\Delta x_1}{x_1} + S_{x_2}^Z \frac{\Delta x_2}{x_2} + \cdots + S_{x_n}^Z \frac{\Delta x_n}{x_n}. \qquad (4.87)$$

This shows that the relative change in a (complex) response is approximately the sum of relative signed changes of all independent variables weighted by the (complex) signed sensitivity numbers. Table 4.5 provides some useful identities for partial derivatives of complex variables.

4.7.2. Approximate Sensitivity. It is essential that the reader feel comfortable about partial derivatives, especially those that are complex. First-order finite differences will be explained because it is a practical method and should convince even the most apprehensive reader that partial derivatives are nice. It is presumed that the connection between real-function slope and derivative can be recalled, particularly as it defines an ordinary real first derivative. The k*th* variable x_k has been discussed; a formal notation of the entire set of variables needs to be introduced; it is called a column vector:

$$\mathbf{x} = \begin{bmatrix} x_1 \\ x_2 \\ \vdots \\ x_k \\ \vdots \\ x_n \end{bmatrix}. \qquad (4.88)$$

It may be written in row form, using the transpose operator that swaps rows and columns:

$$\mathbf{x} = (x_1, x_2, \cdots, x_k, \cdots, x_n)^T. \qquad (4.89)$$

A convenient definition of a finite-difference approximation to a partial derivative is now possible:

$$\Lambda_k Z_{in} \simeq \frac{Z_{in}(\mathbf{x} + \Delta x_k) - Z_{in}(\mathbf{x})}{\Delta x_k}. \qquad (4.90)$$

For instance, suppose that there is a ladder network with n L's and C's. For their nominal values residing in the vector **x** defined by (4.88), the input impedance $Z_{in}(\mathbf{x})$ is computed at a frequency that does not change. Now the k*th* component x_k is changed by a small amount, Δx_k, and the slightly different input impedance $Z_{in}(\mathbf{x} + \Delta x_k)$ is calculated. These three numbers, two being complex, are used in (4.90) to approximate the partial derivative. It requires $n+1$ complete analyses of the ladder network to get all n partial

Table 4.6.   First-Order Finite Differences for the Network in Example b in Figure 4.3

k	$\Delta x = x * \Delta$ Δ	$\Lambda^+ Z_{in}$	$\Lambda^- Z_{in}$
L	0.0001	$-0.001799 + j0.003402$	$-0.001800 + j0.003405$
L	0.01	$-0.001764 + j0.003363$	$-0.001836 + j0.003445$
C	0.0001	$-0.001999 + j0.03691$	$-0.002000 + j0.03692$
C	0.01	$-0.001968 + j0.03651$	$-0.002032 + j0.03733$

derivatives. How much is x_k perturbed? Computers with 7 to 10 decimal-digit mantissas require x_k to be increased by about 0.01% (a 1.0001 factor). If it is much less than that, the change in Z_{in} may fall off the end of the mantissa's digits, and no change is seen. If it is much more than that, this linear approximation of slope is too crude. It is easier to talk about the latter "truncation" problem in terms of the Taylor series, which will be discussed in the next chapter.

The network in Example b in Figure 4.3 was analyzed by Program B4-1; its input impedance was calculated for 0.01 and 1% changes in each variable, namely L and C. The perturbation was tried as an increase and as a decrease. The input impedances were employed in (4.90), which was evaluated using Program A2-1. The results are shown in Table 4.6. These values differ from exact results in the third significant figure.

4.7.3.   *Exact Partial Derivatives by Tellegen's Theorem.*   There are several exact means for finding derivatives of complex network functions. It will be shown in Section 7.1 that the coefficients of bilinear functions, which have the form of (2.1) or (4.18), can be determined by only three independent function evaluations. Because the derivative of the bilinear function can be written easily, its exact value is also available with respect to one of the n variables. Fidler (1976) has given a means to obtain the exact partial derivatives of a bilinear function with respect to n variables in just $2n + 1$ function evaluations.

However, Tellegen's theorem enables the calculation of *exact* partial derivatives of complex responses with respect to all n variables in just *one* or *two* network analyses, depending on whether the response is at only one end of the network or is a transfer function, respectively. This is a spectacular result, and the computer memory requirements for variables and code are not too severe for desktop computers. Branin (1973) and others have observed that the same result is available algebraically with a slight savings in computation; so Tellegen's theorem is not really necessary. Even so, it is worth knowing for its general enlightenment and compactness. Penfield et al. (1970) have neatly derived 101 fundamental theorems in electrical engineering using Tellegen's theorem. They correctly claim that no circuit designer should be without it.

Tellegen's theorem states that for any two entirely different (or identical) linear or nonlinear networks (N and $\hat{N}$) having the same branch topology and obeying at least one Kirchhoff law, the respective branch voltage and current sets (vectors) have null inner products; i.e.,

$$\mathbf{V}^T\hat{\mathbf{I}}=0=\hat{\mathbf{V}}^T\mathbf{I}. \tag{4.91}$$

This also applies in time or frequency domains. The second network, $\hat{\mathbf{N}}$, is called the adjoint network; it may or may not be different. First, observe that an inner product is defined in terms of two vectors such as (4.88) and (4.89); suppose that they are the n-element vectors $\mathbf{x}$ and $\mathbf{y}$. Then the inner product of $\mathbf{x}$ and $\mathbf{y}$ is

$$\mathbf{x}^T\mathbf{y}=x_1y_1+x_2y_2+\cdots+x_ny_n. \tag{4.92}$$

Example 4.5. Apply Tellegen's theorem to the networks from Figure 4.3b and c; they are reproduced in Figure 4.18b and c. The branch-4 current arrow has been reversed so that each branch has its current entering its positive voltage, consistent with each branch in the common topology shown in Figure 4.18a. All branch voltage and current values are shown in Figure 4.18 as found by Programs B4-1 and A2-1. Tellegen's theorem says that

$$\mathbf{V}^T\hat{\mathbf{I}}=V_1\hat{I}_1+V_2\hat{I}_2+V_3\hat{I}_3+V_4\hat{I}_4=0. \tag{4.93}$$

In fact, using Program A2-1, $\mathbf{V}^T\hat{\mathbf{I}}=0.0085+j0.0116$, which is as close to zero as might be expected for the digits carried. There are three more such inner products that should be equal to zero: $\mathbf{V}^T\mathbf{I}$, $\hat{\mathbf{V}}^T\hat{\mathbf{I}}$ and $\hat{\mathbf{V}}^T\mathbf{I}$. Evaluate them using the data from Figure 4.18. The reader should write a brief program that calls Program A2-1 subroutines in order to calculate the inner products of complex numbers; it is much easier, and a lot of time will be saved and errors avoided.

Penfield et al. (1970) generalize the Tellegen theorem statement to include the conjugation and any linear operator; for the partial derivative operator with respect to any variable

$$\Lambda\mathbf{V}^T\hat{\mathbf{I}}=0=\hat{\mathbf{V}}^T\Lambda\mathbf{I}. \tag{4.94}$$

Consider the network in Figure 4.18b to be its own adjoint network N and $\hat{\mathbf{N}}$. The port input impedance is

$$Z_{in}=\frac{V_4}{-I_4}, \tag{4.95}$$

and its partial derivative with respect to L yields

$$-\Lambda V_4=\Lambda Z_{in}\cdot I_4, \tag{4.96}$$

where currents are the independent variables. The branch-2 equation in terms of independent current I_2 is

$$V_2=I_2\cdot\omega L(d+j1), \tag{4.97}$$

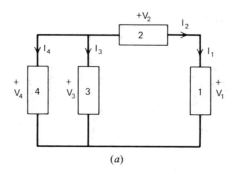

(a)

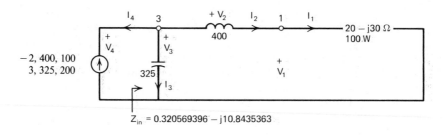

$$Z_{in} = 0.320569396 - j10.8435363$$

50 MHz; nH, pF

$V_1 = 44.7216 - j67.0820$ $I_1 = 2.2361 + j0$

$I_2 = 2.2361 + j0$ $V_2 = 2.809883 + j280.9966$

$V_3 = 47.5313 + j213.9106$ $I_3 = -21.8164 + j4.9623$

$I_4 = +19.5803 - j4.9622$ $V_4 = 47.5313 + j213.9106$

(b)

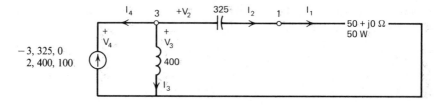

50 MHz; nH, pF

$V_1 = 50 + j0$ $I_1 = 1 + j0$

$I_2 = 1 + j0$ $V_2 = 0 - j9.794127$

$V_3 = 50 - j9.7942$ $I_3 = -0.073954 - j0.398627$

$I_4 = -0.9260 + j0.3986$ $V_4 = 50 - j9.7942$

(c)

Figure 4.18. Two different networks having the same topology. (a) General topology; (b) original network N; (c) adjoint network $\hat{N}$.

106

and its derivative with respect to L is

$$\Lambda V_2 = I_2 \cdot \omega(d + j1). \tag{4.98}$$

Using the differentiation operator on the original network N, the left-hand side of (4.94) is

$$\Lambda V_4 \cdot \hat{I}_4 + \Lambda V_3 \cdot \hat{I}_3 + \Lambda V_2 \cdot \hat{I}_2 + \Lambda V_1 \cdot \hat{I}_1 = 0, \tag{4.99}$$

and the right-hand side of (4.94) is

$$\hat{V}_4 \cdot \Lambda I_4 + \hat{V}_3 \cdot \Lambda I_3 + \hat{V}_2 \cdot \Lambda I_2 + \hat{V}_1 \cdot \Lambda I_1 = 0. \tag{4.100}$$

Then, subtract (4.100) from (4.99), separate the pair of port terms from the rest, and substitute (4.96) and (4.98):

$$\Lambda Z_{\text{in}} I_4 \hat{I}_4 + \hat{V}_4 \Lambda I_4^0 = \Lambda V_3^0 \hat{I}_3 - \hat{V}_3 \Lambda I_3^0$$
$$+ I_2 \omega(d + j1) \hat{I}_2 - \hat{V}_3 \Lambda I_2^0 + \Lambda V_1^0 \hat{I}_1 - \hat{V}_1 \Lambda I_1^0. \tag{4.101}$$

All partial derivatives with respect to currents are zero because currents are independent variables. Partial derivatives with respect to L of branches 1 and 3 are zero because neither branch contains the variable L. Hopefully, these considerations have not hidden the simplicity of the result:

$$\Lambda_L Z_{\text{in}} = \frac{g_k}{I_4^2}, \tag{4.102}$$

where

$$g_L = \omega(d_L + j1)I_2^2. \tag{4.103}$$

This says that the exact partial derivative of input impedance with respect to L (in henrys) is found by denormalizing the unit-source sensitivity term g_k. Similarly, the exact partial derivative of input impedance with respect to C (in farads) in Figure 4.18b is

$$\Lambda_c Z_{\text{in}} = \frac{g_k}{I_4^2}, \tag{4.104}$$

where now

$$g_c = -\omega(d_c + j1)V_3^2. \tag{4.105}$$

Only one analysis at the given frequency is required to evaluate all currents and voltages in (4.102)–(4.105). If these exact answers are to be compared with the approximate values in Table 4.6, then the chain rule will have to be used to account for units:

$$\frac{\partial Z}{nH} = \frac{\partial Z}{\partial H} \frac{\partial H}{\partial nH}, \tag{4.106}$$

where the last term on the right is equal to $1E - 9$.

Figure 4.19 contains the general excitation patterns that depend on the selected response function. Transfer functions require two network analyses, including one in the backward direction. Figure 4.19 also shows general expressions for the unit-source coefficients that depend on the nature of the

Partial derivative expressions   Corresponding excitation patterns

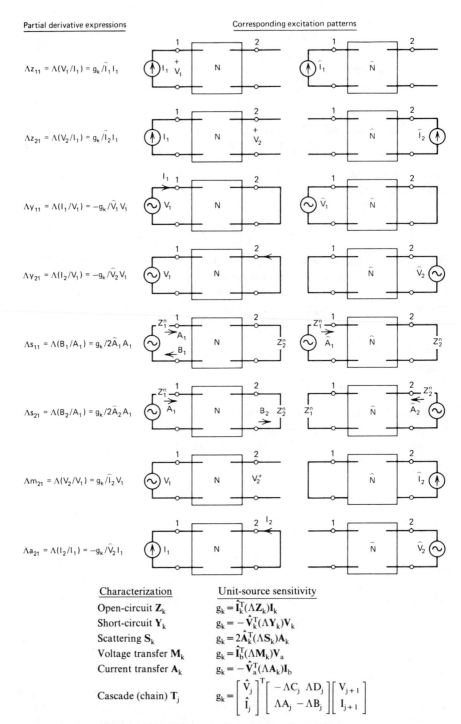

$\Lambda z_{11} = \Lambda(V_1/I_1) = g_k/\hat{I}_1 I_1$

$\Lambda z_{21} = \Lambda(V_2/I_1) = g_k/\hat{I}_2 I_1$

$\Lambda y_{11} = \Lambda(I_1/V_1) = -g_k/\hat{V}_1 V_1$

$\Lambda y_{21} = \Lambda(I_2/V_1) = -g_k/\hat{V}_2 V_1$

$\Lambda s_{11} = \Lambda(B_1/A_1) = g_k/2\hat{A}_1 A_1$

$\Lambda s_{21} = \Lambda(B_2/A_1) = g_k/2\hat{A}_2 A_1$

$\Lambda m_{21} = \Lambda(V_2/V_1) = g_k/\hat{I}_2 V_1$

$\Lambda a_{21} = \Lambda(I_2/I_1) = -g_k/\hat{V}_2 I_1$

Characterization	Unit-source sensitivity
Open-circuit $\mathbf{Z}_k$	$g_k = \hat{\mathbf{I}}_k^T(\Lambda \mathbf{Z}_k)\mathbf{I}_k$
Short-circuit $\mathbf{Y}_k$	$g_k = -\hat{\mathbf{V}}_k^T(\Lambda \mathbf{Y}_k)\mathbf{V}_k$
Scattering $\mathbf{S}_k$	$g_k = 2\hat{\mathbf{A}}_k^T(\Lambda \mathbf{S}_k)\mathbf{A}_k$
Voltage transfer $\mathbf{M}_k$	$g_k = \hat{\mathbf{I}}_b^T(\Lambda \mathbf{M}_k)\mathbf{V}_a$
Current transfer $\mathbf{A}_k$	$g_k = -\hat{\mathbf{V}}_a^T(\Lambda \mathbf{A}_k)\mathbf{I}_b$
Cascade (chain) $\mathbf{T}_j$	$g_k = \begin{bmatrix} \hat{\mathbf{V}}_j \\ \hat{\mathbf{I}}_j \end{bmatrix}^T \begin{bmatrix} -\Lambda\mathbf{C}_j & \Lambda\mathbf{D}_j \\ \Lambda\mathbf{A}_j & -\Lambda\mathbf{B}_j \end{bmatrix} \begin{bmatrix} \mathbf{V}_{j+1} \\ \mathbf{I}_{j+1} \end{bmatrix}$

Figure 4.19. Tellegen excitation and unit sensitivities.

branch immittance; these are in matrix form and may not interest all readers. They do reduce to (4.103) and (4.105) for the two-terminal L and C illustrated. The interested reader is referred to Bandler and Seviora (1970).

4.7.4. Summary of Sensitivities.

Sensitivity has been defined and written in several equivalent forms. Several identities related to partial derivatives were given in Table 4.5 because they are often required in the use of sensitivity relationships. First-order finite differences were explained, so that the reader could become more familiar with partial derivatives. It is also a simple way to obtain reasonably accurate derivatives. Finite differencing reduces complexity and saves computer program steps but runs much slower than some exact methods.

Tellegen's theorem was explained using operator notation according to Penfield et al. (1970). It applies in both time and frequency domains. The time domain partial derivative calculations run a long time because network analysis of a response requires numerical integration of state variable or similar equations. Then the sensitivity calculations often require numerical integration in backward time, using the stored impulse response. This is feasible but perhaps overly ambitious for desktop microcomputers. The frequency domain application of Tellegen's theorem for obtaining exact partial derivatives was explained. It boils down to obtaining the currents through impedance components or voltages across admittance components. These complex numbers are about all that is required to compute exact partial derivatives for sensitivity or optimization purposes. No more than two network analyses are required at each frequency to get the response sensitivity to all variables. This is an amazing result!

Problems

4.1. Write the sequence of topological element-type codes for the "T" interior of the bridged-T network in Figure 4.12, i.e., ignore L_1. Use type$=2$ for the inductor and type$=3$ for the capacitors. The answer should be a string (sequence) of four positive or negative integers.

4.2. Write the topology code table for the following network:

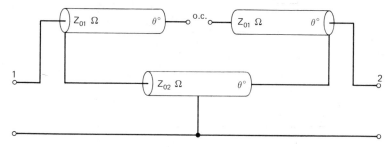

similar to Table 4.3. Assume that stub transmission lines are type 4 and cascade transmission lines are type 5 (also, $Z_{01} = 50$ ohms, $Z_{02} = 25$ ohms, $\theta = 45°$, each line has a 0.06-neper loss, and $\omega_0 = 2.9E10$ radians).

4.3.   Add level-1 and level-2 capability to the flowchart of Program B4-1.

4.4.   Consider the network in Figure 4.1 using only branches Z_L, Y_1, Z_2, Y_3, and Z_4. Suppose that the following data apply at some frequency:

$$P_L = 10 \text{ W}, \qquad Z_L = 30 + j15, \qquad Y_1 = 0.02 - j0.01,$$
$$Z_2 = 150 - j10, \qquad Y_3 = 0.01 + j0.15, \qquad Z_4 = 40 + j65.$$

Construct the table of numbers in the format of Table 4.1. What is the rms current through Z_4 and the rms voltage across Z_2?

4.5.   Show the L-section branch expressions (Figure 4.7) for a lossless cascade transmission line.

4.6.   Find open circuit parameters z_{11} and z_{21} for the lossless network in Problem 4.2. Hints: Find z_{11} by definition; next, find chain parameter C for the entire network by multiplying subnetwork chain parameter matrices; then convert C to z_{21} by identity.

4.7.   Given the two lossless transmission lines:

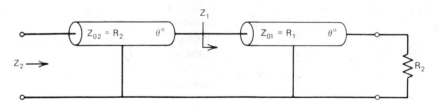

Use (4.18) and (4.22) to
(a)   Express $Z_1(y)$, where $y = \tan\theta$.
(b)   Express $Z_2(y)$.
(c)   Show that when $Z_2 = R_1$, then

$$\theta = \tan^{-1}\left[\left(R + 1 + \frac{1}{R}\right)^{-1/2}\right], \qquad R = \left(\frac{R_1}{R_2}\right)^{\pm 1}.$$

4.8.   Find V_{in} when $V_{out} = 2$ volts for the following network:

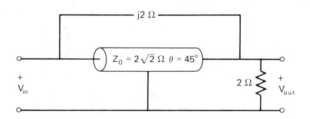

Solve by using (4.40) and by paralleling short-circuit-parameter matrices. Compare the amount of work required for each method.

4.9. Given the following network:

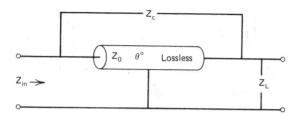

Find Z_{in} when $Z_0 = 50$ ohms, $\theta = 30°$, $Z_c = 5 + j12$, and $Z_L = 10 - j3$.

4.10. Consider the following two-port dissipative (lossy) network:

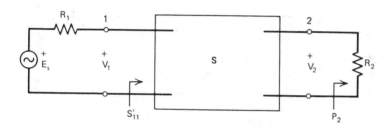

Suppose that the **S** matrix is normalized to the terminating resistors. Show why $S_{21} = 1/E_s$ when $P_2 = 0.25/R_1$. What is the phase reference?

4.11. Derive an expression for the input reflection coefficient S'_{11} of the network in the preceding problem when the load is an arbitrary reflection coefficient, Γ_L.

4.12. Consider the polynomial

$$f(x) = 5x^4 + 2x^3 + 19x + 1.$$

Calculate its exact derivative expression, obtained by calculus, using $x = 5$. Perturb x to the value 5.0005, and use the first-order finite difference to estimate the same derivative.

4.13. Suppose that you have a complex numerical value for the reflection coefficient in (4.57) and also a complex numerical value for its derivative with respect to the network variable x_k.

(a) Give an expression for the partial derivative of the magnitude of the reflection coefficient with respect to x_k.

    (b) Note the SWR definition in (4.59). Give an expression for the partial derivative of SWR with respect to x_k in terms of the reflection coefficient and its derivative found in (a).

4.14. Consider the series dissipation resistance in an inductor:

Find the unit-source sensitivity g_Q using (4.4) and $d = 1/Q$.

4.15. Find the unit-source sensitivities g_{Z_0}, and g_θ for a dissipative cascade transmission line.

4.16. Verify numerically the three Tellegen theory examples suggested below Equation (4.93).

4.17. Again consider the network in Figure 4.1 using only elements Z_L, Y_1, Z_2, and Y_3 (not Z_4 this time). Suppose that at some frequency

$$P_L = 10 \text{ W}, \qquad Z_L = 30 + j15, \qquad Y_1 = 0.02 - j0.01,$$
$$Z_2 = 150 - j10, \qquad \text{and} \qquad Y_3 = 0.01 + j0.15.$$

    (a) Find the exact partial derivative of $Z_{in} = V_4/I_4$ with respect to branch impedance Z_2 using Tellegan's theorem.

    (b) Find an approximate value for the same partial derivative by first-order finite differences. Perturb Z_2 by $5 - j1$, i.e., increase Z_2 to $155 - j11$ ohms.

Chapter Five _____

Gradient Optimization

This chapter shows how design engineers who can write a simple BASIC language subroutine can also use a standard program to select automatically the optimum set of variables for a great variety of mathematical problems, especially for circuit design. The subject of optimization requires more "feel" and art than any other in this book; so it is appropriate to begin by giving the reader some general appreciation of what may and what may not be possible. Intelligently applied optimization frequently provides better answers with less work than belabored, closed-form or approximation theory.

Design or operation of a system ideally involves three steps. First, it is necessary to identify the system's variables and to know how they interact. Second, a single measure of system effectiveness must be formulated in terms of those variables. Only then is the third and last step possible—the choice of system variables that yield optimum effectiveness.

The easiest systems to model are described explicitly by algebraic equations, and these will be the basis of most examples here. But a prime application is the ladder network simulated implicitly by the analysis methods of Chapter Four. An optimizer can automatically adjust some or all component values in networks to improve one or more responses sampled at a number of frequencies. Engineers have always "tweaked" or tuned systems in the laboratory in this way. However, there are compelling and increasingly common technical and economic reasons for eliminating this practice when possible. The synthesis methods in Chapter Three reveal their limited possibilities. Parasitic elements, including dissipation, are usually not considered, and there is no way to deal with element bounds that usually exist. Also, engineers may be unable to assimilate the vast amount of information measurable on systems or encountered during long mathematical procedures such as network synthesis. Optimization often alleviates these difficulties and almost always furnishes insight into how the system variables interact. Sensitivity was considered at a fixed set of system values in Chapter Four. In optimization,

113

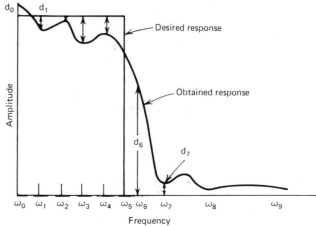

Figure 5.1. A sampled-difference error criterion.

sensitivity is computed at each such "point" in variable space, and this point moves along toward better sets of values.

Regardless of the means for designing and simulating a system, the second step is finding a measure of effectiveness. It usually involves value judgment and is either trivially simple or practically impossible to accomplish. Filter effectiveness in the least-squared-error sense is simple: the differences (d_i) between desired and obtained filter response at each of several significant excitation frequencies may be squared and summed to indicate effectiveness. These differences are shown in Figure 5.1.

Aaron (1956) noted: "As with all models of performance, the shoe has to be tried on each time an application comes along to see whether the fit is tolerable; but, it is well known, in the Military Establishment for instance, that a lot of ground can be covered in shoes that do not fit properly." Such is the case with the least-Pth error criteria, with P being equal to 2 or a larger even integer.

The third step is optimization. The word *optimum*, meaning best, was coined by the mathematician-philosopher Leibniz in 1710 and has an interesting history dating back to the eighth century B.C. Figure 5.2 shows how optimization might proceed for network problems. This amounts to adjusting a certain number of system parameters until the performance satisfies a preassigned requirement. Optimization is a successive approximation procedure, an automated design trade-off, achieving the best in a rational manner. Optimization amounts to handing the computer a set of input values and having the program hand back a set of answers. The computer then automatically re-inputs the adjusted data for many more such "runs" until some defined performance goals have been obtained more closely. The performance error function can be pictured as a surface over many dimensions, such as the two shown in Figure 5.3. Then optimization is a search for a lower elevation

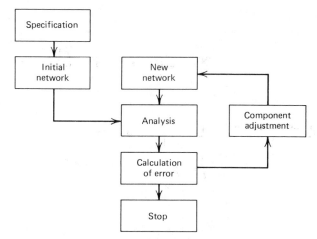

Figure 5.2. Circuit optimization iteration.

on that surface. Facetiously, Hyde (1966) quotes Joseph Petzval as having said in the 1800s that the optimal solution is the best one you have when the money runs out.

It should be mentioned that maximizing some function, say $Q(\mathbf{x})$, is equivalent to minimizing its negative, e.g., $-Q(\mathbf{x})$; the sign just turns the surface upside down. All subsequent discussion refers to minimization without loss of generality.

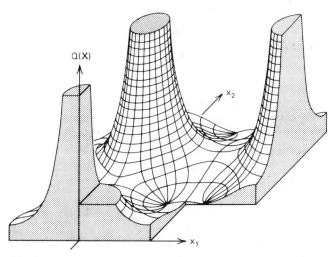

Figure 5.3. A surface over two-variable space; an elliptic function. [From TABLES OF FUNCTIONS WITH FORMULAE AND CURVES by Dr. Eugene Jahnke and Fritz Emde, 1945, Dover Publications, Inc., New York.]

Sooner or later, discussions about optimization turn to the blind-man-on-a-mountain analogy: how does he get down? It is surprisingly informative to exercise the following system function on either hand-held or desktop computers. Program the objective (performance) function

$$Q = \frac{P_2 * P_2 + Q_2 * Q_2}{4}, \tag{5.1}$$

where

$$P_2 = (9 * X_1 * X_1 + 25 * X_2 * X_2 - 36 * X_1 + 50 * X_2 - 164) * 2, \tag{5.2}$$

and

$$Q_2 = (X_1 * X_1 - 4 * X_1 - 3 * X_2 - 8) * 2. \tag{5.3}$$

The "*" indicate multiplication in the BASIC language. Also program the partial derivatives of the objective function Q with respect to the independent variables X_1 and X_2, respectively:

$$G_1 = P_2 * 18 * (X_1 - 2) + Q_2 * 2 * (X_1 - 2), \tag{5.4}$$

$$G_2 = P_2 * 50 * (X_2 + 1) - Q_2 * 3. \tag{5.5}$$

These equations have been programmed in Appendix-A Program A5-1 for HP-67/97 calculators. The reader should try inputting several trial pairs of X_1, X_2 values to minimize Q. Use the derivatives G_1, G_2 to guide your strategy; a necessary condition for a minimum Q value is that both derivatives be equal to zero. A good starting point might be $x = (5, 3)^T$. Examine the points (2, 1.99759808) and (5.84187, 0.92000) and their immediate neighborhoods. This function has three minima and one finite maximum. The need for some background and a reasoned strategy should become evident.

Chapter Five begins with an elementary treatment of quadratic forms, mathematical functions that are ellipsoids in multidimensional variable space. This is shown to be the basis of the conjugate gradient search schemes. The need for a sequence of searches in selected directions will then be clear. A particular linear search, implemented by Fletcher (1972b), will be studied in detail. This is an important part of the Fletcher–Reeves optimizer, which requires less than 1900 bytes of memory. It will be discussed in detail and several examples will be given. Network objective (performance error) functions will be considered next, followed by effective methods for dealing with all sorts of constraints, including variable (component value) bounds. Finally, a brief contrast between gradient and direct-search methods will be drawn. There are reasons for considering the latter, and several sources will be cited for those who may wish to investigate.

5.1. Quadratic Forms and Ellipsoids

When a function is suitably near a minimum, such as shown in Figure 5.3, such a function is approximately a paraboloid that has elliptical cross sections

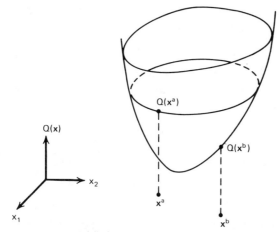

Figure 5.4. A paraboloid function over two-dimensional variable space.

(level curves; see Section 5.1.3). This is similar to the situation illustrated by Figure 3.1 for the root finder. It is shown here in Figure 5.4 for the more usual case where the local minimum value is not zero.

The two-dimensional case illustrates all important properties of the n-dimensional case and will be used in all descriptions. However, it is important to be comfortable using matrix algebra to describe the n-dimensional sets of equations; otherwise, the huge amount of notation would be unmanageable. A little practice with the following examples should overcome the handicap of those not familiar with these slight extensions of the material in Chapter Two.

One central example will be employed to develop many important mathematical and geometrical concepts. A list of some terms that will be of interest is given in Table 5.1. The reader may wish to consult Aoki (1971) during or after working through this chapter; his text is an excellent undergraduate treatment of the topics in Table 5.1, and much more.

Table 5.1. List of Pertinent Matrix Algebra Terminology

Conic section	Matrix	Positive definiteness
Conjugate vectors	Multiply, pre, post	Quadratic form
Eigenvalue	Newton's method	Quadratic function
Eigenvector	Norm	Rotation of axes
Euclidean space	Nonlinear function	Saddle point
Gradient vector	Nonlinear programming	Subspace
Hessian matrix	Orthogonal matrix	Symmetric matrix
Inverse of matrix	Orthogonal vectors	Transpose
Jacobian matrix	Paraboloid	Taylor series
		Unit matrix

5.1.1. Quadratic Functions. A quadratic function of many variables is defined in matrix notation by

$$F(\mathbf{x}) = c + \mathbf{b}^T\mathbf{x} + \tfrac{1}{2}\mathbf{x}^T\mathbf{A}\mathbf{x}. \tag{5.6}$$

Matrix **A** is always real and symmetric (equal to its transpose). Consider what this means for a specific two-dimensional function that will be used as a central example:

$$F = 612 + (-60 - 132)\mathbf{x} + \tfrac{1}{2}\mathbf{x}^T\begin{bmatrix} 26 & -10 \\ -10 & 26 \end{bmatrix}\mathbf{x}. \tag{5.7}$$

Expanding all terms, the equivalent, ordinary algebraic equation is

$$F = 612 - 60x_1 - 132x_2 + (13x_1^2 - 10x_1x_2 + 13x_2^2). \tag{5.8}$$

The reader should be able to obtain (5.8) from (5.7) by applying the skills obtained from Sections 2.2.1 and 4.7.2. The essential feature of a quadratic function is that there are no variables that are raised higher than to the second power and no products composed of more than two variables; i.e., the equation is of second degree.

Level curves are the loci on the variable space where the function value is some constant value. Two level curves for (5.8) are shown in Figure 5.5 on the x_1, x_2 variable space. Level curves in more than two dimensions are harder to visualize, but it is useful to consider a three-variable space (for example, a

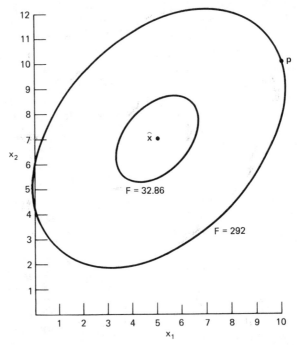

Figure 5.5. Level curves for $F = 32.86$ and $F = 292$ in Equation (5.8).

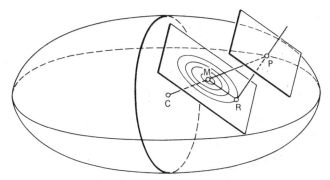

Figure 5.6. Level surface in three variables. A two-dimensional subspace also is shown. [From Acton, 1970.]

quadratic density function). This might appear as in Figure 5.6. Any cutting plane through Figure 5.6 would resemble Figure 5.5. Such reduced degrees of freedom define a subspace, such as the inclined-plane subspace shown in Figure 5.6. A subspace in Figure 5.5 would be a line. One reason subspaces are significant is that many minimization algorithms search in an orderly sequence of subspaces until the minimum is found.

The level curves for the central sample function that are plotted in Figure 5.5 will be studied in more detail. The next two sections deal with finding the center of the loci and the orientation of their axes, respectively. In the process, some concepts of major importance will emerge.

5.1.2. Gradients and Minima.   First recall real functions of real (single) variables. A quadratic function is

$$y(x) = c + bx + \tfrac{1}{2}ax^2. \tag{5.9}$$

The necessary condition for an extreme value or inflection point is that its first derivative be equal to zero:

$$y'(x) = b + ax = 0, \tag{5.10}$$

which produces the coordinate of the extreme value:

$$\hat{x} = \frac{-b}{a}. \tag{5.11}$$

The nature of the function at x is determined by examining the second derivative:

$$y''(\hat{x}) = a. \tag{5.12}$$

If a is strictly positive, (5.11) is the minimum point. If $a = 0$, then (5.11) is an inflection point, being neither a minimum nor a maximum. This familiar analysis extends to multidimensional functions without substantial change.

The matrix algebra rules for differentiation applied to (5.6) produce

$$g(x) = b + Ax, \tag{5.13}$$

where $g(x)$ is often written as $\nabla F(x)$, called grad F. Whichever symbol is used, g is a vector, like x; g is the gradient of F. From the central example in (5.7), the vector b and matrix A can be identified so that (5.13) can be written as:

$$g(x) = \begin{bmatrix} -60 \\ -132 \end{bmatrix} + \begin{bmatrix} 26 & -10 \\ -10 & 26 \end{bmatrix} x. \tag{5.14}$$

Some readers may be more comfortable differentiating (5.8) with respect to both x_1 and x_2:

$$g_1 = \nabla_1 F = 26x_1 - 10x_2 - 60, \tag{5.15}$$

$$g_2 = \nabla_2 F = -10x_1 + 26x_2 - 132. \tag{5.16}$$

Appendix Program A5-2 evaluates the function value and the gradient elements (derivatives) for this particular example. The reader is urged to use that program in conjunction with Figure 5.5. Note that the gradient vectors are always perpendicular to the level curves and point in the direction of steepest ascent.

Finding the minimum of an n-variable quadratic function requires setting each of the n gradient components equal to zero. This means setting (5.13) equal to vector zero; this is equivalent to setting both (5.15) and (5.16) to zero for that particular example. When (5.13) equals zero, then

$$A\hat{x} = -b. \tag{5.17}$$

But the matrix inverse A^{-1} is defined by the relationship

$$A^{-1}A = U, \tag{5.18}$$

where the unit matrix U has all zero elements, except for 1's on the main diagonal. A property of the unit matrix is that when it multiplies a vector, the result is just that vector. Multiplying both sides of (5.17) by A^{-1} yields the $\hat{x}$ values where $g(\hat{x}) = 0$:

$$\hat{x} = -A^{-1}b. \tag{5.19}$$

Identifying b and A by comparing (5.13) with the example in (5.14), (5.19) yields

$$\begin{bmatrix} \hat{x}_1 \\ \hat{x}_2 \end{bmatrix} = \frac{1}{576} \begin{bmatrix} 26 & 10 \\ 10 & 26 \end{bmatrix} \begin{bmatrix} 60 \\ 132 \end{bmatrix} = \begin{bmatrix} 5 \\ 7 \end{bmatrix}, \tag{5.20}$$

where inverse matrix A^{-1} was found by the three conceptual steps for finding inverses: transpose, form the signed cofactors, and divide by the determinant (see any book on matrix algebra, for instance, Noble, 1969). A glance at Figure 5.5 shows that the center of the level curves is indeed at (5, 7), the vector from the origin to the center. The solution (5.19) is the translation of the ellipses from the origin, as shown in Figure 5.7. The rotation of the ellipse with respect to the major axes is discussed in the next section, as well as the issue of whether (5.19) determines a true minimum function point.

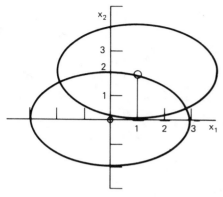

Figure 5.7. Translation of an ellipse. [From Acton, 1970.]

5.1.3.   Quadratic Forms and Graphics.   The preceding section related the coefficient matrix **A** of a set of linear equations to the quadratic function defined by (5.6). The terms at the extreme right in (5.6) are known as the quadratic form $Q(x)$:

$$Q(\mathbf{x}) = \mathbf{x}^T \mathbf{A} \mathbf{x}. \tag{5.21}$$

Matrix **A** was assumed to be a real, symmetric matrix; when **A** is two-dimensional, the quadratic form is

$$Q(\mathbf{x}) = \mathbf{x}^T \begin{bmatrix} a & k \\ k & b \end{bmatrix} \mathbf{x} = ax_1^2 + 2kx_1x_2 + bx_2^2. \tag{5.22}$$

Equation (5.22) is an ellipse centered at the origin. Solving (5.22) for x_2, elliptical level curves for Q can be plotted by

$$x_2 = \frac{-kx_1 \pm \sqrt{k^2x_1^2 - b(ax_1^2 - Q)}}{b}. \tag{5.23}$$

Appendix Program A5-3 uses key B to input values for a, b, and k that define matrix **A** according to (5.22). Key C is used to input the level-curve function value Q. Key A evaluates (5.23) upon entry of various x_1 values. The reader can check Figure 5.5 with Program A5-2, assuming a displaced origin at (5, 7). More important, (5.23) shows that the rotation of the ellipses results from the presence of cross terms such as x_1x_2 in (5.22); if $k=0$ in (5.23) then the x_2 points are symmetric about the x_1 axis.

The type of conic depends on the elements of **A**, namely, a, b, and k, defined by (5.22) (see Figure 5.8). In general, any matrix **A** is said to be positive definite if

$$\mathbf{x}^T \mathbf{A} \mathbf{x} > 0 \qquad \text{for all} \quad \mathbf{x} \neq \mathbf{0}. \tag{5.24}$$

For the two-dimensional case, a little thought shows that $k^2 < ab$ in (5.22) satisfies the positive-definite criterion. Thus the positive-definite matrix in the quadratic form of (5.21) produces the ellipse in Figure 5.8a; a maximum exists, analogous to the real-variable function's second derivative test, as

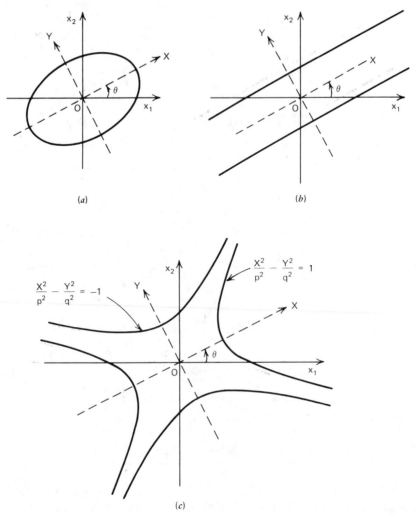

Figure 5.8. Quadratic-form conics. (*a*) Ellipse: ($k^2 < ab$; (*b*) parallel straight lines: $k^2 = ab$; (*c*) hyperbola: $k^2 > ab$. [From B. Noble, 1969.]

discussed previously. Students of the eigenvalue problem

$$Ax = \lambda x \tag{5.25}$$

may be interested in knowing that the eigenvalues λ are inversely proportional to the squared length of the ellipses' axes, and the eigenvectors x give their directions (see Noble, 1969).

Any matrix is said to be singular if its determinant is zero; this would certainly be the case in (5.22) if $k^2 = ab$. Consider the parallel lines in Figure 5.8b in light of the linear equations defined by (5.13). Finally, there is the indefinite matrix case when $k^2 > ab$ in (5.22) associated with the hyperbola in

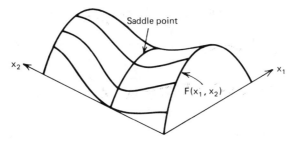

Figure 5.9. A saddle point for a function of two variables. [From Murray, W. (1972). *Numerical Methods for Unconstrained Optimization*. New York: Academic. Reprinted with kind permission from the Institute of Mathematics and Its Applications.]

Figure 5.8c. This is the general situation when the quadratic form in (5.21) may be positive or negative for all **x**. This produces a saddle point, as illustrated in Figure 5.9. A saddle point occurs in function (5.1) at point $(2, 1.99759808)$, as readily determined using Program A5-1 with 0.1% displacements.

5.1.4. Taylor Series. The reader should recall Taylor series of real variables. An expansion of a function about the point $x = a$ is

$$y(x) = y(a) + y'(a)(x-a) + \frac{1}{2!}y''(a)(x-a)^2 + \cdots . \tag{5.26}$$

It is important to define the difference,

$$\Delta x = x - a, \tag{5.27}$$

so that (5.26) reads:

$$y(\Delta x) = y(a) + y'(a)\,\Delta x + \tfrac{1}{2}y''(a)\,\Delta x^2 + \frac{1}{3!}y'''(a)\Delta x^3 + \cdots . \tag{5.28}$$

Figure 5.10 shows the situation for the Taylor series representation of a real variable. Notice the slope and the "neighborhood" at $x = a$, in which a truncated Taylor series might be valid, i.e., when all derivative terms greater than a certain order in (5.28) may be ignored. On the other hand, if the

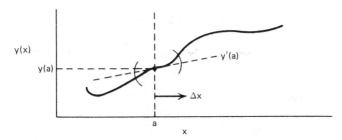

Figure 5.10. Taylor series representation in x or Δx about the point $x = a$.

function is known to be quadratic, as (5.9) for example, then $y'''(x)$ in (5.28) will be zero anyhow. The reader should understand this single-variable case before proceeding. The multivariable case is formulated in exactly the same way.

The multivariable function in (5.6) can be expanded by a Taylor series about point $\mathbf{p}$, where the displacement from $\mathbf{p}$ is

$$\Delta x = x - p. \tag{5.29}$$

The vector $\mathbf{p}$ might be the location of the blind man standing at $\mathbf{p} = (10, 10)^T$ in Figure 5.5. Then the Taylor series for a real function of the vector $\mathbf{x}$ is

$$F(\Delta x) = F(\mathbf{p}) + g(\mathbf{p})^T \Delta x + \tfrac{1}{2}\Delta x^T H(\mathbf{p})\Delta x + \text{h.o.t.}, \tag{5.30}$$

where higher-order terms (h.o.t) are presumed to be insignificant. Matrix $\mathbf{H}$ is known as the Hessian:

$$H \overset{\Delta}{=} \begin{bmatrix} \dfrac{\partial^2 F}{\partial x_1^2} & \dfrac{\partial^2 F}{\partial x_1\,\partial x_2} \\[2ex] \dfrac{\partial^2 F}{\partial x_2\,\partial x_1} & \dfrac{\partial^2 F}{\partial x_2^2} \end{bmatrix}. \tag{5.31}$$

By differentiating (5.13), it is seen that $\mathbf{H} = \mathbf{A}$ for a quadratic function. It is thus possible to expand the quadratic sample function in (5.7) about an arbitrary point, say $\mathbf{p} = (10, 10)^T$. The result in terms of (5.29) is

$$F(\Delta x) = 292 + (100, 28)\Delta x + \tfrac{1}{2}\Delta x^T \begin{bmatrix} 26 & -10 \\ -10 & 26 \end{bmatrix} \Delta x, \tag{5.32}$$

where $\Delta x_1 = x_1 - 10$ and $\Delta x_2 = x_2 - 10$. This describes the function in Figure 5.5 with respect to point $\mathbf{p}$. For quadratic functions, this is the same as shifting the origin; the reader should replace x_1 by $x_1 + 10$ and x_2 by $x_2 + 10$ in (5.8) and confirm that it is equivalent to (5.32).

Analogous to (5.13), the gradient of (5.30) is

$$\nabla F(\Delta x) = g(\mathbf{p}) + H(\mathbf{p})\Delta x. \tag{5.33}$$

So the blind man on a quadratic mountain at point $\mathbf{p}$ (Figure 5.5) could calculate where the minimum should be with respect to that point. In a manner similar to (5.19), the step to the minimum is

$$\Delta \hat{x} = -H(\mathbf{p})^{-1}g(\mathbf{p}). \tag{5.34}$$

Note that the second derivatives in $\mathbf{H}$ must be known. For the central sample function used as an example, the step from point $\mathbf{p} = (10, 10)^T$ to the minimum is

$$\Delta \hat{x} = -H^{-1}g = \frac{1}{576}\begin{pmatrix} 26 & 10 \\ 10 & 26 \end{pmatrix}\begin{pmatrix} -100 \\ -28 \end{pmatrix} = \begin{pmatrix} -5 \\ -3 \end{pmatrix}. \tag{5.35}$$

See Figure 5.5 to confirm this step.

5.1.5. Newton's Method. It is convenient to digress at this point because (5.34) is in fact Newton's method for minimizing a function of many variables. This has been used in Section 3.1.1 in the root finder; it will be used again in Section 6.3 for broadband matching. The Newton, or Newton–Raphson method as it is sometimes called, assumes that the current position (**x** value) is close enough to the minimum so that the higher-order terms in (5.30) are not significant. Another consequence of this assumption is that the partial derivatives in **H** are nearly equal to the values in the quadratic matrix **A** term in (5.13).

Newton's method is usually stated in a somewhat different way. It is said that there are, for example, two functions: $f_1(x) = 0$ and $f_2(x) = 0$, generally nonlinear. Newton's method assumes that they are linear; then they correspond exactly to (5.15) and (5.16), for example. If they were linear, then the step from the current **x** position to the minimum, where $f_1 = 0 = f_2$, would look like (5.34). The Hessian in (5.34) is a matrix of second partial derivatives of $F(x)$ from (5.6), but it is a matrix of *first* partial derivatives from $g(x)$ in (5.14). So the statement of Newton's method usually is: given a vector of functions

$$\mathbf{f} = \begin{bmatrix} f_1 \\ f_2 \end{bmatrix} = \mathbf{0}, \tag{5.36}$$

form the so-called Jacobian matrix of first partial derivatives:

$$\mathbf{J} = \begin{bmatrix} \dfrac{\partial f_1}{\partial x_1} & \dfrac{\partial f_1}{\partial x_2} \\[2mm] \dfrac{\partial f_2}{\partial x_1} & \dfrac{\partial f_2}{\partial x_2} \end{bmatrix}. \tag{5.37}$$

The Jacobian corresponds to the Hessian in the development concerning $F(x)$. Then, an estimated step to the minimum is

$$\Delta \mathbf{x} = -\mathbf{J}^{-1}\mathbf{f}. \tag{5.38}$$

Comparison of (5.38) with (5.34) shows that **J** in (5.37) is analogous to **H** in (5.31), and **f** in (5.36) is analogous to **g** in (5.13).

It is interesting to look back at Moore's root-finder coordinate steps ((3.9) and (3.10) in Section 3.1.1). In Newton's terminology, $f_1 = u$, $f_2 = v$, and the equivalence of the root-finder steps in the variable space to that in (5.38) follows.

5.1.6. Summary of Quadratic Forms and Ellipsoids. This has been a concise look at the matrix algebra crucial to gradient methods for nonlinear programming. It is the foundation of the powerful conjugate gradient method to follow. The subject has been treated by using a central, two-dimensional example and its geometric interpretation. It generalizes to n dimensions, and the fact that the matrix algebra was carried along with the example makes the

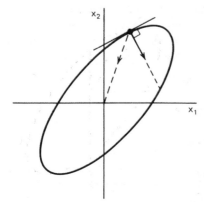

Figure 5.11. The negative gradient and Newton vectors of a quadratic function. [Reprinted with permission of Macmillan Publishing Co., Inc. from *Introduction to Optimization Techniques* by M. Aoki. Copyright © 1971 by Masanao Aoki.]

generalization more easy to follow. The reader should not miss this opportunity to "see" what differential calculus has to say about multidimensional functions, Taylor series representations, and the idea of linearization in the case of Newton's method. The concepts of single-variable functions were stated so that this transition could be related to calculus that every engineer should recall.

Newton's method describes a change in each component of the variable space, which converges to a minimum in just one step for quadratic functions (see Figure 5.11). The Newton vector, or step, can proceed to the minimum (the origin, as shown in Figure 5.11) in just one step. But what if the function $F(\mathbf{x})$ is not quadratic? Also, what if second partial derivatives are not known or inconvenient to compute? Might not a sequence of moves in the direction of steepest descent (negative gradient) lead to the minimum? In how many steps? These are questions that will be considered next.

5.2. Conjugate Gradient Search

Gradient optimization methods assume the availability of partial derivatives. Usually, finding first partial derivatives adds considerable complexity to the programming task or slows program execution time. Second partial derivatives are even less convenient to obtain. Fortunately, there are a number of search methods that do not require second derivatives; the popular conjugate gradient methods belong to this class. Methods that require only function values without any derivatives will be mentioned briefly in Section 5.7.

Almost all optimization methods select a sequence of directions leading to a minimum (or maximum) function value. A minimum in any particular direction is located by varying just one variable, usually some scalar that determines the distance from the last "turning" point, and this procedure is called a linear search. The linear algebra jargon and the special case of linear searches on quadratic surfaces will be described. Several elementary search direction choices will be mentioned, especially the relaxation method (varying each variable in turn) and the steepest descent strategy, which selects the steepest slope direction at each turning point. After considering several more important properties of quadratic functions, conjugate vectors and conjugate direction search methods will be defined. Finally, the Fletcher–Reeves conjugate gradient search direction formula will be discussed with examples.

5.2.1. Linear Search. At the point in variable space (x) where a new search direction (s) has been selected, some clear description of the next linear search must be available. The common notation is

$$x^{i+1} = x^i + \alpha_i s^i, \qquad i = 1, 2, \ldots, \tag{5.39}$$

where the superscript denotes that this is the ith linear search or iteration. The search parameter is the single variable α_i, which determines the distance of x^{i+1} from x^i. For well-posed problems, there will be some optimum $\hat\alpha_i$ that determines the lowest value of $F(x)$ in the s^i direction; in that sense, the linear search is concerned with a function of only a single variable, namely $F(\alpha_i)$.

Consider the nonquadratic surface over two-variable space previously introduced in (5.1)–(5.5). Suppose that the starting point $x = (7, 3)^T$ is selected, where the gradient turns out to have the value $g = (72080, 159976)^T$. Since the gradient is the set of coordinates describing the direction of maximum function increase, a reasonable choice for a linear search might be to the "southwest," i.e., in search direction $s = (-1, -1)^T$. Table 5.2 summarizes a set of moves in this direction according to (5.39) using Program A5-1. A graph of this function of α_i is shown in Figure 5.12. A new turning point is in the vicinity of $x = (5.25, 1.25)^T$, and a new search direction must be obtained, preferably by a more effective procedure than illustrated. Some simple alternatives are considered in the next section. A particular linear search strategy will be considered in some detail in Section 5.3.

Before continuing, examples based on quadratic functions can be implemented much easier if the linear search parameter α is obtained in closed form for these cases. Consider the standard quadratic function defined by (5.6) and write $F(x^{i+1})$ by substituting (5.39):

$$F(x^{i+1}) = c + b^T(x^i + \alpha_i s^i) + \tfrac{1}{2}(x^i + \alpha_i s^i)^T A(x^i + \alpha_i s^i). \tag{5.40}$$

Table 5.2. Searching to the Southwest on (5.1) From $x_1 = 7$, $x_2 = 3$ Using (5.39)

α	αs_1	αs_2	x_1	x_2	F	$\nabla_1 F$	$\nabla_2 F$
0	0	0	7	3	160,016	72,080	159,976
1	-1	-1	6	2	20,740	20,704	43,212
2	-2	-2	5	1	1,972	$-4,824$	$-8,764$
1.5	-1.5	-1.5	5.5	1.5	1,740	5,170	10,400
1.75	-1.75	-7.75	5.25	1.25	38	-462	-728

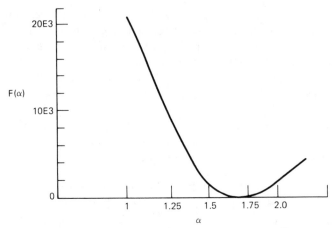

Figure 5.12. A plot of the function in Equation (5.1) in the $(-1, -1)^T$ direction from point $\mathbf{x} = (7, 3)^T$.

But (5.40) is just a function of the single variable α during the linear search. To find the minimum in the search direction $\mathbf{s}$, it is necessary to differentiate $F(\alpha)$ in (5.40) with respect to α and to equate this to zero. The result is:

$$\hat{\alpha}_i = \frac{-(\mathbf{s}^i)^T \mathbf{g}^i}{(\mathbf{s}^i)^T \mathbf{A} \mathbf{s}^i}, \tag{5.41}$$

where $\mathbf{g}^i = \mathbf{g}(\mathbf{x}^i)$. This provides an exact value of the linear search scalar α to a minimum from any point $\mathbf{x}^i$ on a quadratic surface in an arbitrary search direction $\mathbf{s}^i$.

Example 5.1. Consider the central sample quadratic function (5.8) shown in Figure 5.5. Suppose that a linear search is to begin in the negative gradient direction from the point $\mathbf{x}^i = (10, 10)^T$. To find α_i and then the minimum point $\mathbf{x}^{i+1}$ in that direction, (5.41) will require $\mathbf{g}^i$ and the $\mathbf{s}^i = -\mathbf{g}^i$ arbitrarily chosen for this example. The quadratic function gradient vector was defined generally by (5.13) and, for this example, by (5.15) and (5.16). Using Program A5-2 for $x_1 = 10$ and $x_2 = 10$, find $\mathbf{g} = (100, 28)^T$. Appendix Program A5-4 solves real-variable inner products as in the numerator of (5.41) and conjugate forms as in the denominator of (5.41). The significance of the latter will be discussed in Section 5.2.4. As previously noted, the matrix $\mathbf{A}$ is described for this example function by $a = b = 26$ and $k = -10$. The sequence 26, -10, and 26 is entered into an HP-67 calculator with program A5-4 running, and key B is pressed to input these data. The sequence -100, -28, -100, and -28 is input using key A. The (5.41) numerator inner product is found using key D (10784), and the (5.41) quadratic form is found using key E (224384). Then (5.41) yields $\hat{\alpha}_i = 0.048060$. Program A5-4 also evaluates (5.39). Input -100, -28, 10, 10 by using key A. Then input the value above for $\hat{\alpha}_i$ and press key C; x^{i+1} coordinates are: $x_2^{i+1} = 8.65432$ in the X register, and $x_1^{i+1} = 5.194$ in the Y register. Program A5-2 evaluates this as $F^{i+1} = F(x^{i+1}) = 32.86$. The reader should plot this linear search on Figure 5.5. Note that the linear search minimum occurs at the point of tangency to the level curve $F = 32.86$. It is also important to note that the search direction is always orthogonal to the gradient at such points of tangency.

5.2.2. Elementary Search Schemes.

Two obvious schemes for selecting search directions will be discussed. First, there is a relaxation (univariant) scheme by which the coordinate variables are adjusted in sequence, each one obtaining a minimum function value in that coordinate direction. Figure 5.13 shows a typical case for two variables. It is seen that the minima in each coordinate direction are tangent to level curves and that the successive search directions are orthogonal. The behavior in Figure 5.13 is called zigzagging. In sharp valleys such a procedure can fail (hang up), as illustrated in Figure 5.14. Univariant searches on quadratic surfaces without cross terms among the N variables will succeed in exactly N linear searches (iterations), as seen for the two-dimensional function in Figure 5.15.

Another search direction choice is the steepest-descent method Cauchy described in 1847. Each linear search for a minimum is made in the negative-gradient direction, as illustrated in Figure 5.16. Once the first search is made, the result is similar to the univariant method. The underlying reason is that the linear search directions are tangent to the level curves at the minimum

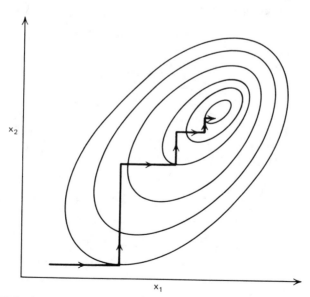

Figure 5.13. Univariant search strategy on a nonquadratic function. [From Box et al., 1969.]

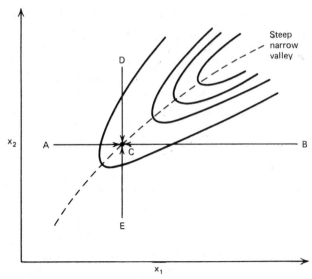

Figure 5.14. Failure of the univariant search in a sharp valley. [From Beveridge and Schechter, 1970.]

point where the gradient is orthogonal. Zigzagging near a minimum in a curving valley results in notoriously slow progress, because all linear search directions are either orthogonal or parallel. An extreme case is shown in Figure 5.17. What is needed is a search direction criterion that breaks this trend and is adaptive in some sense to valleys. Conjugate gradient methods do this and are discussed in Section 5.2.4.

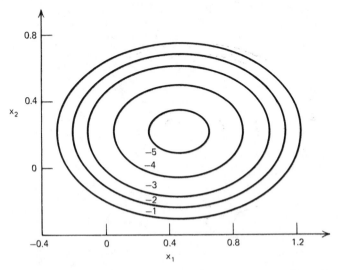

Figure 5.15. Level curves of a quadratic function without cross terms. [From Beveridge and Schechter, 1970.]

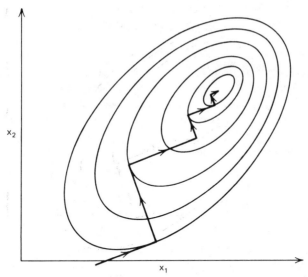

Figure 5.16. Steepest-descent search strategy on a nonquadratic function. [From Box et al., 1969.]

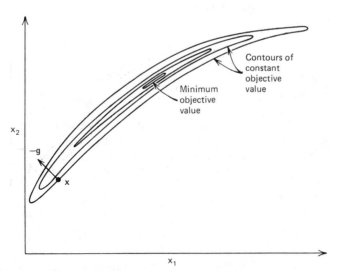

Figure 5.17. A difficult situation for the steepest-descent strategy.

5.2.3. More Quadratic Function Properties. Before proceeding, it is useful to examine three more properties of quadratic functions in N variables, i.e., those structured as in (5.6). First, it is always possible to create N new variables that are linear functions of the original ones so that all cross terms in the new variables disappear. This means that all quadratic functions of N variables can be minimized in exactly N steps in N suitable linear searches (see Figure

5.15, for example). Second, all changes in components of variables are linearly related to the corresponding changes of gradient components between the corresponding points no matter where the two points in question may be located on the functional surface. Thus the mapping of changes in variable values of quadratic functions onto the gradient space is invariant. Third, it will be shown that the altitude above the minimum value of a quadratic surface is equal to a quadratic form composed of the gradient at the point in question and the inverse of its constant Hessian. All of these concepts contribute to a practical understanding of gradient optimization.

Quadratic forms were considered in Section 5.1.3, where it was shown that they are equivalent to quadratic functions, except for a shift of origin to the minimum point. It was also shown that quadratic forms define ellipsoids whose axes are inclined with respect to the coordinate axes if there are cross terms among the variables. It was shown for the two-variable case that a diagonal matrix ($k = 0$) in the quadratic form (5.21) would not produce cross terms; this is true for any number of variables. Therefore, an important issue is how to rotate the coordinate axes to align them with the ellipsoidal axes, i.e., effect a change of variables. The motivation is to eliminate cross terms in N-variable quadratic forms and thus show that the minimum can always be found by no more than N linear searches (see Figure 5.15).

If the matrix in the quadratic form is $\mathbf{A}$ as in (5.21), then what is required is a coordinate-transforming matrix $\mathbf{P}$ such that

$$\mathbf{P}^T\mathbf{A}\mathbf{P} = \Lambda, \tag{5.42}$$

where $\mathbf{P}$ is a so-called orthogonal matrix, and Λ is a diagonal matrix. The eigenvalue problem (5.25), which appears in nearly all branches of engineering and physics, was mentioned in passing in Section 5.1.3. The eigenvectors of matrix $\mathbf{A}$ are geometrically the directions of the related ellipsoid's axes. The columns of $\mathbf{P}$ can be composed of the eigenvectors of $\mathbf{A}$ to produce the result in (5.42). Suppose that the quadratic form $Q(\mathbf{x})$ in (5.21) is to be expressed as $Q(\mathbf{y})$. Then it happens that the change of variable is accomplished by the substitution

$$\mathbf{x} = \mathbf{P}\mathbf{y}. \tag{5.43}$$

This can be confirmed by substituting (5.43) into (5.21) and using (5.42):

$$Q(\mathbf{x}) = (\mathbf{P}\mathbf{y})^T\mathbf{A}(\mathbf{P}\mathbf{y}) = \mathbf{y}^T\Lambda\mathbf{y} = Q(\mathbf{y}), \tag{5.44}$$

where $Q(\mathbf{y})$ has no cross terms, because Λ is a diagonal matrix. The interested reader is referred to Noble (1969) for details.

Example 5.2.   Again working with the $\mathbf{A}$ matrix from the central example (5.7), its eigenvalues turn out to be 36 and 16, and its eigenvectors are $(1, -1)^T$ and $(1, 1)^T$. The important concept is that these eigenvectors can be used as the columns in matrix $\mathbf{P}$; then (5.43) defines the substitutions

$$x_1 = y_1 + y_2, \qquad (5.45)$$

$$x_2 = -y_1 + y_2. \qquad (5.46)$$

Using these in (5.22) produces

$$Q(\mathbf{y}) = 72y_1 + 32y_2, \qquad (5.47)$$

so that the cross terms are indeed removed, and the minimum could be found in no more than two linear searches.

It is straightforward to show that changes in the gradient vectors of a quadratic function are mapped by a constant linear transformation to the corresponding changes in the variable vectors. As Figure 5.18 illustrates, points A and B in the $\mathbf{x}$ space have gradient values (perpendicular to their level curve), and these gradient vectors can be plotted in their own space. There may be more than one $\mathbf{x}$ with the same $\mathbf{g}$. Apply the gradient expression (5.13) of a quadratic function to points $\mathbf{x}^i$ and $\mathbf{x}^{i+1}$ and their corresponding gradients $\mathbf{g}^i$ and $\mathbf{g}^{i+1}$; the two equations may be subtracted to yield

$$(\mathbf{g}^{i+1} - \mathbf{g}^i) = \mathbf{A}(\mathbf{x}^{i+1} - \mathbf{x}^i). \qquad (5.48)$$

Using Δ to indicate the differences and inverting (5.48), the mapping result is

$$\Delta\mathbf{x} = \mathbf{A}^{-1}\Delta\mathbf{g}. \qquad (5.49)$$

This result was anticipated by Newton's step in (5.34), which went to a minimum where $\mathbf{g}^{i+1} = \mathbf{0}$ was required. The importance of (5.49) is that it shows the invariance of that mapping, independent of locations on any quadratic surface.

Finally, it is shown that the altitude above the minimum value of a quadratic surface at some point $\mathbf{p}$ is equal to a quadratic form composed of the gradient at the point in question, $\mathbf{g}(\mathbf{p})$, and the inverse of its constant

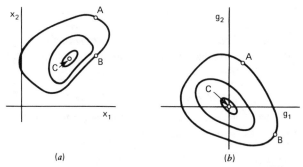

(a)   (b)

Figure 5.18. A mapping of variable space to gradient space. (*a*) Constant objective function curves in the variable space; (*b*) corresponding loci and points in the gradient space. [From Davidon, 1959.]

Hessian, $\mathbf{A}^{-1}$. Consider the function in (5.30) and its gradient in (5.33) when $\mathbf{H} = \mathbf{A}$. When this gradient is zero, $\Delta\mathbf{x}$ in (5.34) corresponds to the location of the minimum value. Substituting this in (5.30) yields

$$F(\mathbf{p}) - F(\Delta\hat{\mathbf{x}}) = \tfrac{1}{2}\mathbf{g}^T(\mathbf{p})\mathbf{A}^{-1}\mathbf{g}(\mathbf{p}). \qquad (5.50)$$

This is the amount by which $F(\mathbf{p})$ exceeds its minimum value.

The most popular optimization algorithm is the Fletcher–Powell method, which was first described by Davidon (1959). It is also known as the variable metric method, and it is worthwhile to observe that the latter name comes directly from (5.50). Davidon noted that the matrix $\mathbf{A}^{-1}$ in (5.50) associates a squared length to any gradient. Therefore, he considered the inverse Hessian matrix for any nonlinear function as its metric or measure of standard length. His optimization method starts with a guess for $\mathbf{H}^{-1}$, usually the unit matrix $\mathbf{U}$. This produces the steepest descent move according to (5.34). Following each iteration, Davidon "updates" the estimate of the inverse Hessian, so that it is exact when a minimum is finally found. In the interim, Davidon's metric varies, thus the name. There is also some statistical significance to the inverse Hessian for least-squares analysis (see Davidon, 1959).

Variable metric methods in N dimensions require the storage of $N(N+1)/2$ elements of the symmetric, estimated inverse Hessian matrix; so they are not considered here for personal computers, although such methods converge rapidly near minima. There are many variable metric algorithms, but Dixon (1971) showed that most of these, which belong to a very large class of algorithms, would produce equivalent results if the linear searches were absolutely accurate. Instead, another kind of conjugate gradient algorithm will be described, because it requires only 3N storage registers; it converges rapidly to good engineering accuracy, but lacks the ultimate convergence properties of variable metric methods. It is the Fletcher–Reeves conjugate gradient algorithm, which was originally suggested for very large problems (e.g., 1000 variables) on large computers. It is very effective for many problems (e.g., up to 25 variables) on desktop computers. The nature of the conjugate gradient search direction is described next, followed by a description of the Fletcher–Reeves algorithm.

5.2.4. Fletcher–Reeves Conjugate Gradient Search Directions.

Two vectors, $\mathbf{x}$ and $\mathbf{y}$, are said to be orthogonal (perpendicular) if their inner product is zero, i.e.,

$$\mathbf{x}^T\mathbf{y} = 0 = \mathbf{x}^T\mathbf{U}\mathbf{y}, \qquad (5.51)$$

where the unit matrix has been introduced to emphasize the following concept. The vectors are said to be conjugate if

$$\mathbf{x}^T\mathbf{A}\mathbf{y} = 0, \qquad (5.52)$$

where $\mathbf{A}$ is a positive-definite matrix. Conjugacy requires that the vectors are not parallel. More remarkably, conjugate vectors relate to A-quadratic forms as depicted in Figure 5.19. Just as illustrated for ellipsoids without cross terms,

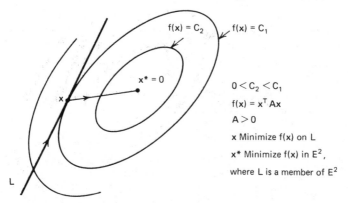

Figure 5.19. Two A-conjugate vectors on a quadratic surface. [Reprinted with permission of Macmillan Publishing Co., Inc. from *Introduction to Optimization Techniques* by M. Aoki. Copyright © 1971 by Masanao Aoki.]

a sequence of N A-conjugate linear searches to minima will terminate at the quadratic function minimum. That is why the two vectors in Figure 5.19 are related as shown; clearly, there are an infinite number of such pairs in two-variable space. Like the previous ellipsoids without cross terms (Figure 5.15), each linear search must find the exact minimum in that direction.

Example 5.3. A negative-gradient line search from $\mathbf{p}=(10, 10)^{\mathrm{T}}$ to a minimum was calculated for the quadratic function in Example 5.1. The minimum in the direction $\mathbf{s}^1=(-100, -28)^{\mathrm{T}}$ was found to be at $x_1=5.1940$ and $x_2=8.6543$. The surface, depicted in Figure 5.5, has its global minimum at $\mathbf{x}=(5, 7)^{\mathrm{T}}$; therefore the vector from the line-search minimum to the global minimum must be in the direction $\mathbf{s}^2=(-0.1940, -1.6543)^{\mathrm{T}}$. The conjugate form, as in (5.52), may be evaluated using Program A5-4:

$$(-0.1940, -1.6543)\begin{bmatrix} 26 & -10 \\ -10 & 26 \end{bmatrix}\begin{bmatrix} -100 \\ -28 \end{bmatrix}=0.1104\dot{=}0. \qquad (5.53)$$

Therefore, directions $\mathbf{s}^1$ and $\mathbf{s}^2$ are conjugate.

What has been illustrated is that conjugacy plus line search (to an exact minimum) equals quadratic termination (no more than N searches to find the minimum). It has been remarked that the sequence of "quasi-Newton" moves in the variable metric scheme results in conjugate search directions (to a sequence of line minima). How else might the sequence of conjugate search directions be generated? Fletcher and Reeves (1964) show that the following recursion generates a sequence of conjugate directions:

$$\mathbf{s}^i = -\mathbf{g}^i + \beta_i \mathbf{s}^{i-1}; \qquad i=1, 2, \ldots, N, \qquad (5.54)$$

$$\beta_1 = 0; \qquad \beta_i = \frac{(\mathbf{g}^i)^{\mathrm{T}}\mathbf{g}^i}{(\mathbf{g}^{i-1})^{\mathrm{T}}\mathbf{g}^{i-1}}; \qquad i=2, 3, \ldots, N. \qquad (5.55)$$

The Fletcher–Reeves formula is quite simple. As is common practice, the first search direction is the negative gradient. Then, each new search direction is a linear combination of the current gradient and the last search direction; the amount of the latter is scaled in proportion to the squared ratio of magnitudes of the current and last gradients. Derivation of the β_i scale factor is given in Appendix C. Only three vectors must be stored at a time: the x variables, the s search direction, and the g gradient components.

Example 5.4. Example 5.1 was a line search in the negative-gradient direction. It will now be shown that Example 5.3 illustrated a second search direction to the global minimum that happens to agree with the Fletcher–Reeves formula. Program A5-2 shows that at the first turning point, $x = (5.1940, 8.6543)^T$, the gradient is $g = (-11.4990, 41.0718)^T$. The last gradient at point $p = (10, 10)^T$ was $(-100, -28)^T$. Equation (5.55) shows that $\beta_2 = 0.1687$; thus (5.54) yields a new search direction: $s^2 = (-5.3675, -45.7929)^T$. A second linear search in this direction would find that $\alpha_2 = 0.0361$, as in (5.39). Thus $\alpha_2 s^2 = (-0.1940, -1.6543)^T$, as already found by other means in Example 5.3.

Convergence will not be achieved in just N linear searches on nonquadratic surfaces. The Fletcher–Reeves policy is to periodically restart the search direction sequence with the current negative gradient direction. An effective choice is to generate N directions by (5.54) and then start over again with the negative gradient. This has been justified experimentally by many researchers.

5.2.5. Summary of Conjugate Gradient Search. Linear searches have been described, and three strategies for selecting their sequence of directions have been discussed. The relaxation (one-at-a-time) method was shown not to be generally effective; however, it is significant because it works well on ellipsoids without cross-variable terms such as $x_1 x_2$, etc. The steepest-descent strategy is effective far from a minimum but tends to zigzag badly in curved valleys. The conjugate gradient method tends to follow curved valleys better, since it uses prior gradient information to moderate zigzagging.

Several additional properties of quadratic functions were discussed to clarify choices and introduce some concepts that are likely to be encountered in the field of nonlinear programming. The concept of diagonalizing a quadratic form, i.e., making a linear change of variables to obtain alignment with the ellipsoidal axes, amounts to justification for the application of A-conjugacy in search direction selection. It also shows the clear possibility for quadratic termination: the sequence of N linear searches to exact minima in N-variable space so that the global quadratic minimum is found. The constant nature of the mapping of variable to gradient space for quadratic functions was mentioned because of its close relationship to Newton's method and the variable metric search scheme. Davidon's use of the inverse Hessian matrix as a metric for gradients leads to a simple expression for quadratic function elevation above the global minimum. It is also the basis for naming the variable metric method, since Davidon (and later Fletcher and Powell) publicized the idea of updating an estimate of the inverse Hessian matrix.

Finally, conjugate directions were defined in comparison to orthogonal directions. The Davidon–Fletcher–Powell variable metric search directions are A-conjugate, but the symmetric, inverse-Hessian matrix estimate requires a substantial amount of memory to store. The Fletcher–Reeves method requires memory for only 3N vectors and works nearly as well, except for final convergence. The Fletcher–Reeves search algorithm works well for engineering accuracy in the memory space provided in desktop computers.

The mechanics of a linear search by Fletcher are discussed next, because of the important assumption that each linear search is stopped at the exact minimum in that direction.

5.3. Linear Search

Nearly all gradient search methods require linear searches, i.e., line searches to minima in a sequence of directions. The single, real variable in such searches has been defined as α in (5.39). A value for α may be calculated according to (5.41) when the surface is known to be a quadratic function in the general form of (5.6). However important a quadratic model may be in formulating search strategies, the usual surface is not at all quadratic except in the immediate vicinity of local minima, so that linear searches must find the minimum as a function of α by a comprehensive procedure.

Figure 5.20 illustrates a typical linear search profile. There are three stages in the linear search for the optimum value $\hat{\alpha}$: (1) estimate the order of magnitude of $\hat{\alpha}$; (2) establish bounds on the vicinity of the minimum; (3) interpolate the value of α within those bounds.

First, the slope in the search direction (directional derivative) will be defined, and an order of magnitude of α will be determined based on the expected quadratic behavior of α near the minimum. The classical cubic interpolation using two function values and two derivatives will be explained,

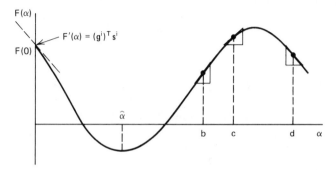

Figure 5.20. Profile of linear search function $F(\alpha)$.

and terminating criteria will be discussed. Finally, the linear search procedure programmed by Fletcher and used in the following Fletcher–Reeves optimizer will be described and illustrated by an example.

5.3.1. Slope in the Linear Search Direction.

The components of the gradient vector at any point $\mathbf{p}$, namely $\mathbf{g}(\mathbf{p})$, indicate the rate of change of function value in each coordinate direction. During linear searches, Figure 5.20 indicates the need to have the function's rate of change in some arbitrary direction $\mathbf{s}$. This may be obtained by recalling the Taylor series expansion about point $\mathbf{p}$, as given in (5.30). In this case, the $\Delta \mathbf{x}$ displacement is conveniently expressed according to (5.39) as $t\mathbf{s}$, where t is some real scalar similar to α. The classical definition of a derivative is then:

$$F'(\alpha) = \lim_{t \to 0} \frac{F(\mathbf{p} + t\mathbf{s}) - F(\mathbf{p})}{t}. \tag{5.56}$$

However, the numerator of (5.56) may be replaced using (5.30). Only the gradient term will remain, since all other higher-order terms will vanish in the limit. For linear search purposes, the point $\mathbf{p}$ will be represented as $\mathbf{p} = \mathbf{x}^i + \alpha_i \mathbf{s}^i$, so that the directional derivative becomes:

$$F'(\alpha) = \mathbf{g}(\mathbf{x}^i + \alpha_i \mathbf{s}^i)^T \mathbf{s}^i. \tag{5.57}$$

This provides the means for determining the slope at any point on the function illustrated in Figure 5.20. This slope will be required for several purposes, such as in estimating the gross magnitude of the first trial α value, as discussed next.

5.3.2. Finding the Order of Magnitude of the First Step.

The issue at the turning point, where a new linear search begins, is the choice of the initial value of α as employed in (5.39): should $\alpha = 0.01$ or $\alpha = 10$ be tried? Fletcher (1972b) reported that extensive testing indicated that the rate of change of function value with respect to iteration (linear search) number was fairly constant, except when close to an optimum solution. Thus he advocated the assumption that $F^{i+1} - F^i = F^i - F^{i-1}$. To develop this concept, he further assumed quadratic behavior for $F(\alpha)$:

$$F(\alpha) = a_0 + a_1\alpha + a_2\alpha^2. \tag{5.58}$$

The slope versus α according to (5.58) is

$$F'(\alpha) = a_1 + 2a_2\alpha, \tag{5.59}$$

and setting this to zero gives the value of $\hat{\alpha}$ at the minimum:

$$\hat{\alpha} = \frac{-a_1}{2a_2}. \tag{5.60}$$

Then, the minimum function value in this direction is

$$F(\hat{\alpha}) = a_0 - \frac{a_1^2}{4a_2}. \tag{5.61}$$

It is now possible to form an estimate for the initial value of $\hat{\alpha}$ when initiating a new linear search. The function decrease between the last and current turning point is $F^{i-1} - F^i = F(0) - F(\hat{\alpha})$, where the right-hand-side function values are seen in Figure 5.20. Using (5.58)–(5.61), it may be confirmed that

$$F^{i-1} - F^i = \frac{-\hat{\alpha}F'(0)}{2}. \tag{5.62}$$

But $F'(0)$ is available from (5.57), so that Fletcher's estimate for the first value of α at a new turning point is

$$\hat{\alpha} = \frac{-2(F^{i-1} - F^i)}{(g^i)^T s^i}. \tag{5.63}$$

In practice, the author has found that approximately a 10% decrease in current function value can be expected during each linear search; therefore, the numerator of (5.63) can be replaced by $-0.2F(0)$. Note that the denominator is negative, since it is the directional slope at the turning point (origin in Figure 5.20).

5.3.3. Extrapolation, Bounding, and Interpolation.

Having taken the first or subsequent step in a linear search, where the new $\alpha = \alpha_1$, several possible conditions may exist. If the slope is still negative and the function value decreased, another step is appropriate. As seen in Figure 5.20, this could result from too short a step. More information is now available, particularly the slopes at two points. Fletcher (1972b) linearly extrapolates these two slopes, again assuming the quadratic behavior of the $F(\alpha)$ function. Figure 5.21 applies where the extrapolation of the slope to zero predicts the necessary

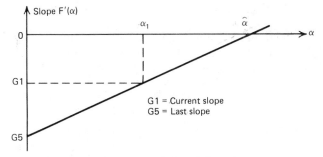

Figure 5.21. Linear extrapolation of the α slope to zero.

condition for a minimum. By similar triangles,

$$\frac{G5 - G1}{G5} = \frac{\alpha_1}{\hat{\alpha}}. \tag{5.64}$$

The increase beyond the α step just taken is $\hat{\alpha} - \alpha_1$, so that

$$\hat{\alpha} - \alpha_1 = \alpha_1 Z, \qquad \text{where} \quad Z = \frac{G1}{G5 - G1}. \tag{5.65}$$

Fletcher limits the extrapolation to be no more than four times the prior step; i.e., Z in (5.65) is limited to 4. The variable names employed correspond to the program code to follow.

Figure 5.20 shows that a minimum has been bounded in α when either $F(\alpha) > F^i$ or when the slope is positive. Suppose that this occurs at $\alpha = \lambda$. There are now four pieces of information: the two function values, $F(0) = F$ and $F(\lambda) = F9$; and the two slopes $F'(0) = G5$ and $F'(\lambda) = G1$. These four items enable the fit of a cubic function, which can interpolate the minimum between the bounds. The cubic function approximates a flat spring fitted to the known function values and slopes, provided that the slopes are small. Davidon (1959) suggested the following formulation, and it has been widely applied since then.

Suppose that the fitting function has the form

$$h(\alpha) = a_0 + a_1\alpha + a_2\alpha^2 + a_3\alpha^3. \tag{5.66}$$

Then, at $\alpha = \lambda$,

$$F9 = F + G5 \cdot \lambda + a_2\lambda^2 + a_3\lambda^3, \tag{5.67}$$

$$G1 = G5 + 2a_2\lambda + 3a_3\lambda^2. \tag{5.68}$$

The last two equations can be solved for coefficients a_2 and a_3:

$$a_2 = \frac{3(F9 - F) - \lambda(2G5 + G1)}{\lambda^2}, \tag{5.69}$$

$$a_3 = \frac{2(F - F9) + \lambda(G5 + G1)}{\lambda^3}. \tag{5.70}$$

It is convenient to define the constant z as

$$z = \frac{3(F - F9)}{\lambda} + G1 + G5. \tag{5.71}$$

The cubic interpolation step in α is then obtained by differentiating (5.66) and equating that to zero. The root of the resulting equation that is between $\alpha = 0$ and $\alpha = \lambda$ is thus obtained after considerable algebra:

$$\hat{\alpha} = \lambda \frac{1 - (G1 + W - z)}{2W + G1 - G5}, \tag{5.72}$$

where an additional defined constant is

$$W = (z^2 - G5 \times G1)^{1/2}. \tag{5.73}$$

The forms of these equations are designed to minimize cancellation by subtraction of nearly equal quantities. As before, the variable names correspond to those appearing in the following BASIC language optimizer program.

Example 5.5. A problem from Dejka and McCall (1969) illustrates the cubic fitting procedure. Given the function

$$F(x) = (x_2 - x_1^2)^2 + (x_1 - 1)^2, \qquad (5.74)$$

estimate the minimum along the line αs, where $s = (1, 1)^T$. Suppose that the minimum is bounded between the points $\alpha = 0$ and $\alpha = 1.5$. The four pieces of information can be obtained from (5.74): at $x = (0, 0)^T$, $F(0) = F = 1$, and $g(0) = (-2, 0)^T$; at $x = (1.5, 1.5)^T$, $F(1.5) = F9 = 0.812500$ and $g(1.5) = (5.5, -1.5)^T$, where g is the gradient vector. To get the slopes at $\alpha = 0$ and $\alpha = 1.5$, (5.57) is employed: $F'(0) = G5 = -2$ and $F'(1.5) = G1 = 4$. Then (5.71) yields $z = 2.3750$, (5.73) yields $W = 3.693322$, and (5.72) predicts that a minimum within the bounds is at $\hat{\alpha} = 0.904071$, where (5.74) yields $F(\hat{x}) = 0.016723$. By inspection of (5.74) the true minimum is at $\hat{\alpha} = 1$, where $F(\hat{x}) = 0$.

5.3.4. Fletcher's Linear Search Strategy.

The three stages of linear searches described above have been applied in the conjugate gradient optimizer Program B5-1 in Appendix B. The general view of this Fletcher–Reeves optimizer will be treated in Section 5.4. The emphasis here is on the linear search strategy as programmed by Fletcher (1972b). A flowchart of this part of the optimizer is shown in Appendix D, as modified for just one variable (line 860 was removed). The features of this chart will be discussed briefly, and an example will be considered.

There are some initial calculations preceding reentry point 490 in Appendix D, the last one estimating the first value of the linear search scalar α according to (5.63). The step size according to (5.39) is $\Delta x = \alpha s$, and this is calculated and tested for an absolute change of less than 0.00001, a stopping criterion. Initially the convergence flag ICON would not be set, so that the algorithm increments x from its value at the beginning of the linear search and then recalculates the function and its gradient values at that point.

Fletcher's algorithm then checks to see if the magnitude of the slope has decreased by more than a factor of 10; if so, the linear search is terminated rather than approach the minimum more closely. Otherwise, a test is made for either of the two conditions that will initiate a cubic fit, namely a function increase or positive slope. When either condition is detected, the program branches to line 710, the last Δx step is withdrawn, and a new step length is computed by cubic interpolation. The program then continues to reentry line 490 to take that chosen step.

The extrapolation based on the linear slope (quadratic function) assumption is indicated in the flowchart in Appendix D when none of the three preceding tests cause branching. The extrapolation factor is calculated according to (5.65), and the program again returns to reentry line 490.

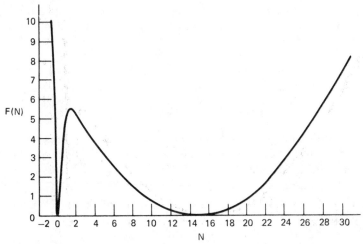

Figure 5.22. Profile of the Fano filter squared-error function in Equation (5.75).

Example 5.6. A highly nonlinear, real function from Section 8.5 will be used to illustrate the linear search algorithm previously described. A squared-error function is:

$$F(N) = \left\{ N - \frac{\sinh^{-1}\left[(\sinh^2 0.8814N)(10^3 - 1)/(10^{0.6} - 1) \right]^{1/2}}{\sinh^{-1}(1.3)} \right\}^2. \quad (5.75)$$

A profile of this function is shown in Figure 5.22. Optimizer Program B5-1 requires a subroutine starting at line 1000 to calculate the error function, in this case (5.75). This BASIC language code is shown in Table 5.3. The derivative of the quantity in the largest brackets in (5.75) is obtained by finite differences in a manner similar to (4.90), as programmed in line 1040 of Table

Table 5.3.   Subroutine for (5.75) in Optimizer Program B5-1

```
1000 REM FANO FILTER SQUARED–ERROR FUNCTION
1002 DEF FNS(X) = (EXP(X) – EXP(– X))/2
1004 DEF FNI(X) = LOG(X + SQR(X*X + 1))
1006 DEF FNQ(N) = N – (FNI((FNS(.8814 * N)**2*335.11)** .5))ᵃ
     /FNI(1.3)
1010 Q = FNQ(X(1))**2
1020 IF Y% = 0 THEN F = Q
1030 IF Y% = 1 THEN F9 = Q
1040 G(1) = (FNQ(1.0001*X(1)) – FNQ(X(1)))/(.0001*X(1))
1045 G(1) = 2*FNQ(X(1))*G(1)
1050 RETURN
9999 END
```

ᵃThe symbol ** indicates exponentiation.

5.3. The derivative of F(N) with respect to N follows by elementary calculus, as programmed in line 1045. The reader should run the modified Program B5-1, starting with several different values of variable N. Be sure to start once with N = 1.9, so that an undesired minimum is obtained, as shown in Figure 5.22. It is also useful to place diagnostic PRINT statements in the optimizer program, using the Appendix D flowchart, so that the program decisions are observable.

5.3.5. Summary of Linear Searches. There are three stages in the linear search for a minimum in a particular direction: (1) estimate the order of magnitude of the search scalar α; (2) establish bounds on the vicinity of the minimum; (3) interpolate the value of α within these bounds. The function to be minimized is usually not quadratic, so that linear searches must have comprehensive features to handle the nonideal circumstances. However, basic strategies are obtainable from some important, ideal assumptions.

The initial value of linear search scalar α is found by assuming a quadratic linear search profile in variable α; that, coupled with the fact that the function usually decreases about the same amount in each linear search, establishes a reasonable first value for the α step. The minimum is considered bounded when either the function value has increased or the slope is found positive after the step is taken. If the step was so small that the slope is still negative, then limited, linear extrapolation of the slope to zero is taken to lengthen the initial step. Once bounded, cubic interpolation is used to locate more closely the minimum in that direction. This process is repeated until convergence is obtained.

The flowchart in Appendix D shows the linear search strategy in the Fletcher–Reeves optimizer program. It was slightly modified for just one variable to illustrate its behavior on a nonlinear, squared-error function of a single variable. Fletcher terminates the linear search whenever the adjustment is very small or when the magnitude of the slope in the direction of linear search has been reduced by a factor larger than 10. The flowchart for Fletcher's linear search is applicable to the linear search in optimizer Program B5-1. In fact, the linear search constitutes most of the program, the remainder involving the choice of search directions, as discussed in Section 5.2.4. The next topic will be the entire Fletcher–Reeves optimizer.

5.4. The Fletcher–Reeves Optimizer

The FORTRAN program written by Fletcher (1972b) some years after the publication of the algorithm by Fletcher and Reeves (1964) has been translated to BASIC and appears in Appendix Program B5-1. A summary of the Fletcher–Reeves strategy is followed by a discussion of the program listing, an example network problem, and mention of potential scaling difficulties.

5.4.1. Summary of Fletcher–Reeves Strategy. The unconstrained, nonlinear programming problem is:

$$\min_{\mathbf{x}} Q(\mathbf{x}) = F(x_1, x_2, \ldots, x_N), \tag{5.76}$$

where $\mathbf{x}$ is a vector composed of N variables. The process is easily visualized by inspection of Figure 5.3. This objective function and its gradient ∇Q must be added to the BASIC language computer code provided. The gradient is:

$$\nabla Q = \mathbf{g}(\mathbf{x}) = \left(\frac{\partial F}{\partial x_1}, \frac{\partial F}{\partial x_2}, \ldots, \frac{\partial F}{\partial x_N} \right)^{\mathrm{T}}. \tag{5.77}$$

The gradient may be described analytically, if available, or found numerically by 0.01% finite differences. The user should consider an "awful warning" concerning excessive numerical noise, such as might occur if a named variable might inadvertently be declared an integer as opposed to a floating-point number. The resulting discontinuous behavior of the objective function will have a disastrous effect on partial derivatives obtained by finite differences. Almost all gradient optimizer programs will appear unacceptably sluggish under these circumstances.

Given an initial starting vector, $\mathbf{x}^0$, a sequence of linear (line) searches,

$$\mathbf{x}^{i+1} = \mathbf{x}^i + \alpha_i \mathbf{s}^i, \tag{5.78}$$

is performed in a calculated direction $\mathbf{s}$ in the variable α_i. Each search terminates when a minimum is approximated so that the directional derivative is nearly zero:

$$F'(\alpha) = (\mathbf{g}^{i+1})^{\mathrm{T}} \mathbf{s}^i \doteq 0. \tag{5.79}$$

The comprehensive procedure to accomplish reasonably accurate line searches on arbitrary functions of α_i was discussed in Section 5.3.

The first linear search direction is the negative gradient (steepest descent), i.e., with $\beta_1 = 0$ in the direction formula

$$\mathbf{s}^i = -\mathbf{g}^i + \beta_i \mathbf{s}^{i-1}; \qquad i = 1, 2, \ldots, N. \tag{5.80}$$

This describes a sequence of directions calculated after estimating each linear search minimum. The new search direction is simply the negative gradient plus a fraction of the just-used search direction. The fraction is:

$$\beta_1 = 0; \qquad \beta_i = \frac{\|\mathbf{g}^i\|^2}{\|\mathbf{g}^{i-1}\|^2}; \qquad i = 2, 3, \ldots, N, \tag{5.81}$$

where the squared-norm notation

$$\|\mathbf{g}^i\|^2 \overset{\Delta}{=} (\mathbf{g}^i)^{\mathrm{T}} (\mathbf{g}^i) \tag{5.82}$$

defines an inner product. It is seen from (5.80) that certain curvature information is accumulated for influencing the choice of subsequent search directions. This strategy was developed on the assumption of quadratic functions where convergence is obtained in exactly N linear searches. Because the objective function is seldom quadratic in practice, (5.80) is restarted in the steepest descent direction ($\beta_0 = 0$) after every N iterations (linear searches).

An important program feature is the criteria for stopping the iterative search for a minimum. This implementation by Fletcher stops when the changes being made in every component of the **x** variable vector are less than 0.00001, or when 100 iterations (linear searches) have been performed. The running time of the algorithm increases dramatically for even smaller changes; engineering problems often allow even earlier termination. Perhaps a better stopping criterion is the relative changes of variables. One advantage of real-time computing is the ability of the user to manually intervene whenever appropriate.

5.4.2.   The BASIC Language Computer Program.   Appendix Program B5-1 is a listing of Fletcher's program VAØ8A as translated into BASIC from FORTRAN. These 114 lines require only 1849 bytes in the Commodore PET computer, and only 15 additional bytes are required for each optimization variable. The program requires the user to define the objective function as subroutine 1000. The particular objective function and the gradient defined in lines 1000–1060 will be discussed in the next section. Unused BASIC names are given in line 60. Each execution of the program requires the user to state the number of variables, which should be consistent with the defined objective function. Then the starting values of the variables are requested. That run-time input is coded in lines 70–145.

Some program control constants are set in lines 150–170; this is less flexible than originally provided by Fletcher (1972b). The number of iterations is limited to 100, the absolute change in each variable must be less than 0.00001 for convergence, and the first step length in each iteration is based on an expected 10% decrease in function value.

The flowchart in Appendix D for a single-variable linear search is very nearly applicable to the entire B5-1 program; the reader should generalize it by reference to the complete program listing. The initial and subsequent setting of search direction to steepest descent is made by lines 230–240. The FOR–NEXT loop, to accomplish N searches before resetting to steepest descent, spans lines 260–850. These directions are calculated in lines 330–400 according to (5.80) and (5.81). Having chosen a search direction, the slope in that direction is computed by lines 410–440 according to (5.57). The linear search occurs as discussed in Section 5.3, except that each variable is increased by line 535 according to (5.78), and lines 850 and 860 implement repeated sequences of N linear searches.

5.4.3.   The Rosenbrock Example.   Lootsma (1972, pp. 29, 67, 68, 74–88, 101, 120, 185) gives many standard nonlinear programming (NLP) test problems, perhaps the most popular being the so-called Rosenbrock banana function, described by

$$Q = 100(x_2 - x_1^2)^2 + (1 - x_1)^2. \tag{5.83}$$

The gradient is

$$g_1 = -400(x_1 x_2 - x_1^3) - 2(1 - x_1), \tag{5.84}$$

$$g_2 = 200(x_2 - x_1^2). \tag{5.85}$$

These equations are programmed in lines 1000–1060 in Program B5-1. There is a required feature in this BASIC language conversion of the original FORTRAN program (see lines 1020 and 1030). The objective function must have the name Q; these two lines then assign this value to either names F or F9, depending on the value of integer flag Y%. This must be included in each different objective function subroutine to replace the subroutine argument list feature found in FORTRAN but missing in BASIC. The shape of this surface, especially the long, curved valley, is illustrated in Figure 5.23.

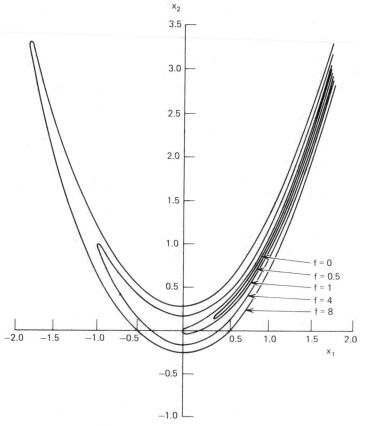

Figure 5.23. Some level curves for the Rosenbrock function in Equation (5.83). [Reprinted with permission of Macmillan Publishing Co., Inc. from *Introduction to Optimization Techniques* by M. Aoki. Copyright © 1971 by Masanao Aoki.]

Table 5.4. Typical Output for the Rosenbrock Problem[a]

```
# VARIABLES, N =?   2
INPUT STARTING VARIABLES X(I):
      1?   1         2?   -1
```

ITN = 0		IFN = 1
F = 400		
I	X(I)	G(I)
1	1	800
2	-1	-400

ITN = 1		IFN = 5
F = 32.3379952		
I	X(I)	G(I)
1	.102183049	21.1560238
2	-.551091525	-112.30658

ITN = 2		IFN = 8
F = 1.11719726		
I	X(I)	G(I)
1	-.0564383475	-2.18894177
2	-1.84100673E-04	-.673877548

ITN = 3		IFN = 18
F = .393823649		
I	X(I)	G(I)
1	.386114878	.784104271
2	.1360583	-2.60527968

ITN = 4		IFN = 21
F = .349279705		
I	X(I)	G(I)
1	.414709915	-2.52994041
	.180178959	1.63892907

ITN = 5		IFN = 23
F = .334620229		
I	X(I)	G(I)
1	.422829231	-.500579577
2	74919157	-.77308031

ITN = 6		IFN = 26
F = .244148841		
I	X(I)	G(I)
1	.542205182	3.11706106
2	.275392712	-3.71874958

ITN = 7		IFN = 28
F = .216721811		
I	X(I)	G(I)
1	.534919786	-.490506925
2	.284084414	-.410952665

ITN = 14		IFN = 47
F = 8.44751143E-03		
I	X(I)	G(I)
1	.925609721	1.84971288
2	.85135558	-1.07955505

ITN = 15		IFN = 49
F = 5.85904863E-03		
I	X(I)	G(I)
1	.923520252	-.0367466431
2	.852575064	-.06291833

ITN = 16		IFN = 52
F = 1.29751824E-03		
I	X(I)	G(I)
1	.975016735	.962065277
2	.948062727	-.518981593

ITN = 17		IFN = 54
F = 6.76262365E-04		
I	X(I)	G(I)
1	.974015938	-.0112750203
2	.948602602	-.0208891644

ITN = 18		IFN = 55
F = 3.53812794E-05		
I	X(I)	G(I)
1	.997877715	.217552957
2	.995204262	-.111134487

ITN = 19		IFN = 57
F = 5.48555021E-06		
I	X(I)	G(I)
1	.997659736	-9.54650997E-04
2	.995315613	-1.86720026E-03

ITN = 20		IFN = 59
F = 7.56015734E-09		
I	X(I)	G(I)
1	1.00001798	3.43919836E-03
2	1.00002745	-1.70140411E-03

ITN = 21		IFN = 61
F = 2.11894212E-10		
I	X(I)	G(I)
1	1.00001454	5.58176544E-06
2	1.00002915	1.18787284E-05

[a] The output for iterations 8–13 has been omitted.

147

The output for the Rosenbrock problem, starting at $x_1 = 1$ and $x_2 = -1$, is shown in Table 5.4 (see Figure 5.24). The data show that ITN = 21 iterations (linear search directions) and IFN = 61 function and gradient evaluations are required to locate the global minimum at $x = (1, 1)^T$ to at least 0.00001 accuracy in each variable.

The reader should run this example to observe the effects of several changes. Several new starting points should be tried. The accuracy set in line 160 can be reduced. The number of variables can be set to 20 instead of 2 by inputting the latter number when asked and setting all but the first two variables to an arbitrary number, e.g., 0. This will illustrate how much of the computing time is in search overhead, because the full 20 variables will be treated by the Fletcher–Reeves algorithm even though only the first two determine the problem defined in subroutine 1000. It is informative to add the statement 392 Z = 0. This causes the search to be of steepest descent at all

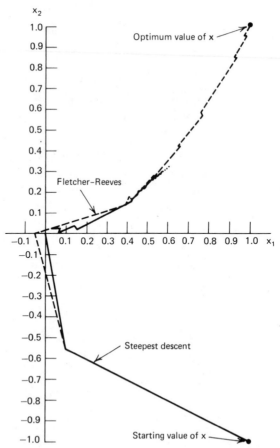

Figure 5.24.   Trajectories in the x plane for the Fletcher–Reeves and steepest-descent algorithms.

times. The trajectory is shown in Figure 5.24. It will not reach the minimum in the specified 100-iteration limit; inputting GOTO150 will cause the program to continue the searches. Delete temporary statement 392 and add 855 GOTO260. This disables the policy of resetting to steepest descent after every N iterations.

5.4.4. Scaling. The new user may construct a problem of his own design only to find that it won't optimize. The difficulty is often in the scaling of the variables, i.e., sensitivity. This is equivalent to partial derivatives; so the user should be aware of a rule of thumb regarding units of the variables. In the context of electrical network problems, suppose that the level curves in Figure 5.23 belong to the two variables in one of the L-section networks of Figure 4.3. For the frequencies of interest, these network L and C design variables make sense in units of nanohenrys and picofarads. A useful rule of thumb is: if any variable is increased by unity, do solutions still make sense? Another symptom is the gradient vector; the magnitude of its elements should be roughly equal and about unity within a factor of $1000^{\pm 1}$. But suppose that the inductance is specified in microhenrys; then an increase from 0.4 to 1.4 microhenrys is a much bigger jump than from 400 to 401 nanohenrys. What is at stake is seen in Figure 5.23; a bad choice of variable units can squeeze the curved valleys into razor-thin slits, so that the optimizer's finite word length search is in fundamental trouble.

An illustration of this effect is easily created using the Rosenbrock example. One or more initial variable values input at the beginning are rescaled, e.g., increased by a factor of 100. Then, at the beginning of subroutine 1000, these variables are decreased by 100 and then increased again before returning from that subroutine. Also, the corresponding derivatives must be decreased by 100 before returning (an application of the chain rule from calculus). Upon trying this, the effect on the gradient is immediately obvious—the number of function evaluations is increased by about half again. The reader is urged to try this on the Rosenbrock function to observe scaling and its effect on search difficulty.

5.4.5. Summary of the Fletcher–Reeves Program. The Fletcher–Reeves search strategy has been reviewed and BASIC language Program B5-1 has been described in the context of previously discussed topics. The Rosenbrock two-variable, nonlinear problem was described, and a number of enlightening, temporary program modifications were suggested. Also, the subject of scaling of variables was mentioned; it is the foremost pitfall the new user is likely to encounter when formulating his own objective function.

In addition to scaling, an "awful warning" was issued to be sure that only smooth functions are modeled for gradient optimization. This is especially true when the gradient vector is obtained by finite differences. Another warning about gradients is that evaluation of analytical expressions should be checked by comparison with finite differences before even trying optimization. Failure in optimization is commonly due to incorrectly formulated or programmed gradients, so that the optimizer is working with bad information.

The great virtue of the Fletcher–Reeves algorithm is that its computer memory requirements are proportional to 3N, where there are N variables. The Fletcher–Powell and other variable metric algorithms require a memory proportional to N^2. They all belong to the class of conjugate gradient algorithms, but the variable metric algorithms, being quasi-Newton, converge more rapidly when very near a minimum. This means that Fletcher–Reeves Program B5-1 should be very satisfactory on small machines employed for engineering applications requiring only moderate accuracy.

5.5. Network Objective Functions

The numerous test problems constructed by mathematicians, such as the preceding Rosenbrock example, are enlightening and provide some measure of effectiveness for various optimization algorithms. But what kind of objective functions are appropriate for automatic adjustment of design variables in electrical networks? The following methods are easy to implement and have an interesting resemblance to weighted-sample integration techniques (Section 2.3). The optimization process can also be viewed as a curve-fitting process. However, as mentioned in Section 2.5, nonlinear programming is often ineffective when compared to methods that are specifically formulated for certain problems.

On the other hand, many network design requirements cannot be solved by existing closed-form methods, as evident by the brief exposure to network synthesis in Chapter Three. Also, the designer may not be aware of more appropriate methods or may not have the time or inclination to implement them. Then optimization of networks is worth trying, especially if there is an approximate design basis to serve as a starting point for both insight and values.

The following sections describe several important kinds of network objective functions and their gradients. An example using Fletcher–Reeves optimizer Program B5-1 is given.

5.5.1. Integral Error Functions. Most cases of optimization in the frequency or time domains amount to curve fitting, as seen in Figure 5.1. The error can be defined as the square of the area between a desired function (the rectangle) and the approximation function. This is expressed as

$$\min_{\mathbf{x}} E = \int_{\omega_1}^{\omega_2} e^2(\mathbf{x}, \omega) \, d\omega = \int_{\omega_1}^{\omega_2} (R - G)^2 \, d\omega, \tag{5.86}$$

where the first integrand emphasizes its dependence on both the variables (**x**) and frequency. The second integrand might represent the difference between a response function (R) and the goal function (G).

Since integration on digital computers is discrete anyhow, the measure of goodness of fit can be a process of frequency sampling. The Euclidean norm (inner product) mentioned in (5.82) applies here as well:

$$\|E\| = \left(e_1^2 + e_2^2 + \cdots + e_N^2\right)^{1/2}. \tag{5.87}$$

This might correspond to sampling at the *i*th frequency, where e_i is the difference between the response and the goal. The next section combines these concepts in a form convenient for optimizing network response functions sampled at several frequencies or times.

5.5.2. Discrete Objective Functions. A typical discrete objective function for network response is shown in Figure 5.25, as described mathematically by

$$E(\mathbf{x}, \omega) = \sum_{i=1}^{M} W_i (R_i - G_i)^P, \tag{5.88}$$

where P is an even integer (the *P*th difference), R_i is the response, G_i is the goal, and W_i is the weight factor at the *i*th frequency. None of these quantities are complex. For example, if a network is to be adjusted so that an impedance approximates some given impedance values at various frequencies, then an approximate response might be SWR, according to (4.59) and (4.54). Compare (5.88), with P=2 and $W_i=1$, to (5.87). Also, (5.88) may be generalized to account for more than one kind of response, $R_{i,k}$, by adding a second, nested summation on k. Two responses might then be SWR and voltage, where the weights $W_{i,k}$ must equalize the scales for the two different kinds of responses. In practice, only very few kinds of responses are successfully considered simultaneously, and there is a good chance for a standoff (over constraint), so that optimization is ineffective.

A "satisfied-when-exceeded" feature can be included in a program for (5.88), so that $W_i=0$ is employed whenever $R_i > G_i$. This feature is useful

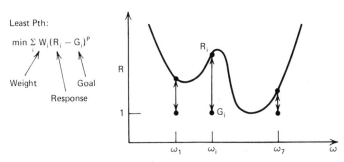

Figure 5.25. Least-*P*th error function with weighted frequency samples.

Floating Pth:

$$\min\left[X_{N+1} + \sum_i (R_i - G_i + \frac{X_{N+1}}{W_i})^P\right] \qquad R$$

$W_i = 1$ illustrated

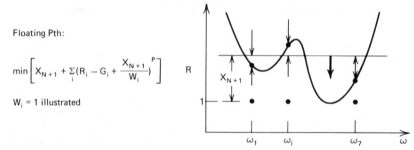

Figure 5.26. Least-*Pth* error function weighted relative to an "extra" floating variable providing slack.

when amplitudes of the response exceed a certain level in filter stopbands; this might be the case for Figure 5.1 if frequency samples 6–9 were required to be equal to or greater than some positive number instead of the unlikely null values illustrated. This approach does not cause discontinuous function behavior, so that the derivatives are still those of a smooth function.

It is also possible to "float" the goal values in an objective function, as illustrated in Figure 5.26. The floating goal requirement is encountered in time delay equalization, where a constant delay is desirable without concern for its absolute value. The function shown in Figure 5.26 is not as well behaved as (5.88); so the user can expect to have some difficulty selecting suitable weights.

Figure 5.27 illustrates the minimax case similar to the curve-fitting result in Section 2.4. The objective function is the maximum difference or residual among all samples. It is easy to program the computer to find what this is, but this approach causes large, discontinuous changes in the function and is thus unsuitable for gradient optimization. Suppose that each sampled difference in Figure 5.25 is greater than unity. Then, as P is made larger and larger, the main contribution to the total error will be the largest difference sample. Temes and Zai (1969) have shown that the minimax (equal differences) case

Minimax:

$$\min\left\{\begin{array}{c}\max\\i\end{array}[W_i(R_i - G_i)]\right\} \qquad R$$

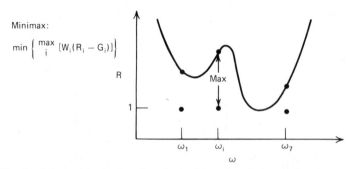

Figure 5.27. A minimax objective function obtained by a least-*Pth* error function when P→∞.

occurs when P→∞, for suitable functions. It is interesting to think of this process in terms of $1/P \to 0$, because the Richardson extrapolation to zero considered in Section 2.3.2 for Romberg integration is also applicable here. Thus, the minimax conditions can be predicted without actually making P all that large. The proper extrapolation variable and other important parameters will not be treated here; satisfactory minimax results often can be obtained by simply setting P = 2, 10, and 30 in a sequence of minimizations. This point will be explored in the network optimization example in Section 5.5.4.

5.5.3.   *Objective Function Gradient.* When finite differencing is used to obtain partial derivatives, then the entire objective function—as in (5.88)—should be employed in the difference functions. However, if partial derivatives of the response function(s) $R_{i,k}$ are available analytically or, more likely, by application of Tellegen's theorem, then (5.88) should be differentiated so that the partial derivatives of the response function may be employed. Differentiation of (5.88) with respect to x_j produces

$$g_i = \frac{\partial E}{\partial x_j} = P \sum_{i=1}^{M} W_i (R_i - G_i)^{P-1} \frac{\partial R_i}{\partial x_i}. \qquad (5.89)$$

Again, note that response R_i is a real quantity; e.g., if it is SWR and derivatives of Z_{in} are available, then identity (5) in Table 4.5 will be required to express the derivative of R_i needed in (5.89).

By the Tellegen method, partial derivatives of complex quantities are also complex; thus 2N registers and additional computer coding will be required to exploit this approach. Of course, the minimization time will be much less than when using finite differences, because there will be no wasted calculations, and the exact partial derivatives will speed convergence.

5.5.4.   *L-Section Optimization Example.* The concepts in Chapter Five are now brought together for a practical network optimization problem, which will illustrate almost all fundamental techniques. The lowpass L section shown in Figure 4.18b will be optimized to match a frequency-dependent load impedance to a resistive source impedance over a band of frequencies. Design methods for this impedance matching problem will be considered in Chapter Six.

Appendix-B Program B5-2 is composed of Fletcher–Reeves optimizer Program B5-1 lines 150–940; lines numbered less than 150 input data, and lines numbered greater than 940 form an error function and its partial derivatives (gradient vector). The general process is flowcharted in Figure 5.28a. Also, the function and gradient computation are shown in Figure 5.28b, and the sampled-error-function formation is shown in Figure 5.28c.

A brief discussion of Program B5-2 code should reveal the simple details. The L and C values (in henrys and farads) are input into X(1) and X(2) by lines 100 and 110, respectively. Line 120 inputs the value of P, which should initially be 2. After minimization, the program is sent to this line (by line 999)

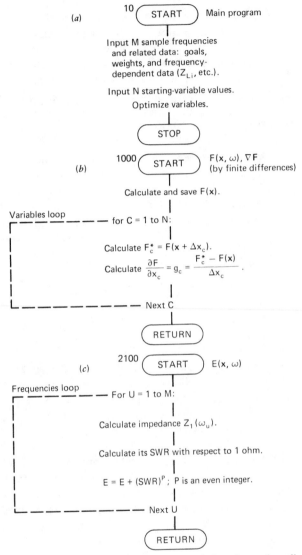

Figure 5.28. Typical network optimization. (*a*) Overall; (*b*) function and gradient; (*c*) sampled-error function.

so that larger values of P may then be specified. The optimizer looks to subroutine 1000 for its objective function (F or F9). Thus Program B5-2 begins the calculation flowcharted in Figure 5.28b at line 1000. Line 1005 is an SWR print control feature utilized in line 2165. More important, the unperturbed function value is obtained by the GOSUB2100 at line 1010, and perturbed values are obtained and used in the FOR–NEXT loop 1040–1090. The flowchart in Figure 5.28c shows the sum of the P*th* errors obtained by

Table 5.5. Some Lowpass L-Section Results for $Z_L = 0.25 + j0$

			SWR at Radian Frequency		
P	L	C	0.8	1.0	1.2
2	1	1	3.1733	4.2656	7.0016
	1.6471	0.4117	1.8626	1.1545	1.5670
10	1.6710	0.4192	1.8122	1.1040	1.6686
30	1.6903	0.4226	1.7817	1.0739	1.7338

subroutine 2100, which ends at line 2190. Line 2170 corresponds to the *ith* term in (5.88), where the response R_i is the standing-wave ratio (SW), weight W_i is fixed at unity, and goal G_i is fixed at zero.

The standing-wave ratio SW is computed at each sample frequency by the GOSUB3000 in line 2160. Subroutine 3000 calculates the input impedance of the network in Figure 4.18b according to the easily obtained expression

$$Z_1 = \frac{[R_L] + j[\omega L + X_L]}{[1 - \omega C(\omega L + X_L)] + j[\omega C R_L]}. \tag{5.90}$$

The SWR calculation is that defined by (4.59) and (4.57) when $R_1 = 1$ is assumed. The four real and imaginary parts of (5.90) are assembled and employed in lines 3010–3070. Note that this lowpass-network SWR function assumes a unit source and is frequency normalized, so that units of henrys, farads, and radians are appropriate. Also, note that network analysis Program B4-1 could have been used for more general networks, especially since the likely variables for optimization appear in the $X(\cdot)$ array in both B4-1 and B5-2.

Table 5.5 shows some results obtained by starting L-section optimizer B5-2 at $L = C = 1$ for $P = 2$ and continuing, after sequential minimizations, with $P = 10$ and $P = 30$. The load impedance was specified as $0.25 + j0$ ohms at each of three sample frequencies, but arbitrary impedances at any number of frequencies could have been employed. The SWR values shown were printed by Program B5-2, line 2165, when the variables were unperturbed (flag variable $C = 0$ set by line 1005). Note the tendency for equal SWR deviations at the band edges for increasing values of P. According to (4.59), SWR can be no less than unity, so that the squared error cannot be less than 3; it started at 77.29 and decreased to 7.26 in ITN = 7 iterations (linear searches) using IFN = 26 function evaluations (not counting the additional 52 perturbed evaluations). Also, each of the 78 error function evaluations required network analyses at three frequencies. It is easy to see why more efficient network response and sensitivity calculations are essential when optimizing more than just a few variables.

5.5.5. Summary of Network Objective Functions and Optimization.

The concept of the area between desired and approximating functions over a range has been viewed as a measure of curve-fitting acceptability. Then the concept of numerical integration as a weighted sampling of a difference function has been applied to the formulation of a weighted, discrete error function over the sample space, usually frequency or time. Several kinds of approximating response functions may be treated in a common summation of sampled errors if the weighting factors of each type function are selected to equalize error contributions to a common scale.

Several variations of this method were mentioned. A "satisfied-when-exceeded" rule applied to each sampled response ignores the contribution to the error function when the response exceeds its goal. This technique is especially useful in obtaining minimum stopband selectivity at the same time that passband requirements are being fulfilled. The method does not upset the continuous function requirement, which must be maintained for use with gradient optimizers. A bias or "float" to a goal was described; it is implemented as an added variable that is minimized along with the error function. The third kind of error function is the minimax; it looks for the worst sampled difference and minimizes it. However, the worst difference can jump from sample to sample during adjustment of variables, causing gross discontinuities in the objective function. This unacceptable behavior may be avoided by using the original, weighted, least-P*th* objective function in a sequence of minimizations, with P = 2 and greater even-integer values.

When derivatives of each sampled response are directly available, it is useful to differentiate the weighted-difference summation analytically and employ the sampled-response derivatives directly. Otherwise, finite differences may be obtained using the weighted-difference summation directly for perturbed and the unperturbed sets of variables. It was emphasized that the error function and its components are real functions. Any complex function and its derivatives (e.g., input impedance) must be transformed by appropriate identities (such as those in Table 4.5).

Finally, a complete network optimization example was added to Fletcher–Reeves optimizer Program B5-1. It can serve as a model for the general technique, and flowcharts of major functions were furnished for this purpose. The optimizer input section was modified to solicit values of sample frequencies and corresponding frequency-dependent load impedances. The objective subroutine 1000 was written for a lowpass, L-section network normalized to 1 ohm and 1 radian; the two variables were L and C (in henrys and farads, respectively). A straightforward expression for input impedance was written for this particular network; it was noted that incorporating ladder analysis Program B4-1 for this purpose is not difficult. The input SWR was raised to the P*th* power and summed at each frequency to constitute an evenly weighted error function with uniform goals of zero. Since SWR ⩾ 1, the minimum possible objective function value is equal to the number of the samples. It would also be easy to have frequency-dependent source impedances. Then the important case of an interstage network connecting two transistors could be optimized, even accounting for gain slope versus frequency.

Table 5.5 summarized some results that showed a $10:1$ reduction in squared error as well as the tendency toward minimax behavior for $P=10$ and $P=30$ minimizations. It is suggested that the stopping criterion in line 160 (0.00001) is probably smaller than need be. Engineering design usually does not require this kind of accuracy, and the Fletcher–Reeves algorithm is known to converge slowly near a minimum. Users might consider a value of $E=0.01$ or use of a 0.1% relative change–stopping criterion instead of the absolute change criterion presently incorporated.

A common rule of thumb is that the number of samples should be at least twice the number of variables. If there are too few samples, the function may oscillate wildly between frequency samples while giving the illusion of a very good fit of sampled response to goals. Which samples to take, how they are weighted, which multiple response types are not conflicting, and many other aspects of network optimization are more a matter of experience and insight than science. This is also true of questions concerning how close to a minimum must one start the variables and whether the minimum is global as opposed to inferior local minima, which trap the search prematurely. In the latter case, the usual advice is to try starting at a variety of points in the variable space. As for starting reasonably near a solution, that is what the rest of this book is all about. The main virtue of an optimizer is its ability to treat significant second-order effects that are too difficult or inconvenient to treat otherwise.

5.6. Constraints

The subject of constraints deals with the explicit or implicit relationships among optimization variables ($\mathbf{x}$). The most elementary constraints are upper and/or lower bounds and linear dependence, such as

$$x_j \geqslant 0, \tag{5.91}$$

$$k_1 \leqslant x_j \leqslant k_2, \tag{5.92}$$

and

$$x_i + x_j = k_3. \tag{5.93}$$

An implicit constraint is one that cannot be stated explicitly, e.g., the requirement that a calculated attenuation function have some specified value at a stopband frequency. However, this example would correspond to (5.93) in that the constraint is always active or "binding" and thus removes at least one degree of freedom from the problem; this is typical of equality constraints. Inequality constraints may not be binding in various subsets of the variable space; this could be the case for the "satisfied-when-exceeded" performance constraint or those in (5.91) and (5.92). For both equality and inequality constraints, points in the variable space where constraints are violated are said to constitute an infeasible region.

The constrained optimization problem can be stated as follows:

$$\min_{\mathbf{x}} E(\mathbf{x}) \text{ such that } \mathbf{c}(\mathbf{x}) \geqslant \mathbf{0} \text{ and } \mathbf{h}(\mathbf{x}) = \mathbf{0}, \tag{5.94}$$

where E is a function of both the variables ($\mathbf{x}$) and the sample parameter (ω). Each component of vector $\mathbf{c}$ constitutes one inequality constraint, such as (5.91). Each component of vector $\mathbf{h}$ constitutes one equality constraint, such as

$$h_1 = x_4 x_7 - k_4. \tag{5.95}$$

An example of inequality constraints is shown in Figure 5.29. Note that the unconstrained minimum is infinitely far out in the first quadrant, but the feasible region causes this problem to have the identified optimum. If an equality constraint were added, it might appear as a line locus in the feasible region.

On small computers, bounds on variables are best incorporated by nonlinear transformation of the variables. For example, letting the optimizer adjust x in $v = x^2$ while computing the function with variable v will ensure a v that is always positive. For other constraints, there are penalty functions that increase

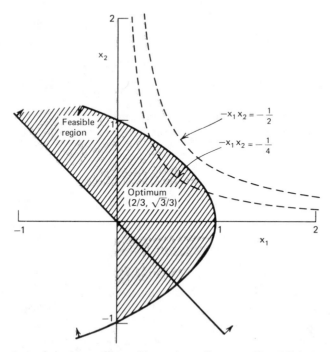

Figure 5.29. An optimization problem with two inequality constraints. Minimize $-x_1 x_2$ such that $-x_1 - x_2^2 + 1 \geqslant 0$ and $x_1 + x_2 \geqslant 0$. [From Fiacco and McCormick, 1968.]

the objective function when constraints are violated, i.e., when the **x** vector is in an infeasible region. The next sections will describe these techniques and provide an evolving example of most of the concepts. A few network applications for constrained optimization will be suggested.

5.6.1. Simple Constraints.

The easiest constraint to maintain is the equality constraint $x_j = k$. Just set the partial derivative with respect to this variable equal to zero. The reader may wish to try this by rewriting line 1050 in Program B5-1, the Rosenbrock function: 1050 G(2)=0. When run, the starting value of X(2) will never change because the optimizer sees no function decrease in coordinate direction X(2). There are many times when such constraints are temporarily useful, such as when a variable tends to go through zero to negative values. The variable can be held at some value by equating the derivative to zero.

An objective function used as an example throughout the rest of the constraints discussion is

$$Q(\mathbf{x}) = 4x_1 + x_2 + \frac{r}{x_1} + \frac{r}{x_2}, \tag{5.96}$$

where r is some fixed, real number, e.g., $r = 1$. An objective subroutine to implement (5.96) in the Fletcher–Reeves optimizer (Program B5-1) is shown in Table 5.6. Note that the derivatives have been written in lines 1040 and 1050 using ** to indicate exponentiation. The function $Q(\mathbf{x})$ in (5.96) is shown in Figure 5.30. The reader should run the optimizer with this function, starting from several points, such as $\mathbf{x} = (1, 2)^T$, $(0.25, 0.7)^T$, and $(0.25, 1.5)^T$. Note that in the first and second cases the program halted with an overflow error. Asking the computer for the values of **x** after this event reveals that the optimizer search has wandered into the second and third quadrants, respectively, where the function descends forever. If linear searches from the starting point never leave quadrant 1, then the minimum at $\mathbf{x} = (0.5, 1)^T$ is found successfully.

Clearly, it is desirable that the variables be bounded positive; this is sometimes necessary in network optimization also. A means for maintaining

Table 5.6.   Objective Subroutine for (5.96) With r = 1

```
1000 REM BARRIER FUNCTION EXAMPLE
1005 R = 1
1010 Q = 4 * X(1) + X(2) + R/X(1) + R/X(2)
1020 IF Y% = 0 THEN F = Q
1030 IF Y% = 1 THEN F9 = Q
1040 G(1) = 4 − R/X(1)**2
1050 G(2) = 1 − R/X(2)**2
1060 RETURN
```

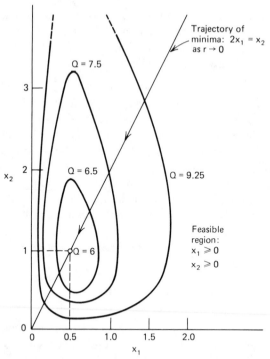

Figure 5.30.   Level curves of $Q(x)$ in Equation (5.96) when $r = 1$.

variables positive is in the transformations

$$x_i = \sqrt{x_i} \tag{5.97}$$

for all N variables. This is performed *inside* the optimizer. The user inputs (positive) values of **x** as before; the optimizer works with values of the square root of **x**; and the objective function subroutine, also being outside the optimizer, converts back to **x** again before making its calculations. So even though the optimizer may make *its* variables negative, there will be no decrease in the objective function subroutine, and the optimizer will therefore withdraw its variables to the first quadrant again. This is implemented for (5.96) by using the code in Table 5.7 instead of that in Table 5.6. Lines 1006–1007 transform the internal variables x to the outside variables v:

$$v = x^2. \tag{5.98}$$

The inverse operation is accomplished in line 135. Since the derivative of (5.98) is

$$\frac{\partial v}{\partial x} = 2x, \tag{5.99}$$

the chain rule yields

$$\frac{\partial F}{\partial x} = \frac{\partial F}{\partial v}\frac{\partial v}{\partial x} = \frac{\partial F}{\partial v}2x \tag{5.100}$$

Table 5.7. Objective Subroutine for (5.96) With Squared-Variable Transformations

```
 135 X(I) = SQR(X(I))
 295 PRINT K; X(K)*X(K); G(K)/(2*X(K))
 935 PRINT K; X(K)*X(K); G(K)/(2*X(K))
1000 REM BARRIER EXAMPLE (5.96) WITH SQUARED-
     VARIABLE TRANSFORMATION
1005 R = 1
1006 V1 = X(1)**2
1007 V2 = X(2)**2
1010 Q = 4*V1 + V2 + R/V1 + R/V2
1020 IF Y% = 0 THEN F = Q
1030 IF Y% = 1 THEN F9 = Q
1040 G(1) = 4 - R/V1**2
1050 G(2) = 1 - R/V2**2
1052 G(1) = G(1)*2*X(1)
1054 G(2) = G(2)*2*X(2)
1060 RETURN
```

for each x and v component. This is employed in lines 1052–1054 to scale the gradients for the optimizer's variable space, and in lines 295 and 935 to scale the gradients to the outside world's variable space. The program should now be run for the three previous cases to note that the positive-variable constraints yield the correct optimum (Figure 5.30) from all starting points in the first quadrant.

A number of bounding constraint transformations are shown in Table 5.8; the first one is that employed above in (5.98). An interesting application of the upper and lower bounds shown in Table 5.8 was suggested by Manaktala (1972) and called "network pessimization." Suppose that a certain lowpass network was constructed with elements having $+/-$ tolerances. At each

Table 5.8. Some Transformations to Impose Simple Constraints on Variables

Constraint	Transformation
$v \geqslant 0$	$v = x^2$
$v > 0$	$v = e^x$
$v \geqslant v_{min}$	$v = v_{min} + x^2$
$v > v_{min}$	$v = v_{min} + e^x$
$-1 \leqslant v \leqslant 1$	$v = \sin x$
$0 \leqslant v \leqslant 1$	$v = \sin^2 x$
$0 < v < 1$	$v = e^x/(1 + e^x)$
$v_{min} \leqslant v \leqslant v_{max}$	$v = v_{min} + (v_{max} - v_{min})\sin^2 x$, or
	$v = \frac{1}{2}(v_{max} + v_{min}) + \frac{1}{2}(v_{max} - v_{min})\sin x$
$v_{min} < v < v_{max}$	$v = v_{min} + (v_{max} - v_{min})e^x/(1 + e^x)$

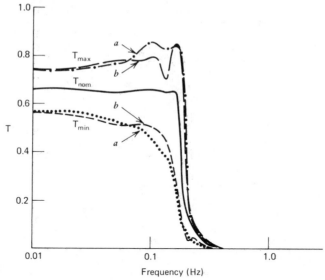

Figure 5.31. Worst-case variations for pessimization of two lowpass network designs. Lowpass $N = 5$ networks: *a*, zeros in left- and right-half planes; *b*, zeros only in left-half plane. [Reprinted with permission from Manaktala, 1972.]

frequency, there must be some adverse combination of tolerances that would produce worst-case selectivity, both maximum and minimum. This is shown in Figure 5.31. Rather than employing the usual time-consuming Monte-Carlo method, it was suggested that a constrained optimizer program could find the minimum and maximum selectivity at each frequency subject to the bounding element tolerance ranges—truly a pessimization problem. The performance of any network would then be contained inside the envelope shown in Figure 5.31.

5.6.2. Barrier Functions for Inequality Constraints. The complete constrained optimization problem was defined by (5.94). This section considers the vector **c** of inequality constraints that are generally nonlinear. It is remarked in passing that a subset would consist of linear constraints of the form

$$\mathbf{Ax} - \mathbf{b} \geqslant \mathbf{0}. \tag{5.101}$$

These boundaries are lines in 2-variable space, otherwise hyperplanes. Minimization with these constraints is like descending on the surface of Figure 5.3, except that it has been placed in a restricting glass box; the descent should conform to these glass walls, or hyperplanes, when encountered. The most common means for doing this is to project linear search directions on such constraining surfaces when encountered. This complicates linear search algorithms and is beyond the scope of the present treatment; the interested reader is referred to Rosen's projection method described by Hadley (1964, pp.

315–325). For small computers, systems of constraints defined by (5.101) may be treated by the following barrier technique.

A barrier function for generally nonlinear constraints in vector **c** is:

$$\min_{\mathbf{x}} Q(\mathbf{x}) = E(\mathbf{x}) + r \sum_{i=1}^{M} \frac{1}{c_i(\mathbf{x})} ; \qquad r \to 0. \tag{5.102}$$

The nature of a barrier function is seen by considering the following objective function:

$$\min E(\mathbf{x}) = 4x_1 + x_2 ; \qquad \mathbf{x} \text{ is positive.} \tag{5.103}$$

This function is shown in Figure 5.32. Clearly, the constrained optimum is at the origin, as indicated. The barrier function corresponding to (5.102) has already been written; it is (5.96), for which figure 5.30 applies. Note that the value $r = 1$ produced a minimum at $\mathbf{x} = (0.5, 1)^T$. The barrier is created by the infinite contours of r/x_1 and r/x_2 or, in general, r/c_i for the ith constraint approaching zero, the edge of its feasible region.

The barrier function is employed in a sequence of unconstrained minimizations, each for a smaller value of parameter r in (5.102). (It can be shown analytically that the limit at $r = 0$ exists.) An expression for these minima can be written for barrier function (5.96) by setting its partial derivatives equal to zero:

$$\frac{\partial Q}{\partial x_1} = 4 - \frac{r}{x_1^2} = 0, \tag{5.104}$$

$$\frac{\partial Q}{\partial x_2} = 1 - \frac{r}{x_2^2} = 0. \tag{5.105}$$

Any particular minimum occurs at $x_1 = \sqrt{r}/2$ and $x_2 = \sqrt{r}$. Eliminating r shows

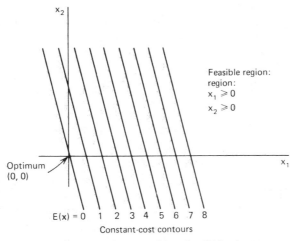

Figure 5.32.   Level curves of function $E(\mathbf{x}) = 4x_1 + x_2$.

Table 5.9.   Additions to Table 5.7 to Implement SUMT

Delete line 1005 and add the following:

145 R = 10
146 FOR JJ = 1 TO 20
147 R = R/10
148 PRINT"**************"
149 PRINT"***   R = ";R,"   ***"
998 NEXT JJ

---

that the trajectory of the sequence of minima in this case is simply $x_2 = 2x_1$, as shown in Figure 5.30. In general, the trajectory is an analytic function of r, Q(r). It is well behaved and its derivatives exist.

Fiacco and McCormick (1968) call this process the Sequential Unconstrained Minimization Technique (SUMT). It is easy to demonstrate this process for the example already programmed in this section. Table 5.9 lists the one deletion and several additions to implement the sequence of unconstrained minimizations. When the program is run, an unconstrained minimization with r = 1 occurs first, then r is reduced by a factor of 10, and the process is repeated. The successive optima are on the trajectory described above and depicted in Figure 5.30. The exact solution will be obtained in the limit as r→0. There is no need to arrive there computationally. It is generally true that a Richardson extrapolation to the limit in the parameter $\sqrt{r}$ (as seen analytically) is valid on the trajectory function Q(r). (Recall the Richardson extrapolation concept introduced in Section 2.3.2 in connection with Romberg integration.) The reduction factor for r is not a critical parameter; values from 4 to 25 usually result in about the same total number of function evaluations in progressing along the trajectory of successive minima.

5.6.3. Penalty Functions for Equality Constraints.

The vector **h** set of equality constraint functions in (5.94) can be enforced by compound functions of the form

$$\min_{\mathbf{x}} Q(\mathbf{x}) = E(\mathbf{x}) + \frac{1}{\sqrt{r}} \sum_{k=1}^{P} h_k^2(\mathbf{x}); \qquad r \to 0, \qquad (5.106)$$

which add a penalty to the total objective function when each and every h_k constraint is not zero. Comparable to the SUMT method, there is a trajectory function Q(r), where r is a sequence of decreasing values. As seen in (5.106), smaller values of r add a larger penalty to unsatisfied constraints h_k. In the limit as r→0, the constraints must all be satisfied; i.e., the unconstrained optimizer has been forced to find the region of **x** space that is feasible, if it exists. The starting values for variables (**x**) usually will be unfeasible, which is the opposite of the barrier (inequality) constraint function.

In fact, one application of penalty functions is to acquire feasibility for inequality constraints. To do so for inequality constraint c_i, use the penalty constraint function

$$h_k(x) = \left\{ \min\left[0, c_i(x) \right] \right\}^2. \tag{5.107}$$

A little thought will show that this is exactly equivalent to the "satisfied-when-exceeded" technique discussed in Section 5.5.2. It is seen that the partial derivatives of h_k in (5.107) exist at the boundary of feasibility.

Penalty functions are usually better behaved in unconstrained optimization than barrier functions. This is usually due to the mechanics of the linear search process, where the infinite barrier may be overstepped by the necessarily finite exploratory moves. A review of the example problem and its treatment by transformation of variables in Section 5.6.1 supports this conclusion.

5.6.4. Mixed Compound Function for All Constraints.
Fiacco and McCormick (1968) derived the necessary conditions for defining a combined barrier and penalty function:

$$\min_{x,r} F(x) = E(x) + r \sum_{i=1}^{M} \frac{1}{c_i(x)} + \frac{1}{\sqrt{r}} \sum_{k=1}^{P} h_k^2(x), \tag{5.108}$$

with derivatives

$$\frac{\partial F}{\partial x_j} = \frac{\partial E}{\partial x_j} - r \sum_{i=1}^{M} \frac{\partial c_i / \partial x_j}{c_i^2(x)} + \frac{2}{\sqrt{r}} \sum_{k=1}^{P} h_k(x) \frac{\partial h_k}{\partial x_j}. \tag{5.109}$$

One practical consideration in (5.108) is the choice of the starting value for r. If it is too small, then the c_i inequality constraint barriers will be too far away and steep, so that the h_k penalty functions will tend to dominate the objective $E(x)$. Difficulties of the opposite nature exist if the initial r is too large. There are fairly sophisticated means for selecting the initial r value, but one way that at least leaves the objective $E(x)$ somewhat in control has been satisfactory. The value of $E(x)$ and of each summation in (5.108) is obtained for the contemplated starting point in variable (x) space. Then the first r value is chosen so that the absolute value of barrier and penalty contributions is just 10% of the $E(x)$ contribution to $F(x, r)$. This procedure requires the solution of a real quadratic equation.

Fiacco and McCormick (1968) also show why and how the Richardson extrapolation to the limit operates. Using this extrapolation for all variables often places the solution inside unfeasible regions. In short, there are some programming complexities to be overcome in applying the Richardson extrapolation to barrier, penalty, and mixed functions. The good news is that personal computer users operate in the loop with program execution. The complicated program features required in a timeshare environment to avoid receiving big bills for unforeseen runaway programs are hardly necessary with personal computer applications.

5.6.5. Summary of Constraints.   Unconstrained optimization will often produce negative variable values that have no physical meaning. Sometime during function minimization, variables may become negative, but return to positive values again at convergence. This is one reason why a trial problem run without constraints is not a bad idea. The most elementary constraints are bounds on elements; these may prevent negative variable values or contain variables within ranges such as component tolerance intervals. The squared or trigonometric transformations of variables are often effective. An example of the former was programmed in this chapter. Another type of constraint is a set of linear inequalities. It was noted that the projection method whereby the minimization is conducted on surfaces bounded by the related hyperplanes ("glass walls") requires rather complicated techniques in linear searches, and thus was not discussed further. Constraints are simply relationships among variables. Those mentioned so far in this paragraph can be stated explicitly. There are many others that cannot be so stated and are therefore implicit constraints.

The barrier method for inequality constraints and the penalty method for equality constraints (including the "satisfied-when-exceeded" constraint) were described. Then the two methods were combined in one mixed compound function. In these cases, the trajectory parameter r was introduced. Assignment of a value to r enabled an unconstrained optimization to occur. A sequence of choices for decreasing r values leads to the constrained solution, the process being called the Sequential Unconstrained Minimization Technique (the well-known SUMT). A lot of computer time is consumed in the process, and failure-proof extrapolation methods for predicting the limit process without closely computing the limit are not easy to program. The interested reader is referred to Gill and Murray (1974) and Lootsma (1972, pp. 313–347).

Despite some complexities, the reader should have knowledge of these methods, because there are many special cases where some of these concepts can be meaningfully applied. This is especially true for personal computing, where an educated observer remains in the driver's seat. Machine time is prepaid, so that programs need not be constructed with the guaranteed performance of robots in space—or limited expense accounts on computer timeshare services.

5.7. Some Final Comments on Optimization

The methods in this chapter were selected because they are practical engineering tools and their explanation involves important mathematical concepts. However, the reader should be aware of an entirely separate kind of optimization, known as direct search. Direct search methods explore the function surface without benefit of gradient information and have the substantial advantage that function smoothness (continuity) is not required. There are many kinds of systematic direct search schemes, and almost all of them are heuristic methods, developed more on the basis of intuition and experience than on an extensive rational basis. They are often less automatic than gradient methods, requiring a number of parameters to be set rather arbitrarily. It should be recognized that the "operator-in-the-loop" nature of personal computers makes direct search more attractive than when used on remotely engaged computer services.

There is at least one direct search method that requires only 3N memory locations, as does the Fletcher–Reeves algorithm. That is the pattern search algorithm of Hooke and Jeeves (1961). A FORTRAN code that can be modified for this purpose has been given by Kuester and Mize (1973). Briefly, function values are computed at a starting (base) point and at a small displacement in one variable. If this is successful (reduced elevation), then a small displacement in the next variable is tried. If this is also successful, the base point is moved along a vector through the second successful point; otherwise, another variable is tried. The strategy is that successful moves are worth trying again. The interested reader can find a useful explanation of the details in Beveridge and Schechter (1970).

Sadly, a strong warning must accompany all claims for optimization; it is, after all, only the last step in engineering design. Some of its advocates have the tendency to use it as an excuse for neglecting the first two steps: identifying design variables and how they interact, and creating a measure of effectiveness. The acronym GIGO is apt: garbage in, garbage out. Optimization does stimulate good modeling of systems. Time and again it has been found that, once optimization problems have been suitably structured, the solution (or lack of one) is then apparent almost by inspection. The author believes that optimization (nonlinear programming) is a major circuit design tool, in the same league with the programmable calculator/computer on which it depends.

Problems

5.1. Shift the origin of the central sample function in (5.8) by the substitutions $x_1 \leftarrow x_1 + 10$ and $x_2 \leftarrow x_2 + 10$. Simplify the resulting equation and compare it to the Taylor series expansion (5.32) about the point $\mathbf{p} = (10, 10)^T$.

5.2. Show that the root-finder steps in (3.9) and (3.10) are identical to Newton steps in (5.38). *Hint*: Let $f_1 = u$, $f_2 = v$, $x_1 = x$, and $x_2 = y$.

5.3. Use Programs A5-2 and A5-4 in Appendix A to verify Table 5.2.

5.4. Verify in two-variable space that

$$(\mathbf{Ax})^T = \mathbf{x}^T \mathbf{A}^T$$

even when $\mathbf{A}$ is not symmetric.

5.5. Evaluate by (5.41) the optimal step for the central sample function (5.8) from $\mathbf{p} = (10, 10)^T$ in direction $\mathbf{s} = (-1, -2)^T$. Find $\mathbf{x}^{i+1}, \mathbf{g}^{i+1}$, and show that the directional derivative (5.57) at $\mathbf{x}^{i+1}$ is zero.

5.6. Make a table for the central sample function (5.8) at $\mathbf{x}^T = (10, 10), (5, 3), (-1, -1)$, and $(5, 7)$. Verify (5.48) and (5.50) using all these data.

5.7. Write a Taylor series expansion about the point $\mathbf{x} = (4, 4)^T$ using first and second partial derivatives of the function

$$F(\mathbf{x}) = \ln(x_1 x_2) + \sqrt{x_1} + 3\sqrt{x_2} .$$

Make a table of $\Delta \mathbf{x}$ varying by ± 0.2 about this point and showing the percent difference between true function values and those estimated by the Taylor series.

5.8. Examine the flowchart in Appendix D for the Fletcher–Reeves linear search scheme; expand it to describe the entire Fletcher–Reeves algorithm (Program B5-1).

5.9. What revised value of $\mathbf{b}$ would cause the minimum of the central sample function (5.8) to be at $\mathbf{x} = (-3, -4)^T$?

5.10. Define a *standard function* as

$$F(\mathbf{x}) = c + \mathbf{b}^T \mathbf{x} + \tfrac{1}{2} \mathbf{x}^T \mathbf{A} \mathbf{x},$$

where

$$\mathbf{A} = \begin{bmatrix} 14 & 2 \\ 2 & 11 \end{bmatrix}$$

and

$$c = 500, \qquad \mathbf{b} = (-94, -67)^T.$$

Is $\mathbf{A}$ positive definite? Why?

5.11. For the standard function in Problem 5.10:
  (a) Find F, $\nabla_1 F = g_1$, $\nabla_2 F = g_2$, and the slope in the direction $\mathbf{s} = (1, -2)^T$, all at the point $\mathbf{x} = (3, 7)^T$.
  (b) What is the value of the metric defined by (5.50) at $\mathbf{x} = (3, 7)^T$?

5.12. For the $\mathbf{A}$ matrix in Problem 5.10:
  (a) Confirm that the eigenvalue problem (5.25) is satisfied by $\lambda_1 = 15$, $\mathbf{x}^1 = (2, 1)^T$; and by $\lambda_2 = 10$, $\mathbf{x}^2 = (-1, 2)^T$.

(b) Find the diagonal matrix $\Lambda = \mathbf{P}^T \mathbf{A} \mathbf{P}/5$, where $\mathbf{P}$ is a suitable orthogonal matrix.

5.13. For the standard function in Problem 5.10:

(a) Write the matrix equation for the Taylor series in $\Delta \mathbf{x}$ about the point $\mathbf{p} = (3, 7)^T$.

(b) Find the $\Delta \mathbf{x}$ step to the minimum F location from $\mathbf{p}$; use matrix calculations.

5.14. Given the two nonlinear functions

$$f_1(x_1, x_2) = x_1^4 - x_1^2 x_2 - 52x_1 + 11x_2 + 23,$$

$$f_2(x_1, x_2) = 51x_1 - x_1 x_2^2 - 94x_2 + x_2^3 + 325,$$

calculate by the Newton method the estimated $\Delta \mathbf{x}$ step to the minimum F location from the coordinates $x_1 = 3$ and $x_2 = 7$.

5.15. For the standard function in Problem 5.10:

(a) Compute the Fletcher–Reeves linear searches to the location of minimum F starting from $\mathbf{x} = (3, 7)^T$. Use (5.41) in all your linear searches. Show all values of $\mathbf{x}$, α, $\mathbf{s}$, $\mathbf{g}$, and β involved.

(b) Show *numerically* that your search directions are A-conjugate and that the gradient at each turning point is orthogonal to the last search direction.

5.16. Consider $\mathbf{A}$ in problem 5.10 and search directions $\mathbf{s}^1 = (1, -2)^T$ and $\mathbf{s}^2 = (2, 1)^T$:

(a) Are $\mathbf{s}^1$ and $\mathbf{s}^2$ A-conjugate? Show work.

(b) Are $\mathbf{s}^1$ and $\mathbf{s}^2$ orthogonal? Show work.

(c) Explain the results in (a) and (b) in terms of minimizing quadratic functions.

5.17. Given any two arbitrary functions $f_1(x_1, x_2) = 0$ and $f_2(x_1, x_2) = 0$ and their partial derivatives, write a discrete, unweighted, least-squared-error optimizer objective function and its gradient equations. These should be in forms so that particular cases could be used in subroutine 1000 in optimizer Program B5-1 to find the solution of x_1 and x_2.

5.18. Write the barrier function equation for the constrained minimization problem in Figure 5.29.

Chapter Six

Impedance Matching

Impedance matching is the design of a network or transducer so that a terminating impedance is transformed exactly to a desired impedance at a frequency, or is transformed approximately over a band of frequencies. Figure 6.1 shows the situation where load impedance Z_2 may be specified as some LC subnetwork terminated by a resistance or by complex numbers associated with arbitrary frequencies. The desired input impedance Z_{in} may be similarly specified or may be contained in a neighborhood described as some maximum standing-wave ratio (4.59). Section 9.6 will consider dissipative network transformations; in this chapter only lossless, passive networks are considered. Impedance transformation is usually desired for control of power transfer from a finite impedance source, and is thus related to the same requirements discussed in Chapter Three for doubly terminated filters. There is one important difference: impedance matching usually is concerned with given terminating impedances that are complex, not simple resistances. This results in simple restrictions for single-frequency transformations. There are complicated constraints when matching complex impedances over a band of frequencies. These problems will be considered in order of increasing generality.

Chapter Six begins with impedance matching at a single steady-state frequency, first with two- and three-element networks composed of a combination of inductors and capacitors, and then with one or two cascaded transmission lines. It is remarkable that these subjects are seldom treated in modern electrical engineering curricula, even though they appear in almost all pertinent texts and handbooks published before 1960. Practicing engineers responsible for radio frequency circuit design invariably query prospective employees about L, T, pi, and perhaps transmission line matching because it is a matter of frequent concern. The treatment here includes tried and true concepts, which will be extended to broadband matching and direct-coupled filters (Chapter Eight) as well, especially the idea of the loaded Q of an impedance. The Smith chart as a means of visualizing the matching process

170

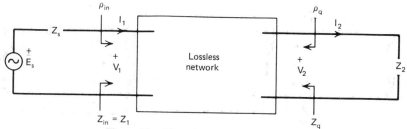

Figure 6.1. The matching network problem.

will be discussed. An innovation is the use of hand-held computers to provide data for plotting, greatly speeding the process and reducing eyestrain.

The rest of the chapter addresses impedance matching over a frequency band, referred to as broadband matching, as opposed to the single-frequency impedance matches, which are often useful over fairly narrow frequency bands (or can be made so by optimization). It might be assumed that if a fairly extensive network is designed for one central frequency in the band of interest, then an optimizer should be able to start with these element values and obtain the best possible match over the band. Unfortunately, this does not work in most cases because of the large number of useless local minima in the objective function's surface. So theoretical methods are necessary, some with major limitations that may be candidates for elimination by optimization of the theoretical result. One limitation always present is the assumption of lossless networks; dissipative effects usually will be compensated by optimization.

The classical method for broadband impedance matching was thoroughly described by R. M. Fano (1950). He extended Bode's integral matching limitation for RC load networks to load networks composed of any number of LC elements terminated in a resistance. The theory becomes too complicated for more than three load reactance elements. The closed formulas by Levy (1964) presented here enable the consideration of a single RLC load branch. He accounts for three types of sources: resistive, a single RLC source branch, and lossless (singly terminated) sources. The subject is invariably presented as the lowpass case; i.e., loads are parallel RC or series RL, and the frequency band begins at dc. Practical applications usually require pass bands above dc, which are obtained by a simple network transformation, and the loads are RLC, as mentioned. Thus the development requires consideration of the transformation that changes the network from lowpass to bandpass. Another feature of classical lowpass theory is that the source resistance is dependent. Usually, the designer must use a particular source resistance. In the case of bandpass networks, Norton transformations enable the replacement of all-L or all-C pairs (L sections) by three-element sections of like kind. An arbitrary impedance transformation within a limited range is possible, and there are no frequency effects. Norton transformations are also a part of the Fano theory in reduction to practice.

A recent application of Fano's integral limitation enables optimal matching of load impedances consisting of a C paralleling a series LR or an L in series with a parallel CR using a lowpass network configuration. The Cottee and Joines (1979) method is described. It employs numerical integration (Section 2.3) and synthesis (Chapter Three). The results are often desirable in practice, and the analysis helps clarify Fano's integral limitations. This is called a pseudobandpass technique, wherein lowpass networks are employed to match over a pass band.

All of the broadband-matching concepts mentioned so far require that the physical load be related to a hypothetical lumped-element terminating network. This subject is called load classification and is mentioned only briefly in this chapter. Carlin (1977) presented a new method for designing lossless matching networks; this method utilizes directly experimental load impedance data sampled at arbitrary frequencies. The arithmetic is well conditioned, so that the required optimization step works well in nearly all cases. Also, the technique is especially well suited for producing arbitrarily shaped power transfer functions versus frequency. This has special application in microwave amplifier network design, where it can compensate for the approximate 6 dB/octave roll-off of transistor gain above the critical frequency. These extra considerations are treated in the optimization step, so that it is equally easy to accommodate amplifier noise figure, stability, or other constraints that can be formulated in impedance terms. Carlin's method is based on a very practical application of the Hilbert transform, which relates reactance frequency behavior to resistance behavior. A separate program for this aspect is provided. The final steps in Carlin's method require fitting a rational polynomial (Section 2.5) and synthesis (Chapter Three).

As in other chapters, there are many personal computer programs furnished. Both programs and concepts are useful design tools for the practicing radio frequency engineer. The ladder network analysis procedures of Chapter Four will be useful for verifying designs produced in this chapter. It is assumed that the reader can write a simple program to convert reactances at a given frequency to L and C values and vice versa.

6.1. Narrow-Band L, T, and Pi Networks

The four reactance configurations considered for the network in Figure 6.1 are shown in Figure 6.2. There are several important conventions adopted. First, the lossless inductors and capacitors are shown as reactances at the one design frequency. Second, an inductor is implied if the reactance is positive, and a capacitor is implied if the reactance is negative. Third, the basic design relationships assume a match from simple load resistance R_2 to input resistance R_1. The two L sections in Figure 6.2 have fundamental constraints on

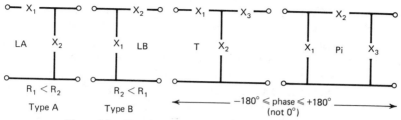

Figure 6.2.   Four lumped-element reactance configurations.

the direction of resistance match, as indicated; i.e., a type-A L section can only decrease the resistance level. Finally, the transfer phase is defined as the angle by which current I_2 lags I_1 in Figure 6.1. For resistive Z_1 and Z_2, this is the same as the phase lag of V_2 with respect to V_1. The phase angles of type-A and type-B L sections are dependent, assuming that the R_1 and R_2 terminations are independent. However, the phase is also independent for the T and pi sections shown in Figure 6.2 within the range indicated.

It has been remarked that the phase is not of interest in many cases; however, it is a convenient parameter and represents a degree of freedom for T and pi networks. (The phase sign is a degree of freedom for L sections.) It is also noted that L, T, and pi sections may be designed by the $1+Q^2$ method described in Section 6.1.3 without consideration of phase. The reader is expected to adopt the techniques most useful for his purpose.

This topic will be developed by first considering the interface impedances resulting from the use of a lossless network, especially the relationship between Z_q and Z_2 and between Z_s and Z_1 in Figure 6.1. Then the basic case for T and pi network matching from resistance R_2 and R_1 will be given. The L sections will be special cases of these, in which branch X_3 in Figure 6.2 is removed. To accommodate complex source and load impedances, series-to-parallel impedance conversions and the opposite case will be developed. Also, the impedance of paralleled impedances will be discussed. These conversions adapt complex terminations to the prior analysis for resistive transformations. Finally, the role of graphic procedures—especially the Smith chart—will be considered in some detail. Programs are provided.

6.1.1. Lossless Network Interface Impedances.

There is an important impedance concept associated with maximum power transfer by a lossless network of any type. Consider the power transferred from the source to the network in Figure 6.1. According to the analysis in Section 3.2.3, maximum available power (P_{as}) is transferred when $Z_1=Z_s^*$. For lossless networks, the maximum available power must arrive undiminished at the load end of the network, where the Thevenin impedance looking back into the network is Z_q. At the load port, then, there must be an equivalent Thevenin source providing the same maximum available power; therefore, $Z_2=Z_q^*$. In fact, the matched, lossless network can be cut at any interface and a conjugate match will exist;

i.e., the pertinent generalized reflection coefficient magnitude (3.46) must be zero at every interface. If there is a mismatch anywhere, then the pertinent reflection coefficients at every interface must all have the same *magnitude*, since the actual and available power are the same everywhere.

The most elementary application occurs when only resistances are anticipated at the lossless network terminals in Figure 6.1. Then $Z_q = R_2 = Z_2$ implies that $Z_s = R_1 = Z_1$. Especially, it is common to talk about resistive terminations where the simple fact that $Z_1 = R_1$ would not necessarily imply that $Z_s = R_1$. This small nuance arises in Section 6.1.4 when matching from a complex load to a possibly complex source.

6.1.2. Real Source and Real Load. The reactance equations for T and pi networks are given in Table 6.1, along with the modifications for the L sections. T-section matching relationships will be verified; the pi-section relationships can be verified similarly.

Consider Figure 6.3. Define

$$Z_0 = R_2 + jX_3 . \tag{6.9}$$

Replacing X_3 in (6.9) with its T-section expression (6.5) from Table 6.1 yields

$$Z_0 = \frac{R_2 e^{j\beta} - \sqrt{R_1 R_2}}{j \sin \beta} . \tag{6.10}$$

If $I_2 = 1 + j0$ in Figure 6.3, then $V' = Z_0$ and

$$I_1 = \frac{Z_0}{jX_2} + 1. \tag{6.11}$$

Table 6.1. T, Pi, and L Reactance Equations

T		Pi	
$X_1 = \dfrac{\sqrt{R_1 R_2} - R_1 \cos \beta}{\sin \beta}$	(6.1)	$X_1 = R_1 R_2 \dfrac{\sin \beta}{R_2 \cos \beta - \sqrt{R_1 R_2}}$	(6.2)
$X_2 = -\dfrac{\sqrt{R_1 R_2}}{\sin \beta}$	(6.3)	$X_2 = R_1 R_2 \dfrac{\sin \beta}{\sqrt{R_1 R_2}}$	(6.4)
$X_3 = \dfrac{\sqrt{R_1 R_2} - R_2 \cos \beta}{\sin \beta}$	(6.5)	$X_3 = R_1 R_2 \dfrac{\sin \beta}{R_1 \cos \beta - \sqrt{R_1 R_2}}$	(6.6)

For LA: use T with $X_3 = 0$; $\beta = \pm \cos^{-1} \sqrt{\dfrac{R_1}{R_2}} = \tan^{-1} \sqrt{\dfrac{R_2}{R_1} - 1}$ (6.7)

For LB: use pi with $X_3 \to \infty$; β as above, with R_1 and R_2 exchanged.   (6.8)

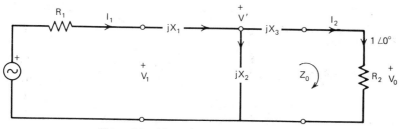

Figure 6.3. Network for T-section analysis.

Then, placing (6.10) and (6.3) from Table 6.1 in (6.11) gives an expression for the input current:

$$I_1 = \sqrt{\frac{R_2}{R_1}}\ e^{j\beta}. \tag{6.12}$$

The current transfer phase into a resistive load is thus

$$\frac{I_2}{I_1} = \sqrt{\frac{R_1}{R_2}}\ e^{-j\beta}, \tag{6.13}$$

where β is the angle by which I_2 lags I_1.

It is now easy to verify the input impedance:

$$Z_1 = jX_1 + \frac{V'}{I_1}. \tag{6.14}$$

Using (6.10) and (6.1) from Table 6.1 in (6.14), a little algebra shows that $Z_1 = R_1$, as required. For a type-A L section, setting $X_3 = 0$ in (6.5) provides

Table 6.2. Sample Problem Data for Appendix Program B6-1[a]

Case	R_1	R_2	β	X_1	X_2	X_3
LA	25	50	45	25	−50	—
LA	25	50	−45	−25	50	—
LB	50	25	45	−50	25	—
LB	50	25	−45	50	−25	—
T	50	50	120	86.60	−50.74	86.60
T	50	50	−120	−86.60	57.74	−86.60
T	50	50	90	50	−50	50
Pi	100	25	150	−17.45	25	−9.15
Pi	100	25	−150	17.45	−25	9.15
Pi	100	25	90	−50	50	−50
Pi	25	100	90	−50	50	−50

[a] Values are in ohms and lagging degrees.

the phase angle expression (6.7); the latter also displays the requirement that $R_1 < R_2$.

The relationships in Table 6.1 have been programmed in BASIC language Program B6-1 (listed in Appendix B together with a flowchart). Table 6.2 contains sample data for program verification and illustration.

Example 6.1. Match a load impedance of $6 + j0$ ohms to an input impedance of $25 + j0$ ohms using L sections. The solutions are shown in Figure 6.4. The required equations appear in Table 6.1. A type-B L section is required, since $R_2 < R_1$. The phase must have a magnitude of 60.67 degrees. Choosing a lagging phase ($\beta = +60.67$) yields the configuration in Figure 6.4a, and the leading phase ($\beta = -60.67$) yields the configuration in Figure 6.4b.

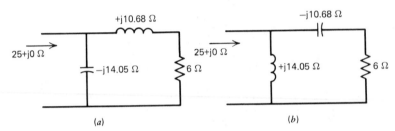

(a)   (b)

Figure 6.4.   Two L sections that match 6 ohms to 25 ohms. (*a*) Lagging phase; (*b*) leading phase.

6.1.3. *Series-Parallel Impedance Conversions.* This section deals with the equivalence at one frequency shown in Figure 6.5. At first glance, it may seem awkward to avoid the impedance-admittance convention by calling the reciprocal conductance and negative reciprocal susceptance "parallel ohms." However, there is a strong tendency to approach problems in familiar units, so that a practical range of values is recognizable as opposed to blind numerical procedures. The need to convert between forms arises when the matching network's series input or output branch faces a parallel impedance termination or vice versa. Then the conversion in Figure 6.5 enables a combination of series (or parallel) reactances in the termination and the network branch. The

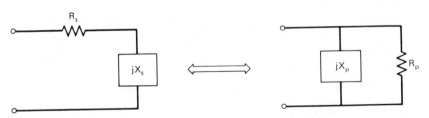

Figure 6.5.   Two impedance forms that are equivalent at a frequency.

concept facilitates complex matching (Section 6.1.4) based on the resistive matching described in Section 6.1.2.

Convenient equations are derived by considering the equivalence of admittance and impedance forms:

$$Y = G + jB = \frac{1}{Z_s} = \frac{R_s}{R_s^2 + X_s^2} + j\frac{-X_s}{R_s^2 + X_s^2}, \qquad (6.15)$$

where $Z_s = R_s + jX_s$. An important definition is

$$Q = \frac{|X_s|}{R_s} = \frac{R_p}{|X_p|}. \qquad (6.16)$$

Then (6.15) shows that

$$R_p = R_s(1 + Q^2), \qquad (6.17)$$

thus the name "$1 + Q^2$" method. The conversion procedure is to solve (6.16) for the appropriate Q and (6.17) for the appropriate R. Then the unused Q relation from (6.16) leads to the unknown reactance X. On hand-held calculators it is tempting to program the conversion in (6.15) using built-in rectangular-to-polar functions. However, they execute much more slowly than the Q relationships. Program A6-1 in Appendix A performs these calculations on function keys A and B using only interchange operations in the four-register stack. The Q concept will be of major importance in this chapter and in Chapter Eight.

Example 6.2.   Suppose that the series impedance $6 + j12$ ohms is required in the parallel form shown in Figure 6.5. Following the data input convention (X before R) given with Program A6-1 listing, key A produces $R_p = 30$ in the X register and $X_p = +15$ (inductive as required) in the Y register. Key B changes the form back to series again.

Another useful relationship is the inverse of (6.17):

$$Q = \sqrt{\frac{R_p}{R_s} - 1}. \qquad (6.18)$$

An alternative to the equations in Table 6.1 is to design L, T, and pi networks by a sequence of $1 + Q^2$ conversions. Example 6.1 could have been worked using (6.18) to find that $Q = 1.7795$. Then (6.16) shows that $X_s = \pm 10.68$ and $X_p = \mp 14.05$. This type-B L section could have been extended by a type-A L section to form a T network. In this approach, the internal parallel resistance level replaces the transfer phase as the arbitrary parameter. An extension of (6.18) involves the L-section branch reactance ranges necessary to match a load impedance of bounded standing-wave ratio (SWR) to a desired source resistance. For the load SWR S_2, defined with respect to a nominal load resistance R_2, the values of the L-section branch Q are bounded by the

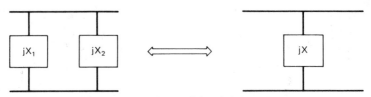

Figure 6.6.   Two reactances in parallel and their equivalent reactance.

extreme values

$$\begin{matrix} \max \\ \min \end{matrix} Q = \sqrt{\frac{R_p}{R_s} S_2^{\pm 1} - 1} \ . \tag{6.19}$$

The Q of the output branch is related to the fixed, nominal load resistance R_2. The extreme values of the two L-section branch reactances do not occur at the same particular load impedance. Derivation of (6.19) is easier to visualize after discussion of the Smith chart in Section 6.1.5.

Combining two reactances in series requires simple addition. Retaining the convenience of calculation in ohms, combining two reactances in parallel requires the relationship

$$X = \frac{X_1 X_2}{X_1 + X_2} = \frac{X_1}{1 + X_1/X_2} \ . \tag{6.20}$$

This is shown in Figure 6.6. Another common requirement is the calculation of one of the paralleled reactances (e.g., X_2 in Figure 6.6) so that the combination with a given X_1 produces the given equivalent X. The relationship can be obtained from (6.20) and put in that functional form as well:

$$X_2 = \frac{(-X_1)}{1 + (-X_1)/X} \ . \tag{6.21}$$

Since the functional forms of (6.20) and (6.21) are identical, programming the functions requires only one algorithm, except for a sign change for reactance X_1. Keys C and D in Program A6-1 evaluate (6.20) and (6.21), respectively. These simple functions are surprisingly useful in practice.

Example 6.3.   Referring to Figure 6.6, suppose that $X_1 = -30$, and $X_2 = 75$ ohms. Entering these into Program A6-1 and pressing key C yields $X = -50$ ohms. Conversely, entering first $X = -50$, then $X_1 = -30$, and pressing key D yields $X_2 = 75$ ohms.

6.1.4. Complex Sources and/or Complex Loads.   The simplicity of the matching relationships discussed so far hides the multiplicity of solutions that may or may not exist in particular cases. The general case of matching a complex load impedance to a source with complex generator impedance will illustrate the subtleties often encountered.

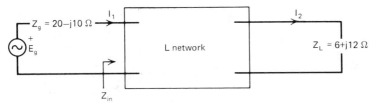

Figure 6.7. Problem for Example 6.4 considering both kinds of L-section networks.

Example 6.4. Consider the matching problem illustrated in Figure 6.7, where $Z_g = 20 - j10$ ohms and $Z_L = 6 + j12$ ohms, and both types of L sections are to be used. The parallel equivalent of the source is $25 \| -j50$, where the symbol $\|$ will be used to mean *in parallel with*. The type-B L section will thus require solutions from a 6-ohm resistance to a 25-ohm resistance, as obtained in Example 6.1 ($X_1 = \pm 14.05$ and $X_2 = \mp 10.68$ ohms reactance; see Figure 6.2). To minimize confusion, the reactances inside the type-B matching network will be designated X_a and X_b, as shown in Figure 6.8. The load was given in series form, and its reactance can become a part of the hypothetical matching element X_2, as shown in Figure 6.8. Then $X_b = -1.32$ ohms by subtraction. Use Program A6-1 to find X_a: enter -14.05, then -50, and press key D. This evaluates (6.21) and yields $X_a = -19.54$ ohms. As a check, convert the load mesh, $6 + j10.68$ ohms, to parallel form ($25.01 \| j14.05$). Then combine parallel reactances $j14.05 \| -j19.54$, using key C, to obtain the equivalent $+j50.01$ ohms. Figure 6.8 shows that this reactance will be canceled by the source reactance, leaving a match to the 25-ohm parallel resistance in the source. Note that the matching network actually used is composed of two capacitors;

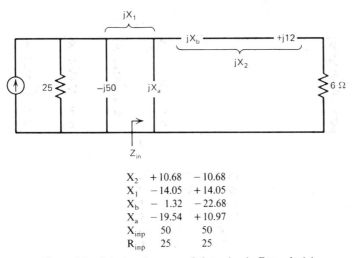

X_2	$+10.68$	-10.68
X_1	-14.05	$+14.05$
X_b	-1.32	-22.68
X_a	-19.54	$+10.97$
X_{inp}	50	50
R_{inp}	25	25

Figure 6.8. Solutions for a type-B L-section in Example 6.4.

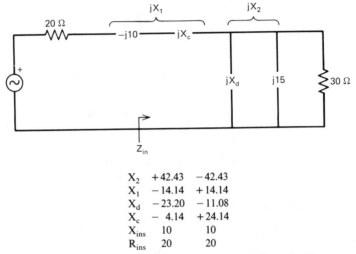

X_2	$+42.43$	-42.43
X_1	-14.14	$+14.14$
X_d	-23.20	-11.08
X_c	-4.14	$+24.14$
X_{ins}	10	10
R_{ins}	20	20

Figure 6.9.   Solutions for a type-A L-section in Example 6.4.

given a particular frequency, their reactances could be changed to farads. The second solution column in Figure 6.8 is obtained by the same procedure.

The type-A L-section solutions are obtained by the analysis recorded in Figure 6.9. The parallel form of the load is required, as obtained in Example 6.2, because the L section ends with a shunt element. Key D is again used to evaluate (6.21), to obtain $X_d = -23.204$ ohms. The rest of this and the second solution are obtained as previously described.

Note that a conjugate match exists at any interface in Figures 6.8 and 6.9. Also, there is no reason to assume that all L-section solutions must exist. The problem in Example 6.4 required a type-A section to decrease the resistance level, and vice versa for type B. T and pi networks may also be used, and they have an extra degree of freedom. Note in Figure 6.7 that phase β is related to terminal currents. Because of the series-to-parallel conversions employed, Figures 6.8 and 6.9 show that phase angle β does not apply to any of the four solutions obtained. This is often the case when there are complex terminations.

6.1.5.   Graphic Methods.
Terman (1943) presented comprehensive design graphs for L, T, and pi networks using the phase parameter. The use of computer programs may not be the last word in design technique; the trends evident in graphic data contribute greatly to problem insight and are highly recommended.

Probably the single most valuable tool in impedance matching is the Smith chart. It is useful in its conventional form in this chapter; it will be applied with much more generality in the next chapter. The Smith chart is the bilinear map of the right-half impedance plane into the unit circle of the reflection plane. Normalizing impedances to resistance R_1, reflection coefficient (4.57) is rewritten as the bilinear function

$$\rho = \frac{Z/R_1 - 1}{Z/R_1 + 1}.$$

(6.22)

The mapping is illustrated in Figure 6.10. Lines of constant resistance map into closed circles of constant, normalized resistance in the Smith chart, and lines of constant reactance map into circular arcs of constant, normalized reactance. On a Smith chart with its center normalized to impedance $1 + j0$ ohms (or mhos) according to (6.22), any complex number has its inverse appear symmetrically about the origin; i.e., given a point Z/R_1, the point $Y \times R_1$ appears on the opposite radial at the same radius, where $Y = 1/Z$. An easily read summary of Smith chart properties has been given by Fisk (1970).

The first result of Example 6.1 is plotted in Figure 6.10. The normalized reactance $(X_2/R_1 = 10.68/25 = 0.43)$ amounts to a displacement of $+0.43$ along the normalized constant resistance circle $(6/25 = 0.24)$. Then, since the X_1 matching reactance is a shunt element, the impedance point is converted to an admittance point by reflection about the origin, as shown in Figure 6.10. This point is necessarily on the normalized unit circle passing through the center of the chart (the center representing $R_1 = 25 + j0$ ohms). Now the Smith chart is considered an admittance chart instead of an impedance chart. Thus the displacement due to $X_1 = -14.05$ ohms is considered a normalized susceptance of $+1/14.05 \times 25 = +1.78$ mhos, which carries the transformation to the chart center, as required. The reader should plot the second solution of

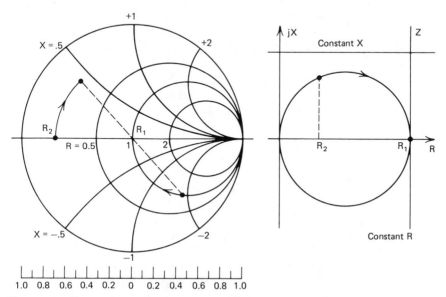

Figure 6.10. The ordinary Smith chart (the unit reflection coefficient circle) on the left is a map of the right-half Z plane on the right.

Example 6.1, which involves negative reactance and susceptance travel on the Smith chart. Negative travel is toward the bottom half of the chart instead of toward the top half (positive half-plane).

It is strongly recommended that the reader write a small program to accomplish the calculation of (6.22) and its inverse, so that Smith chart plotting is simply a matter of locating rectangular and/or polar computed numbers. Cases involving complex sources may be treated as in the analysis above and in Example 6.4. However, a more general treatment will be offered in the next chapter, where the chart center can represent a complex number. Many engineers associate the Smith chart with transmission line solutions; this will be shown in Section 6.2.

6.1.6. Summary of L, T, and Pi Matching.   The topic of L, T, and pi matching began with a comment on the fact that lossless matching networks exhibit conjugate impedance matches at every interface because of conservation of power. Then, functionally similar equations were given and verified for solving T and pi resistive matching network problems in terms of the current-transfer-angle parameter. The two possible L-section configurations were treated as special cases of the T and pi configurations when the output branch was omitted. A small BASIC language program was provided to calculate element reactance at an assumed frequency.

Series-to-parallel impedance conversions and parallel combination of reactances were described in order to always work with impedances as opposed to mixed impedance/admittance units. The former strategy has been found superior because engineers more readily recognize practical ranges of elements in a single unit of measure. A hand-held computer program was provided for these simple relationships, and examples were worked. These tools are vital parts of the complete set of solutions obtained for an example that involves both complex load and complex source, utilizing L-section matching networks.

Finally, a brief comment was provided on the value of graphic visualizations in general and the Smith chart in particular. A much more general treatment of the Smith chart will be furnished in the next chapter.

6.2. Lossless Uniform Transmission Lines

The matching network in these sections will consist of a lossless, uniform transmission line, as shown in Figure 6.11. The load impedance Z_2 and the desired input impedance Z_1 are given; the unknowns are the real transmission line characteristic impedance Z_0 and electrical length θ.

An expression for the input impedance Z_1 will be derived from Chapter Four equations. A related reflection equation will be derived for relationship to the Smith chart. The lossless case will then be examined to produce solutions for a complex source and a complex load. A more simple result will be obtained for the case of a real source and complex load; this will result in a

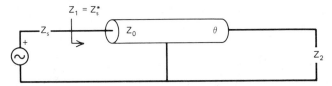

Figure 6.11. The transmission line–matching problem.

convenient graphic design aid for trend analysis. Finally, two important techniques for matching real loads to real sources will be discussed. The inverter (90-degree line) will be a main feature in Chapter Eight.

6.2.1. Input Impedance and Reflection Coefficient. The input impedance of any two-port network was given in terms of its ABCD (chain) parameters in bilinear Equation (4.18). The ABCD parameters for uniform, dissipative transmission lines were given in (4.13)–(4.15). It is quite easy to show that these lead to the following expression for Z_1 in Figure 6.11:

$$Z_1 = \frac{Z_0(Z_2 + Z_0 \tanh \gamma \ell)}{(Z_0 + Z_2 \tanh \gamma \ell)}, \qquad (6.23)$$

where $\gamma \ell = NP + j\theta$; i.e., ℓ is the transmission line length and NP is the loss in nepers for this length.

It is useful to retain the dissipative factor in order to show the general applicability of the Smith chart as a transmission line model. The reflection coefficient in (6.22) can be normalized to Z_0 instead of R_1 and then solved for Z_1/Z_0:

$$\rho_1 = \frac{Z_1/Z_0 - 1}{Z_1/Z_0 + 1} \quad \text{or} \quad \frac{Z_1}{Z_0} = \frac{1 + \rho_1}{1 - \rho_1}. \qquad (6.24)$$

A similar expression relates Z_2/Z_0 and ρ_2. The expressions for Z_1/Z_0, Z_2/Z_0, and the identity

$$\tanh \gamma \ell = \frac{e^{\gamma \ell} - e^{-\gamma \ell}}{e^{\gamma \ell} + e^{-\gamma \ell}} \qquad (6.25)$$

can be substituted into (6.23) to identify the reflection relationship

$$\rho_1 = \rho_2 e^{-2\gamma \ell} = \rho_2 e^{-2NP} e^{-j2\theta}. \qquad (6.26)$$

The Smith chart in Figure 6.10 was described as the reflection plane. Certainly, load impedance Z_2 corresponds to a point ρ_2 on the Smith chart. In polar form, the ρ_2 angle traditionally is measured counterclockwise from the real ρ axis in Figure 6.10. When Z_2 terminates a transmission line of length ℓ as in Figure 6.11, the input reflection coefficient corresponding to impedance Z_1 is computed by (6.26). It shows that the angular length of the line is measured from the ρ_2 radial in a clockwise direction with twice the angular units on the chart plane. For dissipative (lossy) transmission lines, the Smith chart locus from ρ_2 to ρ_1 is plotted with a radius that decays with the angle by the $-2NP$ exponential term in (6.26); i.e., it spirals inward.

Example 6.5. Consider the lossless ($NP=0$) transmission line in Figure 6.11, terminated by $Z_2=40+j30$ ohms. Suppose that the characteristic impedance is 50 ohms. By (6.24), $\rho_2=0.3333\ \underline{/90°}$; this is plotted in Figure 6.12. By (6.26), a 45-degree line rotates ρ_2 clockwise by 90 chart degrees to locate input ρ_{1a}; a 90-degree line rotates ρ_2 clockwise by 180 chart degrees to locate ρ_{1b}. By (6.24), $\rho_{1a}=0.3333\ \underline{/0°}$ corresponds to $Z_1=100+j0$ ohms; similarly, $\rho_{1b}=0.3333\ \underline{/-90°}$ corresponds to $Z_1=40-j30$ ohms. If dissipation loss NP had not been zero, the radius of 0.3333 would have decreased with rotation.

Further considerations will involve only lossless transmission lines; so it is useful to equate $\gamma\ell=j\theta$ in (6.23). This reduces to the input impedance expression for a lossless transmission line:

$$Z_1=Z_0\frac{Z_2+jyZ_0}{Z_0+jyZ_2},\tag{6.27}$$

where definition (4.22) is repeated:

$$y=\tan\theta.\tag{6.28}$$

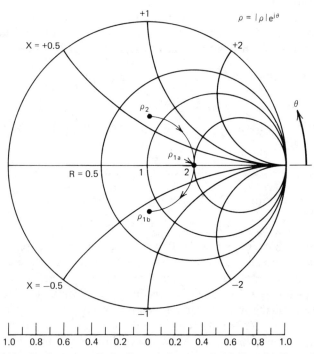

Figure 6.12. Lossless transmission line rotation on a Smith chart for Example 6.5.

This has been programmed efficiently in Program A6-2; given a value of $Z_2 = R_2 + jX_2$, the sequence X_2, R_2, θ degrees, and the real number Z_0 are entered in the stack. Then pressing key A evaluates the input impedance $Z_1 = R_1 + jX_1$, where R_1 is in the X register and X_1 is in the Y register. The two cases in Example 6.5 are easily confirmed by this program.

6.2.2. Complex Sources and Complex Loads.

Jasik (1961) and Milligan (1976) have given expressions for finding the Z_0 and the length of a transmission line that transform complex Z_2 to complex input Z_1 impedances. Day (1975) has described a Smith chart method, which will not be considered here. Moving the denominator of (6.27) to the left side and equating real and imaginary parts, respectively, yields the desired expressions. The real part is

$$R_1 Z_0 - R_1 y X_2 - X_1 y R_2 = Z_0 R_2 . \tag{6.29}$$

The imaginary part is

$$y(R_1 R_2 - X_1 X_2 - Z_0^2) = Z_0 (X_2 - X_1). \tag{6.30}$$

Equation (6.29) yields

$$y = Z_0 q, \tag{6.31}$$

where

$$q = \frac{R_1 - R_2}{R_1 X_2 + X_1 R_2} . \tag{6.32}$$

From (6.31) and (6.28), the electrical length of the required transmission line is

$$\theta = \tan^{-1}(Z_0 q). \tag{6.33}$$

Substitution of (6.31) for y in (6.30) produces an expression for Z_0; further elimination of q, using (6.32) and some algebra, yields the characteristic impedance of the required transmission line:

$$Z_0 = \left(\frac{R_1 |Z_2|^2 - R_2 |Z_1|^2}{R_2 - R_1} \right)^{1/2} , \tag{6.34}$$

when $R_2 \neq R_1$, and the square root exits.

These relationships have been included in Program A6-2 on key B. The desired X_2, R_2, X_1, and R_1 sequence is entered in the stack. Pressing key B provides Z_0 in register X and θ degrees in register Y if a solution exists. Otherwise, an error indication is displayed when the HP-67/97 attempts to compute the square root of a negative number.

Example 6.6. Specifying $Z_2 = 10 + j20$ and $Z_1 = 30 - j40$ ohms requires a matching line with $Z_0 = 22.36$ ohms and $\theta = 65.91$ degrees; this can be checked using the input impedance calculation on key A. If Z_2 is changed to $10 + j30$ with the same Z_1, no match is possible. However, Z_2 can first be rotated by 45 degrees on a 50-ohm line; key A shows the resulting impedance to be

$100 + j150$. Using this as a new Z_2 load will produce the required Z_1 (using $Z_0 = 101.77$ ohms and $\theta = 94.01$ degrees). Thus two cascaded lines can match $10 + j30$ to $30 - j40$ ohms.

6.2.3. Real Sources and Complex Loads.

The case in Figure 6.11, when Z_1 is real and Z_2 remains complex, reduces to a Smith chart design aid that is useful for visualizing ranges of solutions. A Smith chart such as in Figure 6.12 is considered normalized to the desired input resistance R_1 ($X_1 = 0$ is assumed). Load impedance Z_2 is similarly normalized to R_1:

$$\frac{Z_2}{R_1} = \frac{R_2}{R_1} + j\frac{X_2}{R_1} = r + jx. \tag{6.35}$$

Then (6.31) reduces to

$$\theta = \tan^{-1}\frac{(1-r)Z_0/R_1}{x}, \tag{6.36}$$

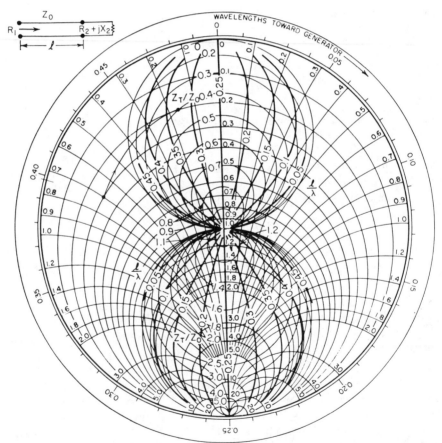

Figure 6.13. Smith chart for transmission line matching of complex loads normalized to a desired real input impedance. [From Jasik, 1961.]

and (6.34) can be written as

$$\left(\frac{Z_0}{R_1}\right)^2 = r - \frac{x^2}{1-r}. \qquad (6.37)$$

It is useful to plot loci of constant θ and constant Z_0/R_1 from (6.36) and (6.37) on the normalized Z_2 plane, i.e., on an $r+jx$ Smith chart grid. Jasik (1961) has done so, as shown in Figure 6.13. This Smith chart is oriented differently from that in Figure 6.12 (90 degrees clockwise), so that the negative half-plane is on the left. Also, the chart perimeter scale for rotation from the load toward the generator is in ℓ/λ—the fraction of a wavelength (360 degrees)—and the electrical length of the transmission line, (6.36), is similarly labeled. Feasible solutions must be within either of the two circular areas.

Example 6.7. Suppose that an impedance of $15-j35$ ohms must be matched to $50+j0$ ohms. Normalizing Z_2 and the Smith chart to $R_1=50$ ohms gives $r+jx=0.3-j0.7$; this is plotted on the left side of Figure 6.13 at the intersection of the two circular coordinates. The corresponding wavelength scale reads 0.398. Using a compass, this point is rotated clockwise until it is within either circular area. Suppose that the initial point is rotated at that radius to the point corresponding to a wavelength scale reading of 0.474; this is the point also corresponding to $Z_0/R_1=0.4$ and $\ell/\lambda=0.325$ as shown. Thus, the 50-ohm line rotation must be $(0.474-0.398)\times360=27.36$ degrees. Then the chart indicates that impedance $(0.2-j0.16)$ can be matched by a $0.4\times50=20$-ohm transmission line that is $0.325\times360=117$ degrees long. This network is shown in Figure 6.14; there is an infinite number of other feasible solutions.

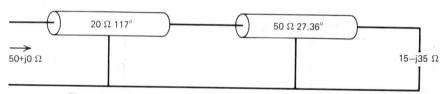

Figure 6.14. One transmission line network that solves Example 6.7.

6.2.4. Real-to-Real Transmission Line Matches. There are two transmission line–matching cases that deserve special mention. The most important is a lossless, 90-degree line called an impedance inverter. As $\theta\rightarrow90$ degrees, (6.28) shows that $y\rightarrow\infty$. Also, (6.27) then shows that the input impedance of a 90-degree line is

$$Z_1 = \frac{Z_0^2}{Z_2}. \qquad (6.38)$$

A 90-degree line is equivalent to 0.25 wavelengths, or half a Smith chart rotation, as shown in Figure 6.13. Although (6.38) is true whether or not Z_2 is

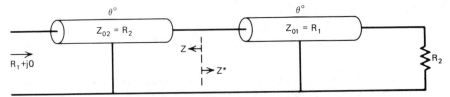

Figure 6.15. Two lossless cascaded transmission lines that match load R_2 to generator R_1; they have the same length and Z_0 values of R_2 and R_1, as shown.

real, Figure 6.12 shows that $Z_1 = 0.5$ when $Z_2 = 2$ and $Z_0 = 1$. One easy means for matching two resistances is to use a 90-degree line having a characteristic impedance that is their geometric mean. Inverters have a much more important role in direct-coupled filter design, as explained in Chapter Eight.

A useful real-to-real transmission line–matching network that is less than 60 degrees long was described by Bramham (1961) and considered in Problem 4.7. It is shown in Figure 6.15. As discussed in Section 6.1.1, conjugate impedances exist at any interface in lossless networks, specifically as shown by Z and Z^* in Figure 6.15. The solution for the common line lengths, obtained by another method in Problem 4.7, can be addressed by this principle. Equate the impedance looking left from the middle of Figure 6.15 to the conjugate of the impedance looking to the right:

$$\frac{R_2(R_1+jyR_2)}{R_2+jyR_1} = \frac{R_1(R_2-jyR_1)}{R_1-jyR_2}.\tag{6.39}$$

The result of cross-multiplying is

$$y^{-2} = \frac{R^3-1}{R^2-R},\tag{6.40}$$

where $R = R_2/R_1$. Long division of the right-hand side yields the final design relation:

$$\theta = \tan^{-1}\left[\left(R+1+\frac{1}{R}\right)^{-1/2}\right].\tag{6.41}$$

Because of symmetry in (6.41),

$$R = \left(\frac{R_2}{R_1}\right)^{\pm 1}.\tag{6.42}$$

Key D in Program A6-2 evaluates (6.41) given a value of R. For small R, analysis shows that the SWR slope versus frequency is only about 15% greater than for the longer 90-degree matching line (inverter). It is easy to use (6.41) to show that $\theta \leqslant 30$ degrees, and this occurs for $R \rightarrow 1$ (see Przedpelski, 1980).

Example 6.8. Suppose that a 50-ohm coaxial cable must be matched to a 75-ohm coaxial cable; i.e., a 50- to 75-ohm resistive match is required over some narrow band of frequencies. According to Figure 6.15, the input line segment should have $Z_0 = 50$ ohms and the output segment should have $Z_0 = 75$ ohms. Evaluation of (6.41) using Program A6-2, key D shows that each line should be 29.33 degrees long at the specified frequency.

6.2.5. *Summary of Transmission Line Matching.* The dissipative transmission line ABCD parameters presented in Chapter Four were employed to show how a clockwise spiral locus on a Smith chart models the input impedance or reflection coefficient of a terminated transmission line as a function of line length. Also, the same function was obtained for lossless lines.

Transformations of load to input impedance when at least one is complex are not always possible with a single line segment. Examples for both complex and real source impedances show that it is possible to rotate a given complex impedance until it can be transformed to the specified resistance by a second transmission line segment. The process was graphically illustrated for real-source situations.

Finally, the 90-degree line transformer (inverter) was mentioned with respect to its Smith chart behavior and importance in direct-coupled filters (Chapter Eight). Then a simple two-segment transmission line–matching network was described that is less than 60 degrees long and matches resistances over narrow frequency bands. Its derivation emphasized the fact that matched, doubly terminated lossless networks of any kind exhibit conjugate impedance matches at any interface. Computer Program A6-1 was provided to evaluate these important lossless transmission line relationships.

6.3. Fano's Broadband-Matching Limitations

Fano (1950) described a complete theory for the design of optimal lowpass matching networks when the load impedance could be specified as that of some LC subnetwork terminated by a resistance (see Z_2 in Figure 6.1). Previously, Bode had given the gain-bandwidth-matching restriction for load impedances consisting of a series LR or parallel CR. The next three topics presented here will involve the most practical of these load networks. They are shown in Figure 6.16 in lowpass form.

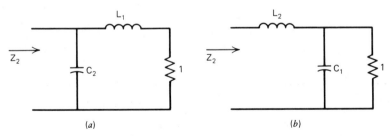

Figure 6.16. Lowpass load impedance forms having two reactances. (*a*) Series resistance; (*b*) parallel resistance.

Although Fano's theory is usually presented in lowpass form, with a passband from dc to some edge frequency (e.g., 1 radian) as a matter of convenience, most practical applications relate to a given pass band between specified frequencies well above dc. Sections 6.5 and 6.6 will include more details of mappings of the lowpass frequency range to an arbitrary bandpass frequency domain. However, it is useful to display the most common bandpass load networks that may be derived directly from the lowpass prototypes shown in Figure 6.16. (The matching network changes in exactly the same way).

Comparison of Figures 6.16 and 6.17 shows that shunt capacitors are replaced by shunt resonators (tanks) and series inductors are replaced by series resonators. For a passband from ω_1 to ω_2 and a geometric band-center frequency $\omega_0 = \sqrt{\omega_1 \omega_2}$, the passband fractional width $w = (\omega_2 - \omega_1)/\omega_0$ mainly determines the bandpass-load element values. For example, $C_1' = C_1/w$ and L_1' resonates C_1' at band-center frequency ω_0. In most cases, the bandpass-load network model is found and translated to its comparable lowpass form for matching network design. Further reference to one- or two-reactance loads will always relate to the lowpass prototype networks in Figure 6.16.

Suppose that the physical problem is matching a short whip antenna to $50 + j0$ ohms over a frequency band. Resonating the (capacitive) antenna at band-center frequency by adding a series inductance often makes the resulting frequency behavior correspond approximately to a series resonator, i.e., the load network in Figure 6.17a without the $C_2' - L_2'$ resonator. If the physical load impedance to be matched is the input to a more sophisticated antenna, the network model probably will be substantially more complicated.

The initial task of deciding which resistively terminated LC network corresponds to the physical load is called load classification. The load data may determine the rational polynomial associated with Figures 6.16 or 6.17 using the method in Section 2.5. Or, an optimizer might be used to repeatedly analyze a network configuration in order to adjust element values to match the known frequency behavior of the load. There have been many sophisti-

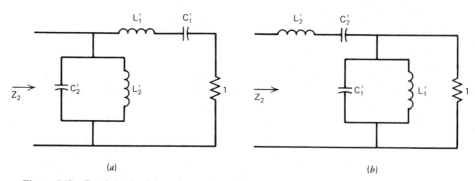

Figure 6.17. Bandpass load impedance forms having two resonators. (a) Series resistance; (b) parallel resistance.

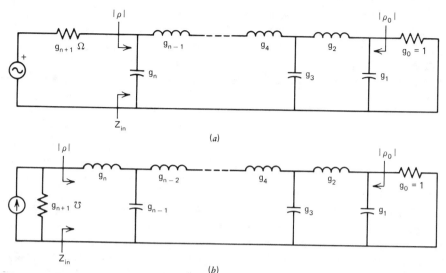

Figure 6.18. Matching networks with load elements g_0, g_1, and perhaps g_2 for (*a*) odd n and (*b*) even n.

cated schemes for load classification. However, it is possible to develop a useful intuition for load models by accumulating experience in Smith chart impedance sketching. Load classification will not be discussed further; for Fano's method, it is assumed that the problem begins with one of the four load networks in Figures 6.16 and 6.17. Carlin's method (Section 6.7) is a means for largely avoiding the load classification problem.

The prototype lowpass matching network will have elements numbered from the load resistance back toward the source, as shown in Figure 6.18. Results for the load configuration in Figure 6.16a will be the same, except that L_1, C_1 and L_2, C_2 are interchanged, respectively.

This section will present and discuss Fano's gain-bandwith integral limitations for the loads in Figure 6.16. First the ultimate limitation for the case of an infinite matching network terminated by a single-reactance load will be described, then a Chebyshev approximation of finite degree will be developed. The single degree of freedom will be identified and used to express an optimal matching relationship. Finally, the Newton–Raphson solution of the transcendental function related to a single-reactance load will be solved, and the optimal matching network performance will be summarized graphically as computed by an included BASIC program.

6.3.1. Fano's Gain-Bandwidth-Integral Limitations.

It has been mentioned in several places that the impedances looking left and right at any interface in lossless, matched, doubly terminated networks are conjugate. For mismatched networks with resistive terminations at each end, any interface presents an equivalent Thevenin generator looking toward the source and its equivalent load impedance looking toward the load. This circumstance exactly fits the discussion of power transfer from a complex source to a complex load in Section 3.2.3. In Figure 6.18, the reflection coefficient ρ_0 is defined with respect to resistance g_0, and ρ is defined with respect to g_{n+1}. For lossless networks, the power available at the source is also available at the load. From Section 3.2.3,

$$|\rho| = |\rho_0|, \tag{6.43}$$

where

$$\rho = \frac{Z_{in} - g_{n+1}}{Z_{in} + g_{n+1}} \qquad (g_{n+1} \text{ in ohms}). \tag{6.44}$$

Clearly, a good impedance match occurs when Z_{in} is nearly equal to the generator resistance g_{n+1}; this is precisely stated as the minimum $|\rho|$. Over a frequency band, a good impedance match would be obtained by minimizing the maximum $|\rho|$. This is shown in Figure 6.19.

Fano (1950) stated the theoretical limitation for load networks representable as resistively terminated LC networks, such as in Figure 6.16. Their lowpass form is

$$\int_0^\infty \ln\frac{1}{|\rho|}\,d\omega = \frac{\pi}{g_1} \tag{6.45}$$

for single-reactance loads, and

$$\int_0^\infty \omega^2 \ln\frac{1}{|\rho|}\,d\omega = \frac{\pi}{g_1^3}\left(\frac{g_1}{g_2} - \frac{1}{3}\right) \tag{6.46}$$

for two-reactance loads. Note that the integrand is essentially the return loss in (4.58). The interpretation for the gain-bandwidth limitation described by (6.45) is illustrated in Figure 6.19 for the bandpass case: the reflection magnitude may be low (good match) over a narrow band or higher (poor match) over a wider band. Fano noted that in no event should the reflection magnitude in the band be zero, as is commonly the case with filters. Making $1/|\rho|$ very large at any point in the pass band necessarily reduces the bandwidth because of the inefficient use of the areas in the integrals above.

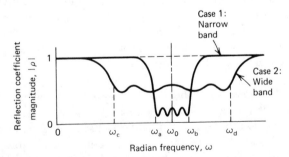

Figure 6.19. Constant gain-bandwidth tradeoffs for a good match over a narrow bandwidth (case 1) and a poor match over a wide bandwidth (case 2). [From Matthaei et al., 1964.]

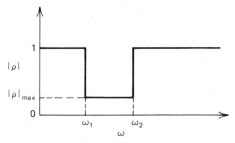

Figure 6.20.   An ideal reflection function for optimal match over a frequency band.

Ideally, the bandpass reflection coefficient function ρ should be in the form of the rectangular box illustrated in Figure 6.20. The box function can be fitted exactly using a polynomial of infinite degree, corresponding to a network having an infinite number of elements. For this upper limit, it is useful to evaluate (6.45) as applicable to a single-reactance load. For a constant reflection value of $|\rho_{\max}|$ from ω_1 to ω_2 and unity (complete reflection) elsewhere, the result for matching networks of great complexity is

$$(\omega_2 - \omega_1)\ln\frac{1}{|\rho_{\max}|} \leqslant \frac{\pi}{RC}. \tag{6.47}$$

The least possible $|\rho_{\max}|$ is thus

$$\min|\rho_{\max}| = e^{-\delta\pi}, \tag{6.48}$$

where the decrement δ is the main matching parameter:

$$\delta \overset{\Delta}{=} \frac{\omega_0}{\omega_2 - \omega_1}\frac{1}{\omega_0 g_1 R} = \frac{Q_{BW}}{Q_L} = \frac{1}{wQ_L}. \tag{6.49}$$

This definition is suitable for both bandpass and lowpass cases. For bandpass networks, $\omega_0 = \sqrt{\omega_1\omega_2}$, the geometric mean frequency. It is convenient to label the first fraction in (6.49) the "Q" of the bandwidth. For lowpass networks, $\omega_0 = \omega_2$ (the upper band edge) and $\omega_1 = 0$, so that $Q_{BW} = 1$. In both bandpass and lowpass cases, the second fraction in (6.49) is clearly a parallel Q, as previously defined by (6.16). A common alternate parameter for Q_{BW} is the fractional bandwidth w:

$$w = \frac{\omega_2 - \omega_1}{\omega_0}. \tag{6.50}$$

It is convenient to express the least possible standing-wave ratio (SWR) in these terms using (4.59):

$$\min SWR_{\max} = \frac{e^{\delta\pi} + 1}{e^{\delta\pi} - 1}. \tag{6.51}$$

Program A6-3 in Appendix A calculates (6.51) using (6.48); this is available on key A.

Example 6.9. What is the least possible input standing-wave ratio over a 50% bandwidth for a matching network terminated by a load with $Q_L = 3$? Enter 3 and then 50 into the HP-67 stack in Program A6-3; key A evaluates (6.48) and (6.51) to yield $\min S_{max} = 1.28 : 1$.

6.3.2. A Chebyshev Approximation of the Ideal Response.

The fact that there is a finite amount of reflection over a band, as illustrated in Figure 6.20, is equivalent to a certain amount of transducer loss, as described by (3.49) in Section 3.2.3. The standard lowpass approximation to box-shaped losses of this sort is illustrated in Figure 6.21. Compare this to the bandpass shapes in Figures 6.19. The function that corresponds to Figure 6.21 (passband edge at 1 radian) is

$$|H(\omega)|^2 = 1 + K^2 + \varepsilon^2 T_n^2(\omega), \tag{6.52}$$

where $T_n(\omega)$ is the Chebyshev function of the first kind described in Section 2.4.1. This is similar to the Chebyshev responses synthesized according to Problems 3.16 and 3.17 (Chapter Three), except for the "flat-loss" term K^2.

Because of the gain-bandwidth integral limitation, the main interest is in the related reflection coefficient. Equations (3.49) and (6.52) yield

$$|\rho|^2 = \frac{(K/\varepsilon)^2 + T_n^2(\omega)}{(1 + K^2)/\varepsilon^2 + T_n^2(\omega)}. \tag{6.53}$$

It is a reasonably straightforward process to obtain an expression for the s-plane poles and zeros of (6.53), considering its squared-magnitude form, $\rho(s)\rho(-s)$, according to (3.50) (see the similar derivation in Guillemin, 1957, pp. 596–598). The s-plane poles of (6.53) are

$$s = \begin{cases} \sinh\left[\pm a \pm j\frac{\pi}{n}\left(m + \frac{1}{2}\right) \right], & n \text{ is even;} \\ \sinh\left[\pm a \pm j\frac{\pi}{n}m \right], & n \text{ is odd,} \end{cases} \tag{6.54}$$

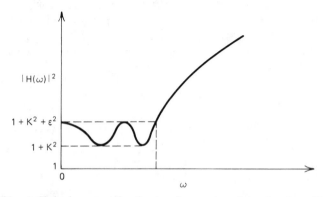

Figure 6.21. A Chebyshev approximation to a lowpass transducer function with flat loss.

where m is an integer. A useful identity is (4.16), which reduces the computations to an evaluation of real sine and cosine functions. The process of expressing the roots results in the defining equations for two important positive parameters, a and b:

$$\sinh^2(na) = \frac{1+K^2}{\varepsilon^2}, \tag{6.55}$$

$$\sinh^2(nb) = \left(\frac{K}{\varepsilon}\right)^2. \tag{6.56}$$

The zeros of the reflection coefficient in (6.53) are given by (6.54), with b substituted for a. With (6.55) and (6.56) substituted into (6.53), useful expressions are obtained for $|\rho|_{max}$ and $|\rho|_{min}$, corresponding to $T_n^2(\omega)$ values 1 and 0, respectively:

$$|\rho|_{max} = \frac{\cosh nb}{\cosh na}, \tag{6.57}$$

$$|\rho|_{min} = \frac{\sinh nb}{\sinh na}. \tag{6.58}$$

The poles and zeros of $\rho(s)$ are available from (6.54); these are significant because $\rho(s) = e(s)/f(s)$ and $p(s) = 1$ according to (3.52)–(3.55) in Section 3.2.4. Choosing $-a$ in (6.54) locates the required left-half-plane poles for e(s). Fano showed that choosing only left-half-plane zeros for ρ by using $-b$ in (6.54) in place of $\pm a$ maximizes the broadband match for ladder networks.

It is now clearly possible to synthesize the network; this could be started from either end. Usually, synthesis is not necessary. However, there is a crucial relationship involving loads with a single reactance (g_1 in Figure 6.18). This relationship turns out to be

$$2\delta_1\sin\frac{\pi}{2n} = \sinh a - \sinh b, \tag{6.59}$$

where the connection to g_1 is through the decrement (6.49).

So far there is one degree of freedom remaining: given the bandwidth and load Q, (6.59) relates parameters a and b; one of them can be chosen arbitrarily. Then the flat loss and ripple in Figure 6.21 are determined by (6.55) and (6.56). Other orders of parameter selection for using the available degree of freedom are possible.

6.3.3. Optimally Matching a Single-Reactance Load.

The objective is to use the one degree of freedom that is available for single-reactance lowpass loads (RC or RL) to minimize the maximum reflection coefficient (6.57) over the band. The constraint is the relationship in (6.59), and the variables are parameters a and b. The number of elements (n) in the networks of Figure 6.18 includes the load reactance g_1. Following Levy (1964), this minimization is determined analytically by employing a Lagrange multiplier, as described in many calculus textbooks.

The functional multiplier λ is defined as

$$b = \lambda a, \tag{6.60}$$

and this is substituted in the $|\rho|_{max}$ expression (6.57). Differentiating the resulting expression with respect to parameter a and setting this to zero, as necessary for a minimum, yields

$$\lambda = \frac{\tanh na}{\tanh nb}. \tag{6.61}$$

Similar substitution of (6.60) into (6.59), followed by differentiation, yields

$$\lambda = \frac{\cosh a}{\cosh b}. \tag{6.62}$$

Now λ may be eliminated from (6.60) and (6.61) to produce the necessary condition for minimum $|\rho|_{max}$:

$$\frac{\tanh na}{\cosh a} = \frac{\tanh nb}{\cosh b}, \tag{6.63}$$

which is still subject to the constraint in (6.59). Note that the integral limitation in (6.45) was not used directly in this case; however, it does indicate that the minimum must exist.

Simultaneous solution of (6.59) and (6.63) produces the values of parameters a and b; thus the ripple parameter ε and flat-loss parameter K are obtained according to (6.55) and (6.56). The selectivity expression (6.52) is then known, and all matching network elements may be found, as shown in Section 6.4.1. The two equations to be solved are transcendental and thus nonlinear. Newton's method from Section 5.1.5 will be applied.

Equations (6.59) and (6.63), respectively, define the functions

$$f_1(a, b) = \sinh a - \sinh b - 2\delta \sin \frac{\pi}{2n} \tag{6.64}$$

and

$$f_2(a, b) = h(a) - h(b), \tag{6.65}$$

where the defined function h with dummy variable x is

$$h(x) = \frac{\tanh nx}{\cosh x}. \tag{6.66}$$

Solutions are obtained by determining the values of a and b that make $f_1 = 0 = f_2$. The Jacobian matrix requires expressions for the partial derivatives of f_1 and f_2 with respect to a and b. It is helpful to employ the derivative expression for (6.66):

$$h'(x) = \frac{n - (\sinh nx)(\cosh nx)(\sinh x)/(\cosh x)}{(\cosh^2 nx)(\cosh x)}. \tag{6.67}$$

Estimated changes in variables a and b to approach a solution are obtained according to (5.37) and (5.38).

Starting values of a and b for the Newton–Raphson method are especially important in obtaining Fano's optimal solution. Good estimates are

$$a = \sinh^{-1}\left[\delta(1.7\delta^{-0.6}+1)\sin\frac{\pi}{2n}\right] \tag{6.68}$$

and

$$b = \sinh^{-1}\left[\delta(1.7\delta^{-0.6}-1)\sin\frac{\pi}{2n}\right] \tag{6.69}$$

for b greater than zero. These were obtained by the author by studying the optimal-solution graphs of Green (1954, pp. 66–69). They will always satisfy the constraints in (6.59) or (6.65). The estimate of the solution for f_2 in (6.65) is usually close enough for engineering work without iterative refinement. This is an important observation when using programmable hand-held calculators.

Program B6-2 in Appendix B implements the Newton–Raphson iterative procedure just described. It is a small BASIC program, and usually converges reliably. For very large values of Q and/or bandwidth, a damping factor of 0.5 in both variable steps (lines 480 and 490) may be necessary to obtain convergence.

Example 6.10. Example 6.9 considered an infinite matching network. Program B6-2 may be used to obtain optimal matching solutions for finite lowpass matching networks. What range of SWR occurs over a 50% passband for $Q_L = 3$ and degree n = 3, 5, 8, and 50? Running Program B6-2 produces the performance data in Table 6.3, as illustrated in Figure 6.22.

Table 6.3. SWR Ripple Over a 50% Passband for Networks of Varying Degrees Terminated by $Q_L = 3$

Q_L	%BW[a]	n	Min SWR	Max SWR
3	50	3	1.3833	1.5486
3	50	5	1.3343	1.4006
3	50	8	1.3095	1.3385
3	50	50	1.2828	1.2840

[a] BW = bandwidth.

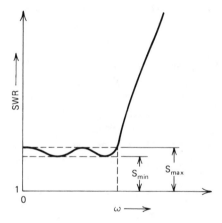

Figure 6.22. Lowpass response showing passband SWR ripple.

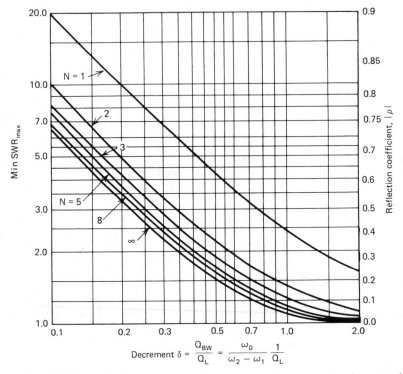

Figure 6.23. SWR/reflection limitations versus decrement of a lowpass or bandpass matching network of degree n.

A graph that plots SWR (and equivalent reflection magnitude) versus decrement (6.49) for various network degrees (n) is a useful design tool. This is easily obtained from Programs B6-2 and A6-3 (see Figure 6.23). Recall that an n = 1 network is the load itself (Figure 6.18); n = 2 represents a single matching element in a lowpass network, and clearly this is the greatest single improvement possible. The data in Example 6.10 and Figure 6.23 show the rapidly diminishing return for increasing the total network degree to greater than n = 5 or 8.

6.3.4. Summary of Fano's Broadband-Matching Limitations.
Fano (1950) published a complete analysis of ideal, lumped-element matching networks that were terminated by load subnetworks of similar structure and ended in a resistance. The usual theoretical extensions of his results have been for lowpass networks, but the common application has been for bandpass cases. There is a simple correspondence between lowpass and bandpass networks that is useful to have in mind during development of the subject; it was introduced here, but its details will be described in Section 6.5. Fano's approach does not deal with the task of load classification, the process of

identifying which lowpass LC network terminated by a resistor corresponds to the physical load being matched. It was suggested that the methods described in Section 2.5 and Chapter Five were applicable; here, it is assumed that the Fano load network is known. For practical applications, his lowpass network structures are terminated by no more than a series L, followed by a parallel RC or a parallel C, followed by a series RL, i.e., one- or two-reactance lowpass loads.

Fano's bandwidth limitations apply to lossless, doubly terminated networks; i.e., they have resistors on both ends. Then the magnitude of the generalized reflection coefficient in Section 3.2.3 must be constant at a particular frequency at all network interfaces, especially at their input and output ports. Certainly, a small input reflection coefficient magnitude corresponds to a good input impedance match. Fano showed that the integral over all real frequencies of the return loss is equal to simple functions of the load components. The ideal reflection coefficient behavior would be some small constant value over the frequency band of interest, and unity (complete reflection) at all other frequencies. Then the integration of this constant provides a simple estimate of the best-possible matching using an infinitely complicated matching network (given the one- or two-reactance-load network). The classical load parameter was defined as the load decrement; it is the ratio Q_{BW}/Q_L, where Q_{BW} is the geometric-mean bandpass frequency divided by the bandwidth, and Q_L is the series X/R or parallel R/X at the band mean frequency. For the lowpass case, the decrement is equal to $1/Q_L$, computed at the band-edge frequency.

An equal-ripple approximation to the ideal "box" shape for the reflection frequency function is obtainable as a Chebyshev function; it was defined as a transducer function and converted into a reflection function according to Section 3.2.3. The expression for the s-plane poles and zeros of the rational reflection function was given in terms of the two defined parameters a and b. The maximum and minimum values of the reflection magnitude were derived from the equal-ripple Chebyshev function. Because the poles and zeros of the reflection coefficient were available, it was noted that matching network synthesis was possible. However, for the present application, this was mentioned only to justify the first stage of such a synthesis, which could produce an algebraic expression for the first load reactance. This expression is a constraint on the reflection relationship, which still leaves one degree of freedom in the matching analysis.

One application for the single degree of freedom of single-reactance loads is to minimize the maximum passband reflection coefficient magnitude while satisfying the load reactance constraint. One function was obtained by using a Lagrange multiplier to minimize analytically the maximum reflection magnitude; the constraint was a second function. Then Newton's method was applied to solve the two nonlinear functions for the values of parameters a and b. Expressions for their starting values were given; these are sufficiently close to a solution so that the Newton iteration may be dispensed with when only hand-held computers are available. BASIC language Program B6-2 was provided to determine optimal matching performance. A comprehensive graph of SWR/reflection versus load decrement for varying network degrees was obtained. It is clear that the number of lowpass network components, including the single-reactance load, should not exceed about eight, because of rapidly diminishing returns. There are other ways to use the single degree of freedom available for single-reactance loads. These will be exploited in Sections 6.4.2, 6.4.3, and 6.6.

6.4. Network Elements for Three Source Conditions

The network LC-element values will be determined in this section by recursive formulas. Three types of sources will be considered (see Figure 6.18). First, the resistive source consisting of g_{n+1} will be considered, as originally assumed; the first matching network element is then g_n. Second, a single-reactance source, consisting of both g_{n+1} and g_n, will be specified so that the first matching network element will be g_{n-1}. Finally, an ideal current source will be considered. In all of these cases, the load will have a single reactance, namely g_1 in Figure 6.18.

The poles and zeros of the reflection coefficient have been given as functions of design parameters a and b (see (6.54)). It was noted that the synthesis of the network element values by the methods described in Chapter Three is straightforward but tedious. However, Green (1954) carried out detailed calculations for cases of low degree and guessed an expression for element values of networks of any degree. It has since been discovered that Takahasi published a complete derivation and proof of general results in Japanese in 1951; the interested reader is referred to Weinberg and Slepian (1960). The closed formulas for element values are easily evaluated once the single degree of freedom is assigned; i.e., parameters a and b are chosen.

6.4.1. Resistive Source Optimally Matched to a Single-Reactance Load. A

resistive source optimally matched to a single-reactance load is illustrated in Figure 6.18, where the source real part is g_{n+1}; the matching network includes g_n through g_2, and the load consists of g_1 and $g_0 = 1$. For the lowpass network in Figure 6.18 with a passband edge at 1 radian, (6.49) and (6.59) yield

$$g_1 = 2\frac{\sin\theta}{\sinh a - \sinh b},\qquad (6.70)$$

where angle θ is

$$\theta = \frac{\pi}{2n}.\qquad (6.71)$$

Parameters a and b are found approximately, from (6.68) and (6.69), or exactly, by the Newton–Raphson iterative procedure in Section 6.3.3.

Green's recursive element formula is

$$g_{r+1} = \frac{4 \sin\left[(2r-1)\theta\right]\sin\left[(2r+1)\theta\right]/g_r}{\sinh^2 a + \sinh^2 b + \sin^2(2r\theta) - 2\sinh a \cdot \sinh b \cdot \cos(2r\theta)}, \quad (6.72)$$

for $r = 1, 2, \ldots, n-1$. The source series resistance or shunt conductance shown in Figure 6.18 is

$$g_{n+1} = \frac{2}{g_n}\frac{\sin\theta}{\sinh a + \sinh b}. \quad (6.73)$$

Note that the source resistance or conductance is dependent. This must be accepted in lowpass networks, but Section 6.5 will show how to provide for fairly arbitrary source resistance levels in corresponding bandpass networks.

Program B6-3 in Appendix B contains Newton's method (Program B6-2) without the print statements; it also performs the prototype element calculations in (6.70)–(6.73).

Example 6.11.   Find the prototype element values for an $n=3$ network that optimally matches a load impedance with $Q_L = 3$ over a 50% bandwidth. The SWR ripple is shown in Table 6.3. Also, as a result of the Newton–Raphson iterative solution, $a = 0.8730$ and $b = 0.3163$. Program B6-3 continues to compute $g_1 = 1.5$, $g_2 = 0.8817$, $g_3 = 1.0561$, and $g_4 = 0.7229$. According to Figure 6.18, when n is odd, the g_{n+1} value is the necessary source resistance. Also, note that g_1 is simply the inverse load decrement according to (6.49).

6.4.2.   Complex Source and Complex Load.   For a complex source and complex load, the given load decrement is $\delta_1 = 1/g_1$. From (6.59),

$$\delta_1 = \frac{\sinh a - \sinh b}{2 \sin\theta}, \quad (6.74)$$

where θ is given by (6.71). The source is now assumed to have a single reactance as well as a resistance. This is another way to assign the single degree of freedom identified in Section 6.3.2. Figure 6.18 shows that the source decrement is

$$\delta_n = \frac{1}{g_n g_{n+1}}. \quad (6.75)$$

However, using (6.73), the source decrement can also be expressed as

$$\delta_n = \frac{\sinh a + \sinh b}{2 \sin\theta}. \quad (6.76)$$

Given the source and load decrements, simultaneous solution of (6.74) and (6.76) for a and b is possible:

$$\sinh a = (\delta_n + \delta_1)\sin\theta, \quad (6.77)$$

$$\sinh b = (\delta_n - \delta_1)\sin\theta. \quad (6.78)$$

Since sinh is an odd function and parameter b must be positive, (6.78) requires that $\delta_n > \delta_1$. The prototype g_r recursion (6.72) was presented as starting at the $g_0 = 1$-ohm end of the network; for this complex source/load case, the 1-ohm end must be the lower decrement end, whether it is the load or the source.

The design procedure is to solve (6.77) and (6.78) and use these values in (6.70)–(6.72) to obtain the element values; start with g_1 equal to the reciprocal of the lesser decrement. The ending real element is again dependent and given by (6.73) or by rewriting (6.75):

$$g_{n+1} = \frac{1}{g_n \delta_n},\tag{6.79}$$

using the greater decrement for δ_n. The central $(n-2)$ elements in Figure 6.18 constitute the matching network. The prototype elements are numbered as shown in Figure 6.18 if the load decrement is less than the source decrement; i.e., the 1-ohm end belongs to the lower decrement. If the source decrement is less than the load decrement, then the source is normalized to 1 ohm and the g_r values from (6.72) are generated from the source end to the load end.

Example 6.12.   Suppose that both source and load terminations included shunt capacitors with decrements of 1.35 and 1.25, respectively. Find the lowest-order matching network and its range of passband SWR. Figure 6.18 shows that only odd-degree networks can have shunt capacitors at both ends. Choosing $n = 3$, (6.77) and (6.78) yield $\sinh a = 1.300$ and $\sinh b = 0.050$. Since load reactance g_1 is a reciprocal decrement, $g_1 = 0.8$. Using (6.72), $g_2 = 1.0514$ and $g_3 = 0.7585$. By (6.79), the source resistance is $g_4 = 0.9766$ ohms. The SWR ranges from 1.0240 to 1.1726 according to (6.57), (6.58), and (4.59).

The networks discussed in this section incorporate single-reactance sources and loads exactly. However, they may not have the least possible SWR_{max} when both given decrements are less than the source decrement obtained by the optimal network in the preceding section. When this is the case, the "optimal" g_n reactance or susceptance may be increased (as part of the matching network) and thereby decrease the decrement to the higher of the given values (see (6.75)). Therefore, given two values for source and load decrement, the lesser of the two should be used first in Program B6-3. The resulting source decrement should then be computed by (6.75); if it is greater than the given decrements, the "optimal" network should be used, with g_n increased as described.

Example 6.13.   The lesser of the two decrements in Example 6.12 was 1.25, which is equivalent to load decrement $g_1 = 0.8$. Using Program B6-3 (with $n = 3$, $Q_L = 0.8$, and $BW = 100\%$, according to (6.49)), obtain $g_1 = 0.8$, $g_2 = 0.9484$, $g_3 = 0.6424$, and $g_4 = 0.9211$. By (6.75), $\delta_3 = 1.6900$, which is greater than the 1.35 decrement given. In fact, (6.75) shows that a 1.35 decrement corresponds to a g_3 value of 0.8042 for $g_4 = 0.9211$. Therefore, the best solution

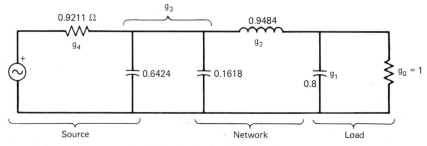

Figure 6.24.   A network solution for Example 6.13 where decrement padding is applicable.

is not that in Example 6.12. If g_3 is padded by adding a matching network shunt capacitor equal to $0.8042-0.6424=0.1618$ farads, then the SWR ripple (from Program B6-3) will be between 1.0856 and 1.1535, which is better than the solution in Example 6.12. This network is shown in Figure 6.24.

6.4.3. Reactive Source and Complex Load.

Referring to Figure 6.18a, source resistance $g_{n+1}\to\infty$ implies an ideal current source excitation by conversion to the Norton form. Conversely, it may be concluded from Figure 6.18b that source conductance $g_{n+1}\to0$ implies an ideal voltage source excitation by conversion to the Thevenin form. However, g_{n+1} in Figure 6.18 is dependent, and g_0 is independent. Therefore, it is convenient to reverse the ends of the network so that g_1 is adjacent to the source. Consider the resultant ideal current source shown in Figure 6.25. The infinite source impedance in parallel with g_1 causes decrement δ_1 to approach zero, corresponding with infinite Q. By (6.74), parameters a and b must be equal. The power available from the source is infinite. However, application of (6.52), (6.55), and (6.56) yields

$$\frac{P_{max}}{P_{min}} = \coth^2 na. \tag{6.80}$$

Recursion (6.72) still applies, conveniently converted to the equivalent form

$$g_r = \frac{4\sin\left[(2r-1)\theta\right]\sin\left[(2r+1)\theta\right]/g_{r+1}}{2(\sinh^2 a)(1-\cos 2r\theta)+\sin^2 2r\theta}, \tag{6.81}$$

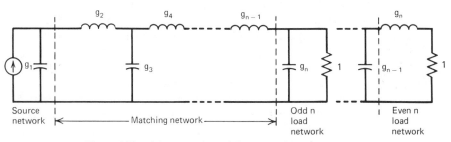

Figure 6.25.   A lowpass network for a reactive current source.

which may be processed using decreasing subscripts: $r = n-1, \ldots, 1$. The g_r starting value comes from (6.75), with $g_{n+1} = 1$, and (6.76):

$$g_n = \frac{\sin \theta}{\sinh a}. \tag{6.82}$$

The design procedure requires that parameter a be determined from either power variation (6.80) or load reactance (6.82). Then (6.81) determines all element values, including the dependent g_1. Often, g_1 is also specified; the problem has no solution if the calculated g_1 is not at least as large. By the duality principle, this method may be extended to the zero-impedance (ideal voltage) source or load problem.

Example 6.14.   Consider the singly terminated network in Figure 6.25 for $n = 5$. Suppose that $g_1 = 0.8$ and $g_5 = 1/\delta_5 = 0.59$. By (6.82), $\sinh a = 0.5238$. Then (6.81) yields $g_4 = 1.2668$, $g_3 = 1.5743$, $g_2 = 1.6014$, and $g_1 = 1.3868$. The computed g_1 is greater than the given g_1 by 0.5868 farads. This shunt padding element is placed at the matching network's input in a manner similar to the arrangement in Figure 6.24. The power variation will be $1.03 : 1$, or 0.11 dB, according to (6.80).

6.4.4.   Summary of Broadband Matching Under Three Source Conditions.

The topic of load impedances consisting of one resistor and one reactance has been considered. The sources considered had just one resistor, or an additional reactance, or a reactance and no resistance. The source condition determined the relationship of parameters a and b. They were found by Fano's transcendental optimal equation, from specified termination decrements, or by equating them so that one decrement was zero. Lowpass prototype element values were obtained for each case by a well-known recursive relationship that avoids network synthesis. This is sometimes called "direct design," since closed forms determine element values.

Program B6-2, which iterated Fano's transcendental solution, was extended by adding the prototype element recursive equation. The dependent source resistance was also calculated. Programs for sources incorporating a single reactance would be quite similar; any one of these would fit in a conventional, hand-held, programmable calculator. The only complexity arises from the order in which prototype elements must be calculated. The resistive source case works from load to source; the load reactance is g_1. The single-reactance-source case works from the end with the lesser decrement associated with g_1. The singly terminated (ideal or lossless source) case works from load to source, but the prototype element g_n is always the load reactance, so that the elements are computed in the order of descending subscripts; the source reactance is dependent. The last two cases involve the possibility that the source reactance may need to be increased to obtain the best solution. This is accomplished by increasing the g_1 value (by making part of g_1 the input element in the matching network).

All three source conditions occur at least as commonly in the bandpass situation. These lowpass results will be extended to bandpass situations in the following section.

6.5. Bandpass Network Transformations

There are standard means for directly transforming lowpass networks to bandpass networks by simple operations performed element by element, as indicated in the introduction to Section 6.3. This will be formalized here. For standardization and numerical conditioning, many network designs are introduced with at least a 1-ohm termination and band-edge or band-center frequency of 1 radian or, occasionally, 1 hertz. Therefore, both impedance level and frequency scaling are commonly required. These will be provided in a simple, hand-held calculator program.

Finally, the resulting bandpass networks obtained from lowpass prototypes require different source and load resistance levels. This is especially true when broadband-matching techniques have left the design with a dependent lowpass generator resistance that is invariably not suitable. There is a method for replacing L sections of inductors or of capacitors in bandpass network structures with pi or T networks of the same component type. These Norton transformations introduce an arbitrary impedance-level change within limits, and are frequency independent. This is the means to change bandpass source and load impedance levels as well as to affect useful changes in impedance level and geometry within the network itself. These techniques will be described and two programs for HP-67/97 hand-held calculators will be provided.

6.5.1. Lowpass-to-Bandpass Transformations.

A lowpass prototype response and a related bandpass response are shown in Figure 6.26. The responses may have flat loss in the passband, similar to Figures 6.20 and 6.21. The most common transformation for lumped-element networks employs the mapping

$$\omega' = Q_{BW}\left(\frac{\omega}{\omega_0} - \frac{\omega_0}{\omega}\right), \tag{6.83}$$

where the inverse fractional bandwidth is

$$Q_{BW} = \frac{\omega_0}{\omega_2 - \omega_1}, \tag{6.84}$$

and the band's geometric-center frequency is

$$\omega_0 = \sqrt{\omega_1 \omega_2} . \tag{6.85}$$

Instead of Q_{BW}, two forms of the fractional bandwidth are often useful:

$$w = \frac{\omega_2 - \omega_1}{\omega_0} = \frac{\omega_2}{\omega_0} - \frac{\omega_0}{\omega_2} . \tag{6.86}$$

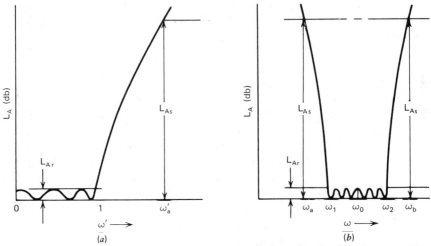

Figure 6.26. A lowpass prototype response (*a*) and a corresponding bandpass response (*b*).

The inverse relationship is also required:

$$\frac{\omega_2}{\omega_0} = \frac{w}{2} + \sqrt{\left(\frac{w}{2}\right)^2 + 1} \ . \tag{6.87}$$

A standard lowpass filter prototype is shown in Figure 6.27. The g_i values are identified with their corresponding L' and C' values. The primes show the relationship to the normalized lowpass frequency scale ω', shown in Figure 6.26. The frequency transformation in (6.83) describes the network in Figure 6.28 with behavior in the ω frequency variable. The conversion of the lowpass network with 1-radian band edge to the bandpass network is quite easy. All bandpass shunt-branch mhos are obtained using

$$\omega_0 C_j = \frac{1}{\omega_0 L_j} = \frac{g_j}{w} \ . \tag{6.88}$$

All bandpass series-branch ohms are obtained using

$$\omega_0 L_k = \frac{1}{\omega_0 C_k} = \frac{g_k}{w} \ . \tag{6.89}$$

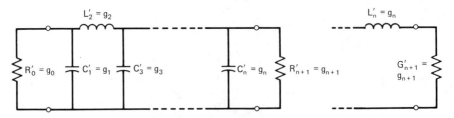

Figure 6.27. A lowpass prototype filter.

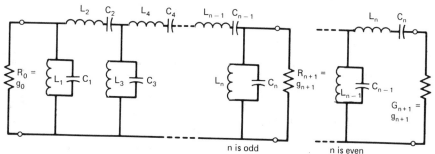

Figure 6.28. A synchronously tuned bandpass filter.

The bandpass filter is called synchronous because all resonators are tuned (resonant) at the same frequency, namely ω_0 according to (6.85).

Example 6.15. Consider the problem of finding the $n = 3$ normalized bandpass matching network in Figure 6.29, where a load consisting of a 20-ohm resistor in parallel with an 11-pF capacitor is to be matched to 50 ohms over a band from 575 to 1000 MHz. The solution is obtained by finding the optimal Fano lowpass network and then transforming this to the corresponding bandpass network. From (6.85), the band geometric-mean frequency is 758.29 MHz; at this frequency, (6.16) yields the load $Q_L = 1.0482$. By (6.86), BW $= 100w = 56.05\%$. These values for n, Q_L, and BW are entered in Program B6-3. The results are shown in the normalized lowpass network in Figure 6.30. The lowpass prototype passband-edge frequency is 1 radian, as shown in Figure 6.26. This will be the geometric-mean frequency of the normalized bandpass network. Equations (6.88) and (6.89) enable the susceptance and

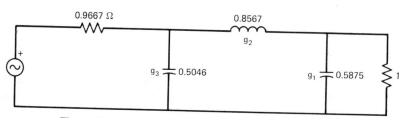

Figure 6.29. A single-reactance-load matching problem.

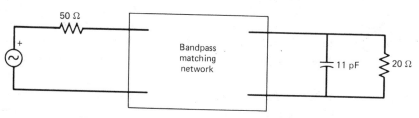

Figure 6.30. Fano optimal lowpass network for Example 6.15.

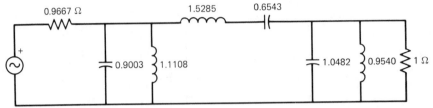

Figure 6.31. Normalized bandpass network for Example 6.15.

reactance of each bandpass network element to be found using the fractional bandwidth w = 0.5605. Since these ohms and mhos are at $\omega = 1$ radian, the values found are also the element values in farads and henrys. The normalized bandpass network is shown in Figure 6.31. By (6.87), its passband extends from 0.7583 to 1.3188, with a geometric-mean band center of 1 radian.

6.5.2. Frequency and Impedance Scaling. The problem posed in Figure 6.29 was not completely solved in Example 6.15 because the bandpass network must be denormalized; i.e., the passband must be centered at 758.29 MHz, and the source resistance must be 50 ohms. These are simple matters of frequency and impedance scaling.

Frequency scaling is based on maintaining the prototype reactances and susceptances of inductances and capacitances, respectively, at some new frequency. Recalling that $X_L = \omega L$ and $B_C = \omega C$, frequency scaling to a higher frequency requires the inverse scaling of *both* L and C values.

Impedance scaling is based on changing the resistance and reactance values throughout the network. Resistances are increased by the desired impedance scaling factor. Recalling that $X_L = \omega L$ and $X_C = -1/(\omega C)$, increasing the impedance level requires increasing the inductances and decreasing the capacitances by the same impedance-scaling factor. Program A6-3 in Appendix A conveniently performs all the simple but crucial scaling operations that convert a lowpass prototype network into the final bandpass network.

Example 6.16. Program A6-3 can be used to go directly from the lowpass prototype network in Figure 6.30 (Example 6.15) to the scaled bandpass network. As the program documentation indicates, values for passband frequencies f_2 and f_1 in hertz (Figure 6.26) and the required impedance-level factor are entered into the stack. Thus 1000E6, 575E6, and (50/0.9667) = 51.722 are entered, and key B is pressed. The passband geometric-mean-center frequency (758.29 MHz) is obtained. The program stores this, the fractional bandwidth w, and the resistance ratio that will get the source resistance up to the desired 50 ohms. Now each prototype g_i value is entered for the series elements (key C) and shunt elements (key D); these evaluate (6.89) and (6.88), respectively. Keys C and D also perform the frequency and impedance scaling

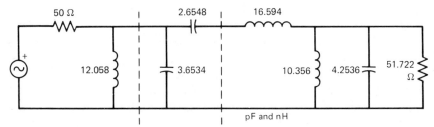

Figure 6.32. Unnormalized bandpass network for Example 6.15.

previously described. Entering $g_3 = 0.5046$ and pressing key D yields the value of the scaled bandpass input shunt ($C = 3.6534$ pF); pressing key E displays its resonating shunt ($L = 12.058$ nH). Similarly, key C is used with g_2 and key D again with g_1. The load resistor is 51.722 ohms. The load still is not in the originally specified RC values, because the dependent Fano source resistance has yet to be compensated. The resulting scaled bandpass network is shown in Figure 6.32.

6.5.3. Norton Transformations.

Example 6.16 in the previous section showed that there is a need for introducing an ideal transformer somewhere in the matching network to provide independence of input and output impedance levels. An easy way to see how this might be accomplished is to derive one case from the set of Norton transformations.

Consider the two networks and the expressions for their open-circuit impedance parameters shown in Figure 6.33. The objective is to equate the sets of z parameters and thus be able to replace the left-hand network with the right-hand network. A case in point is seen in Figure 6.31. There are two adjacent inductors. Incorporating an ideal transformer immediately to their right (and impedance scaling to the right of that) would create a subsection

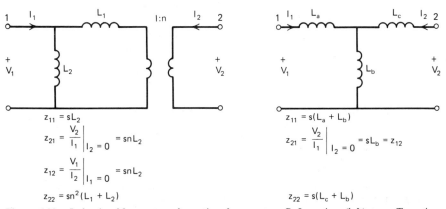

Figure 6.33. Inductive Norton transformation from a type-B L section (left) to a T section (right).

identical to the left-hand side of Figure 6.33. It could then be replaced by its equivalent T network.

For the T network in Figure 6.33, the z parameters may be obtained as functions of the complex frequency s by applying the open-circuit-parameter definitions discussed in Section 3.4.3. The same may be said for the L section in Figure 6.33 if one recalls the rules for the ideal transformer: the current entering the right side increases by turns the ratio n, the voltage on the right appears decreased by n on the left, and the impedance looking in from the right is n^2 times greater than that terminating the left side. These rules lead to the L-section and transformer-combination z parameters shown in Figure 6.33. Then, equating like z parameters for the L section and the T section leads to the following relationships:

$$L_b = nL_2, \tag{6.90}$$

$$L_a = L_2(1-n), \tag{6.91}$$

$$L_c = n^2(L_1 + L_2) - nL_2. \tag{6.92}$$

Also, there are upper and lower bounds on the turns ratio n, which correspond

Table 6.4.   Summary of Norton Transformations

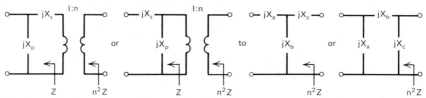

Equations (s = series, p = parallel):
From type-A L to

	T	Pi
	$L_a = L_s + L_p - nL_p$	$L_a = \dfrac{(nL_s)}{n-1}$
	$L_b = nL_p$	$L_b = nL_s$
	$L_c = (nL_p)(n-1)$	$L_c = \dfrac{(nL_s)(nL_p)}{L_s - (n-1)L_p}$

$$1 < n < (1 + \frac{L_s}{L_p}).$$

From type-B L to T and pi, use the equations above, but
1. First replace $n \leftarrow 1/n$.
2. Multiply all answers by the given n^2.
3. Reverse answer order, e.g., use L_c for L_a.
4. $(1 + L_s/L_p)^{-1} < n < 1$.

Note: $L_s = X_s$ and $L_p = X_p$ where both X values are positive numbers. For capacitors, input $X = 1/C$ and convert output by replacing $C \leftarrow 1/C$.

to $L_a = 0$ and $L_c = 0$, respectively. Thus n must be chosen in the closed range

$$\left(1 + \frac{L_1}{L_2}\right)^{-1} \leqslant n \leqslant 1. \tag{6.93}$$

Note that when $L_a = 0$ $(n = 1)$, the T section degenerates into the L section. All possible results for transformations of this type appear in Table 6.4.

Program A6-4 in Appendix A performs all of the preceding calculations for all possible cases in only 80 steps. Compare the operations in Table 6.4 with the L-section matching operations in Figure 6.2. The former are frequency independent and involve only one type of reactance (L or C) at a time, whereas the latter are mixed L and C cases valid only at a single frequency.

Example 6.17. Complete the broadband-matching problem posed in Example 6.15 by replacing the capacitive type-B L section in Figure 6.32 by a pi of capacitors. Use Program A6-4 by entering 1/3.6534 and 1/2.6548 into the HP-67/97 stack and pressing key A. Then select the type-B-to-pi case by pressing key B. The result is the allowable extreme value of n^2 farthest from unity, in this case 0.1771. It is determined from Figures 6.29 and 6.32 that $n^2 = 20/51.722 = 0.3867$ is required, and it is within the allowable range. Entering 0.3867 and pressing the R/S "continue" key produces the first value of reciprocal C in the X register, namely $1/C_a = 0.4905$, or $C_a = 2.0389$. The Y register contains $1/C_b = 0.2342$, or $C_b = 4.2693$. Similarly, the Z register in the stack contains $1/C_c = 0.3852$, or $C_c = 2.5963$. The network to the right of the capacitive type-B L section in Figure 6.32 must be an impedance scaled down by the factor 0.3867, as previously mentioned. Doing this completes the final design shown in Figure 6.34. Observe that the total requirements stated in Example 6.15 and shown in Figure 6.29 have been fulfilled.

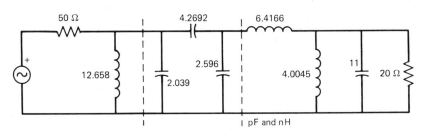

Figure 6.34. Final broadband design for Example 6.15 following the Norton transformation.

6.5.4. Summary of Bandpass Network Transformations.

The standard geometric frequency mapping from lowpass to bandpass response was stated, and the easily remembered design rules for network element conversions were stated. The main parameter is the fractional bandwidth w. The resulting bandpass networks were created by converting all lowpass shunt C's into bandpass shunt resonators and by converting all lowpass series L's into bandpass series resonators. The shunt C's and series L's were inversely scaled by w. The resulting bandpass filter is called synchronous because all of its resonators are tuned to the same geometric band-center frequency ω_0.

Frequency and impedance-level scaling were shown to relate to simple reactance and susceptance concepts. To maintain the same X_L and B_C levels for increased reference frequency, all L's and C's must be decreased by the frequency change factor. To maintain the same X_L and X_C reactance levels for impedance level increase, L's must be increased and C's must be decreased by the impedance-level change factor. The simple but powerful hand-held calculator Program A6-3 was provided to perform lowpass-to-bandpass conversions, frequency scaling, and impedance scaling—all in one quick operation.

Finally, Norton transformations were derived in one case and summarized compactly for all cases. This enabled the introduction of an ideal transformer in a bandpass network adjacent to an L section of two L's or two C's. This subsection may then be replaced by a T or pi section of like-kind elements, eliminating the ideal transformer without changing the frequency response. All of these transformation techniques were illustrated by a broadband-matching example. Another application of Norton transformations is to alter a network topology in order to make element values more reasonable or to avoid parasitic effects. For example, the high impedance that occurs where the series L and C join in Figure 6.31 is often upset by stray capacitance to ground. The incorporated Norton transformer resulted in a topology that does not have such a high impedance point (see Figure 6.34).

6.6. Pseudobandpass Matching Networks

Section 6.1 described the means for designing lumped-element matching L sections at a given frequency. A cascade of such sections could be assembled to match a load resistance to some source resistance, e.g., an even-degree lowpass network such as in Figure 6.35. The transducer loss would be zero at the L-section design frequency, e.g., 1 radian. By (3.49), the dc transducer loss

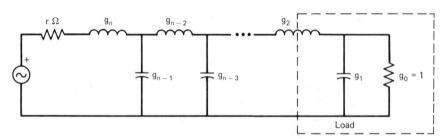

Figure 6.35. An even-degree lowpass matching network.

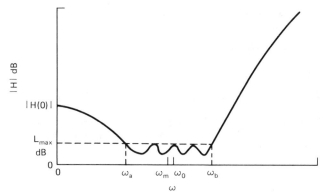

Figure 6.36. A lowpass network transducer function with mismatch at dc.

of the network in Figure 6.35 would be

$$|H(\omega=0)|^2 = \frac{(1+r)^2}{4r} . \tag{6.94}$$

The transducer function of frequency might appear as in Figure 6.36, especially if some of the sections were designed for different frequencies in the vicinity of 1 radian.

The network in Figure 6.35 can be viewed as an impedance transformer with resistance ratio r; its response in Figure 6.36 indicates that the transformation might be valid over a band of frequencies. Also, the two-reactance load indicated in Figure 6.35 coincides with that previously considered in Figure 6.16a. This section deals with the topic of broadband impedance matching by a lowpass network structure over a frequency band above dc, thus the name pseudobandpass impedance matching. Designing individual L sections and then optimizing their response over a band will usually fail because of useless local minima. However, a recent procedure by Cottee and Joines (1979) achieves the desired result.

A frequency transformation that maps a lowpass response into the response in Figure 6.36 will be described. The Fano gain-bandwidth integrals will be evaluated numerically by a BASIC language program so that two-reactance loads can be matched when possible. Finally, the required network synthesis procedure will be described. Norton transformers are not required, since both load and source resistances remain independent. In fact, Norton transformers could not be embedded in the lowpass networks considered. (why?) The penalty in this method is that the number of lowpass LC elements is twice that of an ordinary prototype network.

6.6.1. A Pseudobandpass Frequency Transformation.

The Chebyshev equal-ripple transducer function with flat loss, previously considered in (6.52), is repeated here using identities (2.34) and (2.35). The lowpass prototype frequency variable will be ω' and the degree will be n':

$$|H(\omega')|^2 = 1 + K^2 + \varepsilon^2 \cos^2(n' \cos^{-1} \omega') \qquad (6.95)$$

for $\omega' \leqslant 1$, and

$$|H(\omega')|^2 = 1 + K^2 + \varepsilon^2 \cosh^2(n' \cosh^{-1} \omega') \qquad (6.96)$$

for $\omega' > 1$. These equations define the passband and stopband, respectively, in Figure 6.21. Substituting the following frequency-mapping function into (6.95) and (6.96) produces the response in Figure 6.36:

$$\omega' = \frac{\omega^2 - \omega_0^2}{A}, \qquad (6.97)$$

where the defined constants are

$$A = \frac{\omega_b^2 - \omega_a^2}{2}, \qquad (6.98)$$

$$\omega_0 = \sqrt{\frac{\omega_a^2 + \omega_b^2}{2}}. \qquad (6.99)$$

(See Figure 6.36, which is plotted in terms of the frequency variable ω.) Although the defined constant ω_0 is shown, the band-center frequency is taken as the arithmetic average,

$$\omega_m = \frac{\omega_a + \omega_b}{2}, \qquad (6.100)$$

and is scaled so that $\omega_m = 1$. The relative bandwidth is defined with respect to ω_m:

$$w = \frac{\omega_b - \omega_a}{\omega_m}. \qquad (6.101)$$

Note that both Cottee parameters, ω_0 and w, differ from the parameters with the same names discussed earlier in this chapter.

With (6.97) substituted, the transducer function of ω, defined by (6.95) and (6.96), is a double mapping of the conventional (ω') function shown in Figure 6.21; it maps into Figure 6.36 from ω_0 to 0 and from ω_0 to infinity. It is easy to confirm the mappings of $\omega \to \omega'$ for passband edge frequencies $\omega_b \to 1$ and $\omega_a \to -1$, the ω' image of $\omega' \to +1$. The nature of this mapping is such that the conventional lowpass prototype filter having n' reactive elements corresponds to a new filter having $n = 2n'$ elements, giving the response in Figure 6.36. It will be important to keep track of the complex frequency domains s' and s, corresponding to degrees n' and n and frequency domains ω' and ω, respectively.

6.6.2. Evaluation of Gain-Bandwidth Integrals.
Fano's gain-bandwidth integrals were given in (6.45) and (6.46) for one- and two-reactance lowpass loads,

respectively. An expression for the magnitude of the Chebyshev reflection coefficient was given in (6.53). Thus numerical integration by Romberg Program B2-3, described in Section 2.3, is not difficult. The proper integrand for pseudobandpass networks is

$$\ln\frac{1}{|\rho_1|} = \ln\sqrt{\frac{1 + K^2 + \varepsilon^2 \cos^2\left\{(n/2)\cos^{-1}\left[(\omega^2 - \omega_0^2)/A\right]\right\}}{K^2 + \varepsilon^2 \cos^2\left\{(n/2)\cos^{-1}\left[(\omega^2 - \omega_0^2)/A\right]\right\}}} \tag{6.102}$$

in the pass band and

$$\ln\frac{1}{|\rho_1|} = \ln\sqrt{\frac{1 + K^2 + \varepsilon^2 \cosh^2\left\{(n/2)\cosh^{-1}\left[(\omega^2 - \omega_0^2)/A\right]\right\}}{K^2 + \varepsilon^2 \cosh^2\left\{(n/2)\cosh^{-1}\left[(\omega^2 - \omega_0^2)/A\right]\right\}}} \tag{6.103}$$

in the stopband. The values of constants K and ε will be required; they can be determined as follows.

Assume that the resistance ratio, $r = R_1/R_2$, and the maximum passband ripple, L_{max} in Figure 6.36, are given, where

$$L_{max} = 10\log_{10}(1 + K^2 + \varepsilon^2) \text{ dB}. \tag{6.104}$$

Then (6.94) and (6.96) may be equated for $\omega = 0$:

$$\frac{(1+r)^2}{4r} = 1 + K^2 + \varepsilon^2 \cdot EC, \tag{6.105}$$

where defined constant EC is

$$EC = \cosh^2\left(\frac{n}{2}\cosh^{-1}\frac{\omega_0^2}{A}\right). \tag{6.106}$$

Exponentiating both sides of (6.104) enables its simultaneous solution with (6.105) for the ripple factor:

$$\varepsilon^2 = \frac{10^{L_{max}/10} - (1+r)^2/4r}{1 - EC}. \tag{6.107}$$

The flat-loss factor is now available from (6.104):

$$K^2 = 10^{L_{max}/10} - \varepsilon^2 - 1. \tag{6.108}$$

The only other issue to be resolved before numerically integrating (6.102) and (6.103) is the upper limit of integration. It is well known that the asymptote for the high-frequency attenuation in Figure 6.36 is 6n dB/octave; here the octaves are taken as multiples of passband width above ω_0. The reflection coefficient should be essentially 1 when the attenuation is at least 60 dB. Program B6-4 in Appendix B includes the earlier Romberg integration routine and makes these calculations, including the upper limit of integration in line 250.

Example 6.18. Suppose that the load depicted in Figure 6.35 must be matched over the band from 0.75 to 1.25, where $g_1 = 1.571$ farads, $g_2 = 0.3142$ henrys, $R_1 = 0.25$, and $R_2 = 1$ ohm. Can the load be matched by an $n = 4$ network? Evaluating the right side of Fano's integrals (6.45) and (6.46) yields values of 2.00 and 3.78, respectively. Program B6-4 is used with these data and the trial values of L_{max}, the maximum passband ripple. It can be found that $L_{max} = 0.0924$ dB gives the required integral value of 2.00. Using this L_{max} value and the same program for the two-reactance case, Romberg integration finds the integral value to be 1.82. Since this is less than 3.78, the given load can be matched by an $n = 4$ network, because the effective value of g_2 can be increased (padded) easily enough. Note that (6.46) shows that increasing g_2 can decrease the required integral value to that computed.

6.6.3. Network Synthesis Procedure.

Having determined the Chebyshev parameters K and ε, the first synthesis step is to compute the reflection poles and zeros of the conventional Chebyshev filter in the s' plane according to (6.54)–(6.56). The poles and zeros in the mapped s plane are obtained from an expression resulting from (6.97):

$$s = \sqrt{jAs' - \omega_0^2} . \tag{6.109}$$

Only the left-half-plane poles are used in assembling the reflection coefficient root factors. In the synthesis terminology of Section 3.2.4, the numerator of p(s) is the polynomial f(s), the denominator polynomial is e(s), and p(s) = 1 in this case, since there are no finite zeros of transmission. The network synthesis may then proceed as described in Chapter Three.

Example 6.19. Continue the calculation began in Example 6.18. The s'-plane pole locations from (6.54)–(6.56) and the s-plane pole and zero locations from

**Table 6.5.   Poles and Zeros for Pseudobandpass
Example 6.19**

In the prototype s' plane
  $p' = \pm 1.579 \pm j1.730$
  $z' = \pm 0.3838 \pm j0.8047$
In the double-mapped s plane
  $p' = \pm 0.2787 \pm j1.4158$ and
      $\pm 0.5549 \pm j0.7114$
  $z' = \pm 0.0795 \pm j1.2134$ and
      $\pm 0.1169 \pm j0.8211$
s-plane polynomials:
  $f(s) = s^4 + 0.3927s^3 + 2.204s^2 + 0.4551s + 1.017$
  $e(s) = s^4 + 1.668s^3 + 3.515s^2 + 2.765s + 1.695$
  $p(s) = 1$

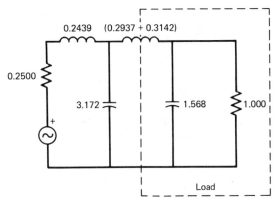

Figure 6.37. Synthesized lumped-element matching network for Examples 6.18 and 6.19. [From Cottee, R. M., and Joines, W. T., *IEEE Trans. Circuits Syst.*, Vol. CAS-26, No. 5, p. 321, May 1979. © 1979 IEEE.]

(6.109) are shown in Table 6.5. The synthesized network is shown in Figure 6.37. Note that the load inductance is a part of the total g_2 element.

6.6.4. Summary of Pseudobandpass Matching.

The narrow bandpass response of matching L sections suggested the use of lowpass networks (cascaded L sections) for broadband matching with resistance transformation. A quadratic frequency mapping function doubled the number of network elements while providing the appropriate correlation between the conventional Chebyshev lowpass, flat-loss function and the pseudobandpass function.

Numerical integration over the frequency axis determined the broadband load-reactance functional values. In practice, the two load-reactance values are given, and trial evaluation of the one-reactance Fano integral determines the flat-loss and ripple factors. Then a solution exists if the two-reactance Fano integral value is less than the corresponding load function requires. If not, a greater number of elements (n) is assumed, and the process repeated.

When the values of conventional Chebyshev constants are found acceptable, the pole/zero locations in the conventional s' plane are computed by formula. The quadratic frequency-mapping function then transforms these n' values into n = 2n' new values. Selection of left-half-plane poles and zeros enables the construction of the Feldtkeller polynomials, and thus network synthesis can proceed.

Cottee and Joines (1979) concede that the integration step can be avoided by proceeding with trial synthesis. However, they claim that the integration approach allows restrictions to be visualized; to that end, they include a dozen design charts. More significantly, their article further considers distributed (transmission line) matching networks terminated by a lumped-element load. The transmission line elements are commensurate—all having the same length —so that a resistively terminated network response would have harmonic passbands extending to infinity on the frequency scale. This does not preclude Fano integration to a finite limit when the load consists of lumped elements that truncate (i.e., band limit) the response.

6.7. Carlin's Broadband-Matching Method

Figure 6.1 pictured the environment for broadband impedance matching: load impedance Z_2 must be transformed by the network to some desired Z_{in} function of frequency. The quality of the match was indicated by a low magnitude of the reflection coefficient ρ_{in} versus frequency. The difference between designing filters and broadband-matching networks is the frequency dependence of load impedance Z_2; it is a resistor in filter design. By Fano's classical method, Z_2 was assumed to represent the input impedance of an LC subnetwork terminated by a resistance. For practical results, the lowpass model of the load impedance must not consist of more than one or two reactances and an end resistor, as shown in Figure 6.16.

Given some arbitrary physical load impedance modeled by impedance data measured at several frequencies, the first—and often difficult—task in applying Fano's method is to classify the actual load, i.e., fit it to the most appropriate lumped-element lowpass model. For loads with bandpass behavior, this usually requires identification of a synchronous bandpass subnetwork and then its corresponding lowpass prototype. Furthermore, the power transfer of the classical method is constant over the band; however, a sloped or other-shaped response often is required.

Fano's method depends on the fact that the magnitudes of the generalized reflection coefficients in (3.46) at any interface in a lossless, doubly terminated network are all equal at a frequency. In fact, his reflection coefficients are conventional, since they are located adjacent to the resistive terminations. Carlin (1977) noted that $|\rho_q|$ is equal to $|\rho_{in}|$ in Figure 6.1. His greater contribution was in noting that a piecewise linear approximation to R_q, the real part of $Z_q = R_q + jX_q$, enables a simple computation of X_q using the Hilbert transform. Furthermore, he showed that power transfer in terms of generalized ρ_q is at most a quadratic function of the R_q piecewise linear function variables. Thus a nonlinear optimization program will usually succeed in obtaining power transfer and/or several other impedance-dependent objectives by a piecewise fit of R_q, the real part of Z_q. The Gewertz method for finding a resistively terminated lowpass network, given the real part of its input impedance, was described in Section 3.5. By Carlin's method, such a network would be the required matching network in Figure 6.1, where the source impedance would be the terminating resistance.

This topic will begin by describing the basis for BASIC language program for finding the imaginary part of a minimum-reactance impedance function from a piecewise linear representation of its real part. The power transfer function will then be derived, and its derivatives will be obtained with respect to the piecewise linear fit parameters. An objective function for the Fletcher–Reeves optimizer (Section 5.4) will be furnished in another BASIC language program. Finally, utilization of the optimal piecewise linear fit to the required Z_q real part to synthesize the matching network will be described. Actually, this last step has been covered completely in Sections 2.5 and 3.5, so that only the connection between these procedures and Carlin's method is required.

6.7.1. Piecewise Hilbert Transform.   Blinchikoff and Zverev (1976, p. 76) give the well-known Hilbert transform that determines the reactance function from a given resistance function:

$$X(\omega) = \frac{1}{\pi} \int_{-\infty}^{+\infty} \frac{R(y)}{y - \omega} \, dy. \tag{6.110}$$

There is a similar function for the inverse transform. Bode (1945, p. 318) gives a more useful form for analysis on linear frequency scales:

$$X(\omega) = \frac{1}{\pi} \int_{0}^{\infty} \frac{dR}{dy} \ln \left| \frac{y + \omega}{y - \omega} \right| \, dy. \tag{6.111}$$

A restriction on these Hilbert transforms is that the function (impedance in this case) must have minimum reactance. Restrictions on transfer functions are similar. Guillemin (1957, p. 301) shows that the phase lag will be least for any transfer function magnitude if its zeros are restricted to the left-half plane. The poles are similarly restricted for passive networks. Such functions are thus called minimum phase; in general, they are associated with ladder (single signal path) networks that do not contain delay equalizer (bridge) sections.

In this case, it is convenient to presume that the equation $R_q(\omega) = Re(Z_q)$ in Figure 6.1 has the piecewise linear form

$$R_q(\omega) = \sum_{k=0}^{n} r_k a_k(\omega), \tag{6.112}$$

where the normalized linear interpolation function is

$$a_k \Big|_{k=1 \text{ to } n} = \begin{cases} 0 & \text{for } \omega \leqslant \omega_{k-1}, \\ \dfrac{\omega - \omega_{k-1}}{\omega_k - \omega_{k-1}} & \text{for } \omega_{k-1} < \omega < \omega_k, \\ 1 & \text{for } \omega \geqslant \omega_k, \end{cases} \tag{6.113}$$

and $a_0 = 1$. This linear interpolation function is easily visualized according to Figure 6.38. The overall form of R_q is illustrated in Figure 6.39. Since this form of resistance (R_q) will be integrated according to (6.111), it must assume a zero value, beginning at some finite frequency. Therefore, an arbitrary but

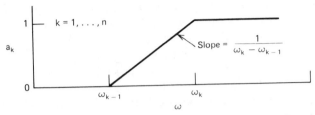

Figure 6.38. Linear interpolation between sample points according to Equation (6.113).

useful choice for the necessary dependent excursion is

$$r_n = - \sum_{k=0}^{n-1} r_k .$$ (6.114)

Using (6.112) in (6.111), the reactance at some frequency ω is

$$X(\omega) = \sum_{k=0}^{n} r_k \frac{1}{\pi} \int_0^{\infty} \frac{da_k}{dy} \ln \left| \frac{y+\omega}{y-\omega} \right| dy.$$ (6.115)

The crux of Carlin's method is a broadly applicable linear combination of the excursions r_k that expresses the reactance function corresponding to (6.112):

$$X_q(\omega) = \sum_{k=0}^{n} r_k b_k(\omega).$$ (6.116)

The reactance contributions, b_k, are

$$b_k(\omega) = \frac{1/\pi}{\omega_k - \omega_{k-1}} \int_{\omega_{k-1}}^{\omega_k} \ln \left| \frac{y+\omega}{y-\omega} \right| dy.$$ (6.117)

Note that the narrow limits of integration result from the single segment of a_k in Figure 6.38 having a nonzero slope. The integral in (6.117) has a simple, closed-form evaluation, as given by Bode (1945, p. 319). Therefore, a final

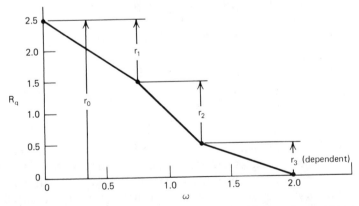

Figure 6.39. A piecewise linear representation of R_q with excursion variables r_k.

expression for the kth reactance contribution is:

$$b_k = \frac{B(\omega, \omega_k) - B(\omega, \omega_{k-1})}{\pi(\omega_k - \omega_{k-1})}, \tag{6.118}$$

where $b_0 = 0$, and the defined function $B(\omega, \bar{\omega})$ is

$$B(\omega, \bar{\omega}) = \bar{\omega}\big[(X+1)\ln(X+1) + (X-1)\ln|X-1| - 2X\ln X\big],$$

$$= 0 \quad \text{if} \quad \bar{\omega} = 0, \tag{6.119}$$

$$X = \frac{\omega}{\bar{\omega}}; \qquad X \neq 1, \quad X \neq 0. \tag{6.120}$$

This remarkable result is easily applied: first, construct a band-limited, piece-wise linear representation of the resistance function; second, calculate the r_k resistance excursions; third, compute the b_k reactance contributions for the desired frequency. The impedance at frequency ω then is

$$Z_q = \sum_{k=0}^{n} r_k(a_k + jb_k), \tag{6.121}$$

using (6.113) with $a_0 = 1$ and (6.118)–(6.120) with $b_0 = 0$. These equations have been coded in BASIC in Program B6-5, (Appendix B), making the last excursion dependent according to (6.114). This computation is equally well suited to hand-held computers.

Example 6.20. The impedance looking into terminals $2-2'$ of the lowpass network in Figure 3.8 can be computed at any frequency using Program B4-1. As a test of the Hilbert transform method, the resistance-versus-frequency curve can be fitted using straight-line segments. Program B6-5 can then be used to compute the related reactance for comparison to the known values from the analysis. Table 6.6 tabulates the input data for Program B6-5 in the first three columns. The program output at these frequencies appears in columns 4 and 5, and columns 6 and 7 show the impedance values obtained

Table 6.6. Hilbert Transform Data for Example 6.20

			Pgm B6-5		Pgm B4-1	
k	r_k	ω	R_q	X_q	R_q	X_q
0	2.2	0	2.2	0	2.2	0
1	−0.0544	0.1	2.1456	−0.3094	2.1465	−0.3208
2	−0.8484	0.7	1.2972	−1.2311	1.2972	−1.2051
3	−0.1676	0.8	1.1296	−1.3435	1.1296	−1.3321
4	−0.7079	1.1	0.4217	−1.3412	0.4217	−1.3444
5	−0.2579	1.3	0.1638	−1.1118	0.1638	−1.0992
6	−0.1205	1.6	0.0433	−0.8362	0.0433	−0.8138
7	−0.0433	2.0	0	−0.6165	0.0102	−0.6028

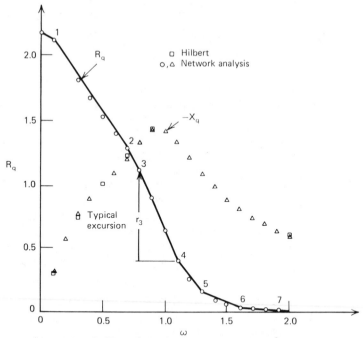

Figure 6.40. R–X graph for the Hilbert transform in Example 6.20.

by analysis in Program B4-1. Also, impedances at frequencies other than the breakpoints may be computed. Figure 6.40 is a graph of these data.

The Hilbert transform calculation works equally well for minimum-phase transfer functions such as

$$H(\omega) = |H(\omega)|e^{j\theta}. \tag{6.122}$$

The appropriate rectangular form for data from an arithmetic frequency scale is:

$$\ln H(\omega) = \ln|H(\omega)| + j\theta, \tag{6.123}$$

where the angle θ is in radians.

6.7.2. Gain Objective Function With Derivatives. Carlin and Komiak (1979) describe a general gain function, which is the inverse transducer function. At the output interface in Figure 6.1, (3.46) and (3.47) yield

$$t(Z_2, Z_q) = \frac{P_2}{P_{a2}} = \frac{4R_2R_q}{\Lambda}, \tag{6.124}$$

where the denominator term is

$$\Lambda = (R_2 + R_q)^2 + (X_2 + X_q)^2. \tag{6.125}$$

Note that gain function t is at most quadratic in r_k because Z_q is linear in r_k according to (6.121). A squared-error objective function is

$$E(\mathbf{r}, \omega) = \sum_{u=1}^{MZ} e_u^2(\mathbf{r}, \omega), \tag{6.126}$$

where the vector of variables is

$$\mathbf{r} = (r_0, r_1, r_2, \ldots, r_{n-1})^T, \tag{6.127}$$

and dependent r_n is computed by (6.114). A well-conditioned error function (residual) at each sample frequency is

$$e_u(\mathbf{r}, \omega_u) = \left[\frac{t(\mathbf{r}, \omega_u)}{g_u(\omega_u)} - 1 \right], \tag{6.128}$$

where $g_u \leqslant 1$ is the arbitrary gain goal (target) value at sample frequency ω_u. The sample frequencies need not coincide with the piecewise linear breakpoint frequencies. These relationships enable the calculation of the objective function in (6.126).

Minimization of the objective function in (6.126) requires its derivatives with respect to the variables in (6.127). Numerical differentiation (finite differencing) is usually unsuitable. Analytically,

$$\frac{\partial E}{\partial r_k} = \sum_{u=1}^{MZ} 2e_u \frac{\partial e_u}{\partial r_k}. \tag{6.129}$$

Note that t in (6.124) is a function of both R_q and X_q, and these are functions of the r_k excursions. Thus the classical chain rule for partial derivatives yields

$$\frac{\partial e_u}{\partial r_k} = \frac{1}{g_u} \left(\frac{\partial t}{\partial R_q} \frac{\partial R_q}{\partial r_k} + \frac{\partial t}{\partial X_q} \frac{\partial X_q}{\partial r_k} \right) \tag{6.130}$$

for use in (6.129). It is a simple matter to write the following derivatives of t from (6.124):

$$\frac{\partial t}{\partial R_q} = 4R_2 \frac{\Lambda - 2R_q(R_2 + R_q)}{\Lambda^2}, \tag{6.131}$$

$$\frac{\partial t}{\partial X_q} = 4R_2 \left[-2R_q(X_2 + X_q) \right] / \Lambda^2. \tag{6.132}$$

Finally, the derivatives of R_q and X_q are required in (6.130). Applying the constraint (6.114) to (6.112) pertinent to this formulation yields

$$R_q = \sum_{k=0}^{n-1} r_k(a_k - a_n), \tag{6.133}$$

so that

$$\frac{\partial R_q}{\partial r_k} = a_k - a_n. \tag{6.134}$$

From (6.116),

$$\frac{\partial X_q}{\partial r_k} = b_k - b_n . \tag{6.135}$$

6.7.3.   Optimization of the Piecewise Resistance Function.   The preceding objective function has been incorporated in the Fletcher–Reeves optimizer in Program B5-1. The result is Program B6-6 in Appendix B. The input section through program line 140 loads the breakpoint data required for the Hilbert transform calculation of reactance from resistance, as in Program B6-5. All but the last resistance excursion become the optimizer variables. The objective function and its gradient are assembled by subroutine 1000 in lines 1000-1260; this requires appeal to subroutine 3000 at every sample frequency to compute $Z_q = R_q + jX_q$. Lines 3020–3040 set constraint (6.114), and lines 3050–3250 perform the Hilbert transform calculations as in Program B6-5.

Example 6.21.   Input the data in Table 6.7 into Program B6-6 to obtain the optimum resistance excursions for a gain of 1.0 at the four sample frequencies. The program output is shown in Table 6.8.

The optimized excursions are $r_0 = 2.2754$, $r_1 = -1.0603$, $r_2 = -1.1167$, and (constrained) excursion $r_3 = -0.0984$. Inspection of Figures 6.1 and 3.8 shows that r_0 is the eventual generator resistance. If it is desirable to hold this at a certain value, e.g., 2.5 ohms in this case, then all that is necessary is to add the statement "1225 G(1)=0" to Program B6-6. A rerun of Example 6.21 shows how the zero gradient holds the first optimization variable at its initial value. The choice of starting excursion values is somewhat arbitrary. Carlin (1977) suggests assuming reactance cancellation and setting the residuals to sustain the dc gain at the in-band breakpoints.

6.7.4.   Rational Approximation and Synthesis.   At this point in Carlin's broadband-matching method, an optimal piecewise linear representation of R_q is known. The remaining task is to realize a network that provides this behavior. This is clearly the subject treated in Section 3.5. The Gewertz method considered there began with a rational function of input resistance in the form of (3.94), or (3.98) in particular. It is always in powers of ω^2 or the equivalent powers of s^2, since resistance is an even function of frequency. The next step in Carlin's method is to fit such a rational function to the piecewise linear representation. This can be accomplished by the method in Section 2.5.

A table of impedance versus frequency and the form of the desired rational polynomial were required in Section 2.5. In the Carlin method, the table of data is created from the piecewise linear resistance function by (1) using symmetric positive and negative frequencies for the even resistance function and by (2) using zero reactance values at every sample. A typical data set is given in Table 6.9. The data in Table 6.9 can be input into Program B2-5 to

Table 6.7. Input to Program B6-6 for Example 6.21

ω^a	R_2	X_2	g_u	k^b	r_k	ω_k
0.1	2.15	0.31	1	0	2.5	0
0.5	1.58	1.01	1	1	−1	0.75
0.9	0.89	1.42	1	2	−1	1.25
1.0	0.65	1.42	1	3	−0.5	2.0

[a] Number of measured Z_2 values is 4.
[b] Number of breakpoints, including $\omega = 0$, is 4.

Table 6.8. Program B6-6 Output for Example 6.21[a]

ITN = 0		IFN = 1	ITN = 29		IFN = 89
F = 2.6976186E − 03			F = 1.22678987E − 08		
I	X(I)	G(I)	I	X(I)	G(I)
1	2.5	.0269056668	1	2.27720383	1.81803783E − 08
2	−1	.0252870452	2	−1.0625151	−2.06528716E − 08
3	−1	8.41058789E − 03	3	−1.1159194	−1.01505479E − 08
ITN = 1		IFN = 5	ITN = 30		IFN = 91
F = 4.61174131E − 05			F = 1.2233111E − 08		
I	X(I)	G(I)	I	X(I)	G(I)
1	2.38714539	1.59244045E − 03	1	2.27546837	4.50758795E − 08
2	−1.10606537	1.54801777E − 03	2	−1.06031197	3.71783939E − 08
3	−1.03527783	6.7775074E − 04	3	−1.11666244	5.39050241E − 09
ITN = 2		IFN = 19	ITN = 31		IFN = 93
F = 8.30479937E − 07			F = 1.22313988E − 08		
I	X(I)	G(I)	I	X(I)	G(I)
1	2.35241762	6.60273984E − 05	1	2.27542275	6.41003251E − 10
2	−1.13975746	6.48530284E − 05	2	−1.0603496	−1.27783317E − 09
3	−1.04982374	4.22180638E − 05	3	−1.11666789	−2.9785896E − 10
ITN = 3		IFN = 32	ITN = 32		IFN = 95
F = 1.79326026E − 07			F = 1.22313053E − 08		
I	X(I)	G(I)	I	X(I)	G(I)
1	2.34265967	3.17542868E − 06	1	2.27541404	3.72293618E − 09
2	−1.14933634	1.22805397E − 06	2	−1.06033326	1.83188975E − 09
3	−1.05596202	6.9709696E − 06	3	−1.11666406	7.02292779E − 10
ITN = 4		IFN = 37	ITN = 33		IFN = 99
F = 1.57306571E − 07			F = 1.22313053E − 08		
I	X(I)	G(I)	I	X(I)	G(I)
1	2.34014065	−3.03374459E − 06	1	2.27541404	3.72293618E − 09
2	−1.15031054	−5.53521328E − 06	2	−1.06033326	1.83188975E − 09
3	−1.06149199	2.40726355E − 06	3	−1.11666406	7.02292779E − 10

[a] The output for iterations 5–28 has been omitted.

Table 6.9. Typical Carlin Piecewise Resistance Data for Fitting Program B2-5

ω^a	R_q	ω	R_q
-1.5	0.0835	0.05	2.1728
-1.0	0.6577	0.10	2.1456
-0.5	1.5800	0.15	2.0749
-0.3	1.8628	0.30	1.8628
-0.15	2.0749	0.50	1.5800
-0.10	2.1456	1.0	0.6577
-0.05	2.1728	1.5	0.0835
0.0	2.2000		

$^a X_q = 0$ for all ω.

obtain the Levy (1959) coefficients for an appropriate lowpass rational polynomial having a constant numerator and a sixth-degree denominator. The pertinent linear system of equations may be solved by Program B2-1. The rational polynomial coefficients of s for the data in Table 6.9 are: $a_0 = 2.1819$, $b_0 = 1$, $b_1 = 0$, $b_2 = -2.0505$, $b_3 = 0$, $b_4 = -2.7689$, $b_5 = 0$, and $b_6 = -3.0330$. Note that the even input data produce the required even fitting function. This polynomial is the basis for the Gewertz procedure in Section 3.5.1, which finds the $Z_q(s) = Z_{RLC}$ impedance function looking into the matching network from the load interface (see Figure 6.1).

The last Carlin step is to synthesize the Z_{RLC} input impedance function obtained by the Gewertz method. This has been described in Sections 3.5.3 and 3.4. The result will be a network like that shown in Figure 3.8; it is similar to an example given by Carlin (1977). Carlin and Komiak (1979) also give a rule of thumb for estimating the required complexity of the rational polynomial used to fit the optimal piecewise linear resistance function; this determines the matching network complexity as well.

6.7.5. Summary of Carlin's Broadband-Matching Method.

Carlin identified at least three important concepts applicable to the broadband-matching problem. First, a piecewise linear representation of a resistance function can be used in a closed-form application of the Hilbert transform to find the corresponding reactance function, assuming minimum phase. The excursions in the piecewise linear representation occur as coefficients in a linear combination of easily computed resistance and reactance contributions. The technique is equally valuable for computing the transfer angle given a piecewise linear fit of transfer magnitude.

Second, the generalized gain function at the load interface is at most a quadratic function of the resistance excursion variables. If a classical least-squared-error solution were employed, a standard quadratic program would suffice. The gain function is well conditioned in any event.

Third, an objective function for optimizing the resistance excursions need not be limited to gain; it might be noise figure, noise measure, or any other function that can be formulated in terms of impedances or admittances. Equally important, an arbitrary goal/target function may be employed so that sloped gain functions may be matched. Like most applications of nonlinear programming (optimization), there are several choices to be made from experience rather than by analysis.

The optimal piecewise linear resistance function must be fitted with an even rational polynomial, so that a matching network may be synthesized. The Gewertz method then provides the input impedance polynomial for the network at the load interface. Standard network synthesis techniques will produce the LC element values.

There are two features of Carlin's broadband-matching method that distinguish it from Fano's classical method. The discrete load impedance versus frequency data set does not have to be identified with a resistively terminated LC load network; i.e., load classification is not required. Also, the well-conditioned optimization process allows sloped-gain or other arbitrary fit of the objective function. Fano's method has been adapted by Mellor (1975) to obtain similar results at the expense of considerable *ad hoc* procedures. Network synthesis is required in these and other methods which are more versatile than the direct design-matching method in Section 6.4.

Problems

6.1. Find four different lossless, lumped-element L-section matching networks that transform a load impedance of $36 - j324$ ohms to match a $50 + j0$ generator impedance.

6.2. Conjugately match a $6 + j25$-ohm load impedance to a $7 + j20$-ohm generator using only capacitors in an L section. Obtain two different solutions.

6.3. Plot a $2:1$ SWR load-locus circle on a Smith chart, and explain why Equation (6.19) is true.

6.4. A T section is composed of the two types of L sections.
   (a) Write an expression for the parallel resistance level across the shunt reactance (X_b) as a function of the T section's terminating resistances and transfer phase angle.
   (b) A conjugately matched T section delivers 1 watt from a 50-ohm source to a 21-ohm load with a lagging current transfer phase of 155 degrees. What is the rms voltage across the shunt reactance?

6.5. (a) Can a single, lossless transmission line transform $6+j25$ ohms to $7-j20$ ohms? If so, give its Z_0 and θ.

   (b) What is the input impedance of a lossless, 50-ohm transmission line 45 degrees long and terminated by a $6+j25$-ohm load?

   (c) Can the input impedance from (b) be transformed to $7-j20$ ohms by a single lossless transmission line? If so, give its Z_0 and θ.

6.6. Rotate load impedance $Z_2 = 100-j150$ ohms on a 50-ohm transmission line that is 45 degrees long.

   (a) What is the input impedance if the line is lossless?

   (b) What is the input impedance if this length of line has a uniform dissipation of 0.25 nepers?

6.7. Suppose that $Z = R + j0$ ohms. Show that SWR with respect to 1 ohm is R for $R > 1$ and is $1/R$ for $R < 1$.

6.8. A two-reactance load (an L section with g_1, g_2, and a 1-ohm resistance) terminates an infinitely complicated bandpass matching network driven by a resistive source. Give an algebraic (containing no integrals) expression for an equality constraint and an inequality constraint on the minimum possible input reflection loss, $\ln(1/|p|)$. The band of interest is ω_1 to ω_2.

6.9. Prove Equations (6.48) and (6.51).

6.10. For a 100% bandwidth, what is the greatest Q_L that can be matched with an SWR $2:1$?

6.11. Derive Equations (6.57) and (6.58).

6.12. Derive Equations (6.61) and (6.62).

6.13. Find the minimum possible decrement of a single-reactance load for optimal broadband match when $\ln|\rho| = 10^{1-\omega}$.

6.14. Estimate Fano's optimal matching solution using Equations (6.68) and (6.69); do not iterate.

   (a) Find the optimal lowpass network that contains two reactances and matches a 2-farad capacitor in parallel with a 1-ohm resistor over the frequency range 0 to 1 radians.

   (b) What is the generator resistance?

   (c) What is the range of SWR in the pass band?

   (d) What is the transducer loss at dc (in dB)?

6.15. Evaluate Equations (6.68) and (6.69) for parameters a and b when $N=3$, $Q_L = 3$, and the bandwidth is 50%.

6.16. Program Equation (6.72) on a hand-held or desktop computer. Start with $g_1 = 1.5000$, 0.7229, and 1.9683. Compute $g_r \cdot g_{r+1}$ and compare the three sequences resulting from the three starting points. Compare the sequences from $r = 1$ to $n - 1$ and from $r = n - 1$ to 1.

6.17. Derive Equation (6.87).

6.18. Transform the lowpass network in Figure 3.8 (Chapter Three) to a bandpass network that is driven by a 50-ohm generator; obtain the 20% bandwidth geometrically centered at 70 MHz. Assume that the passband edges correspond to 1 radian on the lowpass network.

(a) What are the bandpass edge frequencies?

(b) Give all network element values and units.

6.19. Instead of the capacitive L section indicated in Figure 6.32, replace an inductive Norton L section with an inductive T section to obtain an 11-ohm load resistance while retaining the 50-ohm generator resistance. Show all element values in your final network.

6.20. A resistance function versus a linear frequency scale has the form of a straight line from 1 ohm at dc to 0 ohms at 1 radian; it is zero at frequencies greater than 1 radian.

(a) What is the impedance of the associated minimum-reactance function at 0.5 radians?

(b) What is the partial derivative of this impedance with respect to excursion r_1 at 0.5 radians?

6.21. A minimum-reactance impedance function has a piecewise linear real part. This resistance is a constant 2 ohms from 0 to 1.5 radians and a linear function from 1.5 to 2.0 radians. The resistance is zero at all frequencies equal to or greater than 2 radians. Find the reactance at 0.5 and 1.75 radians.

Chapter Seven ─────────────

Linear Amplifier Design Tools

This chapter establishes a basis for many modern amplifier design relationships, especially those related to generalized Smith charts and their bilinear functions. Impedance and power relationships will be investigated in detail. The linear two-port network will be analyzed in terms of Z, Y, and S parameters, as indicated in Figure 7.1. The network may or may not be reciprocal, i.e., y_{12} may not be equal to y_{21}. The simplifying unilateral assumption that $y_{12} = 0$ will be considered only at the end of this chapter. The stability of such networks will be studied. Thus some of these results will be applicable to oscillator design. Further applications of this chapter will appear in Section 9.5, which deals with load effects on passive networks, especially dissipative filters.

Impedance mapping will be the main analytic and computational tool. This technique establishes the position of a small Smith chart image of a branch- or port-terminating impedance plane embedded in a network impedance, admittance, or scattering response plane. For example, all possible values of transducer gain S_{21} as a function of a network branch impedance are easily visualized and calculated. The generalized Smith chart is normalized to a complex number; it will be crucial to the impedance mapping concept.

This chapter begins with the definition of bilinear functions and several methods for determining their three coefficients from a set of characterizing data. The generalized Smith chart bilinear form that maps the right-half plane onto a unit circle will be studied next. Some useful shortcuts and special features in its application will be considered. The bilinear theorem that relates all Z, Y, and S network functions will be derived by obtaining the three-port to two-port reduction formulas. The impedance-mapping relationship will then be derived, including the conversion of bilinear coefficients to the mapping displacement and orientation coefficients.

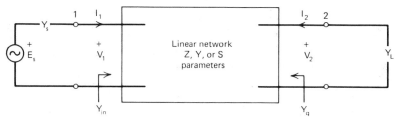

Figure 7.1. A linear two-port network.

Linvill's two-port impedance and power geometric models will be derived for both visualization and subsequent analysis. The per-unit voltage (or current) output power parabola of revolution on its mapped Smith chart base appears in the input immittance plane. Input power per input voltage (or current) is a wedge-shaped surface in the input plane. Thus several gain relationships are easy to see and are the basis for special mathematical development. Finally, the most important computational results will be obtained in terms of scattering parameters. According to current practice, these parameters are usually measured with respect to 50-ohm resistive terminations. A major development tool will be their renormalization to arbitrary, complex impedances. This enables direct consideration of linear, active networks between complex source and complex load; impedance, power, and stability issues are readily considered from that basis. The last subject in this chapter is the specialization of the scattering results to the unilateral case when $S_{12} = 0$. The resulting simplifications allow easier comprehension of some power relationships, although the approximation is often unsatisfactory in practice and unnecessary in the presence of personal computing tools.

7.1. Bilinear Transformations

Almost all complex functions of complex variables associated with linear networks are bilinear, as originally expressed in (2.1):

$$w = \frac{a_1 Z + a_2}{a_3 Z + 1}. \qquad (7.1)$$

Bilinear functions are sometimes called linear fractional transformations (LFT). Dependent function w and independent function Z are usually impedance, admittance, and scattering parameters, i.e., elements of their respective **Z**, **Y**, and **S** matrices that characterize the linear and generally nonreciprocal network. For example, (6.23) expresses the input impedance of a dissipative, uniform transmission line in a bilinear form as a function of the terminating load impedance. These functions are called bilinear because they are linear functions of both w and Z, as seen in an equivalent expression for (7.1):

$$Za_1 + a_2 - wZa_3 = w. \tag{7.2}$$

In this section, two means for determining the three a_i coefficients in (7.1) from characterizing data will be derived. One is easy, fast, and less accurate than the somewhat more complicated, slower, and more accurate second method. A third method for averaging excess, noisy data is mentioned. The generalized Smith chart, normalized to a complex impedance, is described and several useful shortcuts are mentioned.

7.1.1. Determining Bilinear Coefficients. The first and most elementary method works well in numerical practice if at least six decimal digits are carried throughout. It is derived by considering three special values of independent variable Z in (7.1), namely, the triple $(0, 1, \infty)$ and the corresponding dependent w values (w_0, w_1, w_∞). When $Z = 0$, (7.1) yields

$$a_2 = w_0. \tag{7.3}$$

When $Z \to \infty$, (7.1) yields

$$a_1 = w_\infty a_3. \tag{7.4}$$

The required value for a_3 is obtained by letting $Z = 1 + j0$ in (7.1) and substituting (7.3) and (7.4):

$$a_3 = \frac{w_0 - w_1}{w_1 - w_\infty}. \tag{7.5}$$

Infinity is suitably represented by $Z = 1E9 + j0$ in hand-held calculators. A set of consistent data and coefficients is given in Table 7.1.

A slower procedure, which is less vulnerable to poor numerical conditioning, assumes less extreme values for the independent Z triple. Choose the arbitrary Z triple (Z_1, Z_2, Z_3) with the corresponding dependent triple

Table 7.1.   A Set of Bilinear Function Data and the Resulting Coefficients

i	Z_i	w_i
1	0	0.1800 $/-23.0°$
2	1E9	0.4285 $/-55.0°$
3	1	0.5588 $/-26.89°$

$$a_1 = 0.5998 \ /75.0075°$$
$$a_2 = 0.1800 \ /-23.0°$$
$$a_3 = 1.3998 \ /130.0075°$$

(w_1, w_2, w_3). Then (7.2) yields three equations in the three a_i unknowns:

$$Z_1a_1 + a_2 - w_1Z_1a_3 = w_1,$$
$$Z_2a_1 + a_2 - w_2Z_2a_3 = w_2, \qquad (7.6)$$
$$Z_3a_1 + a_2 - w_3Z_3a_3 = w_3.$$

Solving the middle equation for a_2, gives

$$a_2 = a_3P_2 + w_2 - a_1Z_2, \qquad (7.7)$$

where a new constant has been defined:

$$P_i = Z_iw_i, \qquad i = 1, 2, \text{ or } 3. \qquad (7.8)$$

Then a_2 may be eliminated from the first and last equations in (7.6); the result, in matrix form, is

$$\mathbf{M}\begin{bmatrix} a_1 \\ a_3 \end{bmatrix} = \begin{bmatrix} (w_1 - w_2) \\ (w_3 - w_2) \end{bmatrix}, \qquad (7.9)$$

where matrix $\mathbf{M}$ is

$$\mathbf{M} = \begin{bmatrix} (Z_1 - Z_2) & (P_2 - P_1) \\ (Z_3 - Z_2) & (P_2 - P_3) \end{bmatrix}. \qquad (7.10)$$

In order to apply Cramer's rule to solve for a_1 and a_3, the determinant of $\mathbf{M}$ is written as

$$\det \mathbf{M} = Z_1D_{23} + Z_2D_{31} + Z_3D_{12}, \qquad (7.11)$$

where another defined constant is

$$D_{ij} = P_i - P_j. \qquad (7.12)$$

Thus Cramer's rule yields

$$a_1 = \frac{w_1D_{23} + w_2D_{31} + w_3D_{12}}{\det \mathbf{M}}, \qquad (7.13)$$

$$a_3 = \frac{Z_1(w_3 - w_2) + Z_2(w_1 - w_3) + Z_3(w_2 - w_1)}{\det \mathbf{M}}, \qquad (7.14)$$

and a_2 is computed by (7.7).

Program A7-1 in Appendix A performs these computations in 206 steps that run 3 minutes on the HP-67/97. The program is based on the polar four-function complex operations from Program A2-1. The complex numbers are stored in polar form, the magnitude in primary registers and the angle (in degrees) in the corresponding secondary registers. (Calculators without this feature may be programmed with registers similarly paired, with address numbers differing by some constant, e.g., 10). Program A7-1 is based on a programming technique worth remembering for use on small computers. The known sequence of register addresses has been packed into three registers and removed in a sequence of one or two digits. An explanation is based on the

Table 7.2. Register Assignments and Address Sequences for Bilinear Coefficient Program A7-1

R0	R1	R2	R3	R4	R5	R6	R7	R8[a]	R9
$(\text{Det M})^{-1}$	Z_1	w_1	Z_2	w_2	Z_3	w_3	D_{23}	D_{31}	D_{12}
							a_1	a_2	a_3

	Register C	Register D	Register E
For Det **M**	56 8 34 8 1 0	12 8 56 8 3 0	34 8 12 8 5 0
For a_1	56 8 34 8 2 7[b]	12 8 56 8 4 7[b]	34 8 12 8 6 7[b]

For a_3: 6 4 1 9, 2 6 3 9, and 4 2 5 9
For a_2: 34 9 8 7 3 8 4 8

[a] R8 used for scratch during det **M** and a_1 calculations.
[b] Add digits 1 7 by +0.00000017.

register assignments and sequences in Table 7.2. For example, consider the a_3 computation from (7.14) according to the register address sequence shown in Table 7.2. Register 6 (the primary and secondary pair) contains complex w_3, and register 4, containing w_2, is subtracted from that; the result is multiplied by Z_1 from register 1, and this is summed into the register-9 pair. Digit pairs are required when incorporating the P_i values defined in (7.8). The a_2 calculation according to (7.7) requires the sequence shown in Table 7.2; there, $P_2 = Z_2 w_2$ from registers 3 and 4, and this is multiplied by a_3 from register 9, and so on. Table 7.3 contains a consistent set of data to test program operation.

There are also means for determining the bilinear coefficients in (7.1) when the (Z_i, w_i) data are noisy and $i > 3$, as occurs for measured data pairs. In this case, the data pairs require weighting. Suppose that one measured w value is a moderately large impedance and another is a very small impedance. If the measurement error is related in any fixed way to ohms, then the latter value is much less reliable than the former. Kajfez has developed a reasonable weighting scheme in light of the bilinear functions involved. His computation can be

Table 7.3. A Set of Bilinear Function Coefficient Data for Program A7-1

i	Z_i	w_i	P_i	a_i
1	0.1 $\underline{/30°}$	0.1732 $\underline{/-7.8675°}$	0.0173 $\underline{/22.1325°}$	0.6 $\underline{/75°}$
2	0.5 $\underline{/60°}$	0.4473 $\underline{/129.505°}$	0.2237 $\underline{/189.505°}$	0.18 $\underline{/-23°}$
3	1.1 $\underline{/-10°}$	0.5099 $\underline{/-30.324°}$	0.5609 $\underline{/-40.324°}$	1.4 $\underline{/130°}$

$$D_{12} = 0.2406 \;\; \underline{/10.4056°}, \qquad D_{23} = 0.7256 \;\; \underline{/153.3009°}$$
$$D_{31} = 0.5531 \;\; \underline{/-41.9136°}, \qquad \det \mathbf{M} = 0.4627 \;\; \underline{/10.4030°}$$

accomplished easily on desktop computers; the interested reader is referred to Kajfez (1975).

7.1.2. Generalized Smith Chart.

Using methods described by Churchill (1960, pp. 76–77), it is a straightforward matter to show that every bilinear transformation of the closed right-half Z plane onto a closed unit circle must have the form

$$\alpha = e^{j\beta} \frac{Z - Z_c}{Z + Z_c^*}, \qquad (7.15)$$

where β is real and $Re(Z_c) > 0$. The last requirement is especially emphasized. The exponential term merely rotates the unit-circle image and will henceforth be dropped. The generalized Smith chart maps impedances according to

$$\rho = \frac{Z - Z_c}{Z + Z_c^*} = \frac{Z - jX_c - R_c}{Z - jX_c + R_c}, \qquad (7.16)$$

where $Z = R + jX$ and $Z_c = R_c + jX_c$. Clearly, (7.16) could be normalized to R_c by division in the numerator and denominator in the fashion of the ordinary Smith chart relationship given previously in (6.22). The obvious remaining difference is the term jX_c. A little thought shows that it may be combined into a new reactance component $(X - X_c)$ instead of the usual X component. The generalized Smith chart then represents the ordinary chart with center Z_c and constant reactance lines $(X - X_c)$. One practical application concerns power transfer from a complex source to a complex load, as discussed in Section 3.2.3. Thus (7.16) is exactly comparable to (3.46). Note that whether Z_c appears in the numerator or in the denominator is a matter of arbitrary definition. It is convenient here to represent the chart center as Z_c.

It is also important to define the generalized reflection coefficient in admittance form, as follows:

$$\rho = \frac{Y_c - Y}{Y_c^* + Y}. \qquad (7.17)$$

The generalized Smith chart no longer allows substitution of $Z = 1/Y$ in order to change from an impedance to an admittance basis. This does not change (7.16) into (7.17) unless $Z_c = 1/Y_c$ is real.

Example 7.1. Consider a complex source connected directly to a complex load, as in Figure 3.3 in Section 3.2.3. Suppose that $Z_s = 25 - j50$ ohms and Z_L is defined as causing a 2 : 1 standing-wave ratio (SWR) on a 50-ohm transmission line. What is the range of power delivered to the load relative to the power available from the source? The solution will be obtained graphically here and analytically in Section 9.5. The procedure will be to select three or more impedance points on a 2 : 1 SWR circle from a Smith chart normalized to 50 ohms. Then these points will be plotted on a generalized Smith chart normalized to the conjugate of the source impedance in accordance with the

Table 7.4.   Impedance Data for 2 : 1 SWR Renormalized for Example 7.1

Z 2 : 1 SWR	Z wrt[a] 25 ohms	Z wrt 25 + j50 ohms
100 + j0	4 + j0	4 − j2
25 + j0	1 + j0	1 − j2
42.5 + j32.5	1.70 + j1.30	1.70 − j0.70
42.5 − j32.5	1.70 − j1.30	1.70 − j3.30

[a] With respect to.

generalized reflection coefficient in (3.46). Thus the reflection magnitude extreme values can be determined graphically and applied in (3.47) to find the answer. In this case, the complex normalizing impedance is $Z_c = 25 + j50$ ohms. Four convenient 2 : 1 SWR impedance points with respect to 50 ohms are shown in Table 7.4 unnormalized, normalized wrt 25 ohms, and normalized wrt 25 + j50 ohms. These points are plotted on the generalized Smith chart normalized to $Z_c = 50 + j0$ and $Z_c = 25 + j50$ in Figure 7.2, and the required

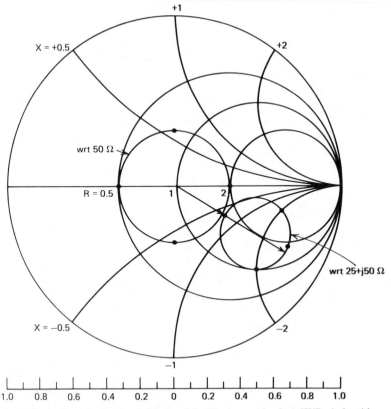

Figure 7.2.   Generalized Smith chart with $Z_c = 25 + j50$ ohms and a 2 : 1 SWR circle with respect to 50 ohms.

circle locus through them has been sketched. The complex reflection coefficient magnitude ranges from about 0.36 to 0.80. Therefore, (3.47) shows that the load power varies from about 0.36 to 0.87 of the maximum available source power.

Since the Smith chart transformations map unique points on a one-to-one basis, it can be concluded that (7.16) maps the left-half Z plane into the region outside the unit reflection circle in the ρ plane. There is a convenient interpretation that avoids having to plot in this region, which is off the Smith chart. Consider the conventional reflection coefficient

$$\rho = \frac{z-1}{z+1} = \frac{(r-1)+jx}{(r+1)+jx} . \tag{7.18}$$

Inverting (7.18) and replacing $r \leftarrow (-r)$ yields

$$\frac{1}{\rho} = \frac{(r-1)+j(-x)}{(r+1)+j(-x)} . \tag{7.19}$$

Comparison of the last two equations reveals that they are the same, except that plotting $1/\rho^*$ requires the constant resistance lines to read as their negatives. Note that the $(-x)$ terms in (7.19) correctly correspond to the $+x$ terms in (7.18) because $1/\rho^*$ results in an offsetting sign change for z. Again, left-half-plane normalizing Z values corresponding to $|\rho|>1$ can be represented directly on the Smith chart by plotting $1/\rho^*$ and reading the reactance values normally and the resistance values as their negatives.

7.1.3. Summary of Bilinear Transformations.

The general form for the standard bilinear (or linear fractional) transformation contains three complex coefficients. Two methods were described for determining these coefficients given three independent and dependent data pairs. The fast and easy method assumes that the independent Z values are all real: 0, 1, and ∞. These can be used with their corresponding dependent w values. In most practical problems, the poor numerical conditioning is tolerable if at least six decimal digits are carried in the elementary calculations. The second method is slower and more accurate, because it allows the selection of three arbitrary, complex values of Z. Computer Program A7-1 was provided for the latter method. A third method was identified in the literature; it applies to cases where a surplus of noisy data pairs are used to find the bilinear coefficients. These must be averaged in a special way. Such a method has been published, and is quite suitable for desktop personal computers.

Generalized Smith charts were shown to result from the unique bilinear transformation that maps the right-half Z plane onto the reflection unit circle. It was emphasized that the complex normalizing impedance that appears as the chart center must have a strictly positive real part. It was also shown that a comparable admittance bilinear form exists to define an admittance generalized Smith chart. However, simple substitution of $Z=1/Y$ does not convert the chart representation from impedance to admittance, as was the case for real normalizing impedances. It was remarked that left-half-plane impedances are represented in the reflection plane region outside the unit circle and thus off the Smith chart. However, plotting $1/\rho^*$ for these impedances, where $|\rho| > 1$, enables the use of the Smith chart in a fairly normal way. The resistance loci must be read as the negative of their usual values and the reactance loci are read normally.

7.2. Impedance Mapping

Impedance mapping is a method that allows a peek into the complex plane associated with any Z, Y, or S response parameter. What one is able to "see" there is a small, rotated, generalized Smith chart representing the entire impedance, admittance, or scattering plane of any network branch. Even more generally, the impedance-mapping formulation enables the restatement of any bilinear function into a form having a complex translation constant and a complex factor that scales and rotates the generalized Smith chart's unique bilinear form (7.16). Impedance mapping is very valuable for visualization, analysis, and computation.

In this section, a linear three-port network will be characterized by its scattering parameters and one port terminated by a fixed reflection coefficient. The equivalent two-port parameters will be derived. This has value in ladder network analysis when a terminated three-port circulator appears in cascade. An HP-67/97 program is provided for this transformation. More generally, this result proves the important bilinear theorem, which states that every Z, Y, or S response of a linear network is a bilinear function of any branch impedance, admittance, or scattering parameter, in any mixed association. For example, response S_{12} must be a bilinear function of any branch impedance, say Z_b. This has many practical applications in neutralization, oscillator, filter, and amplifier design.

Finally, the impedance-mapping relationships will be derived, and a hand-held computer program will be furnished. Many examples will be provided to illustrate these principles and applications.

7.2.1. Three-Port to Two-Port Conversion.
Two-port scattering parameters were considered in Section 4.5.2; the defining system of linear equations was given in (4.46) and (4.47). In general, such systems for any number of ports may be described in matrix notation as

$$\mathbf{b} = \mathbf{Sa}. \tag{7.20}$$

This notation for three-port networks stands for

$$
\begin{aligned}
b_1 &= s_{11}a_1 + s_{12}a_2 + s_{13}a_3, \\
b_2 &= s_{21}a_1 + s_{22}a_2 + s_{23}a_3, \\
b_3 &= s_{31}a_1 + s_{32}a_2 + s_{33}a_3.
\end{aligned}
\tag{7.21}
$$

Three-port scattering parameters will be lower-case s_{ij}, and two-port parame-

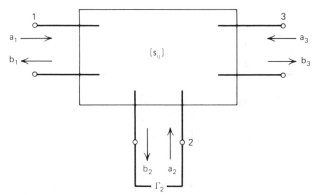

Figure 7.3. A linear, three-port network with scattering parameters s_{ij}.

ters will be upper-case S_{ij} (see the related network in Figure 7.3). Port 2 is constrained by the reflection coefficient Γ_2 so that

$$a_2 = \Gamma_2 b_2. \tag{7.22}$$

Making this substitution in (7.21) yields

$$b_1 = s_{11}a_1 + s_{12}\Gamma_2 b_2 + s_{13}a_3, \tag{7.23}$$

$$b_2 = s_{21}a_1 + s_{22}\Gamma_2 b_2 + s_{23}a_3, \tag{7.24}$$

$$b_3 = s_{31}a_1 + s_{23}\Gamma_2 b_2 + s_{33}a_3. \tag{7.25}$$

Solving (7.24) for b_2 yields

$$b_2 = \frac{s_{21}a_1 + s_{23}a_3}{1 - s_{22}\Gamma_2}. \tag{7.26}$$

Substituting (7.26) into (7.23) and (7.25) yields

$$b_1 = a_1\left(s_{11} + \frac{s_{12}s_{21}}{1/\Gamma_2 - s_{22}}\right) + a_3\left(s_{13} + \frac{s_{12}s_{23}}{1/\Gamma_2 - s_{22}}\right), \tag{7.27}$$

$$b_3 = a_1\left(s_{31} + \frac{s_{21}s_{32}}{1/\Gamma_2 - s_{22}}\right) + a_3\left(s_{33} + \frac{s_{23}s_{32}}{1/\Gamma_2 - s_{22}}\right). \tag{7.28}$$

The equivalent two-port parameters are immediately available by comparison with the original set of two-port equations:

$$S_{11} = s_{11} + \frac{s_{12}s_{21}}{1/\Gamma_2 - s_{22}}, \quad S_{13} = s_{13} + \frac{s_{12}s_{23}}{1/\Gamma_2 - s_{22}},$$
$$\tag{7.29}$$
$$S_{31} = s_{31} + \frac{s_{32}s_{21}}{1/\Gamma_2 - s_{22}}, \quad S_{33} = s_{33} + \frac{s_{32}s_{23}}{1/\Gamma_2 - s_{22}}.$$

In this case the two ports of interest are ports 1 and 3. Also note that as $\Gamma_2 \to 0$, $S_{ij} \to s_{ij}$, as required.

Program A7-2 in Appendix A for the HP-67/97 evaluates the equations in (7.29). The nine three-port scattering parameters are input first. The program then converts any given port-2 terminating impedance (Z_2) to the reflection coefficient Γ_2 with respect to 50 ohms and finds the equivalent two-port parameters.

Example 7.2. Consider the three-port scattering matrix

$$S_3 = \begin{bmatrix} 0.862 \ \underline{/-63°} & 0.800 \ \underline{/160°} & 0.236 \ \underline{/75.3°} \\ 0.050 \ \underline{/20°} & 0.500 \ \underline{/-60°} & 0.300 \ \underline{/-98°} \\ 2.344 \ \underline{/129°} & 0.400 \ \underline{/100°} & 0.708 \ \underline{/-16.1°} \end{bmatrix} \quad (7.30)$$

Load these polar data into Program A7-2 by pressing keys fa and responding to the row/column subscripts displayed. Write these data on a magnetic card for later use. Now terminate port 2 (Figure 7.3) with a resistor of 200 $\underline{/0°}$ ohms by pressing 0, "enter," 200, and key A. The results are

$$S_{11} = 0.8435 \ \underline{/-64.32°}, \qquad S_{13} = 0.3847 \ \underline{/63.03°},$$
$$S_{31} = 2.3561 \ \underline{/128.8°}, \qquad S_{33} = 0.7890 \ \underline{/-15.99°}. \quad (7.31)$$

Since port 2 is normalized to 50 ohms, inputting this value will show that $S_{ij} = s_{ij}$, as expected.

The ladder analysis method from Section 4.2 can incorporate cascaded three-port networks having the third port terminated. It is only necessary to evaluate equations (7.29) and convert the scattering parameters into ABCD parameters.

7.2.2. The Bilinear Theorem.

According to Penfield et al. (1970), the bilinear theorem states that any Z, Y, or S response of a linear network is a bilinear function of any network branch impedance, admittance, or scattering parameter. The response and branch types can be mixed. This is evident from the preceding three-port to two-port conversion results, as will now be shown.

Problem 4.11 asked for the input reflection coefficient in terms of the two-port scattering parameters and an arbitrary load (port-2) reflection coefficient. Figure 4.16 might represent such a network. As in the analysis leading to (3.101) for input impedance, the input reflection parameter is

$$S'_{11} = S_{11} + \frac{S_{12}S_{21}}{1/\Gamma_2 - S_{22}} = \frac{\Gamma_2(-\Delta) + (+S_{11})}{\Gamma_2(-S_{22}) + 1}, \quad (7.32)$$

where the two-port scattering parameter matrix determinant is

$$\Delta = S_{11}S_{22} - S_{12}S_{21}. \quad (7.33)$$

The bilinear theorem can be confirmed by comparison of (7.32) and equations (7.29). Each of the latter have the same bilinear form as seen in the former, at least for reflection parameters. But it is well known that a bilinear function of a bilinear function is itself bilinear. For example, suppose that output port-2 reflection coefficient is

$$\Gamma_2 = \frac{z_2 - 1}{z_2 + 1}, \quad (7.34)$$

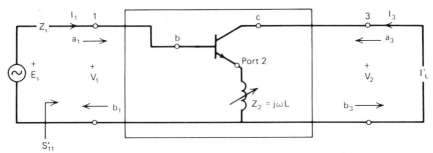

Figure 7.4. Grounded-emitter transistor two-port network.

where normalized $z_2 = Z_2/R_2$. Substituting this into (7.32) yields

$$S'_{11} = \frac{a_1 z_2 + a_2}{a_3 z_2 + 1},\qquad(7.35)$$

where the bilinear coefficients are

$$a_1 = \frac{S_{11} - \Delta}{1 + S_{22}}, \qquad a_2 = \frac{S_{11} + \Delta}{1 + S_{22}}, \qquad a_3 = \frac{1 - S_{22}}{1 + S_{22}}.\qquad(7.36)$$

Thus the bilinear function (7.32) of the bilinear function (7.34) is shown to be bilinear, as in (7.35) with coefficients (7.36).

Consider the transistor in Figure 7.4. The emitter inductor is unavoidable at high frequencies. The reverse transducer gain s_{13} was given in (7.31) when emitter impedance $Z_2 = 200 + j0$; it was $S_{13} = 0.3847 \,\underline{/63.03°}$. For perfect neutralization, $S_{13} = 0$. Setting $S_{13} = 0$ in the (7.29) expression yields the required port-2 termination:

$$\Gamma_2 = \frac{s_{13}}{s_{13}s_{22} - s_{12}s_{23}}.\qquad(7.37)$$

Example 7.3. Use the three-port scattering parameters in (7.30) for the transistor in Figure 7.4. Assume a 50-ohm port normalization. Evaluating (7.37), perfect neutralization occurs when $\Gamma_2 = 1.3055 \,\underline{/164.94°}$. Since the required magnitude is greater than unity, $Re(Z_2)$ would be negative. Therefore, consider setting Γ_2 equal to $1.0 \,\underline{/164.94°}$; using the RTN, R/S feature in Program A7-2 converts Γ_2 to an inductor with a reactance of $+j6.6093$ ohms. Then keys A and D show that $S_{13} = 0.0450 \,\underline{/98.46°}$. Thus a 6.6-ohm inductive emitter reactance produces much better neutralization than does a 200-ohm resistance ($|S_{13}| = 0.3847$) or a 50-ohm resistance ($|S_{13}| = 0.236$).

7.2.3. Mapping. Mapping is the most important single concept and tool in Chapter Seven. The classical analysis of bilinear functions according to Churchill (1960, p. 74) is to express (7.1) in the form of (2.2) in order to show that the bilinear functions amount to linear transformations and inversions. Linear transformations such as in the denominator of (7.1) do not change the shape of curves in the Z plane. Churchill (1960, p. 69) shows that inversions

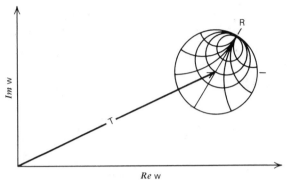

Figure 7.5.   The bilinear transformation $w = T + R\rho$.

$(Y = 1/Z)$ always map circles and lines into circles and lines. For bilinear transformations in linear networks, there is a much more useful decomposition of the standard bilinear form in (7.1); this is

$$w = T + R\frac{Z - Z_c}{Z + Z_c^*}. \tag{7.38}$$

In (7.38), T locates the center of the branch-image circle, the magnitude of R scales its size, and the angle of R determines its rotation with respect to the w-plane coordinate system. The branch-image Smith chart has a complex normalizing impedance (Z_c), as explained in Section 7.1.2. This is illustrated in Figure 7.5.

The significance of (7.38) and Figure 7.5 stems from the bilinear theorem in Section 7.2.2. The small, rotated Smith chart represents the entire right-half plane of any linear network branch impedance, admittance, or scattering parameter. The w plane represents the network's response function, expressed as a scattering, impedance, or admittance parameter. So the w plane might be the Z_{in} plane, the Y_{in} plane, the S_{12} plane, etc. The small branch-image circle may or may not fall within the w-plane unit circle, should that be a scattering parameter and therefore relevant.

It is not difficult to find expressions for the complex constants T, R, and Z_c. These relationships are obtained by putting (7.38) into the form of (7.1) and comparing the coefficients of Z. Thus (7.38) becomes

$$w = \frac{Z[(T+R)/Z_c^*] + [T - R(Z_c/Z_c^*)]}{Z(1/Z_c^*) + 1}. \tag{7.39}$$

It is seen that

$$Z_c = \frac{1}{a_3^*}, \tag{7.40}$$

and

$$R = \frac{a_1}{a_3} - T. \tag{7.41}$$

The right-hand constant expression in the numerator of (7.39) can be equated to a_2; then substitution of (7.41) yields

$$T = \frac{a_2 a_3^* + a_1}{2\,Re\,a_3} = \frac{a_2 a_3^* + a_1}{a_3 + a_3^*}. \tag{7.42}$$

These results will be used to develop many important relationships that would be difficult to formulate otherwise. They are also useful computationally, as will now be shown.

Program A7-3 in Appendix A evaluates the bilinear coefficients in (7.35), using (7.36) and the given two-port scattering parameters. It also continues by evaluating the mapping coefficients in (7.40)–(7.42); this is accomplished by steps 081–180. In addition to the polar, complex four functions (from Program A2-1) on keys B, C, and D, key E computes the generalized reflection coefficient (7.16). The result will be the Smith chart image of a normalized load impedance plane as it appears in the input reflection plane S'_{11}.

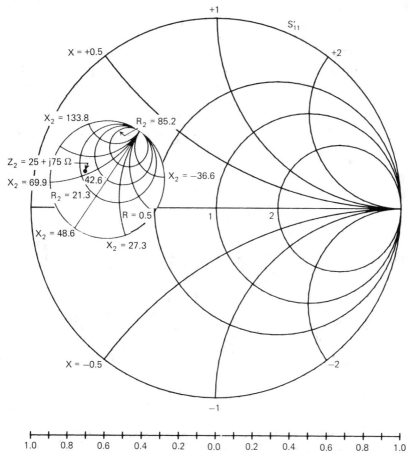

Figure 7.6. A map of the normalized load plane into the S'_{11} plane for transistor Example 7.4.

Example 7.4. Suppose that a transistor's two-port scattering parameters normalized to 50 ohms are

$$S = \begin{pmatrix} 0.47 \ \underline{/161°} & 0.101 \ \underline{/54°} \\ 2.37 \ \underline{/57°} & 0.47 \ \underline{/-71°} \end{pmatrix}. \tag{7.43}$$

Key a in Program A7-3 solicits the input of these values. Pressing key A then computes $T = 0.6070 \ \underline{/165.89°}$, $R = 0.3072 \ \underline{/55.63°}$, and $Z_c = 0.8516 + j0.9715$, which is normalized to 50 ohms. (The bilinear coefficients may be recovered from registers 7–9 using primary and secondary pairing: $a_1 = 0.4471 \ \underline{/184.72°}$, $a_2 = 0.3147 \ \underline{/178.32°}$, and $a_3 = 0.7740 \ \underline{/48.76°}$.) The load impedance point $Z_L = 25 + j75$ ohms in the Smith chart–image circle can be converted to generalized, polar reflection coefficient form using key E: fill the stack with 0.9715, 0.8516, 1.5, and 0.5. The last two values correspond to the stated load reactance and resistance normalized to 50 ohms. Key E then produces $\rho = 0.4374 \ \underline{/102.28°}$. These results are shown in Figure 7.6. The values of T, R, and ρ with respect to Z_c for the given normalized z_L are evident in relation to the scale and angles involved. Note that the input resistance is positive for all possible load values. (Why?)

Example 7.5. Consider the equivalent two-port scattering parameters obtained in Example 7.2 by reduction of the three-port network in Figure 7.3 with port-2 termination $Z_2 = 200 + j0$ ohms. The two-port parameters in (7.31) can be input into Program A7-3. The port-3 to port-1 mapping coefficients are $T = 2.1024 \ \underline{/-128.54°}$, $R = 2.4012 \ \underline{/107.86°}$, and $Z_c = 3.5755 + j4.1174$ (still normalized to 50 ohms as formulated). The input reflection plane S'_{11} is shown

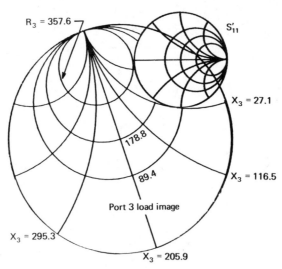

Figure 7.7. A map of the normalized load plane into the S'_{11} plane for the three-port to two-port data from Examples 7.2 and 7.5.

in Figure 7.7; the smaller Smith chart is the S'_{11} unit circle and the larger one is the port-3 load-plane image. It is clear that there is a large region of the Z_3 load plane that causes negative port-1 resistance. This situation is a function of the port-2 termination, which is $Z_2 = 200 + j0$ ohms in this case.

Example 7.6. Consider the pi network shown in Figure 7.8. The capacitive reactances at the frequency of interest are $-j50$ ohms. The bilinear coefficients, related to the scattering transfer function S_{21} as a function of the middle-branch impedance Z_i, are easily found using $Z_i = 0$, $1E9 + j0$, and $1 + j0$ ohms and calculating the corresponding S_{21} values by Program B4-1. The S_{21} values are $0.707107 \underline{/-45.000010°}$, $5E-8 \underline{/-90.000018°}$, and $0.700071 \underline{/-45.567277°}$, respectively. Then (7.3)–(7.5) yield the bilinear coefficients $a_1 = 7.07E-10 \underline{/-45.0026°}$, $a_2 = 0.7071 \underline{/-45.0°}$, and $a_3 = 0.0141 \underline{/44.9975°}$. Finally, (7.40)–(7.42) yield the mapping coefficients

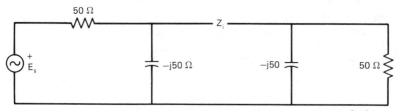

Figure 7.8. A pi network for mapping the Z_i plane into the S_{21} transducer function plane in Example 7.6.

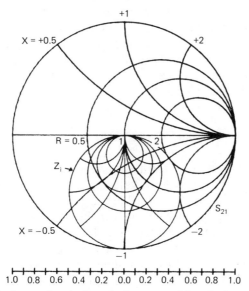

Figure 7.9. A pi-network series-branch plane mapped into the S_{21} response plane.

T = 0.4977 $\underline{/-89.7372°}$, R = 0.4977 $\underline{/90.2628°}$, and $Z_c = 49.9932 + j49.53670$ ohms. These results are shown in Figure 7.9. An application might be tuning the network by varying a reactance in the series branch.

7.2.4. Summary of Impedance Mapping. The linear, three-port network was characterized by its scattering parameters; a set of equivalent two-port parameters was then obtained. Besides applications in ladder analysis, the results clearly show the bilinear effect of any network branch on network response. This is true because any two-port network branch may be "brought out" as a third port. Hand-held computer Program A7-2 was provided to make the three-port to two-port reduction calculations.

Although the three-port to two-port reduction showed the bilinear effect of branch scattering parameters on scattering responses, it was necessary to show that bilinear functions of bilinear functions are bilinear; this was illustrated. The bilinear theorem was thus proved. This theorem states that every Z, Y, or S response of a linear network is a bilinear function of any branch impedance, admittance, or scattering parameter, in any mixed association. A neutralization example was worked. The bilinear theorem has a lot to do with feedback analysis, especially when applied in conjunction with mapping. For example, transistor shunt feedback is easily analyzed by this technique.

It was noted that the standard bilinear form may be decomposed in several different ways. For network behavior, it is especially useful to mold it into a form having a complex constant for translation added to an orientation factor that multiplies the generalized Smith chart function. In this way, the effect of all possible values of a branch impedance on a network response may be visualized. Since bilinear transformations map circles and lines into circles and lines in mixed association, certain critical branch loci can be visualized in the response plane for subsequent analysis. Hand-held computer Program A7-3 was provided to convert two-port scattering parameters into bilinear coefficients that relate normalized load impedance to input reflection. In addition, the mapping coefficients were calculated. Three examples of this technique were provided, and Smith charts illustrated the results.

7.3. Two-Port Impedance and Power Models

The development in this section will be in terms of admittance parameters. One reason for this is that a recently defined power gain is developed in these terms. Impedance could have been used just as readily for the general aspects; in fact, most equations in this section can be expressed using impedance parameters by simply replacing all the y's and Y's by z's and Z's, respectively, and exchanging V's and I's. Ironically, there has been a great emphasis on scattering parameter relationships, and almost all recent design aids involve these parameters. However, many crucial concepts are more readily seen in the impedance or admittance planes; the conversion of expressions using these parameters to those using scattering parameters in no way changes the phenomena. Therefore, this section will utilize admittance parameters, and Figure 7.1 will apply. Section 7.4 will utilize scattering parameters.

The mapping concept plays a critical role in avoiding a tangle of complex algebra that can only obscure significant results. Its embodiment of the generalized Smith chart is important, because the normalized power delivered by a complex source or at the output terminals of any linear two-port network happens to be a parabola of revolution (paraboloid) having the Smith chart as its base. When that Smith chart is mapped into the input plane, the inclined plane that represents input power intersects with the paraboloid of output power. Then efficiency (output power divided by input power) is easy to visualize, as is the point of maximum efficiency, where the plane is tangent to the paraboloid. Thus impedance and power relationships that are far from obvious may be visualized easily.

7.3.1. Output Power Paraboloid.

Power transfer from a complex source to a complex load was considered in Section 3.2.3. The load power, normalized to the source power available, was expressed in (3.47):

$$\frac{P_L}{P_{as}} = 1 - |\alpha|^2. \tag{7.44}$$

This has the form of a parabola,

$$y = 1 - x^2, \tag{7.45}$$

where x is the radius corresponding to a constant reflection magnitude, i.e., constant normalized output power.

To extend these results to the output port of a linear two-port network, consider the model in Figure 7.10; it is consistent with the defining admittance parameter equations, (3.79) and (3.80), in Section 3.4.3. For constant $V_1 = 1$, an equivalent Norton source at output port 2-2' has the available power

$$P_{a0} = \frac{|y_{21}|^2}{4g_{22}}, \tag{7.46}$$

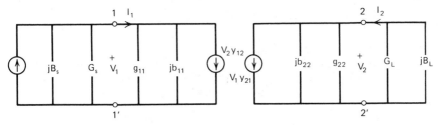

Figure 7.10. An admittance parameter model consistent with defining equations.

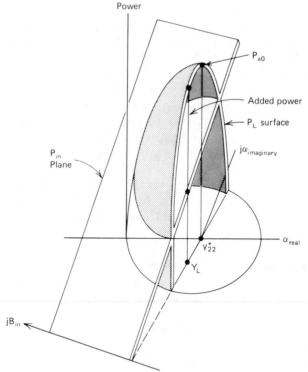

Figure 7.11. Output power paraboloid over a y_{22}^* Smith chart when $V_1 = 1$.

which clearly occurs when $Y_L = y_{22}^*$. The nomenclature

$$y_{ij} = g_{ij} + jb_{ij} \quad \forall \, i \text{ and } j \text{ (where } \forall \text{ means "for all")} \tag{7.47}$$

will be employed consistently. The pertinent load reflection coefficient, according to (7.17), will be

$$\alpha_{22} = \frac{y_{22}^* - Y_L}{y_{22} + Y_L}. \tag{7.48}$$

Thus, for $V_1 = 1$ volt, the Linvill and Gibbons (1961) geometric model of normalized output power is the paraboloid shown in Figure 7.11. The P_{in} plane will be derived in the next section.

7.3.2. Input Admittance Plane. Section 7.2.3 showed how the load plane could be mapped into the input plane. The development here is similar to that for the scattering parameters in Section 7.2.2. It begins by writing the input admittance as a function of load admittance, analogous to (3.101) for impedances:

$$Y_{in} = y_{11} - \frac{y_{21}y_{12}}{y_{22} + Y_L}. \tag{7.49}$$

This may be put in the bilinear form

$$Y_{in} = \frac{a_1 Y_L + a_2}{a_3 Y_L + 1},$$ (7.50)

where the coefficients are

$$a_1 = \frac{y_{11}}{y_{22}}, \qquad a_2 = \frac{\Delta}{y_{22}}, \qquad a_3 = \frac{1}{y_{22}}; \qquad \Delta = y_{11} y_{22} - y_{21} y_{12}.$$ (7.51)

Mapping takes a form similar to (7.38):

$$Y_{in} = T + R \frac{Y_c - Y_L}{Y_c^* + Y_L}.$$ (7.52)

The mapping coefficients are again obtained from (7.40)–(7.42), except that the sign of R in (7.41) must be reversed because of the reversed positions of Z_c and Y_c in the Smith chart definition. The results are

$$T = \frac{2g_{22} y_{11} - y_{21} y_{12}}{2g_{22}},$$ (7.53)

$$R = \frac{-y_{21} y_{12}}{2g_{22}} = r \, exp\left[j(\pi + \theta_{21} + \theta_{12}) \right],$$ (7.54)

$$Y_c = y_{22}^*.$$ (7.55)

The θ_{ij} in (7.54) are the angles of y_{ij}. The input admittance when $Y_L = y_{22}^*$ is identical to the Cartesian coordinates of the T vector. From (7.49) or (7.53),

$$Y_{in}(y_{22}^*) = K|R| - jB_{Ms},$$ (7.56)

where R is from (7.54). The stability factor K will be one of the most important constants in following developments:

$$K = \frac{2g_{11} g_{22} - Re(y_{21} y_{12})}{|y_{21} y_{12}|}.$$ (7.57)

The negative input susceptance is also defined for later use:

$$B_{Ms} = -b_{11} + \frac{Im(y_{21} y_{12})}{2g_{22}}.$$ (7.58)

The input admittance plane geometry is completed by recognizing the input power inclined plane, assuming $V_1 = 1$:

$$P_{in} = G_{in}.$$ (7.59)

The paraboloid and inclined plane in Figure 7.11 thus appear in the input admittance plane, as shown in Figures 7.12 and 7.13. The angles θ_{21} and θ_{12} are the arguments of y_{21} and y_{12}, respectively. It is easy to see from Figure 7.13 that the stability factor K is greater than 1; otherwise there would be a region in the Y_L and α_{22} planes where G_{in} would be less than zero, i.e., the two-port network would be unstable.

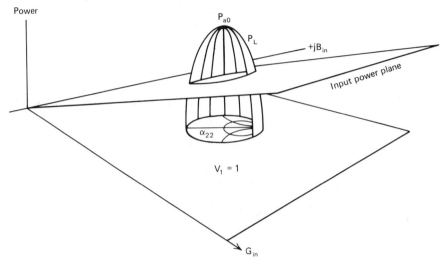

Figure 7.12. Two-port power surfaces over the input admittance plane when $V_1 = 1$.

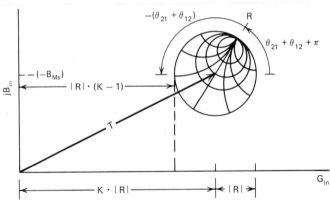

Figure 7.13. Input admittance plane showing the oriented Smith chart.

7.3.3. *Maximum Efficiency.*   For this section and for Section 7.3.5, it will be convenient to solve a minor geometry problem clearly related to the preceding development. Consider Figure 7.14. The difference h is the added power. Differentiating it with respect to x and equating the result to zero shows where h is maximum:

$$(x - b) = \frac{-0.5r^2}{P_{ao}}. \tag{7.60}$$

Intersections η_1 and η_2 in Figure 7.14 are points where the efficiency (power out divided by power in) is unity. Any other line through the origin is related by $P = \eta x$, where η may be less than or greater than unity. Given some value

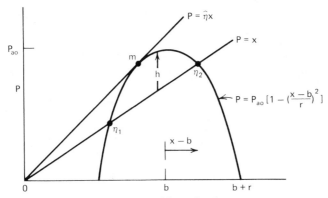

Figure 7.14. A geometry problem related to two-port power.

of η, the two intersections of such a line and the parabola are easily found using the quadratic formula

$$(x-b) = \frac{1}{2}\left[-\frac{\eta r^2}{P_{ao}} \pm \sqrt{\left(\frac{\eta r^2}{P_{ao}}\right)^2 - 4\left(b\frac{\eta r^2}{P_{ao}} - r^2\right)}\, \right]. \qquad (7.61)$$

The extreme value of η, $\hat{\eta}$ in Figure 7.14, is at the point of tangency shown. Note that the tangent point can exist only if $b > r$, as illustrated in Figure 7.14, or if $b < -r$ when the parabola is on the negative axis. This tangency occurs when the radical vanishes in (7.61). A second application of the quadratic formula yields

$$\hat{\eta} = 2P_{ao}\frac{b - \sqrt{b^2 - r^2}}{r^2}, \qquad (7.62)$$

and no solution exists if $b^2 < r^2$. The corresponding coordinate is

$$(\hat{x} - b) = -\left(b - \sqrt{b^2 - r^2}\right). \qquad (7.63)$$

The point of tangency between the plane and the parabola yields an expression for the maximum efficiency of a linear two-port network. Using (7.46) and (7.54) in (7.62) and Figures 7.13 and 7.14, the y-parameter maximum-efficiency expression is

$$\eta_{max} = \left|\frac{y_{21}}{y_{12}}\right|\left(K - \sqrt{K^2 - 1}\,\right). \qquad (7.64)$$

The stability factor K was defined in (7.57). The input admittance corresponding to the maximum efficiency may be found by using (7.63), (7.54), and Figure 7.13:

$$Y_{in}(\eta_{max}) = \frac{|y_{21}y_{12}|}{2g_{22}}\sqrt{K^2 - 1} - jB_{Ms}. \qquad (7.65)$$

The negative input susceptance B_{Ms} was defined in (7.58). Given a network characterized by short-circuit y or other parameters, it is important to realize that the efficiency is a function of only the load and not the source.

Example 7.7. Consider the transistor y parameters

$$y_{11} = 13.008E-3 \underline{/29.46°}, \qquad y_{12} = 1.4000E-3 \underline{/-61.26°},$$
$$y_{21} = 34.4637E-3 \underline{/-90.26°}, \qquad y_{22} = 5.0772E-3 \underline{/86.31°}. \tag{7.66}$$

What are the stability factor and maximum possible efficiency? From (7.57), $K = 1.03253$. From (7.64), $\eta_{max} = 19.088$, or a 12.81-dB gain. Since $K > 1$, the transistor is stable for all possible right-half-plane loads (see also Example 7.10 in Section 7.4.3).

7.3.4. Conjugate Terminations. Roberts (1946) developed the concept of conjugate-image impedances. This is the condition in which a linear two-port network is conjugately matched at both ports. The development is worthwhile, because it will be shown that the load impedance thus defined results in maximum efficiency. Then, if the generator impedance is selected as the conjugate of the corresponding input impedance, the maximum efficiency is also the maximum transducer gain, i.e., maximum P_L/P_{as}.

Referring to Figure 7.15, if

$$Y_{in} = Y_{Ms}^* \quad \text{and} \quad Y_q = Y_{ML}^*, \tag{7.67}$$

then (7.49) yields

$$y_{21}y_{12} = (y_{22} + Y_{ML})(y_{11} - Y_{Ms}^*). \tag{7.68}$$

Note that Y_q may be expressed in terms of the source admittance by using (7.49) with subscripts 1 and 2 interchanged. Thus

$$y_{21}y_{12} = (y_{11} + Y_{Ms})(y_{22} - Y_{ML}^*). \tag{7.69}$$

The last two expressions for $y_{21}y_{12}$ may be equated. The real parts yield

$$\frac{G_{Ms}}{g_{11}} = \frac{G_{ML}}{g_{22}} = \theta_r, \tag{7.70}$$

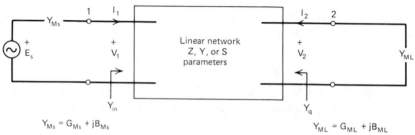

$Y_{Ms} = G_{Ms} + jB_{Ms}$

$Y_{ML} = G_{ML} + jB_{ML}$

Figure 7.15. Conjugate-image admittances $Y_{in} = Y_{Ms}^*$ and $Y_q = Y_{ML}^*$.

and the imaginary parts yield

$$\frac{b_{11} + B_{Ms}}{g_{11}} = \frac{b_{22} + B_{ML}}{g_{22}} = \theta_i. \tag{7.71}$$

Therefore, the terminations must be

$$Y_{Ms} = g_{11}\theta_r + j(-b_{11} + g_{11}\theta_i), \tag{7.72}$$

and

$$Y_{ML} = g_{22}\theta_r + j(-b_{22} + g_{22}\theta_i). \tag{7.73}$$

To find the θ_r and θ_i functions, substitute the last two equations into (7.69):

$$y_{21}y_{12} = g_{11}g_{22}\left[(1 - \theta_r^2 - \theta_i^2) + j2\theta_i\right]. \tag{7.74}$$

Clearly, the imaginary part of (7.74) is

$$\theta_i = \frac{Im(y_{21}y_{12})}{2g_{11}g_{22}}. \tag{7.75}$$

The real part of (7.74) yields

$$\theta_r^2 = 1 - \frac{Re(y_{21}y_{12})}{g_{11}g_{22}} - \frac{Im^2(y_{21}y_{12})}{(2g_{11}g_{22})^2}. \tag{7.76}$$

The following expression for θ_r can be shown to be equivalent to (7.76) by substituting the definition of K from (7.57). A concise expression for θ_r is

$$\theta_r = \frac{|y_{21}y_{12}|\sqrt{K^2 - 1}}{2g_{11}g_{22}}. \tag{7.77}$$

The conjugate-image admittances thus are

$$Y_{Ms} = g_{11}\theta_r + j\left[-b_{11} + \frac{Im(y_{21}y_{12})}{2g_{22}}\right] = g_{11}\theta_r + jB_{Ms}, \tag{7.78}$$

$$Y_{ML} = g_{22}\theta_r + j\left[-b_{22} + \frac{Im(y_{21}y_{12})}{2g_{11}}\right], \tag{7.79}$$

where θ_r and B_{Ms} are defined by (7.77) and (7.58), respectively. Now it is seen that (7.65) and (7.78) are conjugates. The conclusion is that the load admittance that enables a conjugate-image match is also the load admittance that causes the maximum possible efficiency. It is again stated that the maximum possible efficiency is independent of the actual source admittance. However, it is common practice to assume the source admittance Y_{Ms} so that the maximum possible efficiency is also the maximum possible transducer gain. Also, it is repeated that η_{max} has no meaning unless $|K| > 1$.

Finally, it is noted that loci of constant efficiency in the α_{22} plane are an eccentric family of circles. These may be visualized as the intersections of inclined planes and the paraboloid in Figures 7.11 and 7.14 projected onto the

α_{22} plane. However, a generalized reflection coefficient normalized to (7.79) will produce a concentric family of constant-efficiency circles when projected on the different map of the load admittance plane.

7.3.5. Maximum Added Power.

Kotzebue (1976, 1979) has described a method for maximizing the two-port added power for a fixed value of the input port independent variable, e.g., V_1. It has been observed that high-frequency, bipolar junction transistors tend to saturate at their input, while no such clipping is observable at their output. The assumption of constant V_1 in the preceding development enables the extension of the linear design approach to some nonlinear cases, where the so-called large-signal parameters are more appropriate. Kotzebue also argues in favor of the maximum added power approach for the common situation where $K < 1$, and the transistor is potentially unstable.

The added power when V_1 is constant is shown as h in Figure 7.14. The location of the maximum added power was given by (7.60). Figure 7.13, (7.46), and (7.54) enable the expression of the α_{22} reflection coefficient at the point of the maximum added power:

$$\alpha_{22} = \frac{y_{12}^*}{y_{21}}.$$ (7.80)

The corresponding power delivered to the load is available using (3.47) and (7.46):

$$P_L = \frac{|y_{21}|^2 - |y_{12}|^2}{4g_{22}}.$$ (7.81)

Solving (7.48) for Y_L and using (7.80), the load admittance that produces the maximum added power is

$$Y_L = -y_{22} + \frac{2g_{22}y_{21}}{y_{21} + y_{12}^*}.$$ (7.82)

It is interesting that as the reverse parameter $y_{12} \to 0$, $Y_L \to y_{22}^*$.

Kotzebue calls the efficiency when the added power is maximized the "maximally efficient gain." Its expression requires the input power, which is simply G_{in} for $V_1 = 1$ volt. The input admittance for Y_L in (7.82) is obtained by using (7.49):

$$Y_{in} = y_{11} - \frac{y_{21}y_{12} + |y_{12}|^2}{2g_{22}}.$$ (7.83)

Then its real part yields the input power

$$P_{in} = G_{in} = \frac{K|y_{21}y_{12}| - |y_{12}|^2}{2g_{22}},$$ (7.84)

where stability factor K is defined by (7.57). The maximally efficient gain is thus

$$G_{ME} = \frac{|y_{21}/y_{12}|^2 - 1}{2(K|y_{21}/y_{12}| - 1)}. \qquad (7.85)$$

Kotzebue makes the usual assumption that the source is chosen as the conjugate of the input impedance with the unique Y_L in (7.82) in place; then the efficiency is the same as the transducer gain. It is helpful to note that $|y_{21}/y_{12}| = |S_{21}/S_{12}|$. Since $K < 1$ precludes obtaining the maximum possible efficiency, note that the maximally efficient gain is finite for a nonzero denominator in (7.85). The condition for avoiding an infinite G_{ME} is that

$$K > \left| \frac{y_{12}}{y_{21}} \right| = \left| \frac{S_{12}}{S_{21}} \right|. \qquad (7.86)$$

Most practical transistors will satisfy (7.86), so that G_{ME} will be finite even when $K < 1$.

Program A7-4 for the HP-67/97 hand-held calculator computes the essential relationships previously given. It also computes the *overall* stability factor K′:

$$K' = \frac{2(G_s + g_{11})(G_L + g_{22}) - Re(y_{21}y_{12})}{|y_{21}y_{12}|}. \qquad (7.87)$$

Comparison with (7.57) shows that the overall stability factor takes into account the additional damping effect of the source and load conductances. These are seen at the ports in Figure 7.10. The overall stability factor is significant, because bounded source and load admittances may ensure stability for a transistor that otherwise might be unstable.

Example 7.8. Exercise Program A7-4 using the following short-circuit parameters taken from 4-GHz transistor data:

$$Y = \begin{pmatrix} 10.64E-3 \; \underline{/82.13°} & 0.8603E-3 \; \underline{/-88.68°} \\ 34.54E-3 \; \underline{/-16.68°} & 4.549E-3 \; \underline{/34.64°} \end{pmatrix}. \qquad (7.88)$$

These are input individually into Program A7-4 using key a in the manner described in Appendix A. In this case, the transistor is only conditionally stable, since $K = 0.6317$; $G_{ME} = 15.19$ dB, and the load and source reflection coefficients are $0.6869 \; \underline{/16.34°}$ and $0.8532 \; \underline{/71.92°}$, respectively. These reflection coefficients for the load and source correspond to $Y_L = 3.786 - j2.771$ and $Y_s = 2.415 - j14.399$ millimhos. With these terminations in place, the overall stability factor (7.87) is $K' = 2.2237$. A model of the transistor employing the conjugate terminations is useful for developing matching networks. This is shown in Figure 7.16. The conjugate of the source impedance is $11.35 - j67.68$ ohms; the model's series input representation is based on the impedance at 4 GHz. Similarly, the conjugate of the load impedance is $264.13 \| -j360.95$

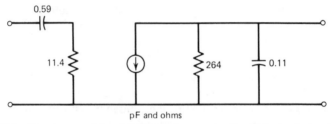

pF and ohms

Figure 7.16. Transistor model at 4 GHz based on maximally efficient gain terminations.

parallel ohms, as shown. Kotzebue argues that this model is more accurate than a model obtained only from S_{11} and S_{22} for the input and output branches, respectively. (See also another example of maximally efficient gain included in Example 7.11 and Figure 7.21.)

7.3.6. *Summary of Two-Port Impedance and Power Models.* A remarkably compact overview of geometric models for linear network behavior is available by the use of mapping concepts. The generalized Smith chart was shown to constitute the base of a parabola of revolution that is the analog of the linear two-port output power normalized to the power available at the output port. The y-parameter model of a two-port network that utilizes two controlled current sources (Figure 7.10) has available output power that is a function of V_1 and two y parameters. Thus it is easy to visualize how the output power per unit V_1 behaves with respect to any load admittance and its corresponding generalized reflection coefficient.

The mapping concept enables the location, orientation, and scaling of the generalized Smith chart in the two-port network's input admittance plane. Location of the load reflection disk completely inside the positive G_{in} plane ensures stability; otherwise, all or part of the possible load admittance region might cause negative input conductance. The stability factor was defined and used as a valuable yardstick to indicate the amount of stability margin in various cases, i.e., $K > 1$. Because input power per unit V_1 is just a plane inclined in the G_{in} direction, there is an easily visualized relationship between load and input impedances and efficiency.

The maximum possible efficiency and the maximum added power conditions were examined. The unique load admittances and corresponding input admittances were obtained for each case. Standard practice specifies a conjugate match at the input port so that efficiency and transducer gain are identical. In fact, a separate development showed that simultaneous conjugate matches at both ports required exactly the load admittance that produces maximum efficiency. In all situations, it is important to remember that efficiency is not a function of source impedance.

Kotzebue's maximally efficient gain was defined as the efficiency when the added power is maximized. It can be interpreted as maximizing the two-port activity. It was also shown to be finite in most cases where the stability is only conditional, i.e., when $K < 1$. An example was worked using Program A7-4 in Appendix A. It calculated the stability factor, maximally efficient gain, both terminations, and the overall stability factor, which includes the damping conductances of these terminations. A transistor model for the input and output branches was derived from the conjugate termination immittances. This provides a starting point for matching network problems.

Developments in this section were written in terms of admittances, although the same development in impedance terms essentially requires only a change in labels. Many of the concepts introduced are more easily visualized in these parameters than in scattering parameters. Ironically, most recent computer design aids are based on scattering parameters because this is the most effective way to characterize physical systems accurately. The next section develops very flexible gain relationships in terms of scattering parameters. All of these concepts are valid in any standard set of characterizing parameters, and conversion from one set to another is simply a matter of running existing short computer programs.

7.4. Bilateral Scattering Stability and Gain

The scattering parameters introduced in Section 4.5.2 are easier to measure than other network-characterizing sets, e.g., open-circuit z and short-circuit y parameters. It is customary to normalize measured scattering parameters to $50 + j0$ ohms. The measuring process then requires port termination by 50-ohm resistors, which can be obtained with considerable accuracy at even very high frequency. Conversely, it is extremely difficult to obtain an open or short circuit at high frequencies. Therefore, the network will now be viewed as illustrated in Figure 7.17.

It will be essential to renormalize the scattering parameter set from one normalizing impedance at each port to other values. This will usually amount to changing a port's normalizing impedance Z_i from $Z_i = 50 + j0$ ohms to some new value, Z_i'. The original set will be designated **S**, and the new set **S'**. The reflection coefficients of the new terminating impedances with respect to the original ones will be r_1 and r_2, as shown in Figure 7.17. This renormalization will provide comprehensive expressions for network behavior between complex sources and complex loads.

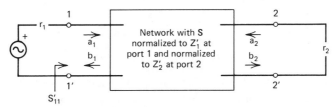

Figure 7.17. A two-port network with scattering notation.

This section begins with the general relationship for converting from one set of port normalizing impedances to another set. This will provide the basis for almost all further development. Then stability will be restated in scattering parameter terminology. The stability factor is equivalent to that previously formulated in y parameters; here, stability circles will be located on input and output reflection planes.

Several gain expressions previously defined will be put in scattering terminology, and several more gains will be defined. In particular, loci of arbitrary gain values will be defined for use when stability is conditional or unconditional.

A number of examples will be worked. However, no programs are provided for this section. They are so common for every major personal computer, desktop and hand-held, that there is little reason for duplication. The emphasis here will be on the origin of the expressions that are commonly employed in current scattering analysis of linear networks. Applications include oscillator as well as amplifier design. Since there are a number of recurring complex constants required throughout this development, they are collected in Appendix E. The equations will be referenced as (E.xx), where xx is the equation number.

7.4.1. *Changing S-Parameter Port Normalization.*   Referring to Figure 7.17, suppose that the scattering matrix S is defined for port 1, normalized to Z_1, and port 2, normalized to Z_2. Port normalization was described in Section 4.5.2. Even though the Z_i constants may be complex, they are usually $50+j0$ ohms, as obtained by automatic measuring equipment, and the following is easier to understand in this context. Suppose that the port normalizing impedances are to be changed from Z_i to Z_i', resulting in the new scattering matrix S'. The new scattering matrix has been related to the original one by Kurokawa (1965); his derivation is too general for the present discussion. The transformation is

$$S' = A^{-1}(S - R^*)(U - RS)^{-1}A^*. \qquad (7.89)$$

Matrices A, R, and U (the unit matrix) are diagonal, and the * superscript is the conjugate operator. The diagonal elements in matrix R are

$$r_i = \frac{Z_i' - Z_i}{Z_i' + Z_i^*}, \qquad (7.90)$$

and the diagonal elements in matrix A are

$$A_i = \frac{1 - r_i^*}{|1 - r_i|}\sqrt{1 - |r_i|^2}, \qquad (7.91)$$

where $i = 1, 2$. Clearly, (7.90) defines the port reflection coefficient r_i of the new normalizing constant, commonly with respect to 50 ohms. Thus r_1 and r_2 will be called the source and load coefficients, respectively.

Although (7.89) is valid for any number of ports, this development will apply only to two-port networks. It is not difficult to use (7.90) and (7.91) in (7.89) to obtain the four expressions for the S'_{ij} parameters. They are:

$$S'_{11} = \frac{A_1^*}{A_1} \frac{\left[(1 - r_2 S_{22})(S_{11} - r_1^*) + r_2 S_{12} S_{21} \right]}{\left[(1 - r_1 S_{11})(1 - r_2 S_{22}) - r_1 r_2 S_{12} S_{21} \right]}, \tag{7.92}$$

$$S'_{12} = \frac{A_2^*}{A_1} \frac{S_{12}(1 - |r_1|^2)}{\left[(1 - r_1 S_{11})(1 - r_2 S_{22}) - r_1 r_2 S_{12} S_{21} \right]}, \tag{7.93}$$

$$S'_{21} = \frac{A_1^*}{A_2} \frac{S_{21}(1 - |r_2|^2)}{\left[(1 - r_1 S_{11})(1 - r_2 S_{22}) - r_1 r_2 S_{12} S_{21} \right]}, \tag{7.94}$$

$$S'_{22} = \frac{A_2^*}{A_2} \frac{\left[(1 - r_1 S_{11})(S_{22} - r_2^*) + r_1 S_{12} S_{21} \right]}{\left[(1 - r_1 S_{11})(1 - r_2 S_{22}) - r_1 r_2 S_{12} S_{21} \right]}. \tag{7.95}$$

These relationships have important analytical and computational applications. For example, if $r_1 = 0$, then (7.92) reduces to (7.32), the input reflection coefficient of a two-port network characterized by 50-ohm scattering parameters and terminated by an arbitrary reflection load referenced to 50 ohms. Also, if the network is operated between complex source Z'_1 and complex load Z'_2, then the forward transducer function is exactly (7.94). Referring to Figure 7.17, the forward power relationship is

$$G_T = |S'_{21}|^2 = \frac{P_L}{P_{as}} = \left| \frac{b_2}{a_1} \right|^2. \tag{7.96}$$

Load terminations that cause instability will cause $|S'_{11}|$ in (7.92) to exceed unity, and this boundary can be located by that relationship; a similar statement can be made for generator r_1 regions that effect the output port.

There is at least one application where normalizing constants Z_i would not be $50 + j0$ ohms. There are advantages to an analysis where Z_i in (7.90) would be the conjugate-image impedances. This usually requires two applications of (7.89), the first one being from a $50 + j0$ normalization to Z_{Ms} and Z_{ML}, as defined for admittances in Section 7.3.4. This will be discussed further in Section 7.4.3.

7.4.2. Stability. The load-image circle in the input admittance plane was described in Section 7.3.2, where it was easy to see that $G_{in} < 0$ was entirely possible, depending on the network's characterizing parameter values at that frequency. The geometry showed that the input power was negative in that region. Taking this to mean instability, a similar analysis will be obtained for scattering characterization. Woods (1976) has shown that a necessary and sufficient condition for unconditional stability is that $|S'_{11}| \leqslant 1$ when $|r_2| \leqslant 1$. His simple criterion for unconditional stability of a linear two-port device, such as a single-chip transistor, is

$$K > 1, \qquad |\Delta| < 1, \tag{7.97}$$

where stability factor K is defined by Equation (E.2), and Δ by Equation (E.1) in Appendix E. These criteria are not sufficient for cascaded amplifiers or active devices embedded in reactive networks, because there may be local instability.

When the network is not unconditionally stable, the regions in the termination planes can be located by imposing the unit reflection magnitude constraint at the opposite port. Consider the fixed source r_1 and locate the values of load r_2 where $|S'_{11}| = 1$. Note that r_1 is a generalized reflection coefficient, as described in Section 7.1.2. An important requirement is that the real part of the normalizing constant be strictly positive. There are no other requirements, so that it is possible to assume that r_1 is normalized to an arbitrary, positive-real generator impedance. Then $r_1 = 0$, and it will still be true that $|S'_{11}| = 1$ locates the r_2 stability boundary. This is a simplifying argument which supports the conclusion that the stability region in the r_2 plane is independent of r_1.

It has been mentioned that $r_1 = 0$ in (7.92) yields (7.32), where $\Gamma_2 = r_2$ in the present analysis. Solving this for r_2 yields

$$r_2 = \frac{S'_{11} - S_{11}}{S'_{11} S_{22} - \Delta}. \tag{7.98}$$

Now S'_{11} is within a unit circle if

$$S'_{11} = \frac{Z - 1}{Z + 1} \tag{7.99}$$

for a hypothetical Z, which is introduced so that impedance mapping may be employed. This is illustrated in Figure 7.18. Substituting (7.99) into (7.98),

$$r_2 = \frac{a_1 Z + a_2}{a_3 Z + 1}, \tag{7.100}$$

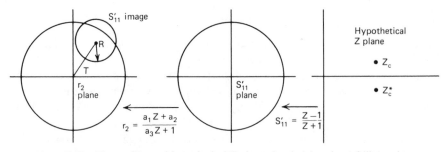

Figure 7.18. The r_2, S'_{11}, and hypothetical Z planes for deriving the stability region.

where the bilinear coefficients are

$$a_1 = \frac{S_{11} - 1}{\Delta + S_{22}}, \qquad a_2 = \frac{S_{11} + 1}{\Delta + S_{22}}, \qquad a_3 = \frac{\Delta - S_{22}}{\Delta + S_{22}}. \tag{7.101}$$

Now r_2 corresponds to w in the standard mapping form, (7.38) in Section 7.2.3. From (7.42),

$$T = \frac{(S_{11} + 1)(\Delta^* - S_{22}^*) + (S_{11} - 1)(\Delta^* + S_{22}^*)}{(\Delta^* + S_{22}^*)(\Delta - S_{22}) + (\Delta + S_{22})(\Delta^* - S_{22}^*)}. \tag{7.102}$$

Further complex algebra reduces this to

$$T = \frac{S_{22}^* - S_{11}\Delta^*}{|S_{22}|^2 - |\Delta|^2}. \tag{7.103}$$

Since only the magnitude of mapping coefficient R will be of interest, (7.41) will be rewritten in the form

$$\frac{a_3}{a_3^*} R = \frac{a_1 - a_2 a_3}{a_3 + a_3^*}. \tag{7.104}$$

Substitution of (7.101) yields

$$\frac{a_3}{a_3^*} R = \frac{|\Delta + S_{22}|^2}{(\Delta + S_{22})^2} \frac{(-S_{12}S_{21})}{|S_{22}|^2 - |\Delta|^2}. \tag{7.105}$$

The stability circles have been located in the r_2 plane, as illustrated in Figure 7.18. Their center is located by mapping coefficient T in (7.103); this is the complex constant expressed by Equation (E.16). The radius of the stability circle is the magnitude of (7.105); this is (E.18). An entirely similar analysis that locates the stability circle in the input (r_1) plane is based on assuming that $r_2 = 0$ and setting $|S_{22}'|$ equal to 1, using (7.95). The same result is obtained, except for interchanging subscripts. The location and radius are given by (E.15) and (E.17).

Bodway (1967) discussed the six possible locations for the stability circle in a port's reflection plane, as shown in Figure 7.19. In each case, the small Smith chart represents the port reflection plane r_i. For discussion, suppose that the stability circles are in the output plane and the Smith charts represent the r_2 output termination plane. Then their interior represents passive terminations having $R_L > 0$. In the left-hand cases (a–c), the port origin is not enclosed by the stability circle; therefore, the device is stable outside the stability circle, corresponding to positive input resistance. In the right-hand cases (d–f), the origin is enclosed by the stability circle; therefore, the device is stable inside the stability circle, especially for the positive R_{in} that results from $Z_L = 50 + j0$ ohms.

In the next section, device power efficiency (or gain) η will be defined in (7.108) as the ratio of power delivered to the load divided by the input power delivered by the source. Amplifiers are usually designed so that both load

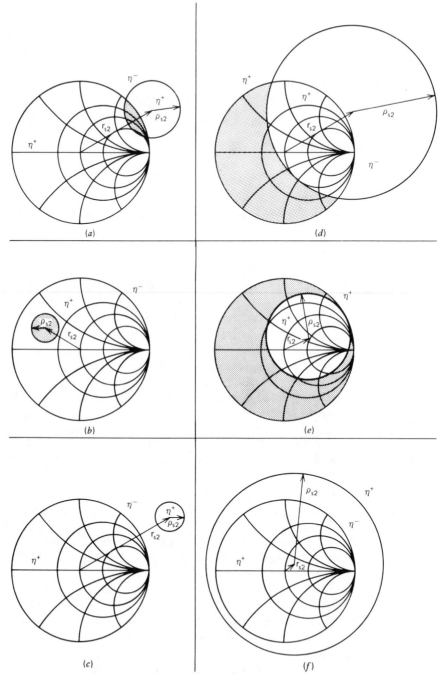

Figure 7.19. Six possible stability circle locations in the output port reflection plane. (*a*) Conditionally stable, $K < 1$; (*b*) conditionally stable, $K > 1$; (*c*) unconditionally stable, $K > 1$; (*d*) conditionally stable, $K < 1$; (*e*) conditionally stable, $K > 1$; (*f*) unconditionally stable, $K > 1$. *Left*: origins not enclosed; *right*: origins are enclosed. Shaded areas: η^-.

power and input power are positive, corresponding to positive R_L and positive R_{in}, respectively. Regions having positive or negative η are shown in Figure 7.19 by superscripts. Note that the region outside the Smith chart in Figure 7.19*b* represents power going into both input and output ports, since $R_L < 0$ and $R_{in} > 0$. Figure 7.19*c* represents a backward amplifier within the small stability circle, because both R_{in} and R_L are negative. The shaded areas represent loci where η is negative and $R_{in} < 0$ when $R_L > 0$; this does not occur in Figure 7.19*c, f*, where the device is unconditionally stable.

The Smith chart origins usually represent $50 + j0$ ohms; the system must be stable for 50-ohm terminations in order to be measurable. When the stability circle does not enclose the origin of the termination's Smith chart (r_i), it defines a region of negative-real network terminal impedance (Z_{in} or Z_q), as shown in Figure 7.19*a–c*. For example, Figure 7.19*c* shows that only certain negative-real load impedances could produce negative-real input impedances. Conversely, when the stability circle does enclose the origin of the termination's Smith chart (r_i), it defines a region of positive-real network terminal impedance (Z_{in} or Z_q), as shown in Figure 7.19*d–f*. For example, Figure 7.19*f* shows that any load with reflection magnitude just slightly greater than unity will cause negative-real input impedance. It is interesting to note that conjugate-image matching is always possible for unconditionally stable networks, but this may or may not be possible for conditionally stable networks.

Example 7.9. Suppose that the device scattering parameters have been measured on a 50-ohm system and found to be

$$S = \begin{bmatrix} 0.385 \ \underline{/-55°} & 0.045 \ \underline{/90°} \\ 2.7 \ \underline{/78°} & 0.89 \ \underline{/-26.5°} \end{bmatrix}. \qquad (7.106)$$

Using (E.2), stability factor $K = 0.909$, so that the stability circles are of interest. Using (E.15)–(E.18), the output plane stability circle is centered at $r_{s2} = 1.178 \ \underline{/29.88°}$, with radius $\rho_{s2} = 0.193$; the input plane stability circle is centered at $r_{s1} = 8.372 \ \underline{/-57.6°}$, with radius $\rho_{r1} = 9.271$. These circles are plotted in Figure 7.21, in the following section.

7.4.3. Bilateral Gains and Terminations. There are several useful gain expressions applicable to Figure 7.17 that are available from the renormalized scattering parameters in (7.92)–(7.95). Transducer gain G_T has been defined in (7.96). Some simplification of the magnitude of (7.94) is available:

$$G_T = |S'_{21}|^2 = \frac{|S_{21}|^2(1 - |r_1|^2)(1 - |r_2|^2)}{|1 - r_1 S_{11} - r_2 S_{22} + r_1 r_2 \Delta|^2}. \qquad (7.107)$$

Using (4.51), it is easy to show that the efficiency may be expressed as

$$\eta = \frac{|S'_{21}|^2}{1 - |S'_{11}|^2}. \qquad (7.108)$$

It can be shown that this reduces to

$$\eta = g_0 g_2 , \tag{7.109}$$

where g_0 in (E.14) is the maximum 50-ohm transducer gain, and g_2 is defined as:

$$g_2 = \frac{1 - |r_2|^2}{\left(1 - |S_{11}|^2\right) + |r_2|^2 D_2 - 2 \, Re(r_2 C_2)} . \tag{7.110}$$

The maximum possible efficiency from (7.64) is similarly expressed in scattering notation by (E.11), because

$$\left| \frac{y_{21}}{y_{12}} \right| = \left| \frac{S_{21}}{S_{12}} \right| . \tag{7.111}$$

The power available at output terminals 2-2' in Figure 7.17 relative to the power available from the source is the available power gain:

$$G_A = \frac{|S'_{21}|^2}{1 - |S'_{22}|^2} = g_0 g_1 , \tag{7.112}$$

where g_1 is defined as:

$$g_1 = \frac{1 - |r_1|^2}{\left(1 - |S_{22}|^2\right) + |r_1|^2 D_1 - 2 \, Re(r_1 C_1)} . \tag{7.113}$$

Conjugate-image matching occurs when the source and load reflection coefficients are given by (E.12) and (E.13). These may be derived by solving the pair of equations obtained by setting the magnitudes of (7.92) and (7.95) to zero. Another approach is to convert the conjugate-image admittance expressions in (7.78) and (7.79). Load reflection coefficient r_{ML} from (E.13) results in maximum efficiency, (E.11). Maximum gain is obtained when the source reflection coefficient is r_{Ms} from (E.12):

$$G_{max} = \eta_{max} \quad \text{when} \quad r_1 = r_{Ms} . \tag{7.114}$$

It was remarked in Section 7.3.4 that generalized reflection coefficients normalized to conjugate-image immittances map constant-efficiency loci onto concentric circles on the generalized Smith charts. It can be shown that equations (7.110) and (7.112) define the related eccentric family of constant g_2 and g_1 circles on the r_2 and r_1 reflection planes, respectively. Bodway (1967) gives the centers of such circles,

$$r_{0i} = \left(\frac{g_i}{1 + D_i g_i} \right) C_i^* , \tag{7.115}$$

with radius

$$\rho_{0i} = \frac{\left(1 - 2K|S_{12}S_{21}|g_i + |S_{12}S_{21}|^2 g_i\right)^{1/2}}{1 + D_i g_i} \tag{7.116}$$

for the r_i planes, where $i = 1, 2$. These are valid for $K < 1$ as well as for $K > 1$.

The value of g_i is always the desired gain divided by the maximum 50-ohm transducer gain $|S_{21}|^2$. In decibels, g_i is the gain with respect to gain g_0. For $g_i \to \infty$, location (7.115) and radius (7.116) approach the stability circle in the r_i plane.

Example 7.10. Suppose that the 50-ohm measured transistor data are:

$$S = \begin{bmatrix} 0.277 \ \underline{/-59°} & 0.078 \ \underline{/93°} \\ 1.92 \ \underline{/64°} & 0.848 \ \underline{/-31°} \end{bmatrix}. \tag{7.117}$$

This is equivalent to the y-parameter data in (7.66) (Example 7.7 in Section 7.3.3), where it was found that $K = 1.033$ and $\eta_{max} = 19.088$ dB. The maximum 50-ohm transducer gain is $g_0 = 3.686$, or 5.666 dB, according to (E.14). To locate the 10-dB efficiency circle in the r_2 reflection plane, calculate $g_2 = 10 - 5.666 = 4.334$ dB. Then (7.115) locates the center at 0.781 $\underline{/33.85°}$, and (7.116) fixes the radius at 0.214. This is shown in Figure 7.20. The 10-dB efficiency

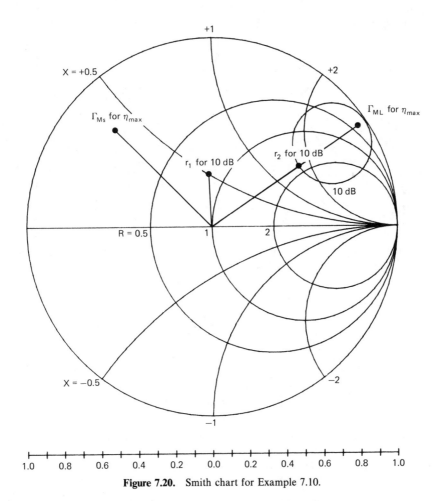

Figure 7.20. Smith chart for Example 7.10.

circle is shown along with r_{ML} and r_{Ms}, according to (E.13) and (E.12). Any load may be selected on the 10-dB circle. Choose the one nearest the center; it is 0.567 $\underline{/33.85°}$. Usually, it is the 10-dB transducer gain locus that is desired; therefore, the source must be the conjugate of the input reflection coefficient. From (7.32), the input reflection coefficient for the selected load is 0.276 $\underline{/-93.33°}$. The source reflection must be the conjugate of this value, and it is obtained by impedance matching, as discussed in Chapter Six. It is also plotted in Figure 7.20.

Example 7.11. Use the scattering data from (7.106) for the conditionally stable device considered in stability Example 7.9. Plot the 12-dB gain circle. The maximum 50-ohm transducer gain is $g_0 = 7.290$ by (E.14). Therefore, since 12 dB corresponds to 15.849, $g_2 = 15.849/7.290 = 2.174$. Then (7.115) locates

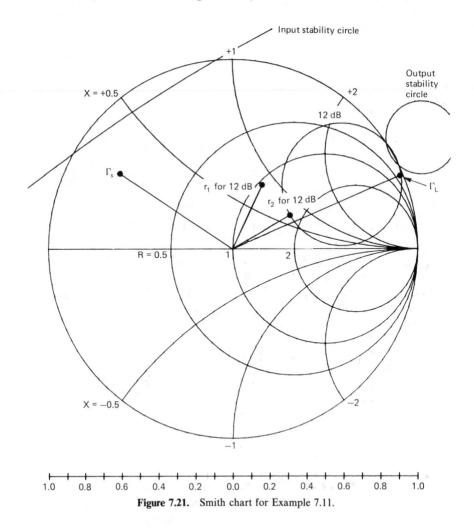

Figure 7.21. Smith chart for Example 7.11.

the 12-dB circle centered at 0.681 $\underline{/29.88°}$ with radius 0.324 from (7.116). This is shown in Figure 7.21. The stability circles from Example 7.9 in Section 7.4.2 are also plotted; it is necessary to select a load on the 12-dB efficiency circle that is in the stable region. Again, arbitrarily selecting the load nearest the center, 0.357 $\underline{/29.88°}$, the input reflection, by (7.32), is 0.373 $\underline{/-64.46°}$, and the source reflection must be the conjugate of this (see Figure 7.21) if the plotted circle is to represent forward transducer gain. Note that the source reflection is outside the input plane stability circle, so that stable operation is assured. Also, the maximally efficient gain source and load reflection coefficients (Γ_s and Γ_L) are plotted in Figure 7.21. For this device, $G_{ME} = 15.26$ dB, according to Section 7.3.5 and Program A7-4.

7.4.4. Summary of Scattering Stability and Gain. It was noted that numerous programs on readily available small computers exist for scattering stability and gain calculations. It is most important to know the territory. This section has explained the origins of the important scattering relationships for bilateral ($S_{12} \neq 0$) networks, which apply to most real devices.

Various significant gain expressions have been identified and several underlying assumptions have been noted. For example, the use of efficiency as forward transducer gain can occur only when the source is chosen to be the conjugate of the input impedance with a selected load in place. Conjugate-image reflection terminations were given, the load reflection enabling the maximum possible efficiency. It was remarked that concentric circles of constant efficiency may be plotted on generalized Smith charts normalized to the conjugate-image impedances; the interested reader is referred to Bodway (1967).

Arbitrary efficiency and gain loci turn out to be eccentric families of circles on the r_1 and r_2 termination Smith charts. Expressions for calculating the location of their centers and their radii were given along with several examples. Similar families for circles of maximally efficient gain (Section 7.3.5) are available in the work of Kotzebue (1976). The most general expression for forward transducer gain was given in (7.94); it may be used without any major assumptions.

7.5. Unilateral Scattering Gain

The bilateral gain equations in Section 7.4 are greatly simplified if it is assumed that the device is unilateral, i.e., $S_{12} = 0$. Also, stability can no longer be considered, because the stability factor, K in (E.2), is no longer defined. A single scalar estimate of validity for this assumption can be calculated to determine the relevance of unilateral analysis. Furthermore, the conceptualization of the entire design process is considerably simplified. However, many high-frequency transistors have substantial S_{12} reverse transducer gain. Readily available computers eliminate the advantage of algebraic simplicity. However, the easier visualization of design factors is still a decided advantage of unilateral design, when it is valid.

7.5.1. *Transducer Gain Simplification.* Setting S_{12} equal to zero in transducer gain G_T in (7.107) yields

$$G_{Tu} = g_s \, g_0 \, g_L, \tag{7.118}$$

where

$$g_s = \frac{1 - |r_1|^2}{|1 - r_1 S_{11}|^2} \tag{7.119}$$

and

$$g_L = \frac{1 - |r_2|^2}{|1 - r_2 S_{22}|^2}. \tag{7.120}$$

Recall that the maximum 50-ohm transducer gain g_0 is defined in (E.14). The maximum unilateral gain occurs when the denominators in g_s and g_L are minimal. For $|S_{ii}| < 1$, this occurs when $r_i = S_{ii}^*$, where $i = 1, 2$. Thus the maximum unilateral transducer gain is

$$G_{Tumax} = g_{smax} \, g_0 \, g_{Lmax}, \tag{7.121}$$

where g_{smax} and g_{Lmax} are defined in (E.5) and (E.6), respectively.

The block diagram depicting the factors in (7.118) and (7.121) will be considered in Section 7.5.3. However, it is appropriate to note that both g_{smax} and g_{Lmax} in (E.5) and (E.6) have the form of (3.49). Each expression constitutes a subtransducer block gain, since $|S_{11}| < 1$ and $|S_{22}| < 1$ were assumed.

7.5.2. *Unilateral Figure of Merit.* Before attempting to utilize the preceding approximation, it is feasible to define a validity factor, or figure of merit, for the unilateral assumption. Bodway (1967) compared the true value of transducer gain G_T and the unilateral approximation G_{Tu}. From (7.107) and (7.118)–(7.120), the gain expressions are related by

$$G_T = \frac{G_{Tu}}{|1 - x|^2}, \tag{7.122}$$

where

$$x = \frac{r_1 r_2 S_{12} S_{21}}{(1 - r_1 S_{11})(1 - r_2 S_{22})}. \tag{7.123}$$

The ratio of true to unilateral gain is bounded by

$$\frac{1}{|1 + |x||^2} < \frac{G_T}{G_{Tu}} < \frac{1}{|1 - |x||^2}. \tag{7.124}$$

The maximum value of $|x|$ when $|S_{ii}| < 1$ leads to the unilateral figure of merit u:

$$u = \frac{|S_{11}||S_{22}||S_{12}S_{21}|}{|1 - |S_{11}|^2| |1 - |S_{22}|^2|}. \qquad (7.125)$$

Therefore, the ratio of true to maximum unilateral transducer gain is bounded as follows:

$$\frac{1}{(1+u)^2} < \frac{G_T}{G_{Tumax}} < \frac{1}{(1-u)^2}. \qquad (7.126)$$

For example, if $u = 0.1$, the ratio of true to approximate transducer gain is bounded between 0.83 and 1.23. Clearly, u must be much less than 0.1 for the unilateral scattering analysis to produce valid results.

7.5.3. Unilateral Gain Circles.

A block diagram interpretation of unilateral gain (7.118)–(7.121) is useful. This is shown in Figure 7.22. The active device's 50-ohm maximum transducer gain (g_0) is the middle block. Once the device and its bias conditions are established, the middle block is invariant. The left-hand block corresponds to the g_s term in (7.118) and (7.119); it is the mismatch between the device's S_{11} and the source reflection coefficient r_1. The load block denotes a similar interpretation. Corresponding to developments leading to maximum unilateral transducer gain (7.121), conjugate matches at the device interfaces maximize power transfer, i.e., when $r_1 = S_{11}^*$ and $r_2 = S_{22}^*$.

Both the g_s and g_L blocks in Figure 7.22 may provide gain, even though they represent passive matching components. This is true because a mismatch (reflection) loss exists between r_1 and S_{11}; the matching network makes up some of this loss, and is a relative gain in this sense. The output network functions similarly.

Families of constant-gain circles may be obtained by setting g_s and g_L to fixed values and solving for r_1 and r_2 in (7.119) and (7.120), respectively. For input gain circles, the centers are located by

$$r_{u1} = \frac{\bar{g}_s |S_{11}|}{1 - |S_{11}|^2 (1 - \bar{g}_s)}, \qquad (7.127)$$

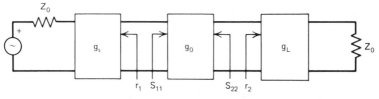

Figure 7.22. Unilateral gain block diagram according to (7.121).

with radius

$$\rho_{u1} = \frac{\sqrt{1-\bar{g}_s}\,(1-|S_{11}|^2)}{1-|S_{11}|^2(1-\bar{g}_s)}, \tag{7.128}$$

where $\bar{g}_s = g_s/g_{smax}$. Locations of output gain circles have similar forms and involve normalized g_L and S_{22}.

When $|S_{11}| > 1$, plot $[(S_{11})^{-1}]^*$ on a Smith chart, using dotted lines for the locus versus frequency. The values may be interpreted as explained in Section 7.1.2. The locations and radii of constant-gain circles are given by (7.127) and (7.128), except that $-\infty < g_s < 0$ and maximum gain g_{smax} is infinite at $r_1 = (S_{11})^{-1}$. Similar relationships apply in the output reflection plane.

Example 7.12. A transistor operated at 4 GHz has the following scattering parameters measured with respect to 50 ohms:

$$S = \begin{bmatrix} 0.51 \ \underline{/154°} & 0.09 \ \underline{/26°} \\ 1.4 \ \underline{/22°} & 0.60 \ \underline{/-58°} \end{bmatrix}. \tag{7.129}$$

Obtain the unilateral design parameters and plot families of input and output gain circles for gains of 0, 0.5, and 1 dB. By (7.125), the unilateral figure of

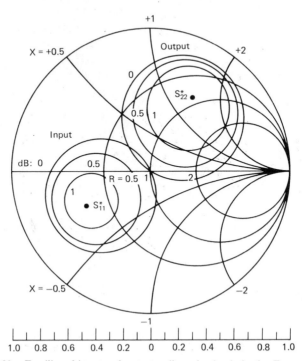

Figure 7.23. Families of input and output unilateral gain circles for Example 7.12.

merit is $u = 0.08$; by (7.126), the ratio of true to maximum unilateral gain is bounded between 0.86 and 1.18 (not dB). Input, device, and output gains by (E.5), (E.14), and (E.6) are 1.31, 2.92, and 1.94 dB, respectively. Therefore, the maximum unilateral transducer gain is 6.17 dB, obtainable when $r_1 = S_{11}^*$ and $r_2 = S_{22}^*$. According to (7.126), the actual transducer gain G_T is bounded between 5.51 and 6.89 dB. The families of circles are shown in Figure 7.23. Note that the 0-dB circles always pass through the chart center. Also, the centers of all input circles lie on the radial to S_{11}^*, and the same is true for output circles and S_{22}^*. The center of the input 0.5-dB circle is 0.44 from the Smith chart center, and its radius is 0.32.

7.5.4. *Summary of Unilateral Scattering Gain.*

Considerable numerical and conceptual simplification is available in unilateral amplifier design when the assumption is valid. The unilateral figure of merit helps make this judgment and provides bounds on the true value of transducer gain.

The optimal matching conditions are the source and load terminations of S_{11}^* and S_{22}^*, respectively. Once device bias has been established, its S_{21} gain term is an invariant factor in the unilateral gain formula. Two other factors represent input and output mismatch. Even less-than-optimal matches can still produce gain relative to the mismatch between 50 ohms and the input or output scattering parameters.

The trend toward readily available computers and maximum required performance reduces the attractiveness of the unilateral design technique. However, there are certain conceptual advantages to recommend it, at least as a first step in amplifier design.

Problems

7.1. Program coefficient equations (7.3)–(7.5) on a hand-held or desktop personal computer. Use the subroutines in Program A2-1 of Appendix A.

7.2. Verify the values of a_i, where $i = 1, 2, 3$, in Table 7.1.

7.3. Find the impedance corresponding to reflection coefficient $S_{11} = 2.2 \ \underline{/-153°}$ by plotting on a Smith chart and by computation. Assume that S_{11} is normalized to 1 ohm.

7.4. Consider the input impedance Z_{in} of the pi network in Figure 7.8. For $Z_i = 1 + j0$ ohms, $Z_{in} = 10.315823 - j20.242819$ ohms.
   (a) Find the bilinear coefficients of Z_{in} as a function of Z_i.
   (b) Use this bilinear relation to show that $Z_{in} = 50 + j0$ when $Z_i = j50$ ohms.

    (c) Find the impedance-mapping parameters T, R, and Z_c for Z_{in} as a function of Z_i.

    (d) Sketch the Z_i-image circle in the cartesian Z_{in} plane.

    (e) Label the resistance locus for $R_i = 25$ ohms and the loci for $X_i = 50$, 75, and 100 ohms, where $Z_i = R_i + jX_i$.

7.5. Find the value of Γ_2 that will make S_{33} equal zero in (7.29) for the 50-ohm S data in (7.30). Is it possible to have a $50 + j0$ Thevenin-generator output impedance at port 3 by terminating port 2 with a passive element and port 1 with a 50-ohm resistance? Why?

7.6. When $V_1 = 1$, $y_{22} = (4.549E-3) - j34.69$ and $y_{21} = (34.54E-3) - j16.68$ mhos. Find the power delivered to the load $Y_L = 0.01 - j0.01$ mhos.

7.7. Obtain the equivalent circuit model for open-circuit z parameters analogous to Figure 7.10. What is the maximum available power, P_{a0}, at the output terminals?

7.8. In Figure 7.6, what value of X_2 allows R_2 to vary over a wide range without causing R_1 to change very much? Why?

7.9. Verify Equation (7.51).

7.10. Show that $Re(T) = K \cdot |R|$.

7.11. Use Equations (7.62), (7.46) and (7.54) to verify the η_{max} expression in Equation (7.64).

7.12. Prove Equations (7.108) and (7.112); in the latter, note that P_{a0} and a_1 are independent (b_2 is dependent).

7.13. An HP GaAs FET transistor has the following S parameters at 12 GHz, measured with respect to 50 ohms: $S_{11} = 0.714 \; \underline{/-124°}$, $S_{12} = 0.073 \; \underline{/39°}$, $S_{21} = 1.112 \; \underline{/69°}$, and $S_{22} = 0.627 \; \underline{/-57°}$.

    (a) Evaluate Equations (E.1)–(E.20) in Appendix E; where appropriate, state whether your answer is numeric or in dB.

    (b) Is the device unconditionally stable? Why?

    (c) What load impedance (in ohms) produces the maximum possible efficiency?

    (d) Give three source impedances that can be used with this load impedance so that maximum possible efficiency will be obtained.

    (e) What is the efficiency when the added power is maximized?

7.14. Derive Equations (7.56) and (7.58).

Chapter Eight

Direct-Coupled Filters

Direct-coupled filters are the most common narrow-band networks in radio frequency engineering. They may be found in very low frequency through microwave applications, often in such different forms as to appear totally unrelated. A large amount of design information exists, the formal basis usually being attributed to Dishal (1949) and Cohn (1957). The design method presented here evolved over the last three decades and is based on the "loaded-Q" concept discussed in Section 6.1.3. It is unique to the extent that loaded Q's of internal resonators are treated as design parameters. It precisely accounts for midband impedance matching and dissipative loss. For passband widths of less than 20%, it provides an accurate estimate of stopband selectivity and an approximation of ideal passband response shapes. Unlike synthesis methods, the loaded-Q design technique enables adjustment of surplus parameters, so that bounds on practical component values may be accommodated.

Practical, direct-coupled filters evolve from an ideal prototype network, shown in Figure 8.1. It is a lumped-element representation, although several

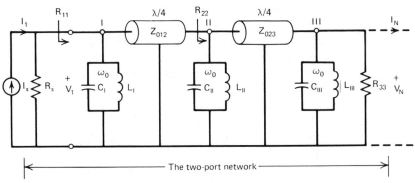

Figure 8.1. The direct-coupled-filter prototype network.

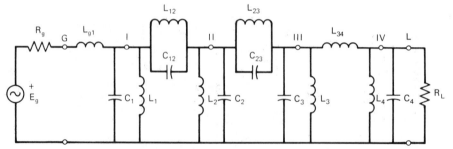

Figure 8.2.   A four-resonator filter with two traps and one L section.

transformations will be given for incorporating distributed (transmission line) elements. The parallel LC resonators (tank circuits) appearing between each network node and ground are coupled by the inverters shown in Figure 8.1. Ideal inverters are lossless, frequency-independent, 90-degree transmission lines. They are assumed to have characteristic impedances and electrical lengths that are frequency independent. Practical inverter networks usually have one of these ideal properties, and the other one is well behaved, with effects that are easily predicted. The entire prototype network impedance level may be elevated by transformers or L sections at one or both ends of the network, so that element values may be located within a suitable range. As far as prototype calculations are concerned, the source and load terminations are simply resistances R_{11} and R_{NN}, respectively.

An example of a practical, direct-coupled network appears in Figure 8.2. The four resonators connected between each node and ground are evident enough, but comparison with the prototype network in Figure 8.1 shows that the inverters have been realized in different ways. The two inverters, between nodes I and II and nodes II and III, are antiresonant "trap" subnetworks that cause zeros of transmission at the corresponding frequencies. Also, the source has been connected by an L section; it could have been connected directly or by a transformer.

The response of the network in Figure 8.2 could be that shown in Figure 8.3. The rigid limitations of classical network approximation and synthesis have been relaxed by accepting reasonable first-order approximations of ideal response shapes, which are both arbitrary and unobtainable using real elements. Therefore, fairly general selectivity specifications, as shown by the barriers in Figure 8.3, may be satisfied by direct-coupled filters, which have great flexibility in both form and component ranges.

This chapter begins with the definition of the prototype network and its main components: resonators and ideal inverters. The selectivity mechanism will be derived, and resonator loaded-Q and inverter impedance parameters will be identified. Next, inductive and capacitive inverters will be introduced, and their impedance-matching and selectivity effects will be identified. Approximate selectivity relationships will be developed so that interactions

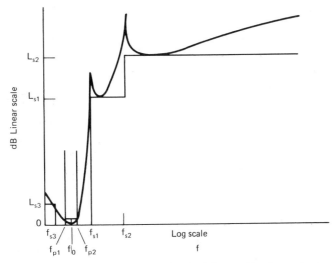

Figure 8.3. Typical direct-coupled-filter selectivity specification and response.

among the several design parameters will be obvious. A practical network will be designed to illustrate these fundamental concepts.

Detailed consideration will be given to practical inverters and resonators. Resonator developments include dissipation and the equivalence between lumped and distributed components. A general basis for analyzing any inverter subnetwork will be presented, and the trap inverter will be analyzed to show why it increases selectivity without increasing midband loss. End (terminal) coupling by both L sections and transformers provides a vital degree of design freedom for elevating filter impedance levels and thus relieving component restrictions; these techniques will be described.

The selectivity behavior in and near the passband is called the response shape and is completely determined by the distribution of loaded-Q values among the resonators in the (ideal) prototype network (Figure 8.1). Four unique shapes will be discussed in detail. The first three are related: Chebyshev (overcoupled), Butterworth (maximally flat), and Fano (undercoupled). The minimum-loss or equal-Q shape will also be considered, because it is practical, simple, and instructive. Useful formulas for predicting the required number of resonators for various passband and stopband selectivity specifications will be furnished. These are well known for all but the Fano (undercoupled) response; it has several valuable characteristics, including good pulse response.

Limitations of this approximate, direct-coupled design technique will be made explicit; it will be shown that they do not eliminate most practical applications. Certain readily available sensitivity relationships will be noted. Also, the well-known tuning procedure for these synchronous filters will be described. Finally, a detailed design procedure, based on a flowchart, will be defined. The required equations have been tabulated, and a particular design example will be related to the pertinent equations. The possibility of further design adjustment using optimization (Section 5.5.4) will be discussed.

8.1. Prototype Network

The lossless network in Figure 8.1 will be considered in this section. The admittance of the resonators appearing from each node to ground will be important for further analysis, and the conductance contributed to end resonators by the load—and perhaps the source—will be included. The impedance-transforming properties of the ideal inverter will be derived from lossless transmission line equations. These developments provide a basis for ABCD chain parameter analysis of the two-port prototype network. It will be shown that there is a recurrence pattern among the parameters as more inverters and resonators are added. Thus, quite general selectivity expressions are available without resorting to numerical analysis of each case.

8.1.1.   Prototype Resonators.   Each resonator in a prototype network, such as Figure 8.1, takes the form shown in Figure 8.4. For purposes of the loaded-Q definition, conductance G_{KK} represents the parallel resistance seen *toward the load*, as presented by that part of the network at the tank resonance frequency. All resonators will be resonant at the geometric midband frequency, and the impedance seen toward the load from each resonator will be real at this frequency. In the case of the N*th* (output) resonator, the parallel resistance is in fact the load resistance R_{NN} indicated in Figure 8.1. The resonator admittance in Figure 8.4 is

$$Y_K = G_{KK} + j\omega_0 C_K F, \qquad (8.1)$$

where

$$F = \frac{\omega}{\omega_0} - \frac{\omega_0}{\omega}, \qquad (8.2)$$

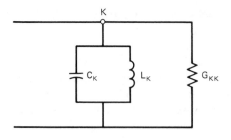

Figure 8.4. The K*th* shunt resonator (tank) with its real or hypothetical resistive load at the tune frequency.

and ω_0 is the midband geometric mean frequency according to (6.85) in Section 6.5.1. It is easy to obtain a similar expression for a series LC network and then to recognize the significance of the lowpass-to-bandpass transformation considered in (6.83). The importance of the bandpass frequency variable F in (8.2) cannot be overemphasized; it will occur in nearly every selectivity expression for direct-coupled filters.

The main parameter is the loaded Q of the K*th* resonator:

$$Q_{LK} = \frac{R_{KK}}{X_K},\qquad(8.3)$$

where $R_{KK} = 1/G_{KK}$. For 1 volt across the resonator in Figure 8.4, it is easy to see that Q is the reactive power divided by the real power. The reactive power is stored in the resonator, and the real power is that which proceeds toward the load as delivered to G_{KK}. The nodal parallel reactance X_K in (8.3) is determined at the midband geometric mean frequency ω_0:

$$X_K = \frac{1}{\omega_0 C_K} = \omega_0 L_K .\qquad(8.4)$$

These definitions are consistent with those in Section 6.1.3. This is singly terminated loaded Q; it does not consider any resistive loading—real or through the intervening circuit—that occurs on the source side of the resonator. Finally, (8.1) may be put in terms of the loaded Q:

$$Y_K = G_{KK}(1 + jQ_{LK}F).\qquad(8.5)$$

The ABCD chain parameters were defined in Section 4.2.1. The ABCD parameters for the K*th* resonator in Figure 8.4 are

$$T_K = \begin{pmatrix} A_K & B_K \\ C_K & D_K \end{pmatrix} = \begin{pmatrix} 1 & 0 \\ Y_K & 1 \end{pmatrix}.\qquad(8.6)$$

These will be used in conjunction with those of the inverters to obtain expressions for the overall ABCD matrix of the prototype network in Figure 8.1.

8.1.2. Ideal Inverters. The hypothetical lossless transmission line segments in Figure 8.1 are defined to have frequency-independent characteristic impedance Z_{0ij} and a constant quarter-wave length, where the ij subscripts denote the adjacent nodes that they connect. These are variously called impedance or admittance inverters because they invert impedances according to (6.38), which is repeated:

$$Z_{in} = \frac{Z_{0ij}^2}{Z_L} .\qquad(8.7)$$

This is shown in Figure 8.5.

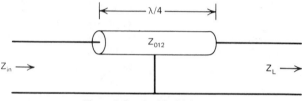

Figure 8.5. An ideal inverter.

It is easy to show why this impedance-inverting behavior occurs. Using the transmission line ABCD parameters from (4.13)–(4.15) for the lossless case yields

$$\mathbf{T}_{12} = \begin{pmatrix} 0 & jZ_{012} \\ jY_{012} & 0 \end{pmatrix},$$ (8.8)

where $Y_{0ij} = 1/Z_{0ij}$. Then bilinear function (4.18), for input impedance as a function of load impedance, yields (8.7). As previously indicated in connection with Figure 8.4, the resonator and inverter designs are established at the midband frequency, where the impedance terminating each resonator is real. Thus, at the midband design frequency ω_0, the inverter impedance is simply

$$Z_{0ij} = \sqrt{R_{ii}R_{jj}} \ ,$$ (8.9)

where the inverter connects the ith node and the jth $(i+1)$ node. This is illustrated for a two-resonator prototype network in Figure 8.6. Note that (8.9) is the means for selecting inverter impedance Z_{0ij}; however, (8.7) is still valid at any frequency, not just ω_0.

The impedance-inverting action of (8.7) causes a given circuit seen through an inverter to look like the dual of that circuit. In particular, the center resonator and the two adjacent inverters in Figure 8.1 have the same transmission characteristics as a series LC resonator. Thus the network in Figure 8.1 with ideal inverters has exactly the same transmission characteristics as the classical bandpass filter in Figure 6.31, assuming compatible choices of

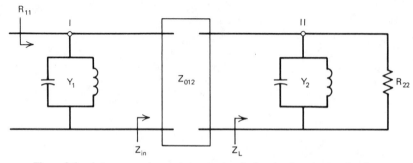

Figure 8.6. A two-resonator prototype network showing inverter terminations.

element values. This equivalence is further developed in the work of Matthaei et al. (1964, pp. 144–149); it will also be apparent from the development in the next section.

8.1.3. Prototype Network Selectivity.

According to (4.9), the ABCD (T) matrix of an entire ladder network can be obtained by multiplying the ABCD matrices of the component subnetworks in order. For example, (8.6) and (8.8) may be applied for the $N = 2$ (two-resonator) network in Figure 8.6. The result is

$$\mathbf{T} = \begin{bmatrix} jZ_{012}Y_2 & jZ_{012} \\ jY_1Z_{012}Y_2 + jY_{012} & jY_1Z_{012} \end{bmatrix}. \tag{8.10}$$

If another inverter matrix and resonator matrix are appended to (8.10), the **T** matrix for the $N = 3$ prototype network corresponding to Figure 8.1 may be obtained. Only the resulting C element is of interest:

$$C = -Y_1Y_2Y_3Z_{012}Z_{023} - Y_3Y_{012}Z_{023} - Y_1Z_{012}Y_{023}. \tag{8.11}$$

Similarly, the C element for $N = 4$ is

$$C = -Y_1Y_2Y_3Y_4Z_{012}Z_{023}Z_{034} - jY_1Y_2Z_{012}Z_{023}Y_{034}$$
$$- jY_1Y_4Z_{012}Y_{023}Z_{034} - jY_3Y_4Y_{012}Z_{023}Z_{034} - jY_{012}Z_{023}Y_{034}. \tag{8.12}$$

There are 52 terms in the C element of the **T** matrix for $N = 5$. The need for only the C element is explained next, followed by the identification of the transfer function's complex polynomial in frequency variable F, defined in (8.2).

Assume that the terminals of the overall two-port network are located as shown in Figure 8.1; i.e., the load resistance and the source shunt resistance (if it exists) are included inside the network as parts of the terminal resonators. Thus output current $I_N = 0$. Then the ABCD-defining equation (4.8) shows that $I_1/V_N = C$. The loss function of interest is

$$L(\omega) = \frac{V_N(\omega_0)}{V_N(\omega)}. \tag{8.13}$$

Since $I_1(\omega) = I_1(\omega_0)$, it follows that the desired loss function is equivalent to

$$L(\omega) = \frac{C(\omega)}{C(\omega_0)}. \tag{8.14}$$

Therefore, only the chain parameter C of an N-resonator filter is required for the loss function.

A general expression for the loss function may be deduced by considering the $N = 3$ case in (8.11). Assume that the source includes a nonzero conductance. From (8.1), there will be two resonator admittance terms that have both real and imaginary parts, namely Y_1 and Y_3. Resonator admittance $Y_2 = jF/X_2$. The first term in (8.11) will produce frequency variable jF with exponents 3 and 2. The second and third terms in (8.11) will produce frequency variable jF with exponents 1 and 0. When $\omega = \omega_0$, $F = 0$, and thus $C(\omega_0)$ is a real number. Clearly, the $N = 3$ loss function has the form

$$L(\omega) = \frac{(jF)^3 + U_2(jF)^2 + U_1(jF) + U_0}{U_0}. \tag{8.15}$$

This confirms that the prototype network in Figure 8.1 can produce the exact polynomial response function of the classical bandpass network in Figure 6.31, because (8.15) is the manifestation of the classical lowpass-to-bandpass mapping in (6.83).

The expressions for chain parameter C, such as (8.11) and (8.12), may be converted to the loaded-Q parameter using definitions (8.3), (8.4), and (8.9). For example, the (8.15) loss function for $N = 3$ is

$$L = \tfrac{1}{2}\Big[(jF)^3 Q_{L1}Q_{L2}Q_{L3} + (jF)^2(Q_{L2}Q_{L3} + Q_{L1}Q_{L2})$$
$$+ (jF)(Q_{L3} + Q_{L2} + Q_{L1}) + 2 \Big]. \tag{8.16}$$

Similarly, the $N = 2$ loss function is

$$L = \tfrac{1}{2}\Big[(jF)^2 Q_{L1}Q_{L2} + (jF)(Q_{L1} + Q_{L2}) + 2 \Big]. \tag{8.17}$$

The important conclusion is that prototype network selectivity is a function of only loaded-Q values. This allows the arbitrary choice of parallel resistance levels throughout the direct-coupled filter to accommodate convenient element values.

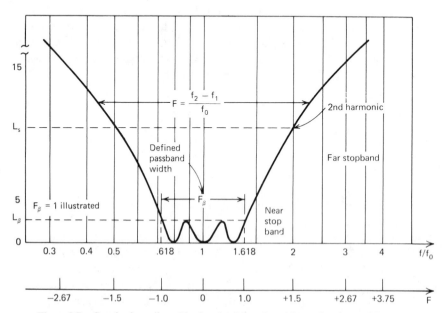

Figure 8.7. Standard semilogarithmic selectivity plot with passband normalization.

8.1.4. Prototype Selectivity Graphs. The N loaded-Q values may be chosen arbitrarily, but an orderly procedure, based on standard response shapes, will be described in Section 8.5. In those cases, passband edge frequencies with an associated loss value will be defined. Suppose that these are labeled F_β and L_β according to (8.2) and $20\log_{10}$ of the magnitude of (8.14), respectively. (The F_β parameter was called w in Chapter Six.) The geometric symmetry of (8.2) provides arithmetic symmetry on a semilogarithmic plot, as shown in Figure 8.7. Note especially the relationship of the normalized frequency f/f_0 and the normalized fractional frequency F/F_β. For example, the indicated F_β and L_β parameters might be defined for $L_\beta = 3$ dB or for a 4.5-dB pass-bandwidth loss value, whichever is appropriate. Any two frequencies, f_1 and f_2, at the same loss value have geometric symmetry with equal arithmetic displacements on the semilogarithmetic plot in Figure 8.7.

A convenient Bode breakpoint analysis results from considering the asymptotic behavior of (8.16) and (8.17), as illustrated in Figure 8.8. Since (8.16) and (8.17) are typical for any number of N resonators, consider the behavior of $20\log_{10}|L|$ for large F when the Nth-degree term dominates:

$$L = -6 + 20\log_{10}\Pi Q_{LK} + N20\log|F| \text{ dB} \quad \text{for large F.} \qquad (8.18)$$

The Π notation indicates a product of Q_L factors, i.e., $Q_{L1}Q_{L2}\cdots Q_{LN}$. Also, (8.2) approaches linearity in f/f_0 for large f or F, where $f = \omega/2\pi$. Clearly, the breakpoint location on the dB loss axis in Figure 8.8 is set by the first two terms in (8.18), and the 6N dB/octave slope is set by the third term. It is useful in the following developments to normalize all Q_{LK} values to the output resonator load Q, Q_{LN}. The definition of the normalized loaded Q is thus

$$\bar{Q}_{LK} = \frac{Q_{LK}}{Q_{LN}}, \qquad K = 1, 2, \ldots, N. \qquad (8.19)$$

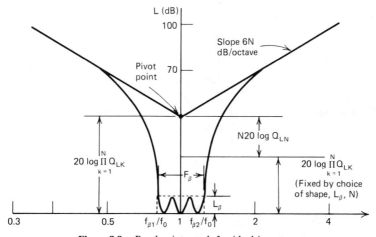

Figure 8.8. Breakpoint graph for ideal inverters.

Then, the loaded-Q product may be written as

$$\Pi Q_{LK} = Q_{LN}^N \Pi \bar{Q}_{LK}. \qquad (8.20)$$

This is also shown as segments of the breakpoint displacement in Figure 8.8. An important denormalizing equation also results from (8.20):

$$Q_{LN} = \left(\frac{\Pi Q_{LK}}{\Pi \bar{Q}_{LK}} \right)^{1/N}. \qquad (8.21)$$

For ideal response shapes, such as maximally flat or equal ripple, there will be two dependent constants and a set of normalized loaded-Q values that result from choosing N resonators and an L_β passband loss level. These constants are the products $Q_{LN} F_\beta$ and $\Pi \bar{Q}_{LK}$. Choosing a value of the fractional passband width F_β determines a particular value of Q_{LN}, using the former constant. However, it is seen from (8.18) or Figure 8.8 that choosing a stopband loss value (L dB) determines the loaded-Q product and, consequently, Q_{LN} by (8.21). Clearly, selectivity scaling may be determined by either passband width or by stopband selectivity, but not by both.

Finally, a review of the collection of terms leading to (8.16) and (8.17) will reveal that the -6-dB term ($20 \log 0.5$) in (8.18) is due to nonzero source conductance (see R_s in Figure 8.1). Therefore, sources shall be considered lossy or lossless (ideal current source). In the latter case, there is no -6-dB term in breakpoint loss expression (8.18).

8.1.5. Summary of Prototype Network.

A prototype direct-coupled filter network is composed of lossless resonators separated by inverters and terminated by a load resistance, and possibly by a source resistance. The inverters are lossless, quarter-wave transmission lines having a frequency-independent electrical length and a characteristic impedance. All resonators are tuned to the passband geometric center frequency and see a real impedance looking toward the load. The resonator's singly loaded Q is the parallel resistance divided by either the inductive or capacitive resonator reactance. The characteristic impedance of the inverter, Z_{0ij}, is the geometric mean of its input and output resistance values at the band-center (tune) frequency.

An expression for direct-coupled filter selectivity may be obtained by forming the product of all ABCD chain matrices for the resonators and inverters. The resonator frequency function F appears in a polynomial for the filter loss function. Therefore, the prototype direct-coupled filter may have exactly the same selectivity as classical bandpass filters obtained from lowpass prototypes by geometric frequency mapping (Section 6.5.1). The coefficients in these polynomials can be reduced to functions of only the loaded Q. Therefore, the loaded Q uniquely determines selectivity and leaves arbitrary the choices of resonator parallel resistance values. It is this flexibility that accommodates bounds on practical filter components.

For large values of F, the N*th*-degree term in the selectivity polynomial dominates. Then the logarithm of the loaded-Q product locates a breakpoint in the selectivity semilogarithmic graph, and the asymptotes have a 6N dB/octave slope versus normalized frequency. It is convenient to normalize resonator loaded-Q values to the output loaded-Q value. It was noted that four standard passband shapes will be analyzed; in each case, the choice of the number of resonators and the passband-edge loss value determine the normalized loaded-Q values (a distribution) and two constants: $Q_{LN}F_\beta$ and $\Pi\bar{Q}_{LK}$. It was concluded that either passband width or stopband loss can be independently specified, but not both.

Shunt parallel LC resonators have been considered. The same development applies for the case of series LC resonators separated by inverters.

8.2. Designing with L and C Inverters

The most important physical inverters are pi networks of inductors or capacitors having negative components in the shunt branches. It will be shown that these correspond exactly to quarter-wave (90-degree) transmission lines at all frequencies. However, the characteristic impedances of inductive and capacitive inverters are linearly or inversely proportional to frequency, respectively. The stopband selectivity estimate can be easily adjusted for that frequency dependence, but it will produce a first-order distortion of the passband shape, which increases with passband width.

The inverter properties of L and C pi networks will be derived, and simple design rules will be obtained for selectivity adjustment and absorption of negative elements into adjacent positive elements. The consequent changes in the breakpoint analysis will be noted, and a practical direct-coupled filter will be designed.

8.2.1. Simple L and C Inverters.   Simple L and C inverters are shown in Figure 8.9. The ABCD matrix for these pi networks may be obtained by premultiplying the matrix result in (4.10) by an additional shunt-branch ABCD matrix; the result is:

$$T = \begin{bmatrix} 1+ZY & Z \\ Y(2+ZY) & 1+ZY \end{bmatrix}. \tag{8.22}$$

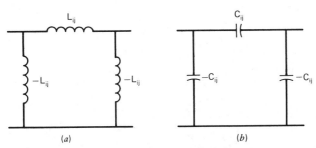

Figure 8.9.   Simple L (*a*) and C (*b*) Inverters.

For the L inverter, $Z = j\omega L_{ij}$ and $Y = -1/j\omega L_{ij}$. Then (8.22) and the ABCD matrix for a lossless transmission line (Section 4.3.1) may be compared to show that the L inverter is 90 degrees long at all frequencies and that $Z_{0ij} = \omega L$. A similar conclusion may be obtained for the C inverter, except that it is -90 degrees long at all frequencies, and $Z_{0ij} = 1/(\omega C)$. This is a remarkable result. The Z_0 of L and C inverters is equal to the reactance of the top (coupling) element at tune frequency ω_0. The design procedure is thus quite elementary: the tune frequency reactance level of the L or C inverter branches is equal to the geometric mean of the parallel resistances on either side of the inverter.

How are the negative elements obtained? Replace the ideal inverter in Figure 8.1 with the capacitive inverter in Figure 8.9a. Clearly, the negative shunt C's in the inverters may be absorbed by the adjacent resonator shunt C's. It turns out that the resonator shunt C's are always larger than the inverter branch C's in all practical cases. Inductive inverters are absorbed in a similar manner.

8.2.2. Magnetically Coupled Inverters. It is easily shown that a transformer contains the inductive inverter in Figure 8.9b. Van Valkenburg (1960, p. 304) gives various equivalent networks for transformers; the relevant cases are shown in Figure 8.10. Figure 8.10a and b represents familiar forms, where

$$L_{12} = \frac{L_p L_s - M^2}{M}, \qquad (8.23)$$

and mutual-coupling M may be positive or negative, depending on the winding orientation. Figure 8.10c shows the shunt inductances divided to

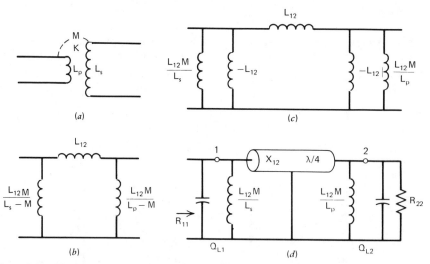

Figure 8.10. Transformer equivalent circuits. (*a*) Physical transformer; (*b*) pi equivalent network; (*c*) pi with divided shunt branches; (*d*) inductive pi inverter replaced by quarter-wave transmission line.

reveal the inductive inverter equivalent to that in Figure 8.10*b*. The inverter $Z_0 = \omega L_{12}$.

Figure 8.10*d* is slightly less general. It assumes that no other inductance is connected to nodes 1 and 2 and that each node is parallel resonated by capacitors. In this special case, (8.9) can be used to show that

$$Q_{L1}Q_{L2} = \frac{R_{11}L_s}{\omega_0 L_{12}M} \frac{R_{22}L_p}{\omega_0 L_{12}M} = \frac{(\omega_0 L_{12})^2 L_s L_p}{(\omega_0 L_{12})^2 M^2} . \tag{8.24}$$

But the coefficient of flux-linkage coupling is defined to be

$$K_{12} = \frac{M}{\sqrt{L_p L_s}} . \tag{8.25}$$

Therefore, when Figure 8.10*d* applies,

$$K_{12} = \frac{1}{\sqrt{Q_{L1}Q_{L2}}} . \tag{8.26}$$

There are many filters involving more than two resonators where (8.26) is valid; otherwise, the coupling coefficient may still be calculated using Figure 8.10*c* and (8.25).

8.2.3. An Accurate Stopband Selectivity Estimate.

The asymptotic behavior of loss function (8.14) is easily modified to account for the Z_0 frequency dependence. Consider the role of the Z's in (8.11) and (8.12) for $N = 3$ and $N = 4$, respectively. In all cases, the Z_0 terms are factors in the coefficient of the N*th*-degree jF frequency variable. The necessary modification to (8.18) for the upper stop band is

$$L = -6 + 20 \log \Pi Q_{LK} + N20 \log|F| + (NMI - NCI)20 \log \frac{f}{f_0} \text{ dB}, \tag{8.27}$$

where $L \geqslant 20$ dB or $f/f_0 \geqslant 1.2$ ensure the validity of the estimate. NMI stands for the number of magnetic inverters, and NCI stands for the number of capacitive inverters. Figure 8.11 illustrates how the prototype selectivity in Figure 8.8 is affected when there are two more inductive than capacitive inverters present. According to (8.27), it is possible to make NMI equal to NCI, so that there is no inverter selectivity tilt in the stop band. However, a certain amount of passband distortion will remain, even in that case. It will be more severe for wider passband widths and for response shapes having ripple.

Program A8-1 in Appendix A computes (8.27) in a very flexible fashion. If all but one variable in (8.27) are given, the remaining variable will be calculated. Frequencies f/f_0 and F are found implicitly by secant search; all others can be solved explicitly. As indicated in the program documentation, the loaded-Q product may be input directly or calculated from the sequence of individual resonator loaded-Q values input one at a time.

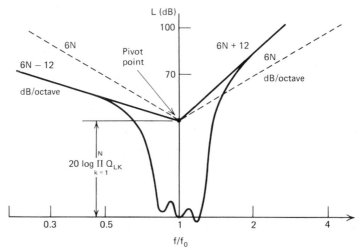

Figure 8.11. Breakpoint graph for a surplus of two inductive inverters.

Example 8.1. Suppose that a doubly terminated, direct-coupled filter will have resonators with loaded Q's of 15, 5, and 4. Two inductive inverters will be used. What will the second-harmonic attenuation be? To obtain the answer, $N = 3$ is entered, and key A is pressed. Next, Key B is pressed, and prompting message "1" is displayed; 15 is entered, and R/S is pressed. Prompting message "2" is displayed, and $Q_{L2} = 5$ is input. Similarly, $Q_{L3} = 4$ is input, with the loaded-Q product 300 being the last display obtained. Inverter counts 2 and 0 are entered into registers Y and X for NMI and NCI, respectively, and key fe is pressed. Doubly terminated key fb is pressed. The second-harmonic frequency ($f/f_0 = 2$) is entered, and key D pressed. Finally, key E is pressed, and $L = 66.15$ dB loss is obtained.

At what frequency will 60 dB be obtained? Enter 60 and press key E, then press key D. After time for iteration, the answer $f/f_0 = 1.81$ is displayed.

Suppose 75 dB is now required at the second harmonic; what loaded-Q product is required? Enter 75 and press key E; enter 2 and press key D. Then, press key C and obtain $\Pi Q_{LK} = 831.12$. Pressing key fc shows that the mean Q_{LK} is 9.40 (i.e., $831.12 = 9.40^3$).

For a loaded-Q product of 600, how many resonators are required to obtain a 75-dB loss at the second harmonic? Enter 600 and press key C; then press key A and obtain $N = 3.8$. Enter the next-higher integer (4), press key A, then key E; this shows that four resonators would provide a 75.69-dB loss at the second harmonic.

There are endless "what-ifs" in this design process; Program A8-1 is a great help in making optimal design trade-offs.

8.2.4. A Design Example.

The following example illustrates all major considerations in direct-coupled filter design. The other techniques that follow only alter these primary relationships by various approximations.

Suppose that the following requirements are given: A three-resonator filter is to be excited by a lossy, 50-ohm source and terminated by a 100-ohm resistor. The filter is to be tuned to 50 MHz, and 60-dB attenuation is required at 90 MHz, using an approximate, maximally flat response shape for the pass band. Inductance values are limited to the range 20–300 nH. Find all component values.

It can be seen from (8.27) that using two inductive inverters will reduce the loaded-Q product and therefore the reactive power in the resonators (and dissipative loss and sensitivity, considered later). Figure 8.12 shows the prototype and evolving topologies. The distribution of loaded-Q values for the maximally flat passband response shape will be shown to be 1, 2, 1 in Section 8.5.2. The fractional frequency corresponding to $f/f_0 = 90/50 = 1.8$ is $F = 1.2444$. Using Program A8-1 to evaluate (8.27), the loaded-Q product must be 319.5424. Using (8.21), $Q_{L3} = 5.4263$; thus $Q_{L1} = 5.4263$ and $Q_{L2} = 10.8529$ according to (8.19).

Converting the inductance limits into reactance at a tune frequency of 50

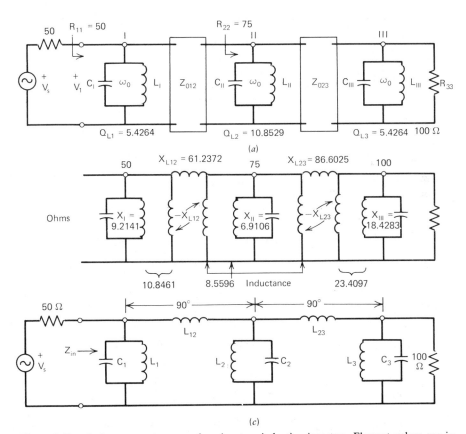

(a)

(c)

Figure 8.12. A three-resonator example using two inductive inverters. Element values are in ohms. (*a*) Prototype inverters between resonators; (*b*) substitution of inductive inverters; (*c*) final network elements from combined positive and negative elements.

MHz, it is found that these reactances must be in the range 6.2832–94.2478 ohms. The synchronous resonator reactances are calculated next using (8.3). Thus $X_I = 9.2141$ and $X_{III} = 18.4283$ ohms. The value of R_{22} shown in Figure 8.12a does not effect selectivity; trying $R_{22} = 75$ ohms yields $X_{II} = 6.9106$ ohms, which is within limits. However, it is still necessary to be sure that the inverter reactances are within limits. Using (8.9), $X_{L12} = \sqrt{50 \cdot 75} = 61.2372$ and $X_{L23} = \sqrt{75 \cdot 100} = 86.6025$ ohms, which are within limits. A new trial value of R_{22} would be necessary if any of these three inductive reactances were out of the allowable range. The circuit reactances are shown in Figure 8.12b.

It is now necessary to absorb the negative reactances. This is conveniently accomplished by Program A6-1, using key C. The final reactance values are indicated in Figure 8.12b. There is a very convenient *rule* illustrated in Figure 8.12b: all the L's touching a node must resonate all the C's touching that node. This is easily seen, because the two inverter branches touching a node cancel when in parallel, i.e., when adjacent nodes are grounded. The final element notation is shown in Figure 8.12c. The rule says that L_1 and L_{12} in parallel will resonate C_1. Similarly, L_{12}, L_2, and L_{23} in parallel will resonate C_2. Designs should always be checked to ensure that this rule is satisfied.

Table 8.1.   Element Values and Data for the Design Example in Figure 8.12 and for Analysis Program B4-1

Units:	Frequency 1E6	Inductance 1E−9	Capacitance 1E−12
Load (ohms):	Resistance 100	Reactance 0	

Power in load $= 1/4R_s = 0.005$ watts

Topology:	Type	Value	Q	Remarks
	3	172.73	0	C_3 pF
	−2	74.52	0	L_3 nH
	2	275.66	0	L_{23}
	3	460.61	0	C_2
	−2	27.25	0	L_2
	2	194.92	0	L_{12}
	2	34.52	0	L_1
	−3	345.46	0	C_1
	1	50.00	0	R_s

Note: Q=0 implies an infinite Q value.
At 50 MHz, $Z_{in} = 49.9994 + j0.0752$ ohms.
At 90 MHz, $|S_{21}| = 60.11$ dB.
There is a loss of 3 dB at 45.0 and 54.3 MHz (18.81% bandwidth).
The ideal, maximally flat case has 3 dB at 45.6 and 54.82 MHz (18.43%).

Note the difference in L_I and L_1. The nomenclature is significant; henceforth, only prototype (synchronous resonator) reactances will have roman subscripts. Table 8.1 shows the final element values for the band-center frequency of 50 MHz. It is in the format for running analysis Program B4-1 (Section 4.1.4). Several results are provided in Table 8.1. The 90-degree nodal phase relationship shown in Figure 8.12c was also confirmed.

8.2.5. Summary of Designing With Simple L and C Inverters.

The ABCD chain matrix for a pi arrangement of L's or C's was obtained for the case where both shunt branches were negative elements. Comparison with the ABCD matrix of a lossless, 90-degree transmission line revealed that these pi inverters are 90 degrees long at all frequencies and have characteristic impedances (Z_0) that are proportional or inversely proportional to frequency for L and C inverters, respectively. It was also shown that the magnetic transformer equivalent circuit incorporates an inductive inverter with shunt inductances left over on each side. When these shunt inductances constitute the total synchronous nodal inductances, then the coupling coefficient K_{ij} (flux linkage) between windings is $(Q_{Li}Q_{Lj})^{-1/2}$ between the coupled nodes.

The prototype asymptotic selectivity estimate that was connected with a breakpoint graph in Section 8.1.4 was modified to account for inverter frequency dependence. It turned out that each inductive inverter added 6 dB/octave to the upper stopband slope and subtracted that amount in the lower stop band. Also, capacitive inverters affected the slope in the opposite manner. This comprehensive estimate of stopband selectivity is valid for frequencies greater than $1.2f_0$ (geometrically symmetric about the tune frequency) or for losses greater than about 20 dB. Program A8-1 was furnished to make the selectivity estimate from any set of dependent variables in order to find the remaining variable. An extended example of program utilization was included.

Finally, a complete design example for a three-resonator filter was furnished. It contained all the major steps in direct-coupled filter design; subsequent refinements will not alter this fundamental procedure. Design success was confirmed by analysis using Program B4-1; the desired selectivity and impedance match were obtained. The choice of loaded-Q distribution in the ratios $1, 2, 1$ anticipated the maximally flat response described in Section 8.5. However, any arbitrary loaded-Q distribution could have been used without affecting the selectivity and impedance-matching outcome. Similarly, an arbitrary choice of resistance level within the filter provided acceptable element values. It was shown in Section 8.1.3 that such choices of resistance levels (and related inverter impedances) do not affect the selectivity or response shape.

A rule was provided for checking any basic direct-coupled design: all the L's and C's touching a node should resonate. This check and the use of an analysis program justify carrying about five significant figures in calculations, even though such accuracy has little meaning in the real world of physical components. Otherwise, numerical or procedural mistakes are very easy to overlook.

It was noted that classical bandpass filters, like that in Figure 6.31 (Section 6.5.1), may be modified by employing one capacitive and one inductive Norton transformer (Section 6.5.3), so that a filter having a direct-coupled appearance is obtained. Even though it appears to have been obtained using one L and one C inverter, the design is not direct coupled because it violates the node-resonance rule. However, such Norton transformer applications are possible for odd N, and there is no passband or stopband distortion in these cases.

8.3.   General Inverters, Resonators, and End Couplings

Design of practical filters requires substantial departure from the ideal case. Inverters may be realized as apertures in waveguide walls, resonators may depart from lumped-element frequency behavior and dissipate energy, and acceptable impedance levels for these may require end couplings.

This section develops the fact that all lossless passive networks contain an inverter with some residual admittances that must become parts of adjacent resonators. The trap top-coupling network will be developed from this principle, and its remarkable ability to improve stopband selectivity will be demonstrated.

It will be shown that resonator dissipation affects tune frequency input impedance much more than inverter dissipation. An expression for input impedance with dissipation will be derived, and a means for compensating for the change will be described. A reasonable amount of dissipation will not seriously affect the stopband attenuation estimate (8.27), because it has offsetting effects. It will also be shown that any resonant two-terminal network may be viewed as an ideal resonator to a first-order approximation, namely with the same resonance frequency and slope versus frequency as the lumped prototype.

End couplings can be L sections, radio frequency (rf) transformers, or direct connections to terminating resistors. Dissipation will be considered for L sections in a treatment that is only a slight extension of Section 6.1. The rf transformer may be realized as actual windings. However, the resonator is often a coaxial or waveguide cavity, and the transformer is just a wire loop that provides coupling to the magnetic field in the cavity. A basis for these more general situations will be provided.

8.3.1.   Inverters in Admittance Parameters.   An equivalent circuit for the defining admittance parameters was given in Section 7.3.1 (Figure 7.10). The defining equations in (3.79) and (3.80) for short-circuit y parameters show that the equivalent circuit in Figure 8.13 is valid for the reciprocal case where $y_{12} = y_{21}$.

Equating $Z = -1/y_{21}$ and $Y = y_{21}$ in (8.22) shows that the inverter characteristic admittance, Y_0, is $Y_0 = B$ when $y_{21} = jB$, i.e., when y_{21} is an imaginary

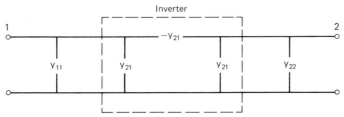

Figure 8.13. A reciprocal short-circuit-parameter equivalent circuit when $y_{12}=y_{21}$. The dashed lines enclose the inverter portion of the network.

number. A sufficient condition for y_{21} being an imaginary number is that the network be lossless. In that case, the maximum possible efficiency is unity, and (7.64) shows that stability factor $K=1$. Then the definition of K in (7.57) reveals that $K=1$ when $g_{11}=0=g_{22}$ and $y_{21}=jB$, thus proving the sufficient condition.

Note that susceptances y_{11} and y_{22} in Figure 8.13 become part of the synchronous reactance in their respective resonators, as seen from the particular case in Figure 8.12*b*. The frequency behavior of the resulting resonators will be equated to the ideal LC resonator in Section 8.3.4.

Example 8.2.   Identify the inverter in a 30-degree length of a 100-ohm lossless transmission line, and resonate each end to make a direct-coupled filter between the 50-ohm terminations. The ABCD parameters of a transmission line in (4.13)–(4.15) can be converted to y parameters according to (3.86); $y_{11}= -jY_0\cot\theta$, and $y_{21}= +jY_0/\sin\theta$. Therefore, inverter $Y_{012}=Y_0/\sin\theta$ and half of each resonator must be composed of admittance $-jY_0\cot\theta$. Figure 8.14 shows the equivalent circuit in impedance terms. The equivalent of a shorted-stub transmission line at each end of the inverter must be resonated to make a direct-coupled filter. In this example, $Z_{012}=50$ ohms, and the shunt reactances associated with y_{11} and y_{22} are $+j57.74$ ohms. Capacitors on either end having the negative of the latter value are required, as shown in Figure 8.15. Programs A6-1 and A6-2 may be used to confirm the impedances noted in Figure 8.15. It can be shown that the voltages across the capacitors have a

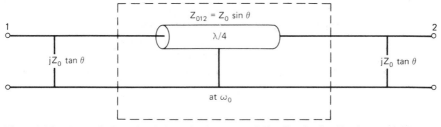

Figure 8.14. An equivalent circuit for a lossless transmission line having Z_0 ohms and θ length.

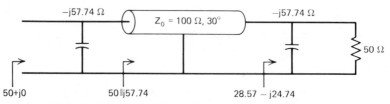

Figure 8.15. A direct-coupled filter using a short length of transmission line.

90-degree phase difference. Also, if the load resistor is changed to 25 ohms, the input impedance becomes $100+j0$ ohms. (Why?)

8.3.2. Trap Inverters. Consider the trap inverter shown in Figure 8.16. If the trap is resonant above the passband-center frequency (i.e., $\omega_n > \omega_0$), then the trap will appear to be an inductance at tune frequency ω_0 where the inverter design is accomplished. The susceptance of a parallel LC network resonant at ω_n is available from (8.1) and (8.2):

$$B(\omega) = \omega_n C\left(\frac{\omega}{\omega_n} - \frac{\omega_n}{\omega}\right), \tag{8.28}$$

and $\omega_n = 1/\sqrt{LC}$. The trap susceptance at $\omega = \omega_0$ is thus

$$B(\omega_0) = \omega_n C\left(\frac{\omega_0}{\omega_n} - \frac{\omega_n}{\omega_0}\right), \tag{8.29}$$

and the adjacent resonators must absorb this equivalent negative inductance, as explained in Section 8.2.4.

A useful conversion of (8.28) is obtained by using the frequency term within the parentheses in (8.29) to both multiply and divide (8.28). After some algebraic manipulation, (8.28) can be restated as:

$$B(\omega) = -B(\omega_0)\frac{\omega}{\omega_0}\left[\frac{1-(\omega_n/\omega)^2}{(\omega_n/\omega_0)^2 - 1}\right]. \tag{8.30}$$

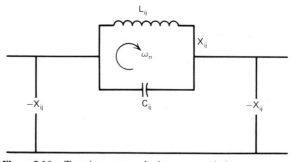

Figure 8.16. Trap inverter producing a transmission zero at ω_n.

It is important to interpret the three factors in (8.30). The first two terms have been extracted in the form typical of the simple capacitive inverter described in Section 8.2.1: $B(\omega_0)$ is an inverter Y_0 and (ω/ω_0) is the linear frequency behavior of capacitive susceptance. Therefore, (8.30) was written with $\omega > \omega_n$ in mind, when the trap appears capacitive.

The analysis in Section 8.1.3 showed that the Z_0 of each inverter was a factor in the coefficient of the highest-degree frequency term in the response polynomial [see (8.10)–(8.12)]. The first two terms in (8.30) produce exactly the effect of a simple capacitive inverter. The effect of the third term in (8.30) is to increase the breakpoint of the asymptote for stopband selectivity by $20 \log$ of its inverse (inverting from admittance to impedance).

The numerator and denominator of the third term are both positive when $\omega > \omega_n$ and $\omega_n > \omega_0$. Thus the third term is unity when

$$\omega/\omega_n = \left[2 - \left(\frac{\omega_n}{\omega_0} \right)^2 \right]^{-1/2}. \tag{8.31}$$

An analysis of the third term in (8.30) in light of (8.31) reveals that the trap increases selectivity above the trap frequency, except for the case where $\omega_n/\omega_0 < \sqrt{2}$, when the benefitting frequencies must be less than (8.31). It is usually desirable to set the trap resonance quite close to the upper passband-edge frequency; so the latter restriction imposed by (8.31) will apply. However, a trap resonance close to the pass band severely aggravates the dissipative effects on the passband edge.

This analysis for $\omega_n > \omega_0$ and $\omega > \omega_n$ leads to the following rule: classify the trap as a simple C inverter in (8.27), and increase the loss estimate by

$$20 \log \left| \frac{(\omega_n/\omega_0)^2 - 1}{1 - (\omega_n/\omega_s)^2} \right|. \tag{8.32}$$

When $\omega_n/\omega_0 < \sqrt{2}$, (8.32) is negative when the frequency is greater than (8.31). A similar analysis for $\omega_n < \omega_0$ and $\omega < \omega_n$ leads to the rule: classify the trap as a simple L inverter in (8.27) and increase the loss estimate in (8.27) by

$$20 \log \left| \frac{(\omega_0/\omega_n)^2 - 1}{1 - (\omega_s/\omega_n)^2} \right|. \tag{8.33}$$

Figure 8.17 shows the breakpoint boost provided by (8.32). Note that the trap appears inductive between ω_0 and ω_n and capacitive above ω_n. The consequent change of 12 dB/octave in the asymptote slope is shown in Figure 8.17. The increase in the breakpoint due to (8.32) or (8.33) can be used to reduce the loaded-Q product while obtaining the same stopband loss at one particular frequency. This reduces resonator reactive power and thus dissipation loss, as described in Section 8.3.3.

There are several caveats regarding traps. The trap null frequency should be no closer to the passband edge than absolutely necessary; this rounds the

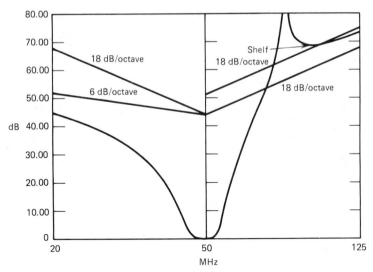

Figure 8.17. Breakpoint boost due to an upper stopband trap.

passband edge in dissipative networks. Selectivity between the null frequency and the passband cannot be guaranteed, because the trap null is extremely sensitive. In fact, the "shelf" indicated in Figure 8.17 is the *only* loss specification that a trap should be expected to fulfill. A reciprocal pole factor can be fit to the loss curve around a trap pole of attenuation, and the shelf loss value can be estimated. The analysis is beyond the scope of this introduction; however, Appendix G contains the necessary equations.

Example 8.3. A three-resonator design was accomplished in Section 8.2.4 and shown in Figure 8.12. Change L_{23} to an 80-MHz trap by adding C_{23} in parallel with L_{23} between nodes II and III. The loss at 90 MHz is still of interest; it was 60 dB before. However, the trap will look capacitive above 80 MHz, not inductive as before. Therefore, the NMI–NCI factor is now 0 instead of 2, which is a 10.21-dB reduction from the filter having two inductive inverters. However, (8.32) is equal to 17.42 dB; so there is a net gain of 7.21 dB as a result of changing the inductor to an 80-MHz trap. This means that the attenuation at 90 MHz would increase to 67.21 dB for the same loaded-Q product of 319.58; analysis showed that the actual attenuation was 68.84 dB. This response curve is plotted in Figure 8.17. The passband distortion due to nonideal inverters caused 3-dB frequencies of 44.90 and 53.77 MHz for an 18.05% passband width. Compare this to the data in Table 8.1. Another option is to reduce the loaded-Q product by a factor of 2.29 (reduce each resonator Q by 32%) and keep the 60-dB attenuation at 90 MHz. The value of trap C_{23} is obtained from (8.29) where $B(\omega_0) = -1/\sqrt{75 \cdot 100}$.

The fractional frequency part within the parenthesis in (8.29) is -0.9750; therefore, trap C_{23} susceptance at the trap null frequency is $1/84.4374$ mhos.

Using ω_n, $C_{23} = 23.5611$ pF; its resonating inductor value is $L_{23} = 167.9829$ nH. Note that no other value of the network in Figure 8.12 was changed. (Why?)

8.3.3. Dissipation Effects.

It will be shown that the only dissipation that matters is in the resonators; the effect of the inverter dissipation is an order of magnitude lower. The power loss attributable to each resonator may be expressed in a very simple relationship. Figure 8.4 showed a resonator with a hypothetical (or real) conductance across it to represent the real power going toward the load. Figure 8.18 shows the same resonator with the unavoidable dissipation conductance G_{Kd}. Loaded-Q definition (8.3) will be retained as the ratio of parallel resistance to parallel reactance without losses.

The unloaded Q of the K*th* resonator is defined as

$$Q_{uK} = \frac{B_K}{G_{Kd}} = \frac{1}{X_K G_{Kd}}, \tag{8.34}$$

where the sign of the reactance is ignored. The efficiency in (4.45) was the ratio of load power to input power; in terms of the power dissipated, P_{Kd}, the efficiency of the K*th* resonator, is

$$\eta_K = 1 - \frac{P_{Kd}}{P_{in}}. \tag{8.35}$$

Assume that $V_K = 1$ in Figure 8.18 without loss of generality. Then $P_{Kd} = G_{Kd}$ and $P_{in} = G_{KK} + G_{Kd}$. Using (8.3) and (8.34), (8.35) becomes

$$\eta_K = 1 - \frac{1}{1 + Q_{uK}/Q_{LK}}. \tag{8.36}$$

Thus, the common expression for the K*th* resonator efficiency is

$$\eta_K = 1 - \frac{Q_{LK}}{Q_{uK}} \qquad \text{for} \quad Q_{uK} \gg Q_{LK}. \tag{8.37}$$

The overall efficiency of the total direct-coupled filter is just the product of each resonator's efficiency. For example, the efficiency of the filter in Figure 8.12c when all inductors have unloaded Q's of 100 is 0.9835 dB using the three factors as in (8.36). Network analysis shows that the actual efficiency is 0.9420 dB.

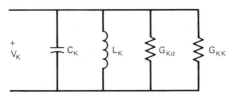

Figure 8.18. A dissipative resonator with conductance G_{Kd} related to its unloaded-Q factor (Q_{uK}).

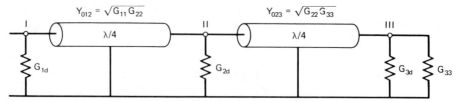

Figure 8.19. An $N=3$ dissipative, direct-coupled filter at tune frequency ω_0.

Figure 8.12a shows that the network at the tune frequency has the form shown in Figure 8.19, because all the resonators are parallel resonant. Inverter relationship (8.9) in its admittance form enables the expression of the input conductance:

$$G_{in}=G_{1d}+\frac{G_{11}G_{22}}{G_{2d}+G_{22}G_{33}/(G_{3d}+G_{33})}. \tag{8.38}$$

The input conductance of the lossless network is G_{11}. Divide both sides of (8.38) by G_{11}, and introduce G_{22} and G_{33} so as not to disturb the equality; this yields

$$\frac{G_{in}}{G_{11}}=\frac{G_{1d}}{G_{11}}+\frac{G_{11}G_{22}/G_{11}G_{22}}{\dfrac{G_{2d}}{G_{22}}+\dfrac{G_{22}G_{33}/G_{22}G_{33}}{G_{3d}/G_{33}+G_{33}/G_{33}}}. \tag{8.39}$$

The left-hand side is inverted to express a resistance ratio, and the loaded and unloaded Q's of (8.3) and (8.34) are incorporated. The result is the continued fraction

$$\frac{R_{in}(\text{lossless})}{R_{in}(\text{lossy})}=\hat{Q}_{L1}+\frac{1}{\hat{Q}_{L2}+1/(\hat{Q}_{L3}+1)}, \tag{8.40}$$

which always ends with 1. The relative Q, $\hat{Q}_{LK}$, is defined as

$$\hat{Q}_{LK}=\frac{Q_{LK}}{Q_{uK}}. \tag{8.41}$$

The continued fraction expansion in (8.41) will be computed by a recursive procedure in Section 8.6. In general, the direct-coupled-filter input resistance tends to change very little as a result of dissipation when there is an even number of resonators. This is due to the "seesaw" effect that each inverter has on its load and input resistances [see (8.7)]. When the inverter's load resistance increases, the inverter input resistance decreases, and vice versa. Whatever the change from the desired input resistance, it may be restored simply by changing some inverter Z_0 value (top-coupling reactance). It is also possible to restore the match by adjusting the input end coupling, when it is employed as desired in Section 8.3.5.

To compare the effect of dissipation on resonators and inverters, consider the dissipative resonator in Figure 8.18. At tune frequency ω_0, the input conductance of the K*th* resonator is

$$G_{Kd} + G_{KK} = G_{KK}\left(1 + \frac{Q_{LK}}{Q_u}\right).$$ (8.42)

On the other hand, the magnitude of a dissipative inverter coupling component is

$$|R_{ij} + jX_{ij}| = X_{ij}\sqrt{1 + \frac{1}{Q_u^2}} \doteq X_{ij}\left(1 + \frac{0.5}{Q_u^2}\right).$$ (8.43)

The last term results from the standard approximation for the square root of a binomial just slightly greater than unity. Clearly, the inverter dissipation disturbance is an order of magnitude less than that in resonators, and this is easily confirmed in practice.

There is a remarkable happenstance concerning the effect of dissipation on stopband selectivity. It turns out in almost every case that the selectivity estimate (8.27) is essentially unaffected by dissipation. What happens is that the selectivity of a lossless network, which is all due to input mismatch reflection, as described in Chapter Three, is replaced by a sum of reflection and efficiency loss that is nearly the same. For example, setting each inductor Q_u equal to 100 in the network in Figure 8.12c produces a 90-MHz loss of 60.13 dB, instead of the lossless network value of 60.12 dB. But the former is the sum of the 25.77-dB mismatch loss in (4.60) and the 34.36-dB efficiency loss in (4.45). This phenomenon can almost always be expected to occur, thus increasing the utility of the selectivity estimate in (8.27).

8.3.4. Equivalent Resonators. The susceptance of the K*th* resonator shown in Figure 8.4 can be determined from (8.1) and (8.2):

$$B_K(\omega) = \omega_0 C_K\left(\frac{\omega}{\omega_0} - \frac{\omega_0}{\omega}\right) = \omega_0 C_K F.$$ (8.44)

Differentiating this with respect to ω and then setting $\omega = \omega_0$ yields an expression for C_K in terms of its slope:

$$C_{Keq} = \frac{1}{2}\frac{dB_K}{d\omega}\bigg|_{\omega = \omega_0}.$$ (8.45)

This is called C_{Keq} because resonators may be employed that are far more complex than simple LC branches. They may still be used as resonators if (1) they are parallel resonant at ω_0, and (2) their slope at $\omega = \omega_0$ is available for use in (8.45). In that situation, there is a first-order correlation between the real resonator and the equivalent C in the lumped-element prototype network, namely (8.45). First-order correlation means that the first two terms in Taylor series (5.26) are known: the constant (zero) and the first derivative (slope).

It is not unusual to encounter discrepancies in the measured resonance bandwidth and the tune frequency reactance of resonated inductors. Often the explanation may be traced to (8.45), where the slope of the resonator does not

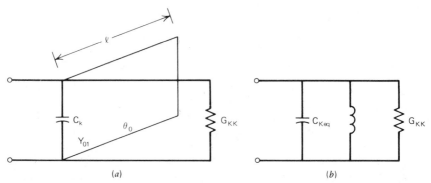

Figure 8.20. A capacitively loaded, short-circuited transmission line (*a*) and its equivalent lumped-element resonator (*b*).

conform to that of the lumped model. Helical resonators are especially subject to this discrepancy because they are coils having approximately 84 degrees electrical length with a small resonating capacitance. The slope equivalence of resonated transmission line segments is often a useful parameter.

Consider the resonated, short-circuited transmission line and its equivalent lumped-element resonator in Figure 8.20. To establish the slope equivalence of C_k and C_{Keq}, the transmission line resonator susceptance may be written as

$$B(\omega) = \omega C_k - Y_0 \cot \theta, \tag{8.46}$$

where the electrical length is

$$\theta = \frac{\omega \ell}{v}. \tag{8.47}$$

The transmission line velocity is v, and the physical length is ℓ. At resonance, $B(\omega_0) = 0$, and

$$\omega_0 C_k = Y_0 \cot \theta_0. \tag{8.48}$$

After differentiating (8.46) with respect to ω, setting $\omega = \omega_0$, and replacing C_k using (8.48), the transmission line resonator slope at the tune frequency is

$$\left.\frac{dB}{d\omega}\right|_{\omega_0} = \frac{Y_0}{\omega_0} \left(\cot \theta_0 + \frac{\theta_0}{\sin^2 \theta_0} \right), \tag{8.49}$$

where θ_0 is the electrical length in radians at tune frequency ω_0.

There are a number of applications where $C_k \to 0$ and $\theta_0 \to \pi/2$ in Figure 8.20. For example, interdigital filters are described by Matthaei et al. (1964). The 90-degree resonators also have a slope equivalence to lumped-element resonators. From (8.45) and (8.49), the equivalent prototype resonator susceptance is

$$\omega_0 C_{Keq} = Y_0 \frac{\pi}{4}. \tag{8.50}$$

Note that the resonator loaded Q computed on this basis will be useful as a passband parameter, but not for determining dissipation. (The equivalent parallel resistance due to transmission line loss may be obtained as described in Section 4.3.2.) The 90-degree resonators have repeating passbands at odd harmonics of the tune frequency, because of periodic resonances.

The capacitive loading of the short-circuited transmission lines in Figure 8.20a spaces the recurring resonances in a nonperiodic manner. Suppose that $B(\omega_0)=0=B(\omega_1)$; from (8-46) and (8.47),

$$\frac{Y_0}{C_k}=\omega_0 \tan\frac{\ell\omega_0}{v}=\omega_1 \tan\frac{\ell\omega_1}{v}. \tag{8.51}$$

Therefore, a second resonance at ω_1 is related to tune frequency ω_0 by

$$\frac{\omega_1}{\omega_0}-\frac{\tan\theta_0}{\tan\left[(\omega_1/\omega_0)\theta_0\right]}=0. \tag{8.52}$$

This transcendental equation may be solved by secant search. For example, when the resonator is 45 degrees long at the tune frequency ω_0, then $\omega_1=4.2915\omega_0$. Resonators may be as short as 10 degrees, which increases the second resonance to about 17.5 times the tune frequency resonance. Incidentally, a useful relationship of transmission line length in free space is

$$\theta_0=\frac{\ell f}{32.81} \quad \text{degrees}, \tag{8.53}$$

where ℓ is in inches and f is in MHz. A graph of (8.52) solutions is shown in Figure 8.21.

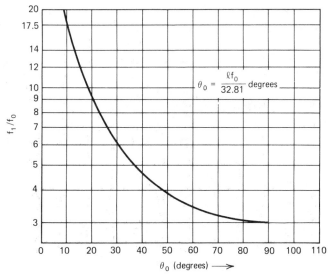

Figure 8.21.   Frequency ratios of recurring pass bands versus short-circuited resonator length at the tune frequency.

The ratio of the two capacitors (and related loaded Q's) in Figure 8.20, based on slope equivalence at the tune frequency, may be found from (8.48) and (8.49):

$$C_{Keq}/C_k = 0.5 + \frac{\theta_0}{\sin 2\theta_0}. \tag{8.54}$$

For example, when $\theta_0 = \pi/4$, $C_{Keq}/C_k = 1.285$. The loaded Q of the resonators in Figure 8.20 is proportional to the resonating capacitors. Therefore, the effective loaded Q that determines passband behavior is 28.5% greater than the apparent loaded Q determined by tune frequency reactance in (8.3). It will be shown that the increased loaded Q decreases the passband width.

Example 8.4. Design an $N=3$ combline filter using capacitive coupling between nodes and 45-degree resonators. Make the effective loaded-Q values and resistance levels equal to those in Figure 8.12. The desired circuit arrangement is shown in Figure 8.22.

The effective loaded-Q values will be 1.285 times the values on a reactance basis, as previously noted for 45-degree resonators. Dividing the Q_L values in Figure 8.12 by 1.285 yields $Q_{L1} = 4.2229 = Q_{L3}$ and $Q_{L2} = 8.4458$. Synchronous node reactance $X_K = R_{KK}/Q_{LK}$, so that $R_{11} = 50$, $R_{22} = 75$, and $R_{33} = 100$ lead to $X_I = 11.8403$, $X_{II} = 8.8801$, and $X_{III} = 23.6805$ ohms, respectively. The shorted-stub input impedance is $jZ_0 \tan \theta$, so that $Z_{0k} = X_K$ when $\theta = 45$ degrees. At 50 MHz, the corresponding synchronous capacitances are: $C_I = 268.84$, $C_{II} = 358.45$, and $C_{III} = 134.42$ pF. The values of X_{12} and X_{23} are the same as shown in Figure 8.12, so that $C_{12} = 51.98$ and $C_{23} = 36.76$ pF. There are two negative inverter capacitances to subtract from C_{II}, and only one each from C_I and C_{III}. The resulting topology code (corresponding to the analysis scheme in Chapter Four) for the combline filter in Figure 8.22 is shown in Table 8.2.

The analysis at 50 MHz showed that the lossless network was tuned to $Z_{in} = 50.0125 + j0.0073$ ohms. The 3-dB loss frequencies were 46.05 and 55.55 MHz; this is an 18.78% bandwidth, compared to the ideal, maximally flat 3-dB bandwidth of 18.43%.

Clearly, Z_{01} and Z_{02} are impractically low because the physical range of transmission line characteristic impedances is about 20–120 ohms. Parallel

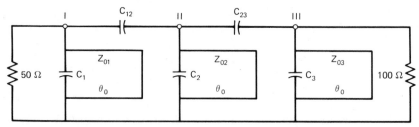

Figure 8.22. A combline filter using two capacitive inverters.

Table 8.2. **Analysis Topology Code for the Combline Filter in Figure 8.22 (Example 8.4)**[a]

Type	Value	Name
4	23.68	Z_{03}
314.16E6	45.00	ω_0 and θ_0
−3	97.66	C_3
3	36.76	C_{23}
4	8.8801	Z_{02}
314.16E6	45.00	ω_0 and θ_0
−3	269.72	C_2
3	51.98	C_{12}
4	11.8403	Z_{01}
314.16E6	45.00	ω_0 and θ_0
−3	216.86	C_1

[a] Units: ohms, pF, and nH.

resistance R_{22} is completely arbitrary; setting R_{22} equal to 600 ohms increases Z_{02} to 71.04 ohms. Changing all capacitances for this circumstance and analyzing the network confirmed that the selectivity was indeed independent of R_{22}. (Try it.) However, it turned out that $C_{23} = 13$ pF, which is near the minimum practical capacitance range. The trade-offs in this procedure are quite visible. The solution for the unacceptably low value of Z_{01} appears in Section 8.3.5.

Slope equivalence (8.45) is a means for estimating passband behavior, and (8.52) estimates spurious passband frequencies for capacitively loaded, short-circuited transmission line resonators. Stopband selectivity estimate (8.27) may be applied to the loaded-line case if the C_{Keq}/C_k ratio in Figure 8.20 is determined for equal prototype and actual resonator susceptance at a given stopband frequency. Replace C_K by C_{Keq} in (8.44) and equate this to (8.46). Substitution of (4.27) and (8.48) yields the requirement for equal stopband susceptances for Figure 8.20:

$$\frac{C_{Keq}}{C_k} = \left[\frac{\omega}{\omega_0} - \frac{\tan \theta_0}{\tan(\theta_0 \omega / \omega_0)} \right] \frac{1}{F}. \tag{8.55}$$

For the second harmonic, this reduces to

$$\frac{C_{Keq}}{C_k} = 1 + \frac{\tan^2 \theta}{3}, \qquad \omega = 2\omega_0, \tag{8.56}$$

by use of a trigonometric identity from Dwight (1961, p. 83).

The second-harmonic selectivity of the combline filter in Example 8.4 may be estimated using (8.56). Since the loaded Q is proportional to the resonator capacitance, (8.56) is found to equal $\frac{4}{3}$ when $\theta_0 = 45$ degrees. The loaded-Q product based on the reactances in Example 8.4 was 150.6130; the effective Q_L of each resonator is one-third more than that in its effect on the second-harmonic selectivity. Thus the loaded-Q product must be increased by 2.3704, or 7.4963 dB, before applying selectivity estimate (8.27); in this case, there are two capacitive inverters and no inductive inverters. Program A8-1 evaluates (8.27) to estimate a 43.58-dB second-harmonic attenuation. Analysis of the combline filter in Figure 8.22 by using values from Table 8.2 yields an actual attenuation of 43.18 dB at the second harmonic (100 MHz).

8.3.5.   End Coupling.   The combline filter in Example 8.4 was left with an input resonator having an extremely low Z_0. Increasing the parallel resistance at node I from 50 to 150 would triple Z_{01} to the acceptable value of 35.52 ohms. A transformer can be placed between the source and node I if the first resonator incorporates the equivalent parallel inductance it presents (see Figure 8.1). The required coupling coefficient is

$$K = \sqrt{\frac{X_p X_s}{R_g R_{11}} \left[1 + \left(\frac{R_g}{X_p} \right)^2 \right]} \,, \tag{8.57}$$

where X_p and X_s are the primary and secondary reactances at ω_0 (see Figure 8.10a). The positive, parallel inductance to be combined into the adjacent resonator is

$$X_{p1} = \frac{R_{11}}{R_{11}/X_s - R_g/X_p} \,, \tag{8.58}$$

where $K^2 \ll 1$ has been assumed.

An alternative means for increasing the end-node parallel resistance is the use of an L section from Section 6.1.2. Any element adjacent to the resonator must become a part of it. However, in this case and in the case of the transformer, the rule that "all L's and C's touching a node must resonate" applies only *before* these coupling sections are combined into the end resonator(s). Element dissipation effects have been considered in Section 8.3.3, where the input resistance disturbance was quantified. It may be offset by the input L section, which also may have dissipative elements, especially inductors. It is straightforward to account for the Q_u of the top-coupling inductor L_{g1} in Figure 8.2. The design relationships for $Q_u \gg 1$ are

$$X_{g1} = \frac{R_{11} - 2R_g}{2Q_u} + \left[R_g(R_{11} - R_g) + \left(\frac{R_{11} - 2R_g}{2Q_u} \right)^2 \right]^{1/2} \,, \tag{8.59}$$

$$X_{p1} = \frac{R_{11}}{\left[R_{11}/(R_g + X_{g1}/Q_u) - 1 \right]^{1/2}} \,, \tag{8.60}$$

where X_{g1} is the top-coupling inductor reactance, and X_{p1} parallels C_I. As before, this coupling may be used on either end or on both ends of the filter.

The selectivity effects of end coupling are similar to those of inverters. When the resistance transformations are 10 or greater (e.g., $R_{11} > 10R_g$), then the end coupling affects selectivity approximation (8.27) like inverters of the same kind. Lesser resistance ratios produce effects of less than 6 dB/octave. A good interpolation formula is included in Appendix G. Its derivation is beyond the scope of the present treatment. Educated guesses between 0 and 6 dB/octave are often satisfactory.

8.3.6. *Summary of Inverters, Resonators, and End Couplings.* Every lossless, reciprocal network contains an inverter; it was identified in terms of its short-circuit y parameters. Inverter $Z_0 = 1/|y_{21}|$, and y_{11} and y_{22} must be incorporated in adjacent resonators. Inverters affect stopband attenuation according to the logarithm of the ratio of Z_0 values at stopband and tune frequencies; so the original breakpoint attenuation estimate may be used for any kind of inverter. The trap inverter is particularly useful because it provides equivalent selectivity with reduced loaded Q's. This effect was expressed as an added term in the selectivity estimate.

One reason for using minimum loaded-Q values is their direct effect on dissipative loss. The efficiency of each resonator at the tune frequency was shown to be a simple function of the loaded-to-unloaded-Q ratio; the product of all such efficiencies is the overall filter efficiency. The tune frequency model of a dissipative, direct-coupled filter is just a set of parallel resistances separated by ideal inverters. The input resistance was shown to be a continued fraction, and that provided an expression for the ratio of input resistances with and without dissipation. The change due to dissipation tends to be greater for an odd number of resonators, but it can be corrected by adjusting the inverter or end-coupling transformation ratios. It was also shown that inverter dissipation effects were an order of magnitude less than resonator effects, and could be safely ignored. An example emphasized the fortuitous effect of dissipation on stopband selectivity: the lossless estimate is still quite accurate because it is usually about equal to the combination of input mismatch loss and efficiency loss in the presence of dissipative elements.

Almost any two-terminal network can be used as a resonator if it is resonant at the tune frequency and has either an acceptable tune frequency slope or stopband susceptance. The equivalent lumped-element, prototype resonator capacitance turned out to be equal to one-half of the resonator slope versus frequency at ω_0. The capacitively loaded, short-circuited transmission line resonator was examined in terms of both its tune frequency slope and stopband susceptance equivalence to the prototype resonator. The equivalence was expressed in terms of the ratio of the lumped resonator C_{Keq} to the actual loading C_k across the transmission line stub. This is the same as the ratio of the equivalent loaded Q to the apparent loaded Q. This type of stub is used in combline filters. Example 8.4 utilized three such resonators, which were capacitively coupled.

Unloaded, 90-degree shorted stubs are used in interdigital filters; their slope equivalence was obtained. In that case, the passband reoccurred at odd harmonics of the tune frequency. It was shown that shortening the line by capacitive loading produces nonperiodic passband reoccurrences that are far removed from the tune frequency for short stubs. A graph of this effect was furnished.

The combline filter in example 8.4 produced two stub Z_0 values that were impractically low. The interior node's parallel resistance can be raised to increase the resonator Z_0 without affecting the selectivity. The need for end coupling to raise the input (and/or output) node's parallel resistance level was made evident. Both the transformer and the L-section end couplings were described; the latter included an inductor unloaded Q_u in the matching formulas provided. The viewpoint for properly combining the end-coupling reactance with the resonator was discussed. Also, the effect on selectivity was described as being similar to the effect on the same type of inverters, especially when the end-coupling resistance ratio exceeded $10:1$. An example of end-coupling design is included in Section 8.6.

8.4. Four Important Passband Shapes

It was shown in Section 8.1.3 that the selectivity function of the prototype network was a function of only the resonator loaded-Q values. When inverters have some frequency dependence in the Z_0 or θ parameters, some distortion occurs, but it may be negligible for bandwidths of less than 20%. The stopband effects of inverter Z_0 are easily anticipated in terms of asymptote tilt in a simple breakpoint analysis. This section concentrates on the loaded-Q distribution. Although loaded-Q values may be selected arbitrarily, it is often useful to have closed formulas for passband characteristics, even if they are only approximately realized by the network.

Four response shapes will be considered; the first three belong to a closely related family based on elliptic loci in the complex frequency plane. These are the Chebyshev (equal-ripple or overcoupled), Butterworth (maximally flat or critically coupled), and Fano (undercoupled) shapes. Fano's undercoupled response shape is not related to his broadband-matching method in Chapter Six: it was suggested in connection with video amplifier design. The fourth shape results from choosing all loaded-Q values to be equal. The resonator efficiency relationship from Section 8.3.3 will show that this produces minimum loss when given a loaded-Q product, i.e., fixed stopband attenuation. It also produces a network having minimum sensitivity, and serves as a simple norm for all other cases.

Transducer functions were described in Section 3.2.4, and their zeros were identified as the natural frequencies of resistively terminated LC networks. The problem set for Chapter Three included both Chebyshev and Butterworth

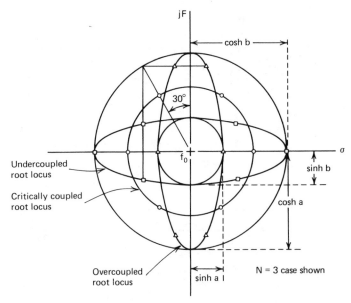

Figure 8.23. A family of bandpass-filter transducer zeros.

polynomials. The zeros of these functions lie on ellipses and a related circle, respectively, as shown in Figure 8.23. Note that the lowpass-to-bandpass-frequency transformation of (6.83) has been assumed. The frequency variable is now F, from (8.2), normalized to a given fractional passband width (6.86), here called F_β in accordance with Figure 8.7. Butterworth transducer zeros lie on the circle, and Chebyshev zeros lie on the vertical ellipse. The elliptic loci may be found from (6.54), but these relationships will be reformulated here. The horizontal ellipse contains the roots of the Fano undercoupled function, which is quite useful but not well known.

This section will include a discussion of each type of response; the first three will be compared. Equations for estimating the number of resonators required for certain selectivity specifications, as well as loaded-Q product and breakpoint analysis information, will be given. The means for tabulating loaded-Q distributions and two normalizing constants will be described. Limited tables and selectivity curves will be provided; it is anticipated that small-computer programs will be written as required. A comprehensive tabulation of design equations is provided in Appendix G.

8.4.1. The Chebyshev Overcoupled Response Shape. The Chebyshev response shape is defined by

$$L(f) = 10\log_{10}\left(1 + \varepsilon^2 T_N^2 \frac{F}{F_p}\right) \text{ dB}, \tag{8.61}$$

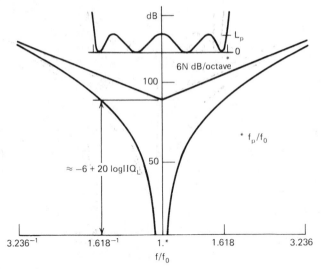

Figure 8.24. Overcoupled breakpoint plot: $N = 4$, $L_p = 1$ dB, $F_p = 0.1$.

where

$$\varepsilon = (10^{L_p/10} - 1)^{1/2} \tag{8.62}$$

and L_p is shown in Figure 8.24. Chebyshev functions of the first kind (T_N) were defined by (2.34) in the passband and by (2.35) in the stopband. As illustrated in Figure 8.24, Chebyshev bandpass functions have N valleys and $N - 1$ peaks located according to

$$\frac{F}{F_\beta} = \cos(\ell\theta), \qquad \ell = 1, 2, \ldots, 2N - 1, \tag{8.63}$$

where $\theta = 2\pi/N$, as previously defined in (6.71). There is a response valley at f_0 for odd values of N and a peak at f_0 for values of N, as shown in Figure 8.24. Direct-coupled Chebyshev networks having an even number of resonators are mismatched at the tune frequency, thus the peak at f_0 in Figure 8.24. The parameter g_{N+1} in Equation (G.10) (Appendix G) is greater than unity for even values of N. Choices of parallel resistances and inverters should be made using R_{NN} decreased by dividing the given R_{NN} by g_{N+1}. After the design is complete, the given value of R_{NN} terminating the network will produce the correct input impedance mismatch.

Chebyshev response (8.61) can be solved for the number of resonators required for given stopband loss L_s at fractional frequency F_s and comparable passband values (L_p and F_p):

$$N = \frac{\cosh^{-1}\left[(10^{0.1L_s} - 1)/(10^{0.1L_p} - 1)\right]^{1/2}}{\cosh^{-1}(F_s/F_p)}. \tag{8.64}$$

There is often a need to locate other points at $F=F_x$ on the curve in Figure 8.24 given the loss value L_x. This may be accomplished by using

$$F_p = \frac{F_x}{\cosh\left[(1/N)\cosh^{-1}(t/\varepsilon)\right]}, \qquad (8.65)$$

where a new parameter, similar to ε in (8.62), is

$$t=(10^{L_x/10}-1)^{1/2}. \qquad (8.66)$$

Example 8.5a. Suppose that the Chebyshev passband ripple is $L_p=0.1$ dB over bandwidth $F_p=0.15$, and the stopband loss is $L_s=30$ dB at $F_s=0.195$. Find N and, using the next higher integer, find the frequencies where $L=3$ dB. Solving (8.64) yields $N=7.9668$. Using $N=8$ in (8.65) yields $F_x=0.158$. Summarizing, when $N=8$ there is a 0.1-dB ripple over the 15% passband width, 3 dB at the edges of a 15.8% bandwidth, and 30 dB at the edges of a 19.5% bandwidth.

The 2N zeros of the Chebyshev function in (8.61) are shown on the vertical ellipse in Figure 8.23, where the dimensioning parameter is

$$a=\frac{1}{N}\sinh^{-1}\frac{1}{\varepsilon}. \qquad (8.67)$$

The location r_m+ji_m of the mth zero is

$$r_m=F_\beta S_N \sin(2m-1)\theta, \qquad (8.68)$$

$$i_m=F_\beta\sqrt{S_N^2+1}\,\cos(2m-1)\theta, \qquad (8.69)$$

and

$$S_N=\sinh\frac{\sinh^{-1}(1/\varepsilon)}{N}. \qquad (8.70)$$

In these equations, $m=1,2,\ldots,N/2$ when N is even, and $m=1,2,\ldots,(N+1)/2$ when N is odd. Using N left-half-plane factors to create a polynomial in (F/F_p) results in forms like (8.15). Green (1954) tabulated the coefficient expressions through $N=5$ and guessed the recursive relationships for the general case. The coefficients of like powers of jF were compared, and a general relationship for successive element values was obtained. The latter result is equivalent to (6.72). In the present case, (8.16) and (8.17) showed that loaded-Q parameters could be identified. The recursion for loaded-Q values for the Chebyshev overcoupled case is given in Appendix G. A typical set of values is shown in Table 8.3 for lossless and lossy sources (see Figure 8.1).

Table 8.3.   Overcoupled Q_L Values for N = 3

L_p (dB)	$\bar{Q}_{L1}$	$\bar{Q}_{L2}$	$\bar{Q}_{L3}$	$\bar{Q}_L$ Product	$Q_{L3}F_p$
		Lossless Source			
0.01	2.542	2.427	1.0	6.169	0.3146
0.10	2.112	2.106	1.0	4.449	0.5158
0.2	1.939	1.937	1.0	3.756	0.6138
0.5	1.687	1.629	1.0	2.748	0.7981
1.0	1.491	1.318	1.0	1.965	1.012
2.0	1.307	0.940	1.0	1.229	1.355
3.0	1.213	0.7011	1.0	0.8501	1.674
6.0	1.087	0.3201	1.0	0.3479	2.708
9.5	1.035	0.1371	1.0	0.1420	4.296
		Lossy Source			
0.01	1.0	1.5420	1.0	1.542	0.6292
0.1	1.0	1.1120	1.0	1.1120	1.0320
0.2	1.0	0.9389	1.0	0.9389	1.2280
0.5	1.0	0.6870	1.0	0.6870	1.5960
1.0	1.0	0.4913	1.0	0.4913	2.0240
2.0	1.0	0.3072	1.0	0.3072	2.711
3.0	1.0	0.2125	1.0	0.2125	3.3490
6.0	1.0	0.0870	1.0	0.0870	5.4150
9.5	1.0	0.0355	1.0	0.0355	8.5910

Example 8.5b.   Reconsider the specifications in Section 8.2.4 for the Chebyshev overcoupled shape (with perfect inverters): the N = 3 lossy-source filter is tuned to 50 MHz, and a 60-dB attenuation is required at 90 MHz. Find the loaded-Q values and the passband width if the passband ripple is to be 0.2 dB. First find output resonator Q_{L3}; (8.18) shows that the loaded-Q product must be 1035.32. Then (8.21) and the normalized loaded-Q product of 0.9389 from Table 8.3 yield $Q_{L3} = 10.3312$. Therefore, $Q_{L2} = 0.9389 \times 10.3312 = 9.7000$, and $Q_{L1} = 10.3312$. Since $Q_{L3}F_{0.2} = 1.2280$, $F_{0.2} = 0.1189$; i.e., the 0.2-dB-ripple passband width is 11.89%.

8.4.2.   The Butterworth Maximally Flat Response Shape.   The Butterworth response shape is defined by

$$L(f) = 10 \log_{10}\left[1 + \varepsilon^2 \left(\frac{F}{F_p} \right)^{2N} \right] \text{ dB;} \tag{8.71}$$

ε was defined in (8.62), and L_p is shown in Figure 8.25. Butterworth response Equation (8.71) can be solved for the number of resonators required for the given stopband loss L_s at fractional frequency F_s and comparable values of L_p

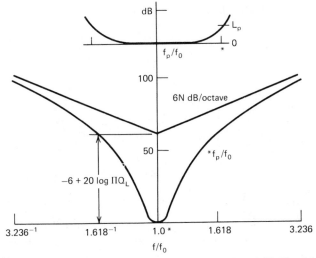

Figure 8.25. Maximally flat breakpoint plot: $N = 4$, $L_p = 1$ dB, $F_p = 0.1$.

and F_p:

$$N = \frac{\log\left[(10^{0.1L_s} - 1)/(10^{0.1L_p} - 1)\right]}{2\log(F_s/F_p)}.\qquad(8.72)$$

To locate other points at $F = F_x$ on the curve in Figure 8.25, given the loss value L_x, use

$$F_p = \frac{F_x}{(q/\varepsilon)^{1/N}},\qquad(8.73)$$

where q was given in (8.66).

Example 8.6. Suppose that the Butterworth defined passband edge is $L_p = 0.1$ dB at bandwidth $F_p = 0.15$, and the stopband loss is $L_s = 30$ dB at $F_s = 0.195$. Find N and, using the next higher integer, find the frequencies where $L = 3$ dB. Solving (8.72) yields $N = 20.3$. Using $N = 21$ in (8.73) yields $F_x = 0.1640$. Summarizing, for $N = 21$ the frequency response curve is 15% wide at the 0.1-dB points, 16.4% wide at the 3 dB points, and 19.5% wide at the 30 dB points. Note that the $N = 8$ Chebyshev filter in Example 8.5 met the same performance specifications.

The $2N$ zeros of the Butterworth function in (8.71) are shown on the circle in Figure 8.23, where the dimensioning parameter is

$$b = \cosh\frac{\cosh^{-1}\left[(\sinh 0.8814N)/\varepsilon\right]}{N}.\qquad(8.74)$$

Table 8.4.   Maximally Flat Q_L Values

N	$\bar{Q}_{L1}$	$\bar{Q}_{L2}$	$\bar{Q}_{L3}$	$\bar{Q}_{L4}$	$\bar{Q}_{L5}$	$\bar{Q}_{L6}$	Q_L Product	$F_p Q_{LN}\varepsilon^{-1/N}$
				Lossless Source				
2	2.	1.000					2.00	0.7071
3	3.	2.667	1.000				8.00	0.500
4	4.	4.121	2.828	1.000			46.63	0.3827
5	5.	5.483	4.472	2.894	1.000		354.90	0.3090
6	6.	6.797	6.000	4.643	2.928	1.000	3327.00	0.2588
				Lossy Source				
2	1.	1.000					1.000	1.4140
3	1.	2.000	1.000				2.000	1.0000
4	1.	2.414	2.414	1.000			5.8280	0.7654
5	1.	2.618	3.236	2.618	1.000		22.1800	0.6180
6	1.	2.732	3.732	3.732	2.732	1.000	104.0000	0.5176

The location $r_m + ji_m$ of the *mth* zero is

$$r_m = F_p \varepsilon^{-1/N}\sin(2m-1)\theta, \qquad (8.75)$$

$$i_m = F_p \varepsilon^{-1/N}\cos(2m-1)\theta. \qquad (8.76)$$

In these equations, $m = 1, 2, ..., N/2$ when N is even, and $m = 1, 2, ...,$ $(N+1)/2$ when N is odd. The recursive relationship for normalized loaded-Q values is given in Appendix G.

Unlike the Chebyshev or undercoupled cases, the Butterworth passband constant $Q_{LN}F_p$ comes from a single, simple expression. The Butterworth Q distributions and constants for values of N from 2 to 6 are given in Table 8.4.

8.4.3.   The Fano Undercoupled Response Shape.   The Fano response shape is defined by

$$L(f) = 10\log_{10}\left[1 + k^2\sinh^2\left(N\sinh^{-1}\frac{F}{F_d}\right)\right] dB, \qquad (8.77)$$

where

$$k = \frac{q}{\sinh(0.8814N)} \qquad (8.78)$$

and

$$q = \left(10^{L_d/10} - 1\right)^{1/2}. \qquad (8.79)$$

Normalizing bandwidth F_d in (8.77) corresponds to loss L_d, the dB droop. It has an upper bound:

$$L_d \leqslant 10\log_{10}\left[1 + \sinh^2(0.8814N)\right]. \qquad (8.80)$$

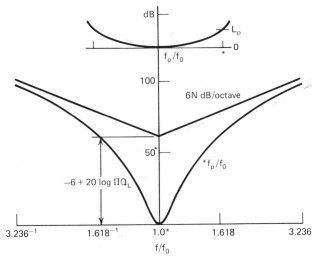

Figure 8.26. Undercoupled breakpoint plot: $N = 4$, $L_p = 1$, $F_p = 0.1$, $L_d = 6$ dB.

This is tabulated in Table 8.6 appearing later in this section. The droop parameter is important in that it determines the shape, but some arbitrary L_p is usually of interest, as shown in Figure 8.26.

Undercoupled response (8.77) can be solved for the number of resonators required for a given stopband loss (L_s) at fractional frequency F_s and comparable values of L_d and F_d:

$$N = \frac{\sinh^{-1}\left[(\sinh^2 0.8814N)(10^{0.1L_s} - 1)/(10^{0.1L_d} - 1)\right]^{1/2}}{\sinh^{-1}(F_s/F_d)}. \qquad (8.81)$$

This equation is implicit in variable N; however, its solution was the subject of Example 5.6 in Section 5.3.4. It can usually be solved by elementary iterative methods, such as a secant search. Since the decibel droop seldom defines the desired passband edge, it is especially important to relate it to an arbitrary loss and related fractional frequency:

$$F_d = \frac{F_x}{\sinh\left[(\sinh^{-1}(t/k))/N\right]}. \qquad (8.82)$$

Example 8.7. Suppose that the undercoupled shape has a 6-dB droop based on $F_d = 0.15$, and the stopband loss is $L_s = 30$ dB at $F_s = 0.195$. Find N and, using the next higher integer, find the frequencies where $L_x = 0.1$ dB. Figure 5.22 in Section 5.3.4 illustrates the solution: $N = 14.7538$ for (8.81) in this case. Using $N = 15$ in (8.82), $F_x = 0.1175$. Summarizing, for $N = 15$ there is an 11.75%, 0.1-dB bandwidth; a 15%, 6-dB bandwidth; and a 19.5%, 30-dB bandwidth.

The 2N zeros of the undercoupled function in (8.77) are shown on the horizontal ellipse in Figure 8.23. The location $r_m + ji_m$ of the mth zero is

$$r_m = F_d C_N \sin(2m-1)\theta, \tag{8.83}$$

$$i_m = F_d \sqrt{C_N^2 - 1}\, \cos(2m-1)\theta, \tag{8.84}$$

and

$$C_N = \cosh \frac{\cosh^{-1}\left[(\sinh 0.8814N)/\varepsilon\right]}{N}. \tag{8.85}$$

In these equations, $m = 1, 2, \ldots, N/2$ when N is even, and $m = 1, 2, \ldots, (N+1)/2$ when N is odd. The recursive relationship for normalized loaded-Q values is given in Appendix G. The lossless-source case produces Q_L distributions similar to those for the Chebyshev and Fano responses. However, for the undercoupled, lossy-source case, there are multiple solutions. For even values of N, there are N/2 distributions, and N/2 more using $Q_{L1} \leftarrow 1/Q_{L1}$. For odd values of N, there is one distribution starting from $\overline{Q}_{L1} = 1$, and $(N-1)/2$ distributions ending with $\overline{Q}_{L1} < 1$. There are $(N-1)/2$ more distributions using $\overline{Q}_{L1} \leftarrow 1/\overline{Q}_{L1}$.

Table 8.5.   Undercoupled, Lossy-Source Q_L Values for N=3

L_d (dB)	$\overline{Q}_{L1}$	$\overline{Q}_{L2}$	$\overline{Q}_{L3}$	$\overline{Q}_L$ Product	$Q_{L3}F_d$
0.1	0.4641	1.268	1.0000	0.5883	0.6668
0.1	1.0000	2.3090	1.0000	2.3090	0.4227
0.1	2.1550	2.7320	1.0000	5.8880	0.3094
1.0	0.3182	0.9656	1.0000	0.3073	1.2370
1.0	1.0000	2.7300	1.0000	2.7300	0.5972
1.0	3.1420	3.0340	1.0000	9.5350	0.3936
3.0	0.2395	0.7730	1.0000	0.1852	1.8330
3.0	1.0000	3.2070	1.0000	3.2070	0.4391
3.0	4.1750	3.2270	1.0000	13.470	0.4391
6.0	0.1813	0.6138	1.0000	0.1113	2.6080
6.0	1.0000	3.8490	1.0000	3.8490	0.8003
6.0	5.5170	3.3860	1.0000	18.680	0.4727
9.5	0.1360	0.4788	1.0000	0.0651	3.6690
9.5	1.0000	4.7450	1.0000	4.7450	0.8783
9.5	7.3550	3.521	1.0000	25.900	0.4988
16.95	0.0720	0.2688	1.0000	0.0194	7.4350
16.95	1.0000	7.9750	1.0000	7.9750	0.9995
16.95	13.880	3.7310	1.0000	51.7800	0.5357

Table 8.6. Fano Minimum Lowpass-Overshoot Droop Values

N	2	3	4	5	6	7	8
L_d (dB)	—	15.40	19.49	27.11	34.76	42.41	50.07
Max L_d	9.54	16.99	24.61	32.26	39.91	47.57	55.22

A typical set of values for the undercoupled, lossy-source case is shown in Table 8.5. Note the available arbitrary choices that do not affect the response shape: not only resistance levels, but also distributions for a given decibel droop. However, two of each set in Table 8.5 simply turn the network end for end. An undercoupled example for $N=4$ and a $6=$ dB droop is worked in complete detail in Section 8.6.

Fano examined the step response of his lowpass filters and recommended certain decibel droop values versus N for minimum overshoot. Although the frequency mapping in (6.83) causes some distortion of group delay, and the inverters will cause more, Fano's criterion for good transient response is a useful guide. His equation for the recommended decibel droop applies for $N > 2$:

$$L_d = 10 \log_{10} \frac{1 + (\sinh^2 0.8814N)}{3.28} \text{ dB.} \tag{8.86}$$

Some of these values are tabulated in Table 8.6 along with the maximum possible droop values from (8.80).

8.4.4. Comparison of Elliptic Family Responses.

The defining response equations for the overcoupled, maximally flat, and undercoupled response shapes may be compared to the ideal selectivity asymptote equation, represented in (8.18), to identify the loaded-Q product (semilogarithmic breakpoint) in each case. Ignoring the -6-dB term, the loaded-Q product for overcoupled filters is

$$20 \log \Pi Q_{LK} = -N20 \log F_p + 20 \log \varepsilon + 6(N-1) \text{ dB.} \tag{8.87}$$

The loaded-Q product for maximally flat filters is

$$20 \log \Pi Q_{LK} = -N20 \log F_p + 20 \log \varepsilon \text{ dB.} \tag{8.88}$$

The loaded-Q product for undercoupled filters is

$$20 \log \Pi Q_{LK} = -N20 \log F_d + 20 \log q + 6(N-1) - 20 \log(\sinh 0.8814N) \text{ dB.} \tag{8.89}$$

Table 8.7 compares the overcoupled and undercoupled cases to the Butterworth case on the basis of (1) relative decibel selectivity for constant passband width and (2) percentage of Butterworth passband width for a fixed stopband selectivity (L_s). It is noted that the defined Fano undercoupled bandwidth is the decibel droop in Table 8.7. However, an arbitrary basis for defining the undercoupled bandwidth is available using (8.82).

**Table 8.7.   Comparisons Relative to Butterworth Passband and
Stopband Characteristics**

N	Relative dB for Same PB[a] Widths		% Butterworth PB Width for Same L_s dB	
	Chebyshev	Fano	Chebyshev	Fano
2	6	−3.03	44.3	−16.0
3	12	−4.90	58.5	−17.1
4	18	−6.59	67.9	−17.3
5	24	−8.26	73.8	−17.3

[a] PB = passband.

8.4.5.   The Minimum-Loss Response Shape.   There is no simple expression for the transfer function that results when all resonator loaded-Q values are equal. However, there is a compact method for obtaining the ABCD chain matrix of a network consisting of iterated sections. It is important in its own right and will be used again in Section 9.1. The strategy will be to cascade N identical subnetworks, each composed of a dissipative resonator followed by a unit inverter. Multiplication of N such chain matrices amounts to raising the typical chain matrix to the N*th* power.

Storch (1954) showed that a real or complex 2×2 matrix, **T**, may be raised to power N by the following identity:

$$\mathbf{T}^N = P_N(y)\mathbf{T} - P_{N-1}(y)\mathbf{U}. \tag{8.90}$$

U is the unit matrix; $P_K(y)$ is a Chebyshev polynomial of the *second* kind, previously mentioned in Problem 2.16; and argument $y = A + D$ from matrix **T**, assuming that **T** is the ABCD matrix. A recursion can be started with $P_{-1} = -1$ and $P_0 = 0$; then

$$P_K = yP_{K-1} - P_{K-2}, \qquad K = 1, 2, \dots, N. \tag{8.91}$$

Table 8.8 shows some of these polynomials.

It is seen from Figure 8.1 that a prototype network without a load resistor is a cascade of N resonator plus inverter subsections, which introduces a superfluous inverter next to the output port. It is clear from (8.16) and (8.17)

**Table 8.8.   Chebyshev Polynomials of
the Second Kind**

$$P_1 = 1$$
$$P_2 = y$$
$$P_3 = y^2 - 1$$
$$P_4 = y^3 - 2y$$
$$P_5 = y^4 - 3y^2 + 1$$
$$P_6 = y^5 - 4y^3 + 3y$$

that the Z_0 values of the inverters do not affect selectivity, so that all of these and the terminating resistance(s) may be set equal to unity without loss of generality. The superfluous inverter next to the load will introduce a 90-degree phase shift, but will have no other effect. Each typical subsection has a chain matrix defined by

$$\begin{bmatrix} 1 & 0 \\ Y & 1 \end{bmatrix}\begin{bmatrix} 0 & j1 \\ j1 & 0 \end{bmatrix} = \begin{bmatrix} 0 & j1 \\ j1 & jY \end{bmatrix}. \tag{8.92}$$

The typical resonator is shown in Figure 8.18, where G_{KK} is approximately unity for low dissipation. To a good approximation under these assumptions,

$$Y = Q_L\left(\frac{1}{Q_u} + jF\right). \tag{8.93}$$

Thus the argument of the Chebyshev polynomials of the second kind is $y = A + D = jY$. Note that Y is complex, and so is jY. Thus (8.90) yields the overall ABCD matrix of a somewhat dissipative, direct-coupled filter ending in one superfluous inverter:

$$T^N = \begin{bmatrix} -P_{N-1}(y) & jP_N(y) \\ jP_N(y) & jYP_N(y) - P_{N-1}(y) \end{bmatrix}, \tag{8.94}$$

where the polynomial argument is

$$y = jY \tag{8.95}$$

for Y in (8.93).

The result in (8.94) may be used to obtain the frequency response of both doubly and singly terminated minimum-loss filters. According to Beatty and Kerns (1964) or (3.68), the transducer-gain scattering parameter normalized to 1 ohm may be expressed in terms of ABCD parameters:

$$S_{21} = \frac{2}{A+B+C+D}. \tag{8.96}$$

Therefore, the transducer gain for a doubly terminated network is

$$L = 10\log_{10}\frac{|M_N|^2}{4} \text{ dB}, \tag{8.97}$$

where

$$M_N = j(2+Y)P_N(y) - 2P_{N-1}(y) \tag{8.98}$$

using (8.94). This may be evaluated recursively using (8.91); however, the M_N values in (8.98) may be expanded as polynomials in complex Y, as tabulated in Table 8.9.

Figure 8.27 shows the $N=4$ response in the passband and stopband, respectively. Note the Q_L/Q_u parameter. Two design graphs for four-resonator, doubly terminated, minimum-loss filters are provided in Appendix F. Note that it is convenient to use $Q_L F$ as the independent variable.

Table 8.9. Doubly Terminated Minimum-Loss-Filter Polynomials in Resonator Admittance Y

$$|M_1| = |2+Y|$$
$$|M_2| = |2+2Y+Y^2|$$
$$|M_3| = |2+3Y+2Y^2+Y^3|$$
$$|M_4| = |2+4Y+4Y^2+2Y^3+Y^4|$$
$$|M_5| = |2+5Y+6Y^2+5Y^3+2Y^4+Y^5|$$
$$|M_6| = |2+6Y+9Y^2+8Y^3+6Y^4+2Y^5+Y^6|$$

The stopband selectivity displayed in Figure 8.27 conforms to (8.18); for this case it becomes:

$$L_s = -6 + N20\log Q_L + N20\log|F_s| \text{ dB}, \qquad L_s \geqslant 20 \text{ dB}. \qquad (8.99)$$

A useful second form of (8.99) expresses the required loaded Q given the attenuation specification:

$$Q_L = (F_s)^{-1}10^{(L_s+6)/20N}, \qquad L_s \geqslant 20 \text{ dB}. \qquad (8.100)$$

A practical, well-known expression for the midband loss is available from

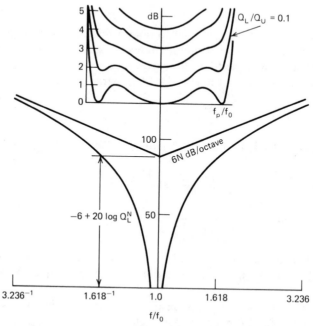

Figure 8.27. Doubly terminated, minimum-loss breakpoint plot: $N=4$, $L_p=1$, $F_p=0.1$.

the product of the resonator efficiencies, (8.37). In this case,

$$L(\omega_0) = -10\log_{10}\left(1 - \frac{Q_L}{Q_u}\right)^N \text{ dB.} \qquad (8.101)$$

However, $\log_{10} x = (\log_{10} e)(\ln x)$, and Dwight (1961, p. 137) give the series

$$\ln(1-x) = -x - \frac{x^2}{2} - \frac{x^3}{3} - \cdots - . \qquad (8.102)$$

In this case, $x = Q_L/Q_u$ is assumed to be small, so that

$$L_0 = L(\omega_0) \doteq 4.3429N\frac{Q_L}{Q_u} \text{ dB.} \qquad (8.103)$$

This valuable relationship is simple to compute and shows that the decibel midband loss is inversely proportional to the resonator unloaded Q.

Furthermore, (8.99) may be solved for N,

$$N = \frac{6 + L_s}{20\log(Q_L F_s)}, \qquad (8.104)$$

and (8.103) may be substituted to obtain

$$L_0 = \frac{4.34(L_s + 6)(Q_L F_s)}{(Q_L F_s)20\log(Q_L F_s)} . \qquad (8.105)$$

Differentiating (8.105) with respect to $Q_L F_s$ and setting this to zero gives a minimum midband-loss condition of

$$\log_{10}(Q_L F_s) = 0.434. \qquad (8.106)$$

Putting this back into (8.104) yields

$$N_o = \frac{L_s + 6}{8.686} . \qquad (8.107)$$

This is the optimal number of resonators for a minimum midband loss when the stopband attenuation (L_s) is specified. The nearest integer value would be used, of course.

The singly terminated response function can be obtained using (8.14); it applies to Figure 8.1 and includes the load conductance. There is no source conductance in the singly terminated case. The overall ABCD matrix consists of (8.94) for the resonators and inverters, but it must be postmultiplied by the ABCD matrix for the unity-load conductance (see Section 4.2.1). The resulting C element of the overall ABCD matrix for this case is

$$C(\omega) = j(1 + Y)P_N(y) - P_{N-1}(y), \qquad (8.108)$$

Table 8.10. Singly Terminated Minimum-Loss-Filter Polynomials in Resonator Admittance Y

$$|C_1| = |1 + Y|$$
$$|C_2| = |1 + Y + Y^2|$$
$$|C_3| = |1 + 2Y + Y^2 + Y^3|$$
$$|C_4| = |1 + 2Y + 3Y^2 + Y^3 + Y^4|$$
$$|C_5| = |1 + 3Y + 3Y^2 + 4Y^3 + Y^4 + Y^5|$$
$$|C_6| = |1 + 3Y + 6Y^2 + 4Y^3 + 5Y^4 + Y^5 + Y^6|$$

where $y = jY$, and Y is in (8.93). Recall that this transfer function is exactly 90 degrees longer than the actual case, because of a superfluous output inverter. The value of $C(\omega)$ in (8.108) may be calculated recursively using (8.91), but it is useful to expand (8.108) into polynomials in admittance Y. The magnitude of (8.108) is tabulated in Table 8.10 using the polynomials from Table 8.8.

The singly terminated response function from (8.13) and (8.14) is

$$L(\omega) = 20 \log_{10} \left| \frac{C(\omega_0)}{C(\omega)} \right| \, dB, \tag{8.109}$$

where C is in (8.108). Usually, it is more practical to compute the term

$$L(\omega) = L_0 - 20 \log_{10} |C(\omega)| \, dB, \tag{8.110}$$

where L_0 is the approximate midband dissipative loss in (8.103).

Programs A8-2 and A8-3 in Appendix A compute the selectivity functions described previously for the doubly and singly terminated minimum-loss filters, respectively. The polynomial coefficients from Tables 8.8 or 8.10 are stored in reverse order (as a string of integers in register 2) upon inputting the desired value of $N \leqslant 6$. The coefficient of the highest-degree term is not stored; also, 0.1 is added for convenience. The unload resonator Q_u factor must be manually stored in register 3. Then input of the loaded-Q value (Q_L) initiates the calculation of the approximate midband dissipative loss (L_0) according to (8.103). Key A is used to store the fractional frequency F. The approximate stopband selectivity may then be computed according to (8.99); the -6 constant is omitted for the singly terminated case. Subroutine C (key C) computes the approximate value of $L_0 - L_s$; this is valid only for values greater than about 20 dB. It computes rapidly, compared to the exact calculation (provided in subroutine E) programmed from the admittance polynomial expressions. A means for systematically adding an arbitrary positive or negative increment to the loaded-Q (Q_L) value is incorporated in the approximate $L_0 - L_s$ calculation, so that a value of Q_L may be found for some desired stopband attenuation relative to the midband loss. (The reader may wish to add an automatic secant search for convergence.) The use of this program is easier to understand in the context of the selectivity curves in Appendix F. The following example helps clarify the necessary design decisions.

Example 8.8. An $N = 4$, doubly terminated minimum-loss filter has the values $Q_u = 100$ and $Q_L = 50$ for unloaded and loaded Q's, respectively. Find the midband loss and the attenuation relative to the midband loss at a 10% bandwidth. Also, find the fractional bandwidth for a 3-dB relative attenuation. What must the loaded Q be to obtain a 45-dB attenuation at a 10% fractional bandwidth? Solutions from Program A8-2 can be located on the selectivity curves in Appendix F. The midband loss is obtained by evaluating (8.103). Entering $N = 4$ by key D, storing $Q_u = 100$ in register 3, entering $Q_L = 50$, and pressing key B yields $L_0 = 8.6859$ dB. Enter $F = 0.1$ and press key A. Key C promptly yields the approximate stopband attenuation 41.2317 dB according to (8.99); key E requires more time to yield the exact value 41.0577 dB according to (8.97).

The 3-dB relative attenuation cannot be found by using key C, because this attenuation is valid only in the stopband. By glancing at the first figure in Appendix F or by trial and error, enter $F = 1.3333/50 = 0.0267$, and press key A and then key E. The attenuation relative to $L_0 = 8.6859$ dB is 2.9945 dB. So $Q_L F = 1.3333$ on the first graph in Appendix F for a 2.67% bandwidth at the relative 3-dB points.

To increase the stopband loss requires a greater Q_L. Program A8-2 includes a feature to enable arbitrary incrementing of Q_L. Store $F = 0.1$ by using key A again, and store $\Delta Q_L = +0.5$ in register I. Pressing key C and then R/S shows that a 41.2317-dB relative attenuation is obtained when $Q_L = 50$. Pressing R/S again shows a new L_0, and pressing keys C and R/S again yields 41.4906 dB using $Q_L = 50.5$ (the chosen increment was added automatically). Successive cycles of keys C, R/S, and R/S lead to $Q_L = 58.0$, with a 44.9986-dB approximate relative attenuation (44.9507 dB by exact subroutine E). Program A8-3 for the singly terminated (lossless-source) case works in the same way.

8.4.6. Summary of Four Important Passband Shapes. There is a well-known family of response shapes based on the root loci of orthogonal ellipses and their common circle. This section included graphic and relational comparisons of their selectivity characteristics. The Chebyshev overcoupled shape has equal ripples and the steepest attenuation at the sides of the passband. It is affected to a greater extend by moderate resonator dissipation and inverter Z_0 tilt than are more rounded shapes. Equations for estimating the required number of resonators for various selectivity requirements were provided. Design charts and sets of various response curves (selectivity, time delay, etc.) are widely available. A sample table of normalized, resonator loaded-Q values was furnished (the generating equations are contained in Appendix G). Even-N Chebyshev cases require an input mismatch at the tune frequency, causing a slight alteration of the standard loaded-Q filter design procedure (see Appendix G).

The Butterworth maximally flat shape is also well known but easier to calculate, since it represents the common ground between the two sets of ellipses. An included table of loaded-Q values for $N \leqslant 6$ is adequate for all Butterworth direct-coupled filter designs (the generating equations are also included in Appendix G). Design charts and response curves are readily available in the literature.

Fano's undercoupled response shape is not well known; it is a "droopy" response with reduced selectivity but improved group time delay. Also, some of the required loaded-Q distributions have high values for the input resonator Q_L; this often enables the accommodation of high-power vacuum tube capacitances. Since these devices are often operated in class C, the generated pulses tend to have reduced overshoot. The equations for the undercoupled response are somewhat more involved, but are still suitable for hand-held calculators. This is especially convenient, because design charts and various response curves are not ordinarily available, although it is easy to generate them on automatic plotters. A sample table of loaded-Q distributions was provided (the generating equations are provided in Appendix G).

The minimum-loss-filter response shape is not a member of the family; it is obtained by making all loaded-Q values equal. It is fairly obvious that the resulting equal resonator efficiencies maximize the overall efficiency, given a certain loaded-Q product as a selectivity constraint. The asymptotic stopband selectivity approximation is even more simple than when resonator loaded-Q values are distinct. In many ways, the minimum-loss case serves as a geometric-mean representative of all other filter response shapes. For example, the approximate midband–dissipation loss expression shows that such loss in decibels is inversely proportional to the unloaded resonator Q_u. Therefore, in nearly any bandpass filter, it is safe to estimate that the midband decibel loss can be halved by doubling the Q_u values of the unloaded resonators. An approximate equation was derived for the optimal number of resonators that minimizes midband loss for a specified stopband attenuation; this is also a valuable rule of thumb for many bandpass-filter response shapes.

Minimum-loss selectivity shapes were described by fairly compact expressions, which included resonator dissipation (unlike most other response shapes). Therefore, the passband width can be estimated accurately for real filter elements. Both doubly terminated (lossy-source) and singly terminated (lossless-source) response shape equations were derived, and two hand-held calculator programs were furnished for generating and using attenuation (loss) curves. Taub (1963) and Taub and Hindin (1964) have published selectivity, time delay, and other graphs. These have a resemblance to Chebyshev overcoupled responses, with small ripple values that increase with filter degree.

Storch's method for raising second-order complex matrices to integer powers was used to obtain the general response equations. It is an important tool for filter designers, because it applies to any ABCD expression for a subnetwork iterated N times.

8.5. Comments on a Design Procedure

A complete design procedure will be described with the aid of a flowchart. The 13 major steps will be described, especially several topics that have not been discussed so far. These are design limitations, optimization of shunt inductance values, sensitivities, and filter tuning.

8.5.1. Design Flowchart. A step-by-step design procedure suitable for manual computation or computer programming is shown in Figure 8.28. The design step numbers correspond to the numbers of the main section headings in Appendix G and to the last number of the subsection headings in Section 8.6. The following discussion emphasizes each step as previously described or currently introduced.

The choice of a passband shape in step 1 is really a choice of loaded-Q distribution. Element constraints are sometimes sufficiently severe to dictate the use of an arbitrary set of loaded-Q values; the passband shape then is nameless and can be determined only by analysis. The stopband estimates are still viable, but step 3 has been preempted. Otherwise, one of the four shapes discussed in Section 8.4 is selected, or a new prototype shape is developed to provide a (normalized) loaded-Q distribution. Conventional shape specifications are: (1) the number of resonators (poles); (2) the passband parameter (such as decibel ripple, droop, or Q_L/Q_u); (3) the fractional passband width (F_p) and loss (L_p).

Step 2 in Figure 8.28 is to choose the configuration (e.g., Figure 8.2), the midband frequency (e.g., Figure 8.3), and the allowable ranges of component values (i.e., $\underline{L} \leqslant L \leqslant \overline{L}$ and $\underline{C} \leqslant C \leqslant \overline{C}$, where the underlines and overlines indicate lower and upper bounds, respectively). Generally, the shunt inductor Q_u values are the resonator Q_u values, because resonator capacitors usually have relatively little loss and inverter dissipation has little effect.

A specific passband shape is obtained by a unique loaded-Q distribution among the resonators in the absence of inverter frequency effects. The actual level to which this distribution is scaled can be determined by knowledge of the "$Q_{LN}F_p$" product for the shape. These values are easily calculated and tabulated, so that specification of F_p determines Q_{LN}. Thus, the N $\overline{Q}_L$ values normalized to Q_{LN} (also tabulated versus shape) can be unnormalized. Step 3 in Figure 8.28 concerns this calculation. The stopband specifications may result in greater Q_{LN} values than those resulting from the passband specification. Step 3 records the decision that the pass band may be more important, because selectivity increases and passband width decreases with increasing loaded Q (increased stored energy). Loss effects in the passband must also be considered in step 3. Dissipation tends to mask ripples, especially at the passband edges. However, the minimum-loss shape is known in the presence of uniform resonator dissipation.

Step 4 is based on the fact that stopband selectivity is completely specified

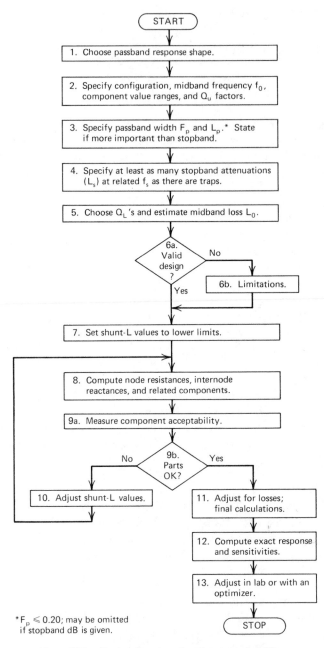

START

1. Choose passband response shape.

2. Specify configuration, midband frequency f_0, component value ranges, and Q_u factors.

3. Specify passband width F_p and L_p.* State if more important than stopband.

4. Specify at least as many stopband attenuations (L_s) at related f_s as there are traps.

5. Choose Q_L's and estimate midband loss L_0.

6a. Valid design?

No

6b. Limitations.

Yes

7. Set shunt-L values to lower limits.

8. Compute node resistances, internode reactances, and related components.

9a. Measure component acceptability.

No

9b. Parts OK?

Yes

10. Adjust shunt-L values.

11. Adjust for losses; final calculations.

12. Compute exact response and sensitivities.

13. Adjust in lab or with an optimizer.

*$F_p \leqslant 0.20$; may be omitted if stopband dB is given.

STOP

Figure 8.28. Design flowchart for direct-coupled filters.

by locating the Bode-plot breakpoint using the loaded-Q product and determining the asymptote tilt due to the surplus of inductive or capacitive inverters. The asymptote slope is 6N dB/octave when inverters are ideal (see Figures 8.11 and 8.17). End-coupling L sections affect stopband selectivity in the same way as like-kind inverters when their loaded Q exceeds 3; otherwise, there is an interpolating formula given in Appendix G. Trap inverter Z_0 and the electrical length both depend on frequency. However, trap inverters give a boost to the breakpoint when notch frequencies are between the passband edge and the second (or half) harmonic. The specific steps to determine the trap "shelf" attenuation levels are given in Appendix G. Step 4 in Figure 8.28 concerns the inverse calculation: if one or more selectivities are required at given frequencies, then the corresponding loaded-Q products and Q_{LN} values must be determined. The greatest Q_{LN} value thus found will satisfy all the stopband specifications.

Step 5 in Figure 8.28 requires a choice of the output resonator (Q_{LN}) from those calculated in step 3 (pass band) and step 4 (stop band). The greatest value of Q_{LN} is selected, unless the minimum passband width is most important, in which case that Q_{LN} is selected. The normalized loaded-Q values are then multiplied by the selected Q_{LN} value, and the midband loss is estimated by using each resonator's efficiency.

8.5.2. Design Limitations.

8.5.2. Design Limitations. Design step 6 requires two tests to indicate whether the specifications have resulted in a design potentially unsuited for this narrow-band method. The first test is to see whether

$$\Pi Q_{LK} > 3^N, \tag{8.111}$$

i.e., whether the geometric-mean, loaded-Q value exceeds 3. Excessively low loaded-Q values may make it impossible to absorb the negative inverter shunt branches into the adjacent resonator components of like kind. Also, the breakpoint loaded-Q-product approximation may fail.

The second test concerns the approximate midband insertion loss:

$$L_0 \leqslant 2N \text{ dB}. \tag{8.112}$$

This limitation results from the assumption that Q_L/Q_u is sufficiently small; it was used in several parts of the analysis of dissipation effects.

There are many useful cases that do not pass either or both of these tests; however, the designer should be alerted to potential difficulties.

8.5.3. Adjustment of Shunt-L Values. Design step 7 in Figure 8.28 sets resonator inductances to their lower limit, and design step 8 uses the loaded-Q values to determine parallel resistances at each node. The end-coupling configuration selected in step 2 will determine the resistance levels of the end resonators. The resonator capacitances and inverter reactances are thus determined. The latter are often too high. Step 9 calls for an examination of these dependent values to see if any are out of bounds. A convenient and well-conditioned measure of violations is the sum of squared differences between squares of top-coupling reactance (in kilohms). The constraint on the variables, namely the shunt inductors, is on their net values after combination of adjacent shunt inverter inductances and possibly a transformer or L-section inductance.

It may be possible to improve the design by adjustment of the shunt-L values if there are violations of element bounds. Step 10 in Figure 8.28 represents this process, which involves N variables in the usual situation. Adjustment of a particular shunt L, therefore that node's parallel resistance, sometimes succeeds on a cut-and-try basis because of the designer's insight into the seesaw impedance reaction of ideal (quarter-wave) inverters. Sometimes, it is useful to construct a constrained optimization (nonlinear-programming) problem. This procedure can minimize the sum of squared differences over the prototype shunt-L space with constraints on the final shunt-L reactance values after combining adjacent inverters and, perhaps, an L-section inductor. This may be accomplished by a conjugate gradient algorithm incorporating the nonlinear constraints by penalty functions, as described in Chapter Five. A simple enumeration search scheme may suffice when only discrete-valued sets of inductors are available.

8.5.4. Sensitivities.   Design step 11 in Figure 8.28 requires design adjustment for dissipative effects, as described in Section 8.3.3. The physical suitability of the resulting filter should then be determined by computing performance and sensitivities according to step 12. This will determine if the component and load tolerances are compatible with the performance expectations. Some important sensitivities of direct-coupled filters are remarkably simple.

To derive the sensitivity of filter input impedance with respect to resonator capacitance, consider the lossless prototype network at the tune frequency. For $N = 3$,

$$Z_{in} = \frac{Z_{012}^2}{R_{22}} = Z_{012}^2 Y_{023}^2 Z_L .$$
(8.113)

Similarly, for $N = 4$,

$$Z_{in} = Z_{012}^2 Y_{023}^2 Z_{034}^2 Y_L = Z_0^2 Y_L .$$
(8.114)

Now consider a filter that has more than four resonators, but the interest is in the fourth resonator $(K = 4)$. Suppose that there is a slight surplus of node capacity, say δC_K. Looking at the (8.114) result,

$$Y_L = G_{KK} + j\omega_0 \delta C_K = G_{KK}(1 + j\delta Q_{LK}).$$
(8.115)

Then the perturbed input impedance is

$$\delta Z_{in} = jZ_0^2 G_{KK} \delta Q_{LK} .$$
(8.116)

Therefore, (4.82) yields the direct-coupled-filter sensitivity of input impedance with respect to the synchronous capacitance of the $K\mathit{th}$ node:

$$S_{C_K}^{Z_{in}} = \pm jQ_{LK}, \qquad (8.117)$$

where the $+$ sign is for even values of N. The $\pm$ signs come from the identity

$$S_{C_K}^{Y_{in}} = -S_{C_K}^{Z_{in}}, \qquad (8.118)$$

which must be considered by comparing (8.113) and (8.114). It should now be clear why minimum-loss filters also have minimum sensitivity.

Similar consideration of (8.113) with respect to any Z_{0ij} yields the surprising result

$$S_{Z_{0ij}}^{Z_{in}} = \pm 2. \qquad (8.119)$$

Also, a detuning analysis of simple LC resonance shows that

$$\frac{\delta\omega_0}{\omega_0} = -\frac{\delta L/L + \delta C/C}{2} \qquad (8.120)$$

for the small perturbation factor δ.

The effect of small changes in the load reflection coefficient Γ_L on transfer function S_{21} is often of interest, especially the effect of changes in the transfer angle θ_{21}. It can be shown that

$$\Delta|S_{21}| = \pm 20\log_{10}(1 + |S_{22}||\Gamma_L|) \text{ dB}, \quad \text{and} \qquad (8.121)$$

$$\Delta\theta_{21} = \pm 57.3|S_{22}||\Gamma_L| \text{ degrees}, \qquad (8.122)$$

where S_{22} is the output scattering coefficient.

8.5.5.   Tuning.   Refer to Figure 8.1 and recall that an inverter changes a short circuit to an open circuit, and vice versa, according to (8.7). A synchronous filter–tuning procedure was described by Dishal (1951). Suppose that there are no end couplings. The following procedure is then used:

1. Set trap frequencies with a grid-dip meter and verify all inverter reactances on an impedance bridge.

2. Short-circuit to ground (or severely detune) the second node, *lightly* couple excitation to the first resonator with a small loop or small capacitive coupling, and then tune the first resonator for maximum voltage across it. Measure voltage V_1 with a very high-impedance voltmeter.

3. Remove the short circuit from node 2 and place it across node 3. Then tune resonator 2 for minimum V_1 (yes, V_1!).

4. Continue moving the short circuit to nodes nearer the load, consecutively tuning the newly unshorted odd (even) resonators for maximum (minimum) voltage V_1 across the input node.

5. Tune the output (load) node with the load resistance disconnected.
6. Connect the load resistance to the output port and a swept-frequency generator having the correct source impedance to the input port. Connect a synchronized oscilloscope to the filter output port, and fine tune the first and last resonators for final response shape.

When end couplings are employed, the following procedure is used:

1. Connect the proper source and use this for excitation.
2. Complete the preceding procedure, except that the last node with end coupling should be treated as an input node when it is adjusted (i.e., connect a matched generator to that end).

Voltage phases and magnitudes are easily confirmed at the tune frequency ω_0

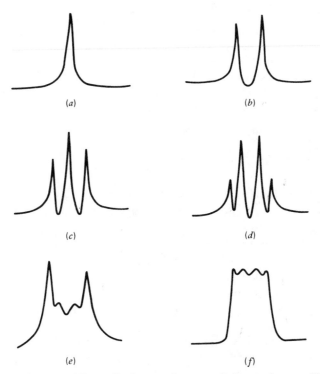

Figure 8.29. Oscillograms of V_1 amplitude versus frequency during synchronous filter tuning. (*a*) Resonator 2 is short-circuited, and resonator 1 is tuned for maximum voltage. (*b*) Only resonator 3 is short-circuited, and the second resonator is tuned for minimum V_1. (*c*), (*d*) The continued procedure. (*e*) Swept-frequency display when the unterminated output node is tuned. (*f*) Final V_1 Chebyshev response when the load resistance is reconnected. [From Dishal, M., *Proc. IRE*, Vol. 39, No. 11, p. 1451, November 1951. © 1951 IRE (now IEEE).]

using a high-impedance vector voltmeter:

1. Adjacent node voltages should differ in phase by ± 90 degrees.
2. Node voltage magnitudes are easily related to the node's parallel resistance, based on power.
3. The L-section end-coupling phase should agree with (6.7) or (6.8), as appropriate.

The swept-frequency display of input node voltage V_1 during the node-tuning sequence should appear as in Figure 8.29. Waveguide and other distributed filters are tuned according to the same principles, namely ideal resonator and inverter behavior. However, slotted lines and lightly coupled probes are often required. This is discussed by Matthaei et al. (1964, p. 668).

Dishal (1951) gives relationships that enable the calculation of coupling coefficients as in (8.26), based on the frequencies of the response peaks in Figure 8.29. These are often useful for evaluating aperture and other inaccessible coupling implementations.

8.5.6.   Summary of Comments on a Design Procedure.   A step-by-step design procedure for narrow-band, direct-coupled filters has been discussed in terms of the flowchart in Figure 8.28. The next section will proceed through almost all of these steps to illustrate the procedure in terms of the tabulated equations in Appendix G.

Design step 13 is the observation of filter behavior in the laboratory. The intent of the design procedure is to produce a first-time design success. The physical properties of many sophisticated filters, especially microwave filters, make final tweaking in the laboratory very difficult and expensive. Besides the analysis to determine proper coupling coefficients just suggested, an optimizer used in conjunction with a network analysis program often can confirm the presence and values of parasitic elements not previously considered. Such optimization can then be used to predict new filter component values that may offset the undesirable effects of the unexpected parasitic elements.

Finally, the great virtue of the loaded-Q filter design method should now be apparent: it is a flexible procedure that accommodates component limitations in nearly every stage of the design.

8.6.   A Complete Design Example

The design steps in Figure 8.28 will be calculated for the network in Figure 8.2 using an undercoupled response shape with selectivity, as in Figure 8.3. The step numbers appear after the last decimal in the following paragraph numbers. The equation numbers correspond to those in Appendix G.

8.6.1. Response Shapes

Input: Undercoupled, $N = 4$, $L_p = 1$ dB, $F_p = 0.1$, $L_d = 6$ dB.

$$\theta = 22.5 \text{ degees.} \tag{5}$$

$$\varepsilon = 0.5088. \tag{6}$$

$$\max L_d = 24.61 \text{ dB.} \tag{18}$$

$$q = 1.727. \tag{19}$$

$$k = 0.1017. \tag{20}$$

$$F_d = 0.1637. \tag{21}$$

$$C_N = 1.290. \tag{23}$$

N	L_d	$\overline{Q}_{L1}$	$\overline{Q}_{L2}$	$\overline{Q}_{L3}$	$\overline{Q}_{L4}$	$\Pi\overline{Q}_L$	$Q_{L4}F_d$
4	6	0.1653	0.5281	0.9738	1.	0.0850	2.092
		0.5424	1.824	3.246	1.	3.211	0.8438*
		1.844	5.985	3.363	1.	37.11	0.4576
		6.050	5.892	3.195	1.	113.9	0.3458

*Arbitrarily selected for this example.

8.6.2. Physical Data

Input: $f_0 = 100$ MHz, $90 \leqslant L \leqslant 900$ nH, $Q_{uL} = 200$, $1 \leqslant C \leqslant 100$ pF.

$$\hat{X}_{Lij} = 311 \text{ ohms,} \qquad \hat{X}_{Cij} = 803 \text{ ohms.} \tag{32}$$

8.6.3. Pass Band

Input: See Section 8.6.1.

$$Q_{L4} = 0.8438/0.1637 = 5.155. \tag{34}$$

8.6.4. Stopbands

Input: $f_{s1} = 140$ MHz, $L(f_{s1}) = 60$ dB, $f_{s2} = 200$ MHz, $L(f_{s2}) = 80$ dB, $NLINV = 1$,   L sections $= 1$ L, $NTRAPS = 2$, $NCINV = 0$.

$$1. \quad \bar{f}_s = 1.4 < 2, \qquad \bar{f}_K = 1.704, \qquad a = .4854. \tag{51}$$

$$m = 6(4 + 3 - 1) = 36 \text{ dB/octave.} \tag{50}$$

$$L_K = 61.69 \text{ dB.} \tag{49}$$

$$DB5 = 9.408 + 8.743 = 18.16. \tag{48}$$

$$DB4 = 4.629 \text{ dB,} \qquad \text{assuming } Q_0 > 3. \tag{47}$$

$$DB3 = 4.629 \text{ dB.} \tag{45}$$

$$DB2 = 3.849. \tag{44}$$

$$DB1 = 36.40. \tag{43}$$

$$\Pi Q_L = 66.00. \tag{42}$$

$$Q_{L4} = 2.129. \tag{41}$$

2. $\bar{f}_s = 2 \geqslant 2, \qquad \bar{f}_K = 3, \qquad a = 1.$ (51)

$$m = 6(4 + 2 - 2) = 24 \text{ dB/octave.} \tag{50}$$

$$L_K = 82.39 \text{ dB.} \tag{49}$$

$$DB5 = 1.779 + 14.64 = 16.43. \tag{48}$$

$$DB4 = 9.54. \qquad \text{(See comment in Section 8.6.12.)} \tag{47}$$

$$DB3 = -9.54. \tag{45}$$

$$DB2 = 34.08. \tag{44}$$

$$DB1 = 37.89. \tag{43}$$

$$\Pi Q_L = 78.42. \tag{42}$$

$$Q_{L4} = 2.223. \tag{41}$$

8.6.5. Q Effects

$Q_{L4} = \max(2.129, 2.223) = 2.223;$

$$\text{also, } 5.155 > 2.223 \therefore Q_{L4} = 2.223. \tag{52}$$

$$Q_{L1} = 1.206, \qquad Q_{L2} = 4.055, \qquad Q_{L3} = 7.216, \qquad Q_{L4} = 2.223. \tag{3}$$

$$\hat{Q}_{L1} = 6.029\text{E} - 3, \qquad \hat{Q}_{L2} = 0.0203, \qquad \hat{Q}_{L3} = 0.0361, \qquad \hat{Q}_{L4} = 0.0111. \tag{53}$$

$$L_0 \doteq 0.323 \text{ dB.} \tag{54}$$

8.6.6. Design Limitations

$$\Pi Q_L = 78.42 \overset{?}{\geqslant} 3^4 = 81 \qquad \text{Not quite.} \tag{55}$$

$$L_0 = 0.323 \overset{?}{\leqslant} 2 \times 4 = 8 \qquad \text{Yes.} \tag{56}$$

8.6.7. Minimum Shunt Inductance

$$L_k = 90 \text{ nH}, \qquad K = 1, 2, 3 \qquad (L_4 \text{ dependent).} \tag{57}$$

8.6.8. Prototype Ohmic Values

$$R_{11} = 68.19, \qquad R_{22} = 229.29, \qquad R_{33} = 408.05,$$

$$R_{44} \overset{\Delta}{=} 50, \qquad L_{IV} = 35.79 \text{ nH.} \tag{58}$$

$$X_{12} = 125.04, \qquad X_{23} = 305.88, \qquad X_{34} = 142.83 \text{ ohms.} \tag{59}$$

$$X_{g1} = 30.15 \text{ ohms} \qquad \text{(lossless network).} \tag{60}$$

8.6.9. Component Acceptability

$$E = 0.1714. \tag{64}$$

8.6.10. Shunt Inductance Adjustment.   For possible optimization (not used):

$$K_L = 10^{-9}, \tag{66}$$

$$\frac{\partial E}{\partial L_1} = -0.1419, \tag{67}$$

$$\frac{\partial E}{\partial L_2} = -0.2977, \tag{68}$$

$$\frac{\partial E}{\partial L_3} = -0.1976, \tag{68}$$

$$\frac{\partial E}{\partial L_4} = 0. \tag{69}$$

8.6.11. Final Component Values

$$\hat{R} = 0.9898. \tag{70}$$

K	Y_K	X_K
1	1.	$6.029E-3$
2	0.02027	1.0001
3	1.0007	0.04211
4	0.03139	1.0006

$R_{11}(\text{lossy}) = 67.49.$ (72)

$X_{g1} = 29.49.$ (73)

$X_{p1} = 114.83.$ (74)

$L_{g1} = 46.93, \quad C_1 = 42.01, \quad L_1 = 164.31, \quad L_{12} = 97.47,$

$C_{12} = 13.26, \quad L_2 = 248.01, \quad C_2 = 28.14, \quad L_{23} = 365.12,$ (77)–(79)

$C_{23} = 1.734, \quad C_3 = 28.14, \quad L_3 = 214.67, \quad L_{34} = 227.34,$

$L_4 = 42.49, \quad C_4 = 70.77 \quad (\text{in pF, nH}).$

8.6.12. Performance and Sensitivity Analysis. The preceding component values, applicable to Figure 8.2, were used to obtain an exact S_{21} frequency response by analysis; the result is equivalent to the graph in Figure 8.3, except that the midband loss was 0.400 dB and the minimum selectivity response above $f_{s2} = 200$ MHz was 83.75 dB instead of the specified $L_{s2} = 80$ dB. This excess selectivity is partly attributable to the 9.54-dB contribution of (G.47), as opposed to the more exact contribution of 3.54 dB of (G.46) using $Q_0 \doteq 0.603$. (The value of Q_0 is known only near the end of the procedure.) This 6-dB difference would give an estimate 2.25 dB low, which is attributable to the marginal condition in (G.55) and the approximation in (G.49). As computed, the selectivity slightly exceeded specifications, and the predicted midband loss of 0.322 dB was slightly exceeded.

As indicated in (G.80)–(G.83), the important parameter in the sensitivity of S_{21} to load SWR is the value of $|S_{22}|$; the typical behavior of this value is shown in Figure 8.30. These data, used in (G.81) and (G.82), indicate that in the passband the transfer response can vary by as much as ± 0.8 dB, and the transfer phase can vary by as much as ± 1.2 degrees for a 1.2 : 1 load SWR of arbitrary reflection phase.

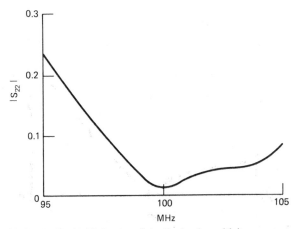

Figure 8.30. Typical behavior of the S_{22} load sensitivity parameter.

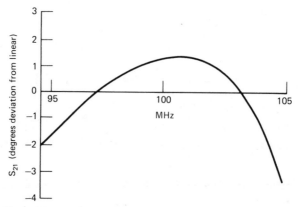

Figure 8.31.   Typical undercoupled deviation from linear phase.

Using a least-squares fitted line as a phase reference, a frequency analysis provided the typical deviation from linear phase shown in Figure 8.31. The gentle rate of change of the undercoupled response (Figure 8.26) is directly related to good phase linearity. Its moderate selectivity has been supplemented in the upper stop band, at the expense of the lower stopband in this case, without seriously disturbing the passband.

The maximum sensitivity of input impedance with respect to capacitance occurs at node 3 with C_{III}; according to (G.85), it is $\pm j7.216$.

8.6.13.   Design Adjustment.   The exact analysis in the preceding section also confirmed the phase angles between nodes at the midband frequency ($f_0 = 100$ MHz). As in (G.87) and Figure 8.2, angle θ_{G1} was 30.6 degrees, and the other node voltage angles differed by the expected 90 degrees, according to the prototype network (Figure 8.1).

Problems

8.1.   Use ABCD parameters to prove that $Z_{in} = Z_0^2/Z_L$ for an ideal inverter.

8.2.   Consider the following network:

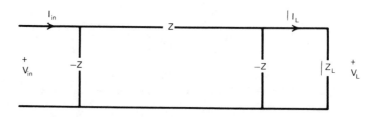

For $Z=0+jX$, find (1) Z_{in}, (2) V_L/I_{in}, and (3) I_L/V_{in}. Assume $I_L=1$ ampere.

8.3. Use ABCD parameters to show that a series impedance (Z) between two unit inverters is equivalent to a shunt admittance (Y). In particular, assume that $Z=j\omega_0 LF$; what is the equivalent Y?

8.4. A 12-element lowpass network is composed of six identical subsections, each as in Figure 4.6 (Chapter Four), with $Z=j\omega$ and $Y=j2\omega$. The overall network is terminated by 1-ohm resistors. Calculate the overall ABCD matrix, S_{21}, and $L=-20\log|S_{21}|$ at $\omega=1/\sqrt{2}$, 1, and $\sqrt{2}$ radians.

8.5. Derive the C expression in the overall ABCD matrix of an $N=3$ ideal prototype network. Compare your result to Equation (8.11).

8.6. Derive $L(\omega)$ in Equation (8.14) in terms of the loaded Q_L values of an $N=3$ ideal prototype network. Compare your result to Equation (8.16).

8.7. A lowpass pi network has an equivalent two-port network that contains an inverter, as shown below.

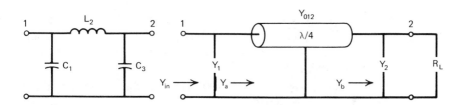

(a) For $\omega=1$ radian, find expressions for Y_{012}, Y_1, and Y_2.
(b) For $R_L=1$, $C_1=4$, $L_2=\frac{1}{2}$, and $C_3=3$, find Y_b, Y_a, and Y_{in}.

8.8. Suppose that $N=3$, $\overline{Q}_{L1}=\overline{Q}_{L3}=1$, $\overline{Q}_{L2}=2$, $f_0=50$ MHz, and 60 dB is required at 90 MHz using a doubly terminated network. For $R_{11}=50$ and $R_{33}=100$ ohms, find the element reactances at f_0 and the lumped-element values using (1) L_{12}, L_{23}, and $R_{22}=150$ ohms; (2) L_{12}, C_{23}, and $R_{22}=75$ ohms; and (3) C_{12}, C_{23}, and $R_{22}=75$ ohms.

8.9. Design a direct-coupled filter that
(a) Uses only L and C components and no traps or end coupling.
(b) Has three resonators, and $R_{22}=200$ ohms.
(c) Is tuned to 100 MHz.
(d) Provides at least 60-dB attenuation at 150 MHz.
(e) Provides at least 52-dB attenuation at 70 MHz.
(f) Operates between two 50-ohm terminations.
(g) Has minimum midband loss for inductor $Q_{uL}=200$ and lossless capacitors.

Find (1) the loaded-Q values, (2) element values for a lossless network, and (3) the approximate dissipative loss in dB at 100 MHz. Describe changes/additions that would raise the entire impedance level without affecting the required selectivity. Could the number of resonators be changed to reduce midband loss while meeting the same stopband loss? If so, how many, and what is the midband loss?

8.10. Derive Equations (8.64) and (8.65).

8.11. Suppose that the filter in Figure 8.12 is to have an overcoupled, 0.5-dB-ripple passband and the original stopband selectivity. Find the new Q_L values and the 0.5-dB passband width. Ignoring inverter tilt, what is the 3-dB bandwidth?

8.12. In Example 8.3, confirm the given values of C_{23} and L_{23} using Equation (8.29).

8.13. Replace the shunt inductors in the filter in Example 8.3 with capacitively loaded, short-circuited transmission line (stub) resonators. Recalling that the stub resonator reactance at tune frequency is $X = Z_0 \tan \theta$, find Z_{01}, Z_{02}, and Z_{03} for $\theta = 45$ degrees, $Q_{L1} = Q_{L3} = 4.2229$, $Q_{L2} = 8.4458$, and $R_{22} = 600$ ohms.

8.14. Suppose that short-circuited, 30-degree stubs are used in resonators.
   (a) What increases or decreases in loaded Q, defined by Equation (8.3), account for the changes in passband width or the attenuation at $f = 1.8f_0$?
   (b) If the desired passband is at 100 MHz, at what frequency is the first spurious passband?
   (c) If these stubs are also used as coupling elements between nodes, how much will each one affect the attenuation at $f = 1.8f_0$?

8.15. Use Storch's method to obtain the first three expressions for minimum-loss, doubly terminated filters in Table 8.9.

8.16. Obtain an expression for the minimum possible loss by solving Equation (8.106) for Q_L and putting this into Equation (8.103).

8.17. An rf transformer has primary and secondary reactances of 30 and 55 nH, respectively. What is its inverter Z_0 at 100 MHz when (coupling coefficient) $K = 0.1$?

Chapter Nine

Other Direct Filter Design Methods

This chapter contains a variety of filter design methods that extend previous methods in several ways. Equal-stub admittance filters introduce another frequency transformation and apply the Storch method for replicating a typical network section into the overall ABCD transfer function. The result is a microwave filter that has important practical relationships to edge-coupled filters, according to a recent article. It is a direct design method because the synthesis techniques in Chapter Three are not required. A program for a hand-held computer is furnished for evaluating the selectivity function.

The concept of traps that cause selectivity notches, introduced in the last chapter, is formalized by considering classical elliptic (Cauer) filter theory. A computer program and nomograph are provided to assist in estimating the elliptic filter order required to meet design specifications. However, synthesis is avoided by Amstutz's spectacular direct design method for accurately determining lowpass filter element values. The family of Butterworth and Chebyshev functions is extended by addition of the inverse Chebyshev and elliptic functions. The concept of double periodicity and equal ripple in both passband and stopband is described in terms of polynomial roots and filter specifications. The basis of Amstutz's permuted-trap, equivalent two-port networks is developed, and two Amstutz programs have been translated into BASIC and included in this treatment.

Several useful transformations of lumped-element branches into other lumped-element configurations are described. First-order (slope) approximations are included to eliminate the redundant traps which result from conventional lowpass-to-bandpass reactance transformations. A compilation of exact transformations from Zverev (and several more from Dishal) are included.

The geometric description of two-port network behavior, contained in Chapter Seven, is applied to the passive network case, with special attention to load effects on mapping and input standing-wave ratio. A theorem concerning composite bilinear functions between different unit circles is provided as a fundamental analysis tool for obtaining concise design results.

The load effect analysis is extended to derive a design basis for filters that absorb rather than reflect energy. These may be viewed as frequency-selective attenuators (pads). They are substantially less selective than reflection filters, but there are many important applications for impedance control and damping of spurious oscillations in system stopbands. Design graphs for several of these invulnerable (load-independent) filters will be provided.

9.1. Equal-Stub Admittance Filters

The direct-coupled filter design method in Chapter Eight provided a means for designing transmission line filters, as shown in Figure 9.1. For 1-ohm terminations, the shorted-circuited, quarter-wave transmission line resonators have loaded Q values

$$Q_{LK} = Y_{0K} \frac{\pi}{4},$$ (9.1)

according to (8.50). Note that the internal resonators in Figure 9.1c have been divided into two parallelled resonators so that each may have half the required Y_0 (and to minimize the generation of extraneous electromagnetic modes). Even so, the Q is too low and the bandwidth too wide for direct-coupled-filter predictions to apply (see Section 8.5.2), since the transmission line normalized admittance is within the range $0.1 \leqslant Y_0 \leqslant 5.1$. The filter can still be constructed, however, and its periodic selectivity curve will appear as in Figure 9.2. The term "commensurate" refers to distributed-element filters composed of components that have the same electrical length.

This section will apply Storch's result to raise a 2×2 complex ABCD matrix to the N*th* power and thereby obtain the response of filters like those in Figure 9.1c. A design basis will result through the choice of a few parameters. In the process, a frequency transformation will be introduced; it is typical of all commensurate-network selectivity functions. A program for the HP-67/97 hand-held calculator will be provided to evaluate the filter selectivity function. Also, the equivalent network for edge-coupled conductors on planar dielectric sheets (printed circuits) will be described. It is equivalent to the dual of the network in Figure 9.1c, using open-circuited stubs in series. These developments can be applied to that case according to the treatment in a very recent article, which circumvents some severe coupling restrictions.

9.1.1. Equal-Stub-Filter Development.   According to Wenzel (1964), all commensurate line networks have transducer gain functions in the form

$$S_{21} = \frac{P(q^2)}{\sin^2 \theta},$$ (9.2)

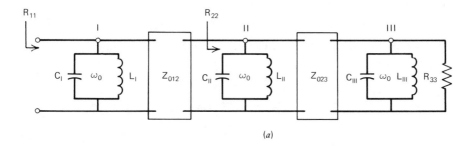

(a)

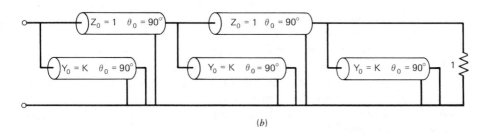

(b)

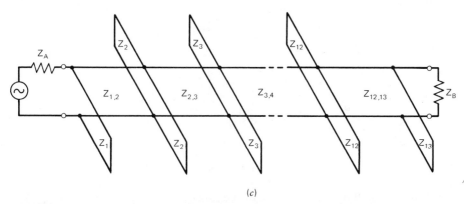

(c)

Figure 9.1. A direct-coupled filter using physical quarter-wave resonators separated by physical quarter-wave lines. (a) Prototype inverters between resonators; (b) coaxial transmission line resonators and inverters; (c) open-wire transmission lines used as inverters and paralleled resonators.

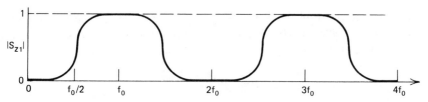

Figure 9.2. Quarter-wave commensurate filter periodic selectivity.

where P is a polynomial in variable q^2:

$$q = \cos\theta. \tag{9.3}$$

Angle θ is the filter component electrical length at any frequency f:

$$\theta = \frac{\pi}{2}\frac{f}{f_0}. \tag{9.4}$$

It is easy to show that such functions have arithmetic symmetry according to

$$\frac{f_1}{f_0} = 2 - \frac{f_2}{f_0} \tag{9.5}$$

between any two frequencies (f_1 and f_2) having the same selectivity. The polynomial P must be obtained for each particular case; it will be obtained here for the network in Figure 9.1c.

Storch's matrix result was applied to replicated filter subsections in Section 8.4.5. In the case shown in Figure 9.1c, the typical section to be replicated is shown in Figure 9.3. To find the ABCD matrix for the typical subnetwork, consider the two cascade lines on both sides of the shunt resonator as three cascaded sections. Because of their symmetry, the ABCD matrix of the subsection in Figure 9.3 is:

$$T_i = \begin{pmatrix} A_i & B_i \\ C_i & D_i \end{pmatrix} = \begin{pmatrix} A & B \\ B & A \end{pmatrix}\begin{pmatrix} 1 & 0 \\ Y & 1 \end{pmatrix}\begin{pmatrix} A & B \\ B & A \end{pmatrix}, \tag{9.6}$$

where

$$A = \cos\frac{\theta}{2}, \tag{9.7}$$

$$B = j\sin\frac{\theta}{2}, \tag{9.8}$$

and

$$Y = -jK\cot\theta. \tag{9.9}$$

K is the characteristic admittance of the shorted-stub resonator. Therefore, the ABCD matrix of the ith subsection is

$$T_i = \begin{bmatrix} (ABY + A^2 + B^2) & (B^2Y + 2AB) \\ (A^2Y + 2AB) & (ABY + A^2 + B^2) \end{bmatrix}. \tag{9.10}$$

Storch's matrix result in (8.90) (Section 8.4.5) can be applied to obtain the elements of the overall ABCD matrix T_N. For lossless, reciprocal, symmetric networks, (8.96) reduces to

$$|S_{21}|^2 = 1 - \frac{(B_N - C_N)^2}{4}. \tag{9.11}$$

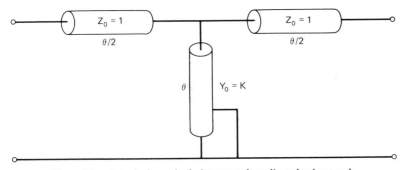

Figure 9.3. A typical equal-admittance-stub replicated subnetwork.

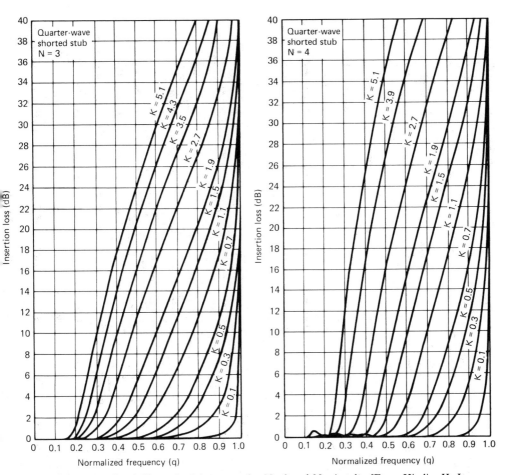

Figure 9.4. Equal-stub-filter selectivity curves for N = 3 and N = 4 stubs. [From Hindin, H. J., and Taub, J J. *IEEE Trans. Microwave Theory Tech.*, Vol. MTT-15, No. 9, pp. 526–527, September 1967. © 1967 IEEE.]

Therefore, only two ABCD parameters are required; (8.90) yields

$$B_N = P_N(y)B_i \tag{9.12}$$

$$C_N = P_N(y)C_i, \tag{9.13}$$

and

$$y = A_i + D_i = 2A_i. \tag{9.14}$$

Two double-angle trigonometric identities reduce (9.14) to

$$y = 2(\cos\theta)\left(1 + \frac{K}{2}\right). \tag{9.15}$$

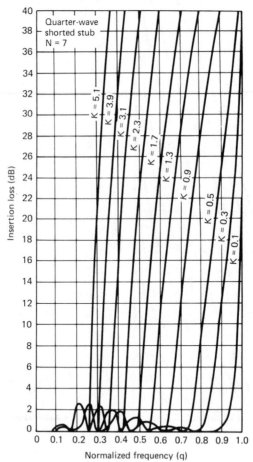

Figure 9.5. Equal-stub-filter selectivity curves for N=7. [From Hindin, H. J., and Taub, J. J. *IEEE Trans. Microwave Theory Tech.*, Vol. MTT-15, No. 9, pp. 526–527, September 1967. © 1967 IEEE.]

Another trigonometric identity can be applied to show that

$$B_N - C_N = P_N(y)(B_i - C_i) = -YP_N(y).$$ (9.16)

Therefore, (9.11) yields the transducer function

$$|S_{21}|^2 = 1 + \frac{K^2 \cos^2\theta}{4\sin^2\theta} P_N^2 \left[2(\cos\theta)\left(1 + \frac{K}{2}\right) \right].$$ (9.17)

The desired result is obtained using frequency variable (9.3):

$$L = 10\log_{10}\left\{ 1 + \frac{K^2 q^2}{4(1-q^2)} P_N^2 \left[2q\left(1 + \frac{K}{2}\right) \right] \right\} \text{ dB}.$$ (9.18)

Selectivity curves for three and four stubs are shown in Figure 9.4. The seven-stub selectivity curve is shown in Figure 9.5.

9.1.2. Equal-Stub-Filter Design Procedure. The frequency variable q is defined in (9.3) and (9.4). Program A9-1 (Appendix A) for the HP-67/97 computer evaluates the Chebyshev polynomials of the second kind according to (8.91) and computes the selectivity function described by (9.18). It also implements the optional input of several equivalent bandwidth parameters. Since there is arithmetic symmetry according to (9.5), the midband frequency is

$$f_0 = \frac{f_1 + f_2}{2}.$$ (9.19)

Therefore, selectivity may be evaluated using

$$q = \cos\left[\left(1 - \frac{w}{2}\right) \cdot \frac{\pi}{2} \right],$$ (9.20)

where bandwidth w is

$$w = \frac{f_2 - f_1}{f_0}.$$ (9.21)

Program A9-1 accommodates these various ways for describing the bandwidth. Also, a useful relationship for wavelength in free space is

$$\lambda = 11802.8/\text{MHz} \quad \text{(in inches)}.$$ (9.22)

Example 9.1. Design a bandpass filter with a minimum, 3-dB bandwidth of 630 MHz, a center frequency of 900 MHZ, and a minimum, 20-dB rejection at 900 ± 500 MHz. From (9.22), each stub must be 3.28 inches long in air and separated by the same distance. The 3-dB, $w = 0.7$ bandwidth ($\frac{630}{900}$) corresponds to $q = 0.522$, according to (9.20) and (9.21). The 20-dB, $w = 1.11$ bandwidth corresponds to $q = 0.766$. The four-stub curves in Figure 9.4 show that the normalized admittance $K = 1.4$ (35.7-ohm stubs in a 50-ohm system) will be satisfactory. By the graph in Figure 9.4 or Program A9-1, the passband ripple will not exceed 0.3 dB.

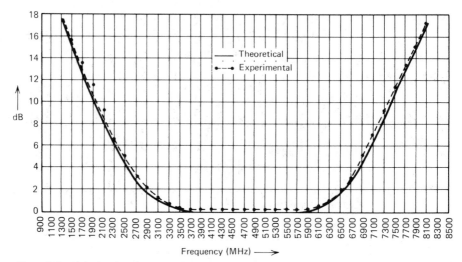

Figure 9.6. Calculated and measured results for N = 3, K = 1. [From Hindin, H. J., and Taub, J. J. *IEEE Trans. Microwave Theory Tech.*, Vol. MTT-15, No. 9, pp. 526–527, September 1967. © 1967 IEEE.]

Example 9.2. Design, analyze, and test a filter having three 50-ohm stubs in a 50-ohm system by using a center frequency of 4700 MHz. Hindin and Taub (1967) constructed this filter. The calculated and measured results are shown in Figure 9.6. The midband loss is $L_0 = 0.2$ dB.

9.1.3. *Variations for Printed-Circuit Filters.* The spaced-stub implementation shown in Figure 9.1c probably is not the main application of this filter design method. Because of several important network equivalence relations, the same response may be obtained in filters that are conveniently implemented in stripline or microstrip configurations. Consider the network in Figure 9.7, which is dual to that in Figure 9.1c. Changing admittance K to impedance Z in (9.9) adapts the entire development to open-circuited stubs, instead of the original short-circuited stubs. Thus selectivity function (9.18) and the selectivity graphs in Figures 9.4 and 9.5 apply when K is replaced

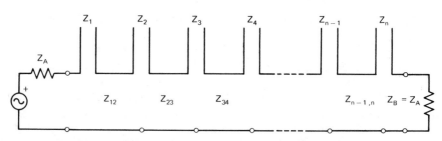

Figure 9.7. Quarter-wave open-circuited stubs with quarter-wave spacing.

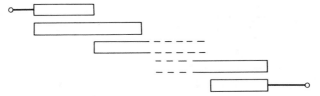

Figure 9.8.   Edge-coupled printed-circuit filter arrangement.

with Z. This dual case is considered because the edge-coupled filter in Figure 9.8 is exactly equivalent to the open-circuited stub filter in Figure 9.7. This equivalence results from the more fundamental equivalence given by Matthaei et al. (1964) and shown in Figure 9.9.

The two coupled bars in Figure 9.9 are over a ground plane (microstrip) or between ground planes (stripline). They have a distributed per-unit capacitance (and inductance) between them and to ground, as do all transverse electromagnetic (TEM) systems. The distributed capacitance is exactly analogous to the static capacitance. Because superposition applies to linear systems, it is convenient to formulate the system description in terms of the even-mode capacitance (when both bars are at the same potential) and the odd-mode capacitance (when the bars have opposite potentials with respect to their ground plane). The equivalent open-circuited stubs and the separating cascade line in Figure 9.9 have wave impedances that are linear combinations of the even- and odd-mode wave impedances.

The point is that an equivalent network does exist. The interested reader may consult Matthaei et al. (1964) for the various means of calculating even- and odd-mode capacitances of various structures. There is also a dual printed-circuit, edge-coupled filter structure corresponding to the original shorted stubs, but it requires parallel bars with ends short-circuited to the ground plane. This is difficult to implement in practice.

Another difficulty that arises is due to a basic incompatibility between the stub filters and the edge-coupled realization: the former is generally a wide band filter, and the coupling between adjacent parallel bars in the latter is often insufficient for this purpose. Minnis (1981) has employed a well-known physical transformation to avoid this problem. There are many other variations of the stub design; for example, stub impedance may be varied to obtain

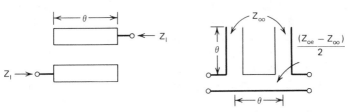

Figure 9.9.   Schematic and equivalent circuit of two coupled bars. [From Matthaei et al., 1964.]

Chebyshev and Butterworth response shapes. (For the latter, see Mumford, 1965; also, see Minnis, 1981, and references therein.)

9.1.4. Summary of Equal-Stub Admittance Filters.

9.1.4.   Summary of Equal-Stub Admittance Filters.   Commensurate filters are composed of distributed elements having equal electrical lengths. It has been shown that there is a standard form for all commensurate-filter transducer functions. Also, the frequency variable is a cosine function, and the resulting frequency behavior has arithmetic symmetry. Commensurate filters composed of quarter-wave, short-circuited shunt stubs separated by quarter-wave, cascaded lines were considered; the same development applied to their duals, employing open-circuited stubs connected in series. The development was based on a typical subsection consisting of one-eighth-wavelength, cascade-connected lines on either side of the shunt stub resonator. Storch's method was used to exponentiate the ABCD matrix of the typical subsection, leading to a compact expression for the filter's transducer gain function.

Typical selectivity curves were provided, and a hand-held computer program enabled computation of the selectivity function. Two examples of the shorted-stub realization were included. However, the filter is often implemented in a printed-circuit form, which is made possible by an equivalence between parallel-coupled bars and separated wireline stubs. In that case, the physical parameters are the distributed even- and odd-mode capacitances or equivalent impedances. This stub filter design is intrinsically wide band. However, the edge-coupled bars are loosely coupled for even close spacing. The introduction of redundant commensurate elements, as recently described by Minnis (1981), solves this problem.

The limited range of characteristic impedances in transmission lines differs drastically from the greater ranges of lumped-element component values that are available. Microwave filter design generally is constrained by these limitations, and an important part of these techniques involves means to circumvent the physical limitations.

9.2. Introduction to Cauer Elliptic Filters

Filters having all poles of attenuation at infinite frequency are called "all-pole filters." Chebyshev all-pole filters were considered in Section 8.4.1. They had ripples of equal magnitude in the passband and a monotonic stopband. Elliptic filters exhibit equal-ripple behavior in both the passband and the stopband. The finite frequencies of attenuation poles are of considerable interest because they are crucial to selectivity relationships. It was shown by (2.34) that the Chebyshev passband behavior may be described by trigonometric functions; these are periodic in one variable, of course. Elliptic filter design requires functions that are doubly periodic; these are the Jacobian elliptic functions.

This section will introduce the background and identify the parameters necessary to design Cauer elliptic filters. The progression of Butterworth, Chebyshev, inverse Chebyshev (only stopband ripples), and Cauer elliptic filter functions will be reviewed. The application of Jacobian elliptic functions will be discussed. However, details of their origin and formulation would obscure the important design results to follow, and are omitted. The traditional belaboring of this subject will be bypassed in the interest of clarity.

This section will deal with filter characteristics in an increasingly complex manner, up to and including the five kinds of elliptic filters to be designed by the computer programs described in Section 9.3.

9.2.1. From Butterworth to Elliptic Filter Functions. In order to conform to practice in the elliptic filter field, certain redefinitions of familiar parameters follow. For passband edge ω_B, the following lowpass filter functions are expressed in terms of normalized frequency:

$$x = \frac{\omega}{\omega_B}. \tag{9.23}$$

Some familiar functions in this variable will be reviewed. For example, the Butterworth function from (8.71) is

$$A(\omega) = 10\log(1 + \varepsilon^2 x^{2N})\,\mathrm{dB}, \tag{9.24}$$

where

$$\varepsilon = (10^{0.1A_{max}} - 1)^{1/2}. \tag{9.25}$$

A_{max} (in dB) is defined as the passband-edge attenuation, as shown in Figure 9.10a. The Chebyshev function from (8.61) is

$$A(\omega) = 10\log\left[1 + \varepsilon^2 T_N^2(x)\right]\,\mathrm{dB}. \tag{9.26}$$

$T_N(x)$ is the Nth-degree Chebyshev polynomial of the first kind, described in Section 2.4.1; it has equal-ripple unit amplitude in the range $-1 \leqslant x \leqslant +1$. This is the familiar response shape in Figure 9.10b.

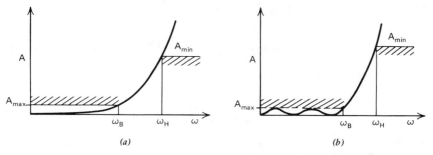

Figure 9.10. Attenuation characteristics of all-pole (a) Butterworth and (b) Chebyshev filters.

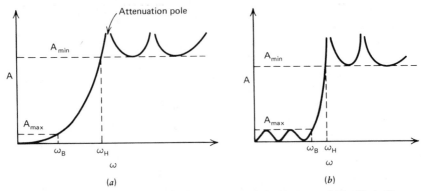

Figure 9.11. Attenuation characteristics of (*a*) inverse Chebyshev and (*b*) elliptic filters.

Two new response shapes are shown in Figure 9.11. The inverse Chebyshev function is

$$A(\omega) = 10\log_{10}\left[1 + \left\{\varepsilon\frac{T_N(\omega_H/\omega_B)}{T_N[(\omega_H/\omega_B)/x]}\right\}^2\right] \text{dB}. \qquad (9.27)$$

This response is shown in Figure 9.11*a*; the passband edge is still ω_B at A_{max} dB. Now there is a stop band beginning at ω_H with shelf attenuation A_{min} dB:

$$A_{min} = 10\log\left[1 + \left(\varepsilon T_N\frac{\omega_H}{\omega_B}\right)^2\right] \text{dB}. \qquad (9.28)$$

The name of this filter comes from the inverse structure of the response of its argument in (9.27). It is not obvious, but the passband is maximally flat at the origin. Furthermore, the filter degree relationship to A_{min} and A_{max} is exactly the same as for the Chebyshev filter, namely (8.64). Also, the frequencies of the stopband poles of attenuation are easily determined by (8.63) and inspection of (9.27).

The most complicated function to be considered is that of the elliptic filter, also known by the names Cauer, Darlington, and Zolotarev. The elliptic filter function is

$$A(\omega) = 10\log\left\{1 + \left[\varepsilon R_N(x, L)\right]^2\right\} \text{dB}, \qquad (9.29)$$

where R_N is a ratio of even polynomials of degree N, and

$$L = \left(\frac{10^{0.1A_{min}} - 1}{10^{0.1A_{max}} - 1}\right)^{1/2}. \qquad (9.30)$$

The elliptic filter response shape is shown in Figure 9.11*b*. In contrast to the familiar lowpass network structure (e.g., Figure 6.18 in Section 6.3), lowpass elliptic filters take one of the three forms shown in Figure 9.12.

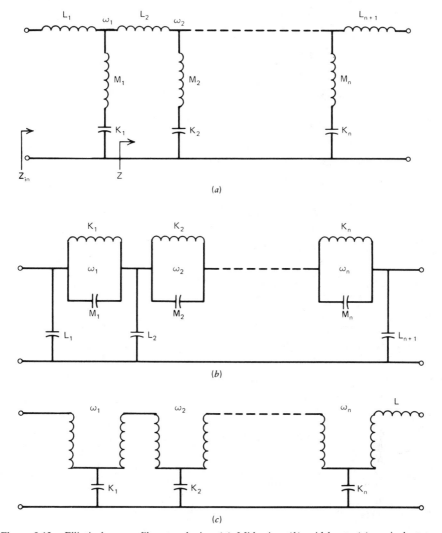

Figure 9.12. Elliptic lowpass filter topologies. (*a*) Midseries; (*b*) midshunt; (*c*) equivalent to midseries using transformers. [From Amstutz, P., *IEEE Trans. Circuits Syst.*, Vol. CAS-25, No. 12, p. 1002, December 1978. © 1978 IEEE].

The rational function in (9.29) can be expressed in its factored form in terms of complex frequency variable s:

$$R(s) = \frac{(s^2 - \omega_1^2)(s^2 - \omega_3^2)(s^2 - \omega_5^2)\cdots}{(s^2 - \omega_2^2)(s^2 - \omega_4^2)(s^2 - \omega_6^2)\cdots}. \tag{9.31}$$

The poles and zeros of (9.31) are computed by a simple calculation involving

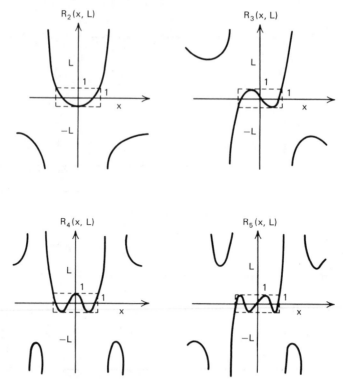

Figure 9.13. Low-order Chebyshev rational functions. [From Daniels, 1974. Copyright 1974 Bell Telephone Laboratories. Reprinted by permission.]

the Jacobian elliptic sine function, which is doubly periodic and only slightly harder to evaluate than the trigonometric sine function. This calculation is included in BASIC Program B9-1. It requires less than 11 lines of code. The rational functions so obtained behave as indicated in Figure 9.13. It is easy to imagine how the square of these functions produces the frequency response in Figure 9.11*b*.

9.2.2. Elliptic Filter Degree, Attenuation, and Pole Frequencies. Zverev (1967) published the well-known nomogram in Figure 9.14, which relates the elliptic filter degree and attenuation to the bandwidth parameters. Figures 9.11*b* and 9.14 show that the passband ripple ($A_{max}=0.1$ dB) and the stopband shelf attenuation ($A_{min}=43$ dB) require $N=5$ for a stopband/passband transition ratio (ω_H/ω_B) of 1.5. Program B9-1 has been translated to BASIC from a program by Daniels (1974). It duplicates the calculations illustrated in Figure 9.14 and computes the attenuation pole (trap) frequencies. A word of caution: Program B9-1 is inaccurate for the solution in the upper-left corner of Figure 9.14 because of the round-off error in the complete elliptic integral calculation (another Jacobian function). A more accurate

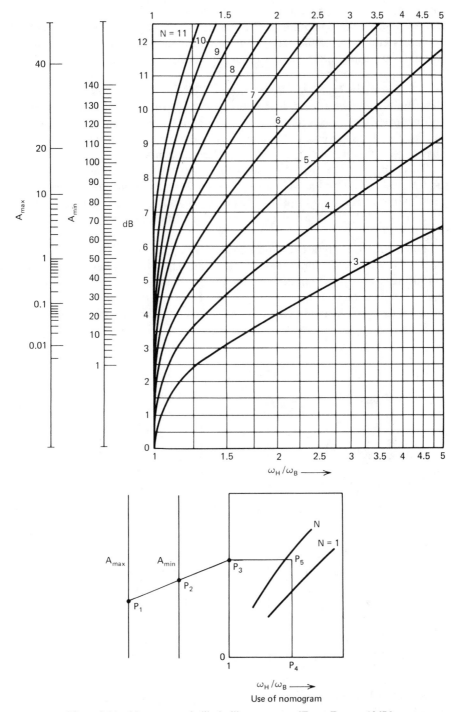

Figure 9.14. Nomogram of elliptic filter response. [From Zverev, 1967.]

algorithm has been programmed by Murdock (1979). However, it is instructive to observe the operation of Program B9-1, especially the computation of the poles and zeros of the rational elliptic function (9.31). Program B9-1 also evaluates the attenuation at any frequency according to (9.29).

9.2.3. The Four Types of Elliptic Filters.

Elliptic filter tables and literature describe filter types a, b, c, and s, often without further elaboration on what distinguishes these cases. Usually, remarks concerning symmetric (type s) and antimetric filters are encountered. Probably the most disconcerting problem is that degree N is not a simple count of filter elements.

This can be put in order by noting that filter type s is symmetric and is the odd-N case. The number of traps is $(N-1)/2$, and the filter must have equal terminating resistances. Midshunt cases $N=3$ and $N=5$ are shown in Figure 9.15. Type s elliptic filters may also occur in the midseries forms shown in Figure 9.12a.

There are three types of antimetric filters, this class having only even-degree N. Antimetric filters have $N/2$ traps. The type-a filter has the exact elliptic response function defined by (9.29). However, the realization must include the ideal transformers shown in Figure 9.16 or have at least one negative element. Also, the type-a filter has unequal termination resistances, as indicated.

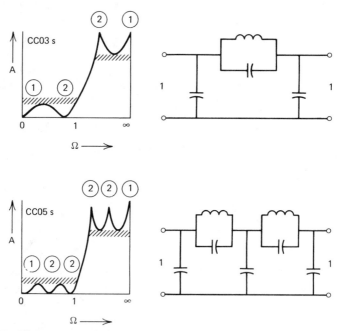

Figure 9.15. Symmetric (type-s) elliptic filters showing transmission pole/zero locations and count (in circles). [From Zverev, 1967.]

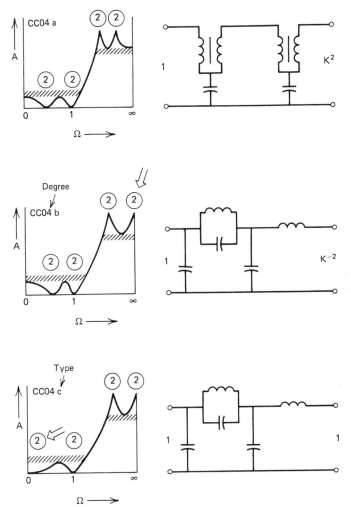

Figure 9.16. Antimetric elliptic filters (types a, b, and c) showing transmission pole/zero locations and count (in circles). [From Zverev, 1967.]

To eliminate the need for a negative element or ideal transformers, the highest two finite-frequency stopband attenuation poles can be moved to infinity. This causes ω_H to increase slightly, so that the type-b response does not cut off quite as fast as the ideal elliptic response function (9.29).

The type-c elliptic filter is a further aberration of the type-b filter: the lowest two passband attenuation zeros can be moved to the origin, so that the termination resistances are equal. The filter cutoff rate is further degraded. The nomogram in Figure 9.14 is only approximately correct for filter types b and c; however, it is usually accurate enough, because the next-higher-integer filter degree is required in any case.

9.2.4. *Summary of Introduction to Cauer Elliptic Filters.* The lowpass response functions of Butterworth, Chebyshev, inverse Chebyshev, and Cauer elliptic filters have certain similarities and increasing complexity in the order given. The rational elliptic function may be written as the ratio of two even polynomials of the same even degree. Their roots are easily computed using the doubly periodic Jacobian elliptic sine function.

Selectivity estimates related to choice of filter degree can be accomplished using a nomogram or a computer program. Program B9-1 does that; it also computes the trap frequencies and the attenuation at any given frequency.

Symmetric elliptic filters have odd-degree N, have $(N-1)/2$ traps, and have equal terminating resistances. Even-degree elliptic filters are called antimetric (from synthesis terminology, which is irrelevant to these purposes). They have $N/2$ traps, have unequal terminating resistances, and require either ideal transformers or one negative element. The even-degree filter just described has the exact response shape and is called a type-a filter. By arbitrarily moving two traps to infinity, the negative element or the ideal transformers are eliminated; this filter, with an approximate elliptic response, is called a type-b filter. A further aberration moves two passband zeros to the origin, to obtain equal terminating resistances; this is called a type-c filter. Both type-b and type-c filters have slightly reduced selectivity cutoff rates, but usually this is not a serious departure from predictions based on the nomogram.

9.3. Doubly Terminated Elliptic Filters

Amstutz (1978) published a procedure and two FORTRAN computer programs for calculating the elements of doubly terminated elliptic filters. The basis of his method will be described, and BASIC language translations of his programs, with examples, will be furnished. As seen in Program B9-1, there is little difficulty in computing the doubly periodic Jacobian elliptic functions. However, it was also noted that round-off error can be a problem. Another Amstutz contribution was a more accurate computation of these functions, especially the elliptic sine function; his method utilizes infinite products instead of infinite summations.

The Amstutz method is quite straightforward, although it incorporates one subtle step. He computes the poles and zeros of elliptic filter transducer function H and characteristic function K, as described in Section 3.2.4. As in (3.52), he obtains the rational polynomial of the reflection coefficient in factor form. Attention is then restricted to the trap frequencies. It is clear from Figure 9.12 that the input coefficient magnitude at these frequencies must be unity (complete reflection). Thus Amstutz finds very simple expressions for both the input impedance and its derivative with respect to frequency Z'_{in} (i.e., group delay) at the trap frequencies. Again referring to Figure 9.12a, the input impedance at the frequency where M_1 and K_1 are resonant is clearly $Z_{in} = sL_1$. Also, it can be shown that M_1 and K_1 can be found from Z'_{in}.

Amstutz's subtle step depends on the equivalence of any two networks having the same topology but permuted trap positions. The key to his method is to permute each of the traps into the input position and apply a recursive algorithm that yields all of the element values. An additional feature of his type-s symmetric filter program is the calculation of the critical unloaded Q (uniform in both inductors and capacitors) that just erases the trap notch nearest the passband.

This section begins with the application of the input impedance and its derivative with respect to frequency. The basis of the trap permutation scheme is then described. Finally, examples using the symmetric and the antimetric programs are discussed.

9.3.1. Input Impedance Relationships.

The analysis will employ the mid-series topology in Figure 9.12a without loss of generality. It was remarked in Section 8.1.1 that the impedance of a series LC branch resonant at ω_n is simply $j\omega_n LF$ (for example, see the admittance case in (8.28), Section 8.3.2). Suppose that the (M_1, K_1) branch in Figure 9.12a is resonant at $s_1 = j\omega_1$. Then the input impedance in the neighborhood of that frequency is

$$Z_{in}(s) = sL_1 + M_1\left(s - \frac{s_1^2}{s}\right). \tag{9.32}$$

Note that impedance Z, seen to the right of the (M_1, K_1) branch in Figure 9.12a, is not zero in practical cases. Differentiating (9.32) with respect to complex frequency s yields

$$Z'_{in} = L_1 + M_1\left(1 + \frac{s_1^2}{s^2}\right). \tag{9.33}$$

From (5.26), a Taylor series approximation valid in the neighborhood of trap frequency s_1 must be

$$Z_{in}(s) = L_1 s_1 + (L_1 + 2M_1)(s - s_1) + \cdots +. \tag{9.34}$$

Now it is possible to identify element values of the first three elements in Figure 9.12a that cause the notch at ω_1. Setting s equal to s_1 in (9.34) yields

$$L_1 = \frac{Z_{in}(s_1)}{s_1}. \tag{9.35}$$

Setting s equal to s_1 in the derivative of (9.34) with respect to s yields

$$M_1 = \frac{Z'_{in}(s_1) - L_1}{2}. \tag{9.36}$$

Since M_1 and K_1 are resonant at ω_1,

$$K_1 = \frac{1}{M_1 \omega_1^2}. \tag{9.37}$$

The significance of results (9.35)–(9.37) comes from the independent knowledge of $Z_{in}(\omega_1)$ and $Z'_{in}(\omega_1)$, as outlined in the previous section (i.e., by knowing the impedance function in pole/zero form, especially at the trap (notch) frequencies).

9.3.2. The Permutation Method to Calculate Trap-Section Elements.

Amstutz (1978) described the equivalence pictured in Figure 9.17: for any two-port network having the structure in (*a*), there exists an equivalent two-port network with the same structure, shown in (*b*), but in which the places of the three elements that produce attenuation peaks at ω_1 and ω_2 have been permuted. This is the genius of his method, because it determines element values without the loss of accuracy that plagues every other known method. Although subtle, it is far easier to compute.

To develop this equivalence according to Lin and Tokad (1968), write the continued fraction expansion for the network input impedance in Figure 9.17*a*:

$$Z_1 = L_{11}s + \cfrac{1}{\cfrac{1}{M_{11}s + \cfrac{1}{K_{11}s}} + \cfrac{1}{L_{22}s + \cfrac{1}{\cfrac{1}{M_{22}s + \cfrac{1}{K_{22}s} + \cdots}}}} \qquad (9.38)$$

It is convenient to form Z_1/s by dividing every impedance phrase in (9.38) by s. Also, define the inverse squared-frequency variables

$$h = \frac{-1}{s^2} \qquad (9.39)$$

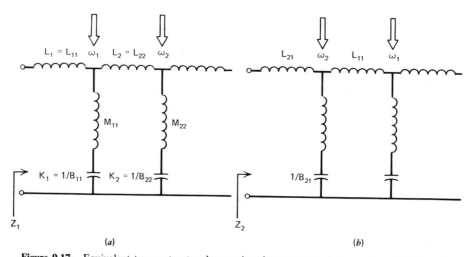

Figure 9.17. Equivalent two-port networks creating the same attenuation poles. (*a*) Pole order ω_1, ω_2; (*b*) pole order ω_2, ω_1. [From Amstutz, P., *IEEE Trans. Circuits Syst.* Vol. CAS-25, No. 12, p. 1003, December 1978. © 1978 IEEE.]

and

$$h_i = \frac{1}{\omega_i^2} = M_{ii} K_{ii} \qquad (9.40)$$

for the *ith* trap. Then (9.38) takes the form

$$\frac{Z_1}{s} = L_{11} + \cfrac{1}{\cfrac{K_{11}}{h_1 - h} + \cfrac{1}{L_{22} + \cfrac{1}{\cfrac{K_{22}}{h_2 - h} + \cdots}}} \qquad (9.41)$$

A similar expression may be written for Z_2/s in Figure 9.17b:

$$\frac{Z_2}{s} = L_{21} + \cfrac{1}{\cfrac{K_{21}}{h_2 - h} + \cdots} \qquad (9.42)$$

The claim is that $Z_2 = Z_1$ for all s. Choose $s = s_2$, and equate (9.42) to (9.41):

$$L_{21} = L_{11} + \cfrac{1}{\cfrac{K_{11}}{h_1 - h_2} + \cfrac{1}{L_{22}}} . \qquad (9.43)$$

Amstutz (1978) defined the constant

$$U = L_{11} - L_{21} . \qquad (9.44)$$

Manipulation of (9.43) results in the solution

$$L_{22} = UV, \qquad (9.45)$$

where a second defined constant is

$$\frac{1}{V} = \frac{U}{(h_2 - h_1) B_{11}} - 1. \qquad (9.46)$$

One additional relationship is required; it comes from the equivalence of the derivatives of (9.41) and (9.42) with respect to h in (9.39). The resulting expressions will again be evaluated at $h = h_2$. In this context, the network in Figure 9.17b does not involve L_{21} and elements to its right, so that (9.42) is truncated to yield the derivative

$$\left| \frac{d(Z_2/s)}{dh} \right|_{h=h_2} = \frac{-1}{K_{21}} . \qquad (9.47)$$

Similarly, the derivative of Z_1/s with respect to h does not involve anything to the right of trap 2 in Figure 9.17a. Therefore, (9.41) is truncated, and its derivative is equated to (9.47). After some algebra, a solution for K_{22} results:

$$K_{22} = \frac{\left[1/K_{11} + (L_{21} - L_{11})/(h_2 - h_1) \right]^2}{1/(K_{11}^2 K_{21}) - 1/K_{11} \left[(L_{21} - L_{11})/(h_2 - h_1) \right]^2} , \qquad (9.48)$$

which further reduces to

$$K_{22} = \frac{[(h_2 - h_1) + (L_{21} - L_{11})K_{11}]^2}{(h_2 - h_1)/K_{21} - (L_{21} - L_{11})^2 K_{11}}. \qquad (9.49)$$

Amstutz (1978) rearranges (9.49) into the last of four main expressions:

$$B_{22} = V^2 B_{21} - (V+1)^2 B_{11}. \qquad (9.50)$$

The calculation illustrated in Figure 9.17 proceeds as follows: obtain L_{11} and L_{21} according to (9.35) and then compute U from (9.44). Compute B_{11} and B_{21} according to (9.37) and then compute V from (9.46). Thus L_{22} may be obtained from (9.45), and B_{22} from (9.50). Repeating, the input trap-section values in Figure 9.17 are determined from the known input impedances and slopes at the trap frequencies. Then (9.44), (9.46), (9.45), and (9.50) are used, in that order, to determine the three values of the second trap section in Figure 9.17a.

Example 9.3.   The element values of a type-c (antimetric) filter are shown in Figure 9.18. A ladder network analysis program similar to Program B4-1 was used to compute the input impedance and its approximate frequency derivative (group delay) using 0.01% finite differences, as described in Section 4.7.2. Table 9.1 shows these values at the three trap frequencies. To find L_4, M_4, and C_4, begin by using (9.35)–(9.37) and the ω_2 data in Table 9.1 to compute $L_2 = 1.0915$, $M_2 = 0.118981$, and $C_2 = 1/B_2 = 1.377177$. A similar calculation using (9.35)–(9.37) and the ω_4 data in Table 9.1 yields $L_{41} = Z_4/s_4 = 0.842907$, $M_{41} = (Z_1' - L_{41})/2 = 0.455362$, and $B_{41} = M_{41}\omega_2 = 0.990429$. Therefore, with appropriate changes in subscripts, (9.44) yields

$$U = L_2 - L_{41} = 0.248593. \qquad (9.51)$$

Then (9.46) yields

$$V^{-1} = \frac{U}{(\omega_4^{-2} - \omega_2^{-2})B_2} - 1 = 0.156978, \qquad (9.52)$$

or $V = 6.370313$. Finally, (9.45) yields $L_4 = UV = 1.5836$, and (9.50) yields

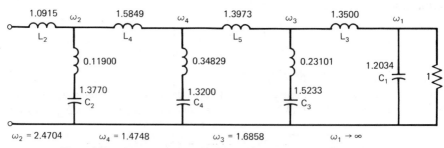

Figure 9.18.   A type-c antimetric elliptic filter appearing in Example 9.3.

Table 9.1. Input Impedance and Its Time Delay at the Three Trap Frequencies for the Elliptic Filter in Figure 9.18

ω	Z_{in}	Z'_{in}
ω_2	$j2.69645206$	1.329461
ω_3	$j1.58927603$	1.553162
ω_4	$j1.24311958$	1.753632

$B_{44} = 0.748384$, which produces $C_4 = 1/B_{44} = 1.3362$. The last value is comparable to the 1.3200 value shown in Figure 9.18, the difference being attributable to the approximate derivatives utilized in this example. (In practice, the derivatives are known exactly from the pole/zero factors.)

9.3.3. The Complete Permutation Algorithm. The algorithm in Table 9.2 generates the elements for M traps. Then the actual element values are $L_i = L_{ii}$, $K_i = 1/B_{ii}$, and $M_i = B_{ii}/\omega_i^2$ for $i = 1, 2, \ldots, M$. The new element values L_{ii} and

Table 9.2. Array of Permutation Algorithm Coefficients for M = 4 Traps

Input:				
Trap k	$i = 1$	$i = 2$	$i = 3$	$i = 4$
ω_1	L_{11}, B_{11}			
ω_2	L_{21}, B_{21}	L_{22}, B_{22}		
ω_3	L_{31}, B_{31}	L_{32}, B_{32}	L_{33}, B_{33}	
ω_4	L_{41}, B_{41}	L_{42}, B_{42}	L_{43}, B_{43}	L_{44}, B_{44}

For $i = 1$ (first column):

For $k = 1, 2, \ldots, M$:

$$L_{K1} = X_1(\omega_K)/\omega_K, \quad \text{where} \quad \tilde{Z}_1 = R_1 + jX_1$$

$$M_{K1} = [Z'(\omega_K) - L_{K1}]/2$$

$$B_{K1} = M_{K1}\omega_K^2$$

For $i = 2, 3, \ldots, M$:

For $j = i, i+1, \ldots, M$:

$$U = L_{i-1,i-1} - L_{j,i-1}$$

$$1/V = U/\left[\left(\omega_j^{-2} - \omega_{i-1}^{-2}\right)B_{i-1,i-1}\right] - 1$$

$$L_{ji} = UV$$

$$B_{ji} = V^2 B_{j,i-1} - (V+1)^2 B_{i-1,i-1}$$

B_{ii} require four parameters, computed in the previous $i-1st$ step; only $L_{i-1,i-1}$ and $B_{i-1,i-1}$ are actual element values. The other two parameters, $L_{i,i-1}$ and $B_{i,i-1}$, can be considered element values when ω_{i-1} is interchanged with ω_i in the trap sequence. Of course, this process may be accomplished from either end of the doubly terminated filter, especially to test accuracy.

It is well known that not every sequence of traps frequencies will yield positive elements. Lin and Tokad (1968) describe the minor tests to be added to the preceding algorithm so that all elements are positive.

9.3.4. Symmetric Type-s Filter Program.   Symmetric filters have odd-degree N, and the number of traps is $M=(N-1)/2$. Figure 9.19 shows the lowpass and highpass midshunt and midseries topologies, respectively, that Program B9-2 (Appendix B) designs. The program is a translation of Amstutz' (1978) FORTRAN into BASIC. The original program utilized double precision, although most eight-bit-microprocessor personal computers should be able to design elliptic filters with as many as seven traps without difficulty.

Two typical runs are illustrated by the computer output in Figures 9.20 and 9.21 for lowpass and highpass networks, respectively. The program discrimi-

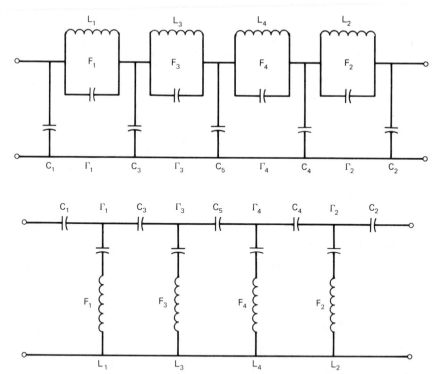

Figure 9.19. Lowpass, midshunt and highpass, midseries topologies computed by Program B9-2 for symmetric type-s filters. [From Amstutz, P., *IEEE Trans. Circuits Syst.*, Vol. CAS-25, No. 12, p. 1011, December 1978. © 1978 IEEE.]

```
SYMMETRICAL ELLIPTIC FLTR,C&S12/78,1009
STBND EDGE (KHZ)= 1.4142136
PASSBAND EDGE(KHZ)= 1
NUMBER OF PEAKS(1-15)= 4
CRITICAL Q= 36.5432758
STBND REJECTION (DB)= 96.95
PSBND RIPPLE (DB)= .177265239
3 DB (KHZ) ABOUT = 1.0223406
NOMINAL OHMS RESISTANCE= 1000
      ** LOW-PASS FILTER **
      KHZ          FARAD              HENRY
                 2.06723216E-07
  3.57749695  1   9.26042224E-09   .213722837
                 3.10090083E-07
  1.57095192  3   5.48128475E-08   .187254436
                 2.72796269E-07
  1.42945989  4   7.18533913E-08   .172523812
                 2.81714209E-07
  2.0034475   2   3.36478463E-08   .187554315
                 1.85128739E-07
PRECISION TEST:-4.93031228E-04
```

Figure 9.20. Lowpass, midshunt, symmetric type-s filter design example.

nates between these two cases by the relative values of the passband- and stopband-edge frequencies in the input data. The number of peaks input by the user is the number of trap notch frequencies. Note that the frequencies are in Kilohertz. Program B9-2 computes the critical Q, which is the uniform unloaded-Q value at which the notch nearest the passband is erased. The uniform, unloaded Q applies equally to all inductors and capacitors; it is defined as

$$Q = \frac{Q_L Q_C}{Q_L + Q_C}. \tag{9.53}$$

```
SYMMETRICAL ELLIPTIC FLTR,C&S12/78,1009
STBND EDGE (KHZ)= 1
PASSBAND EDGE(KHZ)= 1.4142136
NUMBER OF PEAKS(1-15)= 4
CRITICAL Q= 36.5432758
STBND REJECTION (DB)= 96.95
PSBND RIPPLE (DB)= .177265239
3 DB (KHZ) ABOUT = 1.38330963
NOMINAL OHMS RESISTANCE= 1000
      ** HIGH-PASS FILTER **
      KHZ          FARAD              HENRY
                 8.66435038E-08
  .395308117  1   1.93416923E-06   .0838058488
                 5.77613563E-08
  .900227159  3   3.2677054E-07    .0956517997
                 6.56578767E-08
  .989334217  4   2.49274578E-07   .10381885
                 6.35794121E-08
  .705890014  2   5.32314123E-07   .0954988627
                 9.6750099E-08
PRECISION TEST:-4.93031228E-04
```

Figure 9.21. Highpass, midseries, symmetric type-s filter design example.

The integers in the element data refer to the trap positions. Note that they alternate from end to end in the filter so that only positive element values are obtained, as mentioned in connection with the permutation algorithm (Section 9.3.3). The precision test number is the relative discrepancy between the values of the central element calculated from either port.

9.3.5.   *Antimetric Type-a, Type-b, and Type-c Filter Program.*   Antimetric filters have even-degree N, and the number of traps is $M = N/2$. Figure 9.22 shows the lowpass and highpass midshunt and midseries topologies, respectively, that Program B9-3 (Appendix B) designs.

Typical outputs for antimetric type-a, type-b, and type-c filters are shown in Figures 9.23, 9.24, and 9.25, respectively. Observe that the input data include the "half degree"; this is also the number of traps. Frequencies are in radians. The initial output from U through the last RF variable consists of data concerning the natural frequencies; interested readers are referred to Amstutz (1978). The trap sequence alternates from one end of the network to the other, as seen by comparing the integers in the output data to Figure 9.22.

The first precision test number is the relative discrepancy between the

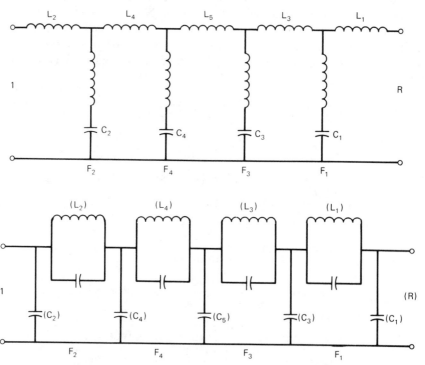

Figure 9.22.   Lowpass, midshunt and highpass, midseries topologies computed by Program B9-3 for antimetric type-a, type-b, and type-c filters. [From Amstutz, P., *IEEE Trans. Circuits Syst.*, Vol. CAS-25, No. 12, p. 1009, December 1978. © 1978 IEEE.]

```
ANTIMET ELLIP FLTR,CT12/78,1008

REJECTION,RIPPLE(DB),1/2-DEG(2-15),TYPE(A,B,OR C):
 83.3068  .17728767   4 A
U= .392699067 AO= .253168244 EP= .414213539
E= .383100151
E= .293985074
E= .159524088
RE=-.0941605422 SE= .447245616
RE=-.232749695 SE= .187680214
RF=-.0386088741 SF= 1.01584498
RF=-.12804916 SF= .909415957
RF=-.239606029 SF= .661764809
RF=-.333472309 SF= .248605769
LD RESIS= 1.50000003  ( .666666651   )
        L(C)        C(L)       PEAK
  1 -.0294237911  .876331629  6.18390915
  2  1.20559776  1.2296171   2.27723343
  3  1.8230507   1.15516466  1.62104523
  4  1.79644462  1.07285916  1.43359222
  5  1.70529984 STPBD EDGE= 1.41421359
TESTS-1.11149158E-04 -7.5083226E-06
```

Figure 9.23. Sample run for an antimetric type-a filter.

values of the central capacitor calculated from each port. The second test is the difference between the sum of all series inductors and a theoretical value. Most eight-bit-microprocessor personal computers should be accurate for "half degrees" (number of traps) up to 8. Note the negative element value, which will always appear in the type-a filter. The schematic representation of the type-c filter, corresponding to the data in Figure 9.25, is shown in Figure 9.18.

```
ANTIMET ELLIP FLTR,CT12/78,1008

REJECTION,RIPPLE(DB),1/2-DEG(2-15),TYPE(A,B,OR C):
 83.3068  .17728767   4 B
U= .392699067 AO= .253168244 EP= .414213539
E= .383100151
E= .293985074
E= .159524088
RE=-.0941605422 SE= .447245616
RE=-.232749695 SE= .187680214
RF=-.0396957664 SF= 1.01621815
RF=-.130545074 SF= .906702149
RF=-.240380105 SF= .655316027
RF=-.329399428 SF= .244461632
LD RESIS= 1.50000002  ( .666666658   )
        L(C)        C(L)       PEAK
  1              .900876092
  2  1.22249445  1.25009281  2.41712146
  3  1.84099295  1.16996665  1.65767848
  4  1.82095152  1.08728732  1.45434413
  5  1.72790256 STPBD EDGE= 1.43359223
TESTS-7.84639269E-05 -5.01144678E-06
```

Figure 9.24. Sample run for an antimetric type-b filter.

```
ANTIMET ELLIP FLTR,CT12/78,1008

REJECTION,RIPPLE(DB),1/2-DEG(2-15),TYPE(A,B,OR C):
 83.3068   .17728767   4 C
U= .392699067 AO= .253168244 EP= .414213539
E= .383100151
E= .293985074
E= .159524088
RE=-.0941605422 SE= .447245616
RE=-.232749695 SE= .187680214
RF=-.0417918631 SF= 1.01712478
RF=-.138269216 SF= .902058076
RF=-.260576346 SF= .637014303
RF=-.390335577 SF= .21738535
LD RESIS= 1      ( 1    )
            L(C)          C(L)          PEAK
    1                  1.20340853
    2   1.09152826   1.37701951   2.47036986
    3   1.35005027   1.5232667    1.68577457
    4   1.58486991   1.32003532   1.47480406
    5   1.39727972 STPBD EDGE= 1.45323639
TESTS 1.2381468E-05   1.28149986E-06
```

Figure 9.25. Sample run for an antimetric type-c filter.

9.3.6.   Summary of Doubly Terminated Elliptic Filters.   The effect of trap resonance on elliptic filter input impedance is such that it is easy to determine the first lumped-element and the next trap-branch-element values. All that is required is the input impedance and its time delay at that trap's frequency; both are available from the poles and zeros of the elliptic transfer function. The computation is made easier by the fact that the input reflection coefficient is necessarily unity at any trap frequency. The poles and zeros are Jacobian elliptic functions that are easy to compute. Improved accuracy is obtained by using the infinite-product formulation of Amstutz.

Amstutz's remarkable permutation algorithm determines all element values without the round-off error typical of all other realization methods. It is based on the fact that two-port networks having the same geometry may have the same response even when the three-element sets responsible for the notches are permuted. This fact was developed by writing continued fraction expansions for two such equivalent networks, each having a different trap subsection at their input port. By equating expressions for their impedances and for their time delays at all frequencies, and at the second trap frequency in particular, a method was determined for finding the element values of the trap subsection that is once removed from the input port.

The development for permuting just two trap subsections enabled the construction of a table and its generating algorithm to determine all element values in a network. Once again, the only required data are the input impedance and time delay at each trap frequency.

Two Amstutz (1978) FORTRAN programs were translated into BASIC and included for use on personal computers. Reasonably sized elliptic filters may be designed promptly and accurately using these programs for all types of doubly terminated elliptic filters. The more precise, symmetric elliptic filter program also included Amstutz's calculation of the critical, uniform unloaded-Q value that would just erase the notch adjacent to the passband. These are extremely valuable tools to have for applied circuit design.

9.4. Some Lumped-Element Transformations

There are many situations in which the network topology obtained by network synthesis or direct design is not satisfactory for practical applications. For example, the lowpass elliptic filters designed in Section 9.3 may be transformed to their bandpass equivalent circuits by the reactance transformation in Section 6.5.1. Each lowpass trap branch produces two bandpass traps, one above and one below the pass band. The topology of the single bandpass branch that produces these two attenuation notches may be altered by exact transformation to another form. Also, one of the notches may not be required at all. An element may be eliminated along with one notch that is approximately equivalent in a Taylor series sense to the transformed bandpass network. This section reproduces some well-known equivalent branch relationships and applies the Norton transformation to obtain several others. Approximate branch relationships that delete one element are derived.

9.4.1. Exact Transformations. Figure 9.26 shows four equivalent bandpass networks that correspond to a lowpass elliptic filter. Figure 9.12*b* may be transformed, as described in Section 6.5.1, into the network in Figure 9.26*a*. In this process, lowpass capacitors become bandpass parallel LC branches, and lowpass inductors become series LC branches. The correspondence between Figure 9.12*b* and the network in Figure 9.26*a* can be recognized by these transformations. It is not so obvious that the remaining networks in Figure 9.26 are exactly equivalent at all frequencies.

Zverev (1967) has compiled a set of equivalent three- and four-element networks that are cataloged according to pole-zero impedance characteristics. These are valid at all frequencies. It is straightforward, but tedious, to verify these by writing the branch impedances in complex frequency s. For example, the network in Figure 9.26*a* may be converted into the topology shown in Figure 9.26*b* by using transformation 12 in Appendix H. The type-IV impedance characteristics graphed in Appendix H indicate that there are two notches (poles) at ω_2 and ω_4 and one zero (corresponding to the band-center frequency ω_3). These frequencies are related to element values as indicated in the separate table in Appendix H for pole-zero frequencies. Note that parameters W, X, Y, Z, Q, and S, which appear in the equivalence tables, are defined in the last section of Appendix H.

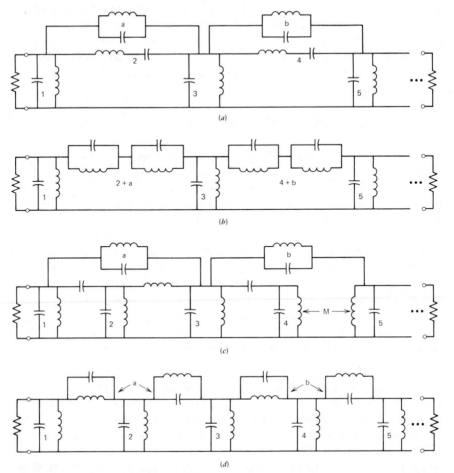

Figure 9.26. Node-type equivalent networks. (*a*) Paralleled, dual coupling branches; (*b*) coupling branches composed of parallel LC traps in series; (*c*) bridged-T coupling; (*d*) single parallel LC trap coupling branches. [From Dishal, M., *IRE Trans. Veh. Commun.*, Vol. PGVC-3, No. 1, p. 115, June 1953. © 1953 IRE (now IEEE).]

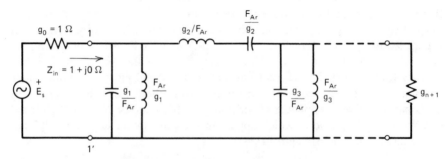

Figure 9.27. A bandpass prototype network.

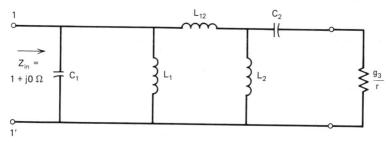

$$C_1 = \frac{g_1}{F_{Ar}} \quad L_1 = \frac{F_{Ar}g_2}{g_1g_2 - (\sqrt{r}-1)F_{Ar}^2} \quad L_{12} = \frac{g_2}{\sqrt{r}\,F_{Ar}} \quad L_2 = \frac{g_2}{(\sqrt{r}-1)\sqrt{r}\,F_{Ar}} \quad C_2 = \frac{rF_{Ar}}{g_2}$$

Figure 9.28. Two-pole bandpass equivalent network using L_{12}.

Another useful set of equivalent networks is available by using the Norton transformations described in Section 6.5.3. Consider the bandpass prototype network in Figure 9.27. Parameter F_{Ar} is the passband fractional frequency, and the g_i are the corresponding lowpass prototype element values for a 1-ohm source resistance. An inductive Norton transformer replacement produces the network in Figure 9.28, which is equivalent at all frequencies. Note that the impedance-scaling parameter r must be greater than unity. A capacitive Norton transformer replacement produces the network in Figure 9.29.

A more flexible transformation applies to a three-pole bandpass prototype network that is modified by an inductive and a capacitive Norton transformation. The results appear to be direct-coupled filters, but they are not, because all the L's and C's touching each node are not resonant. Two possible topologies appear in Figures 9.30 and 9.31. Now there are two impedance scaling parameters, N_a and r; their ranges and relationships may be seen in the equations included in the figures. Various combinations of Figures 9.28 through 9.31 may be employed to obtain suitable element values.

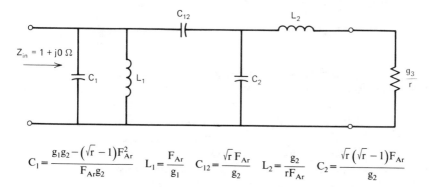

$$C_1 = \frac{g_1g_2 - (\sqrt{r}-1)F_{Ar}^2}{F_{Ar}g_2} \quad L_1 = \frac{F_{Ar}}{g_1} \quad C_{12} = \frac{\sqrt{r}\,F_{Ar}}{g_2} \quad L_2 = \frac{g_2}{rF_{Ar}} \quad C_2 = \frac{\sqrt{r}(\sqrt{r}-1)F_{Ar}}{g_2}$$

Figure 9.29. Two-pole bandpass equivalent network using C_{12}.

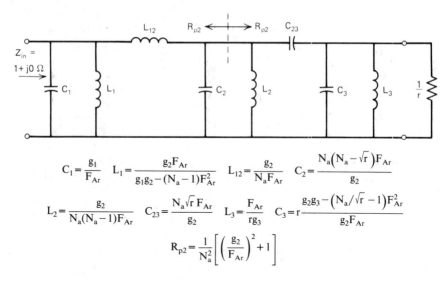

$$C_1 = \frac{g_1}{F_{Ar}} \quad L_1 = \frac{g_2 F_{Ar}}{g_1 g_2 - (N_a - 1)F_{Ar}^2} \quad L_{12} = \frac{g_2}{N_a F_{Ar}} \quad C_2 = \frac{N_a(N_a - \sqrt{r})F_{Ar}}{g_2}$$

$$L_2 = \frac{g_2}{N_a(N_a - 1)F_{Ar}} \quad C_{23} = \frac{N_a \sqrt{r}\, F_{Ar}}{g_2} \quad L_3 = \frac{F_{Ar}}{rg_3} \quad C_3 = r\frac{g_2 g_3 - (N_a/\sqrt{r} - 1)F_{Ar}^2}{g_2 F_{Ar}}$$

$$R_{p2} = \frac{1}{N_a^2}\left[\left(\frac{g_2}{F_{Ar}}\right)^2 + 1\right]$$

Figure 9.30. Three-pole bandpass equivalent network using L_{12} and C_{23}.

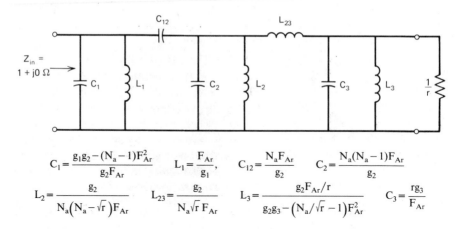

$$C_1 = \frac{g_1 g_2 - (N_a - 1)F_{Ar}^2}{g_2 F_{Ar}} \quad L_1 = \frac{F_{Ar}}{g_1}, \quad C_{12} = \frac{N_a F_{Ar}}{g_2} \quad C_2 = \frac{N_a(N_a - 1)F_{Ar}}{g_2}$$

$$L_2 = \frac{g_2}{N_a(N_a - \sqrt{r})F_{Ar}} \quad L_{23} = \frac{g_2}{N_a \sqrt{r}\, F_{Ar}} \quad L_3 = \frac{g_2 F_{Ar}/r}{g_2 g_3 - (N_a/\sqrt{r} - 1)F_{Ar}^2} \quad C_3 = \frac{rg_3}{F_{Ar}}$$

Figure 9.31. Three-pole bandpass equivalent network using C_{12} and L_{23}.

9.4.2. Trap Approximations. Taylor series were described in Section 5.1.4. A first-order approximation of susceptance by value and slope at a frequency was employed for resonators in Section 8.3.4. The difference here is that the susceptance at the frequency of interest will not be zero. In the following, it will be assumed that trap notch frequencies are greater than any reference frequency. Pole-zero branches having two, three, or four elements, similar to those discussed earlier in Section 9.4.1, will be considered.

It is informative to note that two elements cannot replace one in a first-order equivalence. Consider the two branches shown in Figure 9.32. The

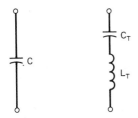

Figure 9.32. An example of two elements that cannot be replaced by one element by use of a first-order approximation.

susceptance of the single capacitor is

$$B_C = \omega C, \tag{9.54}$$

and its slope with respect to radian frequency is

$$\dot{B}_C = \frac{dB}{d\omega} = C. \tag{9.55}$$

The susceptance for the trap in Figure 9.32 is

$$B_T = \frac{-1}{\omega_n L_T(\omega/\omega_n - \omega_n/\omega)}, \tag{9.56}$$

where the trap resonance frequency is ω_n. Its slope with respect to radian frequency is

$$\dot{B}_T = \frac{(\omega/\omega_n + \omega_n/\omega)}{\omega\omega_n L_T(\omega/\omega_n - \omega_n/\omega)^2}. \tag{9.57}$$

Equating susceptances (9.54) and (9.56), the trap slope must be

$$\dot{B}_T = \dot{B}_C \left[\frac{(\omega_n/\omega)^2 + 1}{(\omega_n/\omega)^2 - 1} \right] > \dot{B}_C \quad \text{for} \quad \omega_n > \omega. \tag{9.58}$$

This shows that if the susceptances are equal, the slopes cannot be equal.

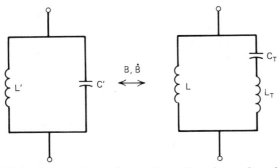

Figure 9.33. First-order equivalence of a two-element branch to a three-element branch.

Equating slopes (9.55) and (9.57), the trap susceptance must be

$$B_T = B_C \left[\frac{(\omega_n/\omega)^2 - 1}{(\omega_n/\omega) + 1} \right] < B_C . \tag{9.59}$$

This shows the converse limitation. Therefore, two elements cannot replace one.

Three elements can replace two elements using a first-order approximation. Figure 9.33 illustrates the replacement of two elements by three, introducing a notch in the process. The relationship among element values may be obtained by writing relationships for the susceptance and its frequency slope, as follows:

$$L = \frac{L'}{1 - (\omega^4/\omega_n^2)L'C'} , \tag{9.60}$$

$$L_T = \frac{1}{\omega^2 C'(\omega/\omega_n - \omega_n/\omega)^2} , \tag{9.61}$$

$$C_T = \frac{1}{\omega_n^2 L_T} . \tag{9.62}$$

Three elements can replace four elements using a first-order approximation. Figure 9.34 illustrates the replacement of four elements by three, removing a notch in the process. The relationship among element values may be obtained by writing relationships for the susceptance and its frequency slope, as

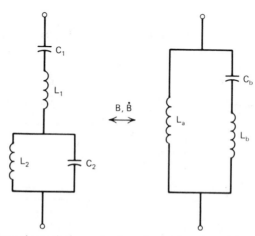

Figure 9.34.   First-order equivalence of a four-element branch to a three-element branch.

follows:

$$L_b = \frac{\omega_n^2}{C_2(\omega_n^2 - \omega_0^2)^2},$$　(9.63)

$$L_a = L_b\left[\left(\frac{\omega_n}{\omega_0}\right)^2 - 1\right],$$　(9.64)

$$C_b = \frac{1}{\omega_n^2 L_b}.$$　(9.65)

9.4.3. *Summary of Some Lumped-Element Transformations.*

There is a large collection of lumped-element equivalent branch topologies; a set by Zverev (1967) is contained in Appendix H. In addition, Norton transformations may be used to produce two- and three-pole equivalent bandpass networks that can radically alter topologies and element values without changing frequency response.

A first-order equivalence based on Taylor series provided several approximate transformations involving traps that produced attenuation poles (notches). It was shown that two elements cannot replace one on the basis of equal susceptance and equal slope at a frequency. This amounts to a statement of Foster's reactance theorem (see Van Valkenburg, 1960, p. 123). However, element values were given for (1) replacing two elements by three and adding a notch, and (2) replacing four elements by three and removing a notch.

9.5. Load Effects on Passive Networks

The image of the right-half load plane as seen at the input terminals of a linear, active two-port network was the circle described in Section 7.3.2. The Y_L plane appeared as a Smith chart with complex normalization in the Y_{in} plane. Filters usually have a high efficiency in the pass band and thus do not shrink the load-plane image in the input plane at these frequencies. However, the input image of a more limited load-plane neighborhood, typically a constant SWR circle, is still of interest. This case occurs in such questions as the effect of a spacecraft antenna SWR on transmitter output in the passband.

Another important question arises at filter stopband frequencies. The load impedance often is specified at passband frequencies. However, both filter and system designers make the assumption that the load impedance in the stopband is either 50-ohms or is so reactive that the filter selectivity will be increased if changed at all. The fact is that an antenna SWR in the stopband is usually very large, and it is equally likely that filter selectivity will be seriously degraded at certain stopband frequencies. Complaints about transmitter excessive harmonic output in the field are common.

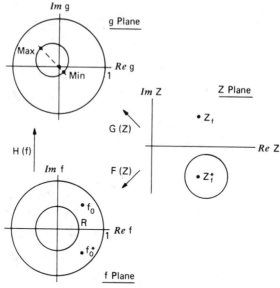

Figure 9.35. A composite bilinear transformation between unit circles and a cartesian plane.

This section applies the impedance-mapping results from Chapter Seven to passive, reciprocal networks. The results are somewhat more compact than if applied to linear active networks; they will be useful in Section 9.6 as well. Parts of Chapter Seven will be recalled, and a new unit-circle to unit-circle bilinear transformation will be described. It will be applied to obtain a simple analytic solution to the power transfer example (Example 7.1 in Section 7.1.2). The maximum efficiency is the key parameter in predicting bounds on impedance and efficiency behavior. Practical bounds will be obtained, and a basis for invulnerable filter design will be developed for use in the next section.

9.5.1. Unit-Circle to Unit-Circle Bilinear Mapping. The generalized Smith chart was described in Section 7.1.2 as the bilinear function that mapped the right-half cartesian plane onto a unit circle. Problem 6.7 showed that concentric circles inside the unit circle were SWR circles in the conventional sense when the unit-circle Smith chart was normalized to a real transmission line characteristic impedance. Otherwise, the generalized Smith chart development showed that concentric Smith chart circles appear as "SWR" circles geometrically centered on the normalizing constant in the Z plane (see Figure 9.35). The "SWR" circle has reflection magnitude R in the f plane; its image in the Z plane is centered on Z_f^*. The F(Z) transformation has been described in (7.16); in terms of Figure 9.35 it is

$$f = F(Z) = \frac{Z - Z_f^*}{Z + Z_f} \triangleq Re^{j\theta}. \qquad (9.66)$$

This form is unique for any bilinear transformation that maps the right-half plane onto a unit circle. Another such form shown in Figure 9.35 is

$$g = G(Z) = \frac{Z - Z_g^*}{Z + Z_g}. \tag{9.67}$$

The purpose of this section is to introduce the fact that bilinear transformations that map unit circles onto unit circles have certain elementary properties. For instance, they must have the form

$$g = H(f) = e^{j\gamma} \frac{f - f_0}{1 - ff_0^*}, \tag{9.68}$$

where $|f_0| < 1$ (see Cuthbert, 1980). According to (9.67), the g-plane origin is the image of $Z = Z_g^*$. But (9.68) shows that it also corresponds to $f = f_0$. Therefore,

$$f_0 = F(Z_g^*) = \frac{Z_g^* - Z_f^*}{Z_g^* + Z_f} \triangleq Me^{j\phi}. \tag{9.69}$$

Cuthbert (1980) shows that, for a fixed R in (9.66) and Figure 9.35, the constant-$|f|$-circle image in the g plane has a maximum radius, defined by

$$|g|_{\max} = \frac{M + R}{1 + MR}, \qquad R \text{ is fixed}, \tag{9.70}$$

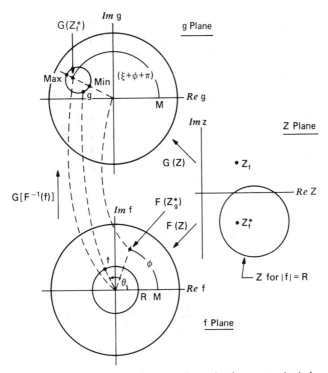

Figure 9.36. Some details of the transformation between unit circles.

and a minimum radius, defined by

$$|g|_{min} = \frac{M - R}{1 - MR}, \qquad \text{R is fixed.} \tag{9.71}$$

The image encircles the origin when $M < R$. This is illustrated in Figures 9.35 and 9.36. Although it is incidental to the following application, angle ξ in Figure 9.36 is

$$\xi = 2 \arg(Z_g^* + Z_f). \tag{9.72}$$

9.5.2. Power Bounds Between a Complex Source and Loads.

Figure 3.3 in Section 3.2.3 illustrated the connection of a fixed, complex source to an arbitrary load impedance. Example 7.1 in Section 7.1.2 illustrated the calculation of power transferred to load impedances contained within and on a $2:1$ SWR circle. The solution was based on the generalized Smith chart situation shown in Figure 7.2. The circuit is reproduced in Figure 9.37.

Compact expressions bounding the power delivered may be obtained by applying the bilinear mapping result in Section 9.5.1. Suppose that both the source and load standing-wave ratios S_s and S_L, respectively, are defined with respect to resistance R_0. Then the f plane in Figure 9.35 becomes the load Smith chart when $Z = Z_L$, $Z_f = R_0$, and $Z_g = Z_s$. The reflection magnitude is related to the standing-wave ratio S by

$$|\Gamma| = \frac{S - 1}{S + 1}. \tag{9.73}$$

Applying (9.66) to this case yields

$$R = |\Gamma_L| = \frac{S_L - 1}{S_L + 1}, \tag{9.74}$$

and (9.69) yields

$$M = |\Gamma_s| = \frac{S_s - 1}{S_s + 1}. \tag{9.75}$$

According to (9.67) and (3.47), when $|g|$ is maximum, load power is minimum, and vice versa. So (9.70) yields

$$\min \frac{P_L}{P_{as}} = \frac{4 S_L S_s}{(S_s S_L + 1)^2}. \tag{9.76}$$

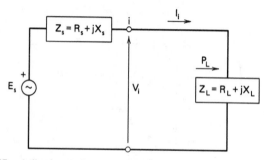

Figure 9.37. A fixed complex source connected to arbitrary complex loads.

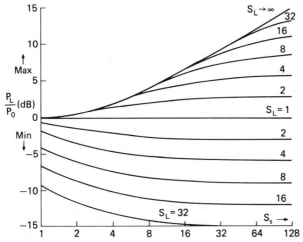

Figure 9.38. Power from a complex source to a complex load relative to the 1 : 1 SWR load power.

Reference the load power to that delivered to the normalizing resistance R_0. Call this P_0 and set S_L equal to 1 in (9.76); this yields

$$\frac{P_0}{P_{as}} = \frac{4S_s}{(S_s+1)^2}. \tag{9.77}$$

Therefore, dividing (9.76) by (9.77) yields the minimum relative power:

$$\min \frac{P_L}{P_0} = \frac{S_L(S_s+1)^2}{(S_sS_L+1)^2}. \tag{9.78}$$

Note that (9.73) is a strictly increasing function of S. By (9.74) and (9.75), $S_s \leqslant S_L$ implies that $M \leqslant R$, so that $|g|=0$ and $P_L = P_{as}$ (because of encirclement of the origin). Otherwise, (9.71) yields

$$\max \frac{P_L}{P_{as}} = \frac{4S_LS_s}{(S_s+S_L)^2}, \qquad S_s > S_L. \tag{9.79}$$

Normalizing to (9.77), the last result is:

$$\max \frac{P_L}{P_0} = \begin{cases} \dfrac{(S_s+1)^2}{4S_s}, & S_L \geqslant S_s, \\[3mm] \dfrac{S_L(S_s+1)^2}{(S_s+S_L)^2}, & S_L < S_s. \end{cases} \tag{9.80}$$

A graph of (9.78), (9.79), and (9.80) is shown in Figure 9.38.

Example 9.4. A transmitter is connected to an antenna as shown in Figure 9.39. The second-harmonic power into a 50-ohm resistive test antenna (load) is

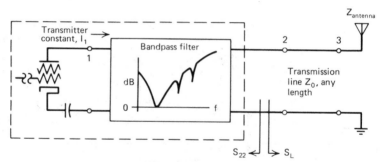

Figure 9.39.   A transmitter connected to a narrow-band antenna.

60 dB below the fundamental power. Also, at the second-harmonic frequency, SWR $S_{22} = 32$ is measured with the power off, and the antenna SWR is $S_L = 16$. Assuming $S_L = 1$ at the fundamental power, what is the possible range of the second-harmonic, spurious output power? According to (9.80) and (9.78), the second-harmonic power can increase by 8.8 dB and decrease by 11.8 dB, respectively. Therefore, the second-harmonic output power can range between 51.8 and 71.8 dB below the fundamental power.

9.5.3. Bounds on Input Impedance and SWR. Maximum efficiency will be the primary parameter for the analysis of bounds on input impedance and SWR, and it is convenient to simplify its expression for passive reciprocal networks. Maximum efficiency (7.64) was derived in terms of admittance parameters in Section 7.3.3. Its definition was in terms of Rollett's stability factor K in (7.57). A different constant is defined here:

$$\mathcal{R} = \frac{g_{11}g_{22} - g_{21}^2}{|y_{21}|^2}. \tag{9.81}$$

This constant is always positive because its numerator is the real-part requirement for passive networks (according to Van Valkenburg, 1960, p. 312). It is easily related to the stability factor:

$$K = 2\mathcal{R} + 1. \tag{9.82}$$

The maximum-efficiency expression in (7.64) may then be formulated for the passive reciprocal network:

$$\eta_{\max}^{\pm 1} = \left(\sqrt{\mathcal{R} + 1} \mp \sqrt{\mathcal{R}}\right)^2. \tag{9.83}$$

This is a monotonic function of $\mathcal{R}$; therefore, many aspects of $\eta_{\max}$ may be analyzed in terms of $\mathcal{R}$ instead. The corresponding choices of $+/-$ signs will be useful in Section 9.6.

It will now be shown that maximum efficiency is the single parameter that controls the relative size of the input-plane image of the load plane. Consider Figure 7.13 in Section 7.3.2. The ratio of image-circle diameter to the distance

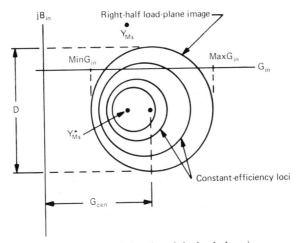

Figure 9.40. The relative size of the load-plane image.

of its center from the imaginary axis is

$$\frac{D}{G_{cen}} = \frac{2}{K}.$$ (9.84)

This is equivalent to the desired expression

$$\frac{D}{G_{cen}} = \frac{4\eta_{max}}{1 + \eta_{max}^2},$$ (9.85)

which may be reduced to (9.84) by using $y_{21} = y_{12}$ in (7.64). This is illustrated in Figure 9.40. The important result in (9.85) states that the relative size of the load-plane image in the input plane is a function only of n_{max}. In the extreme, a lossless network maps the right-half plane onto the right-half plane. Bounds on input admittance may be obtained only by bounds on η_{max}, i.e., on $\mathcal{R}$ in (9.81).

Some results from Cuthbert (1980) are stated. The exact bounds on the input SWR for any load impedance in the right-half plane are

$$S_{in,max} = S_{c1} S_{max}$$ (9.86)

$$S_{in,min} = \begin{cases} \dfrac{S_{c1}}{S_{max}}, & S_{c1} > S_{max}, \\ 1, & S_{c1} \leqslant S_{max}, \end{cases}$$ (9.87)

where S_{c1} is the SWR of Y_{Ms} in (7.78) with respect to the nominal resistance, and

$$S_{max} = \frac{1 + \eta_{max}}{1 - \eta_{max}}.$$ (9.88)

The bounds of S_{in} are easy to measure: use a lossless LC resonator as the load admittance and tune it over its entire range while observing the input SWR.

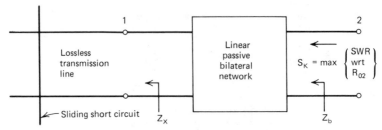

Figure 9.41. The measurement of S_K, the maximum SWR of the input-plane image.

Usually, load admittances are limited to small regions of the right-half plane. A useful upper bound on input SWR can be stated for a given load SWR region:

$$S_{in} \leqslant S_{in0} \frac{S_L S_K + 1}{S_L + S_K}, \qquad (9.89)$$

where S_{in0} is S_{in} when load $S_L = 1$, and S_K is the maximum SWR observed by looking into the output terminals for all values of pure resistance connected to the input terminals. Except for turning the network end for end, the latter is the same as (9.86), i.e.,

$$S_K = S_{c2} S_{max}, \qquad (9.90)$$

where S_{c2} is the SWR of Y_{ML} in (7.79) with respect to the nominal resistance, and S_{max} was defined in (9.88). S_K may easily be measured, instead of being calculated (see Figure 9.41). The value of S_K is found by connecting all possible reactance values to the input port. This may be done, as illustrated, with a sliding short circuit on a transmission line or by a lossless LC resonator tuned over an infinite range. Either way, S_K is the greatest SWR value observed at the output port.

9.5.4. *Summary of Load Effects on Passive Networks.*

The geometric models of linear two-port-network impedance and power behavior were recalled from Chapter Seven. Bilinear mapping was extended by considering the form and particular constants that map unit circles onto unit circles. The reflection-plane bounds of a concentric image from another reflection plane were given. One application was an analytic solution for the range of power delivered from a fixed complex source to an arbitrary complex load.

Passive reciprocal networks were shown to have a simplified expression for maximum efficiency in terms of the real-part parameter $\mathscr{R}$. It was also shown that the relative size of the load-plane image in the input plane is a function only of the maximum efficiency. Also, expressions for the exact SWR extreme values of that load-plane image were stated. More often, the load-plane region is limited to a stated SWR value. An upper bound for the input SWR of that image was stated. Measurement methods were described for obtaining these bounds.

9.6. Invulnerable Filters

Microwave literature contains descriptions of a number of waveguide and other devices connected to a dissipative medium by frequency-selective means. One implementation is to have side-coupling holes that excite chambers containing resistance sheets. These devices usually are designed to absorb unwanted stopband energy, especially in high-power circuits, where the source cannot tolerate unwanted reflected energy. At frequencies above 400 MHz, it is common practice to use three-port circulators to dissipate the reflected energy in a load attached to the extra port. Another application occurs in oscillators when stopband reactive terminations cause unwanted oscillations. Circulators are too large at low frequencies, and many dissipative designs have been *ad hoc* and difficult to model and adapt.

This section describes an organized approach to the design of filters that absorb energy rather than reflect it; i.e., the reflection basis described in Section 3.2.3 is not employed. Instead, the concept of designing a filter having a limited maximum efficiency is exploited by direct design techniques. Two specific filters will be described, and the basis for designing other absorptive filters will be evident. The invulnerable part of the filter capability comes from the fact that maximum efficiency may be achieved for only one unique load impedance; any other load impedance will result in lower efficiency. In this sense, the designer need not have any information whatever about the load impedance; indeed, this is precisely the case in power lines and many other environments.

The only additional development is the expression of the minimum loss associated with the maximum efficiency. This is

$$L_{min} = 10 \log \frac{1}{\eta_{max}} \text{ dB}. \qquad (9.91)$$

Using (9.83) and Equation (G.4) in Appendix G, a surprisingly simple expression for L_{min} may be obtained:

$$L_{min} = 8.6858 \sinh^{-1} \sqrt{\mathcal{R}} \text{ dB}, \qquad (9.92)$$

where $\mathcal{R}$ is defined by (9.81). Open-circuit impedance parameters can also be used by replacing y, g, and b with z, r, and x, respectively, in these equations.

9.6.1. *Invulnerable Bridged-T Network.* Consider the network in Figure 9.42 and the following general definitions:

$$\omega_1 = \frac{1}{\sqrt{LC}}, \qquad (9.93)$$

$$\bar{\omega} = \frac{\omega}{\omega_1}, \qquad (9.94)$$

$$d = R\omega_1 C. \qquad (9.95)$$

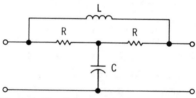

Figure 9.42. A lowpass bridged-T network.

The normalizing frequency ω_1 and the decrement d are the main design parameters. No resistive termination values need be specified for the invulnerable filter in Figure 9.42 because L_{min} is not a function of the source or load impedance or of the impedance scale.

It can be shown (Guillemin, 1957) that the z parameters of the bridged-T network are

$$z_{11} = R\left[\frac{2d^2 + \bar{\omega}^2}{4d^2 + \bar{\omega}^2} - j\frac{\bar{\omega}^2(1 - d^2) + 4d^2}{\bar{\omega}d(4d^2 + \bar{\omega}^2)} \right],$$

$$\tag{9.96}$$

$$z_{21} = R\left[\frac{2d^2}{4d^2 + \bar{\omega}^2} - j\frac{\bar{\omega}^2(1 + d^2) + 4d^2}{\bar{\omega}d(4d^2 + \bar{\omega}^2)} \right],$$

where $z_{22} = z_{11}$ by symmetry. Substitution into

$$\mathcal{R} = \frac{r_{11}r_{22} - r_{21}^2}{|z_{21}|^2}, \tag{9.97}$$

analogous to (9.81), yields

$$\mathcal{R} = \frac{\bar{\omega}^4}{\bar{\omega}^2(1/d + d)^2 + 4} \tag{9.98}$$

for the bridged-T network. From (9.83), small η_{max} or large L_{min} values validate the approximation

$$\eta_{max} \doteq \frac{0.25}{\mathcal{R}}. \tag{9.99}$$

Therefore, (9.98) shows that L_{min} increases at a rate of only 6 dB/octave for large ω.

Also note that L_{min} is a strictly increasing function of $\mathcal{R}$. So, equating the derivative of $\mathcal{R}$ with respect to d to zero shows that $d = 1$ produces the maximum possible value of L_{min} for any given ω. It also can be shown that the condition $d = 1$ causes the conjugate match terminations to equal R for all ω. This constant-resistance condition may also be confirmed by employing

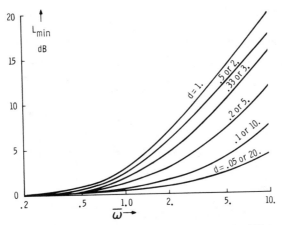

Figure 9.43.   Selectivity curves for the invulnerable bridged-T network.

Bartlett's bisection theorem to convert the bridged-T network in Figure 9.41 to a symmetric lattice, which is a constant resistance when $d = 1$ (see Guillemin, 1957). When $d = 1$, implying that $R = \sqrt{L/C}$, bridged-T networks may be cascaded so that the overall L_{min} is just the sum of the individual L_{min} values.

Equations (9.92) and (9.98) provided the design curves in Figure 9.43. Note that (9.98) gives the same result for d and $1/d$; the curves in Figure 9.43 are

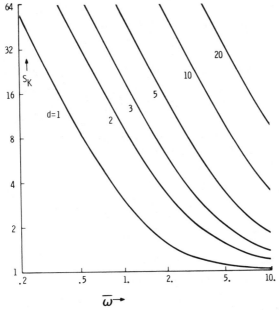

Figure 9.44.   Maximum possible SWR for bridged-T invulnerable filters.

marked accordingly. Of course, the lowpass network may be transformed into bandpass networks, as discussed in Section 6.5.

The maximum possible input SWR, S_K, may be defined with respect to $R_0 = R$, the value of the resistors in Figure 9.42. Then (9.86) yields the maximum possible input SWR. This is plotted in Figure 9.44.

9.6.2. Three-Pole Invulnerable Filter.

The invulnerable filter in Figure 9.45 is considered because it is easily compared to a classical design when $R = 0$. The parameters defined in (9.93)–(9.95) also apply for this network. The development is accomplished using y parameters:

$$y_{11} = \frac{1}{R}\left[\frac{\bar{\omega}^2 d^2}{1+\bar{\omega}^2 d^2} + j\frac{d}{\bar{\omega}} \frac{\bar{\omega}^2 - (1+\bar{\omega}^2 d^2)}{1+\bar{\omega}^2 d^2} \right],$$ (9.100)

$$y_{21} = \frac{1}{R}\left(j\frac{d}{\bar{\omega}} \right),$$ (9.101)

where $y_{22} = y_{11}$ by symmetry. Then the real-part parameter is

$$\mathcal{R} = \left(\frac{d\bar{\omega}^3}{1+\bar{\omega}^2 d^2} \right)^2.$$ (9.102)

Again, the minimum attenuation of this invulnerable filter, according to (9.92), increases at a rate of only 6 dB/octave for large ω. Proceeding as before for the derivative of $\mathcal{R}$ with respect to decrement d, yields the condition for maximizing L_{min} at a given ω. The result is $d = 1/\bar{\omega}$, a function of frequency. According to Guillemin (1957, p. 196), the symmetric lattice for the network in Figure 9.45 is not reciprocal under any conditions. Therefore, this three-pole network cannot have the constant-resistance property; this will complicate the calculation of L_{min} for cascaded sections.

The selectivity response according to (9.92) and (9.102) is shown in Figure 9.46. The dashed line bounds the family of decrement curves according to the requirement of maximum L_{min}, namely $d = 1/\bar{\omega}$.

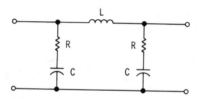

Figure 9.45. A three-pole lowpass invulnerable filter.

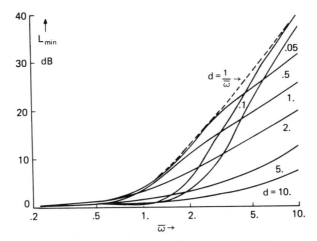

Figure 9.46. Three-pole invulnerable filter selectivity.

Example 9.5. A conventional, doubly terminated bandpass filter tuned to 1 MHz and having a 10% Butterworth passband width will be compared to its invulnerable counterpart. For the conventional Butterworth filter, the lowpass, lossless prototype network, normalized to 1-ohm terminations and a 3-dB passband edge of 1 radian, has element values $C = 1F$ and $L = 2H$. This leads to the network in Figure 9.47 when $R = 0$. The values indicated were obtained by frequency and impedance scaling and resonating each lowpass prototype element according to Section 6.5.1. The Butterworth response is obtained between 50-ohm terminations. The second harmonic on the prototype frequency scale is $\omega = 15$ radians; selectivity curves or (9.24) show that the second-harmonic (2-MHz) attenuation is 71 dB.

The value of R must be selected to design the invulnerable filter with the L and C values fixed as above. One choice is to obtain the maximum possible value of L_{min} at the second harmonic. For the invulnerable lowpass network, (9.93) yields the normalizing frequency $\omega_1 = 1/\sqrt{2}$. The optimum decrement was shown to be $d = 1/\overline{\omega}$. Using the calculated ω_1 value and $\omega = 15$ in (9.94) and (9.95), it is found that $R = 1/15$ in the lowpass network. Therefore,

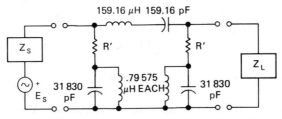

Figure 9.47. Three-pole Butterworth/invulnerable filter in Example 9.5.

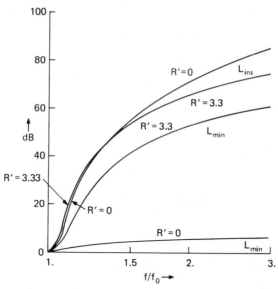

Figure 9.48.   Three-pole Butterworth/invulnerable filter computed selectivity.

$R' = 50/15$ in the scaled bandpass filter in Figure 9.47. Setting $d = 1/\bar{\omega}$ and using $\bar{\omega} = 15\sqrt{2}$ in (9.102) and (9.92), the invulnerable attenuation at 2 MHz will be $L_{min} = 53$ dB. Note that this attenuation exists without regard to the values of Z_s, Z_L, or the impedance-scaling factor utilized.

The network of Figure 9.46 was analyzed with unloaded-Q values assigned to each inductor to provide a realistic comparison between the conventional ($R' = 0$) and invulnerable ($R' = 3.33$) filters. The results are shown in Figure 9.48, where only the upper-half passband is pictured relative to tune frequency $f_0 = 1$ MHz. The worst-case (conjugate-image match) curves are shown for L_{min}; they will be better than that for most loads. Both filters had a transducer loss of 0.8682 dB at 1 MHz. At the second harmonic, the Butterworth filter ($R' = 0$) had an insertion loss of 70.57 dB, and the invulnerable filter ($R' = 3.33$) had an insertion loss of 65.65 dB. If conjugate-image matches had occurred at the second harmonic, the invulnerable filter would deliver a load power 53.06 dB below the available source power; the Butterworth filter load power would be only 6.067 dB down. Of course, such an exact combination of both terminal impedances is unlikely, but L_{min} does constitute an invulnerable lower limit to the infinite set of degrading terminations that are as likely as not to occur in practice.

As developed in Section 9.5.3, significant differences in the maximum input SWR of the two filter types may be expected when Z_L may assume any value in the right-half plane. At $f = 1.05f_0$ in the passband, the maximum SWR (S_K) was 23.10 for the conventional filter and 10.49 for the invulnerable filter. The corresponding numbers at $f = 2f_0$ were 8922.0 and 14.98. The 3-dB bandwidths were about equal.

9.6.3. Summary of Invulnerable Filters. Lowpass bridged-T and three-pole invulnerable filters were described, selectivity curves were derived, and a design example with analysis was provided. Both open-circuit z and short-circuit y parameters were utilized when convenient, because maximum-efficiency relationships are valid under a simple change of nomenclature.

These invulnerable filters were shown to have a minimum-possible attenuation rate of only 6 dB/octave at high frequency. In effect, they are selective "pads" that absorb energy on a frequency-selective basis. Like resistive pads, they reduce the load SWR that appears at the input port, a valuable property in many critical applications. They may be used in conjunction with reflection filters to obtain the best characteristics of the two techniques.

Problems

9.1. Since commensurate-network selectivity is a function of $q^2 = \cos^2\theta$, use $\cos^2\theta = \cos^2(\pi - \theta)$ to show that any two frequencies having the same selectivity are related by Equation (9.5): $f_1/f_0 = 2 - (f_2/f_0)$.

9.2. Which types of elliptic filters have
   (a) The exact elliptic filter response?
   (b) A negative element or perfect transformers?
   (c) Equal terminating resistances?
   (d) $(N-1)/2$ traps, where N is the filter degree?
   (e) $N/2$ traps?

9.3. An elliptic filter of what degree is required to produce a 0.3-dB-ripple pass band to 1.2 MHz and a 53-dB stopband shelf starting at 2.4 MHz?

9.4. The circuit below has $Z_{in} = 0 + j6$ ohms and $dZ_{in}/d\omega = 4$ at $\omega = 2$. M_1 and K_1 are resonant at 2 radians. Find L_1, M_1, and K_1.

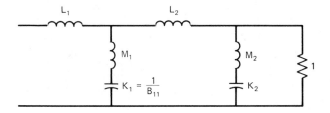

9.5. In the permuted circuit below, $L_{21} = \frac{22}{9}$, $K_{21} = \frac{1}{9}$, and M_{21} and K_{21} are resonant at $\omega = \frac{3}{2}$ radians. Find L_2, M_2, and K_2 in the figure in Problem 9.4.

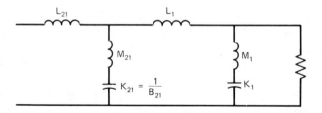

9.6. Use Equations (9.67)–(9.69) to find $e^{j\gamma}$ in Equation (9.68).

9.7. A variable-length transmission line having $Z_0 = R_1$ has input SWR S with respect to (wrt) R_1. The line is terminated by an impedance that produces an SWR S_0 wrt R_0, as pictured below. Set $|g|$ equal to $(S-1)/(S+1)$ and use Equations (9.70)–(9.71) to show that the maximum S is $S = S_1 \times S_0$, where S_1 is the SWR of R_1 wrt R_0, i.e., $(R_1 - R_0)/(R_1 + R_0) = (S_1 - 1)/(S_1 + 1)$. Also, find the minimum S, including any special conditions.

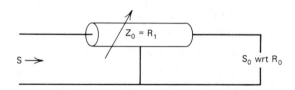

9.8. Work Example 7.1 in Chapter Seven using formulas from Chapter Nine.

9.9. Derive $K = 2\mathfrak{R} + 1$ from Equation (9.82) and $\eta_{max} = \sqrt{\mathfrak{R} + 1} - \sqrt{\mathfrak{R}}$ from Equation (9.83) using Equations (7.57), (7.64), and (9.81).

9.10. At a stopband frequency, an invulnerable filter has short-circuit parameters $y_{11} = y_{22} = (13.68 + j5.883) \times 10^{-3}$ and $y_{21} = y_{12} = (-0.02 + j3.333) \times 10^{-3}$ mhos.

(a) What is the relative size of the load-plane image in the input plane?

(b) What is the value of L_{min}, the minimum possible loss?

(c) How many different load admittances can produce this L_{min} value?

9.11. A passive network has open-circuit z parameters $z_{11} = 4 + j25$, $z_{21} = 1 - j15$, and $z_{22} = 2 + j37$ ohms. Find

(a) The relative size of the load-plane image in the input plane.

(b) The value of the minimum possible loss (L_{min}) by two different equations.

9.12. Derive Equation (9.92): $L_{min} = 8.68 \sinh^{-1} \sqrt{\mathfrak{R}}$.

Appendix A

HP-67/97 Programs

Program A2-1. Polar Complex Four Functions

Label:	X $\circlearrowright$ Y	+	1/Z	*
Key:	A	B	C	D

Examples: Let $Z_1 = 5 \underline{/53.1°} = 3 + j4$ and $Z_2 = 2 \underline{/30} = \sqrt{3} + j1$ in degree mode.

1. $Z_1 + Z_2 = 6.88 \underline{/46.58°}$; 53.13, ↑, 5, ↑, 30, ↑, 2, B.
2. $Z_1 - Z_2 = 3.26 \underline{/67.09°}$; 53.13, ↑, 5, ↑, 30, ↑, 2, CHS, B.
3. $1/Z_1 = 0.20 \underline{/-53.13°}$; 53.13, ↑, 5, C.
4. $Z_1 * Z_2 = 10.00 \underline{/83.13°}$; 53.13, ↑, 5, ↑, 30, ↑, 2, D.
5. $Z_1/Z_2 = 2.50 \underline{/23.13°}$; 53.13, ↑, 5, ↑, 30, ↑, 2, C, D.

```
001   *LBLC  Start cmplx 1/Z
002    1/X
003    X≥Y
004    CHS
005   *LBLA  Start swap RX,RY
006    X≥Y
007    RTN  End 1/Z & swap
008   *LBLD  Start cmplx mult
009    X≥Y
010    R↓
011    X
012    R↓
013    +
014    R↑
015    RTN  End cmplx mult
016   *LBLB  Start cmplx add
017    →R
018    R↑
019    R↑
020    →R
021    X≥Y
022    R↓
023    +
024    R↓
025    +
026    R↑
027    →P
028    RTN  End cmplx add
```

Program A5-1. Swain's Surface

See Equations (5.1)–(5.5).

Label:	$X_1 \rightarrow$	$X_2 \rightarrow$	F	$\nabla_1 F$	$\nabla_2 F$
Key:	A	B	C	D	E

Examples: See Table 5.2.

001	*LBLA	Process X1		051	+	
002	STO1	X1 into R1		052	x	
003	X²			053	STO0	Temp STO
004	STO2	Sqrd into R2		054	GSB2	Calc Q2
005	LSTX			055	3	
006	RTN	End process X1		056	x	
007	*LBLB	Process X2		057	CHS	
008	STO3	X2 into R3		058	RCL0	RCL Temp
009	X²			059	+	
010	STO4	Sqrd into R4		060	RTN	End ∇2F Calc
011	LSTX			061	*LBL1	Start P2 Calc
012	RTN	End process X2		062	RCL2	
013	*LBLC	Start Q calc		063	9	
014	GSB1	Calc P2		064	x	
015	X²			065	2	
016	STO0	Temp STO		066	5	
017	GSB2	Calc Q2		067	RCL4	
018	X²			068	x	
019	RCL0	RCL Temp		069	+	
020	+			070	3	
021	4			071	6	
022	÷			072	RCL1	
023	RTN	End Q2 calc		073	x	
024	*LBLD	Calc ∇1F		074	-	
025	GSB1	Calc P2		075	5	
026	1			076	0	
027	8			077	RCL3	
028	x			078	x	
029	RCL1			079	+	
030	2			080	1	
031	-			081	6	
032	x			082	4	
033	STO0	Temp STO		083	-	
034	GSB2	Calc Q2		084	2	
035	2			085	x	
036	x			086	RTN	End P2 Calc
037	RCL1			087	*LBL2	Start Q2 Calc
038	2			088	RCL2	
039	-			089	4	
040	x			090	RCL1	
041	RCL0	RCL Temp		091	x	
042	+			092	-	
043	RTN	End ∇1F Calc		093	3	
044	*LBLE	∇2F Calc		094	RCL3	
045	GSB1	Calc P2		095	x	
046	5			096	-	
047	0			097	8	
048	x			098	-	
049	RCL3			099	2	
050	1			100	x	
				101	RTN	End Q2 Calc

387

Program A5-2. Central Quadratic Function

See Equations (5.8), (5.15), and (5.16).

Label:	$X_1 \rightarrow$	$X_2 \rightarrow$	F	$\nabla_1 F$	$\nabla_2 F$
Key:	A	B	C	D	E

Example: $X_1 = 10$, $X_2 = 10$, $F = 292$, $\nabla_1 F = 100$, $\nabla_2 F = 28$.

```
001  *LBLC  Start F Calc
002  RCL2      (5.8)
003  RCL4
004   +
005   1
006   3
007   x
008  RCL1
009  RCL3
010   x
011   1
012   0
013   x
014   -
015  RCL3
016   1
017   3
018   2
019   x
020   -
021  RCL1
022   6
023   0
024   x
025   -
026   6
027   1
028   2
029   +
030  RTN  End F Calc
031  *LBLD  Start ∇1F Calc
032  RCL1      (5.15)
033   2
034   6
035   x
036  RCL3
037   1
038   0
039   x
040   -
041   6
042   0
043   -
044  RTN  End ∇1F Calc
```

```
045  *LBLE  Start ∇2F Calc
046  RCL3      (5.16)
047   2
048   6
049   x
050  RCL1
051   1
052   0
053   x
054   -
055   1
056   3
057   2
058   -
059  RTN  End ∇2F Calc
060  *LBLA  Process X1
061  STO1  X1 into R1
062   X²
063  STO2  Sqrd into R2
064  LSTX
065  RTN  End Process X1
066  *LBLB  Process X2
067  STO3  X2 into R3
068   X²
069  STO4  Sqrd into R4
070  LSTX
071  RTN  End Process X2
```

Program A5-3. Calculate Quadratic-Form Level Curves

See Equation (5.23).

Label: $X_2(X_1)$ $b, k, a \rightarrow$ $Q \rightarrow$
Key: A   B   C

Example: $b = 13 = a$, $k = -5$, $Q = 292$, $X_1 = 5$;   then, $X_2 = 3$ and 0.8462
in RX and RY, respectively.

```
001  *LBLA   Calc X2 from X1 per (5.23)
002  STO6    X1 into R6
003   X²
004  RCL1    a
005   ×
006  RCL4    Q
007   -
008  RCL3    b
009   ×
010  CHS
011  RCL2    k
012  RCL6    X1
013   ×
014  STO5    Temp k·X1
015   X²
016   +      Arg radical
017   √X
018  STO0    Temp STO radical
019  RCL5    k·X1
020   +
021  CHS
022  RCL3
023   ÷      Soln with "-" signs
024  RCL0    RCL radical
025  RCL5    k·X1
026   -
027  RCL3
028   ÷      2ND soln
029  RTN     End calc X2 from X1
030  *LBLB   Process b,k,a:RZ,ZY,RX
031  STO1    STO a
032  R↓
033  STO2    STO k
034  R↓
035  STO3    STO b
036  R↓
037  R↓
038  RTN     End process b,k,a
039  *LBLC   Process Q
040  STO4    Q into R4
041  RTN     End process Q
```

Program A5-4. Linear Search, Inner Products, and Conjugate Forms

See Equations (5.39), (4.92), and (5.52).

	Input				
RT:	s_1^i				
RZ:	s_2^i	a	Input α		
RY:	x_1^i	k	$\overline{x_1^{i+1}(\alpha)}$		
RX:	x_2^i	b	$x_2^{i+1}(\alpha)$	$x^T s$	$x^T As$
Key:	A	B	C	D	E

Example: $s_1^i = 1.1$, $s_2^i = -1.2$, $x_1^i = -1.3$, $x_2^i = 1.4$, $a = 4.1$, $k = 5.2$, $b = 6.3$; then $x^{i+1} = (-0.4750, 0.50)^T$ for $\alpha = 0.75$, and $x^T s = -3.1100$, $x^T As = -0.3270$.

```
001  *LBLE  Start Conjg. Form      033  *LBLC  Calc x(i+1) from α
002  RCL1                          034  ST08   α into R8
003  RCL4                          035  RCL1   s1
004   x                            036   x
005  RCL3                          037  RCL3   x1
006  RCL2                          038   +     x1(i+1)
007   x                            039  RCL2   s2
008   +                            040  RCL8   α
009  RCL6                          041   x
010   x                            042  RCL4   x2
011  RCL5                          043   +     x2(i+1)
012  RCL1                          044  RTN    End calc x(i+1)
013   x                            045  *LBLB  Process a,k,b
014  RCL3                          046  ST07   b into R7
015   x                            047  R↓
016   +                            048  ST06   k into R6
017  RCL7                          049  R↓
018  RCL2                          050  ST05   a into R5
019   x                            051  R↓
020  RCL4                          052  R↓
021   x                            053  RTN    End process a,k,b
022   +                            054  *LBLA  Process 2 points
023  RTN  End Conjg. Form          055  ST04   x2 into R4
024  *LBLD  Start Inner Prod.      056  R↓
025  RCL1                          057  ST03   x1 into R3
026  RCL3                          058  R↓
027   x                            059  ST02   s2 into R2
028  RCL2                          060  R↓
029  RCL4                          061  ST01   s1 into R1
030   x                            062  R↓
031   +                            063  RTN    End process 2 points
032  RTN  Enf Inner Prod.
```

Program A6-1. $1 + Q^2$ Series-Parallel and Parallel Reactances

See Equations (6.16), (6.17), (6.20), and (6.21).

Input convention:

RY: X_s X_p X_2 X

       or   or   or

RX: R_s R_p X_1 X_1

   $Z_s {\to} Z_p$ $Z_p {\to} Z_s$ $X_1 \| X_2$ $X_2(X, X_1)$ $X \backsim Y$

Key: A   B   C   D   E

Output convention:

RY: X_p X_s

       or   or   or

RX: R_p R_s X X_2

```
001   *LBLB   Calc Zs(Zp)
002   GSB1    L1 stack shuffle
003   GSB0    L0 stack shuffle
004   X≠Y     Set up for Rs
005   ÷       Calc Rs
006   GSB1    L1 stack shuffle
007   x       Q·Rs=Xs
008   GTOE    To swap Rs&Xs
009   *LBLA   Calc Zp(Zs)
010   GSB1    L1 stack shuffle
011   X≠Y     Swap Rs & Xs
012   GSB0    L0 stack shuffle
013   x       Get Rp
014   GSB1    L1 stack shuffle
015   ÷       Get Xp
016   *LBLE   Swap RX & RY
017   X≠Y
018   RTN     End Zs or Zp or swap
019   *LBLD   Calc X2(X,X1)
020   CHS     per (6.21)
021   *LBLC   Calc (6.20)
022   GSB1    L1 stack shuffle
023   ÷       X2/X1 or X1/X2
024   1
025   +
026   X=0?    If X1=-X2, don't div
027   RTN
028   ÷
029   RTN     End calc (6.20)
```

```
030   *LBL1   Start a stack shuffle
031   ENT↑
032   ENT↑
033   R↑          See shuffle table
034   RTN     End shuffle L1
035   *LBL0   Start a stack shuffle
036   ÷
037   ENT↑
038   ENT↑        See shuffle table
039   x
040   1
041   +
042   R↑
043   RTN     End shuffle L0
```

Stack	Start	LBL0	LBL1
RT	d	c	a
RZ	c	b/a	a
RY	b	$(b/a)^2+1$	a
RX	a	c	b

Program A6-2. Transmission Line Matching

See Equations (6.27), (6.28), (6.32)–(6.34), (6.41), and (6.42).
Stack and key identification:

RT:	X_2	X_2			
RZ:	R_2	R_2			
RY:	$\theta \rightarrow X_1$	$X_1 \rightarrow \theta$		R_2	
RX:	$Z_0 \rightarrow R_1$	$R_1 \rightarrow Z_0$	$R\downarrow^2$	$R_1 \rightarrow \theta$	$X \subset Y$
Key:	A	B	C	D	E

```
001  *LBLE  Swap X & Y
002   X⇄Y
003   RTN   End Swap
004  *LBLC  Start roll down twice
005   R↓
006   R↓
007   RTN   End roll down
008  *LBLB  Z0,θ from Z1 & Z2
009   STO1  R1 into reg1
010   R↓
011   R↓
012   STO2  R2 into reg2
013   R↓
014   ×
015   STO0  X2·R1 into reg0
016   CLX
017   RCL1  R1
018   LSTX
019   R↓
020   R↓
021   ×
022   ST+0  R1·X2+R2·X1
023   CLX
024   RCL2  R2
025   LSTX
026   R↓
027   →P
028   X²    |Z2|²
029   X⇄Y
030   CLX
```

```
031   RCL1  R1
032   ×
033   R↓
034   →P
035   X²    |Z1|²
036   R↑
037   X⇄Y
038   RCL2  R2
039   ×
040   -     num of (6.34)
041   RCL2
042   RCL1
043   -
044   STO1  R2-R1
045   ÷
046   √X    Z0
047   STO2  into reg2
048   EEX
049   CHS
050   8
051   ST+0  so can't be 0
052   1
053   8
054   0
055   RCL1
056   CHS   R1-R2
057   RCL0
058   ÷     q
059   RCL2  Z0
060   ×
```

061	TAN⁻¹	θ
062	X<0?	if negative
063	+	add 180°
064	RCL2	Z0
065	RTN	End Z0, θ
066	*LBLA	Calc Zin per (6.27)
067	STO3	Z0 into reg3
068	R↓	
069	EEX	
070	CHS	
071	8	
072	-	maybe θ = 90°
073	TAN	
074	STO4	y into reg4
075	R↓	
076	→P	Z2
077	RCL3	Z0
078	÷	
079	STO5	$\|Z2\|/Z0=\|\overline{Z}2\|$
080	1/X	
081	X⇄Y	θ2
082	STO6	into reg6
083	CHS	
084	X⇄Y	$\overline{Y}2$
085	GSB0	add jy
086	ST÷3	Zo/(denom)
087	X⇄Y	
088	CHS	
089	RCL6	θ2
090	RCL5	$\|\overline{Z}2\|$

091	GSB0	add jy
092	R↓	
093	+	add angles
094	RCL6	θ2
095	-	
096	R↑	
097	RCL5	$\|\overline{Z}2\|$
098	÷	
099	RCL3	
100	X	scale
101	→R	
102	RTN	End Zin
103	*LBL0	Add jy to a given polar Z
104	→R	
105	X⇄Y	
106	RCL4	
107	+	
108	X⇄Y	
109	→P	
110	RTN	End add jy
111	*LBLD	Start θ per (6.41)
112	÷	R per (6.42)
113	ENT↑	
114	1/X	
115	+	
116	1	
117	+	
118	1/X	
119	√X	
120	TAN⁻¹	
121	RTN	End (6.41) for θ

Note: rewrite (6.27) for LBLA: $\quad Z_1 = \dfrac{Z_0(\overline{Z}_2+jy)}{\overline{Z}_2(\overline{Y}_2+jy)}, \quad \overline{Z}_2 = Z_2/Z_0, \quad \overline{Y}=1/\overline{Z}$

Program A6-3. Min SWR$_{max}$ and Lowpass-to-Bandpass Transformations

See Equations (6.51), (6.85), (6.86), (6.88), and (6.89).

Stack and key identification:

RZ:		f_2Hz		
RY:	Q_L	f_1Hz	$\rightarrow C_{ser}$	$\rightarrow L_{sh}$
RX:	%BW$\rightarrow$SWR	$R_{norm}\rightarrow f_0$	$g_i\rightarrow L_{ser}$	$g_i\rightarrow C_{sh}$	X$\subset$Y
Key:	A	B	C	D	E

Note: Record with DSP 4 ENG.

001	*LBLE	Swap X & Y		030	*LBLC	Start series calc
002	X≠Y			031	RCL3	Rnorm
003	RTN	End swap		032	x	Z-scaled gi
004	*LBLB	Start f0 calc		033	*LBL0	Start common calc
005	STO3	Rnorm into R3		034	RCL0	w
006	R↓			035	÷	scaled gi/w
007	STO0	f1 into R0		036	RCL1	ω0
008	X≠Y			037	÷	Lser or Csh
009	-	f1-f2		038	RCL2	ω0²
010	LSTX	f2		039	X≠Y	
011	RCL0	f1		040	x	
012	x			041	LSTX	Rcl Lser or Csh
013	√X	f0		042	X≠Y	
014	STO1	f0 into R1		043	1/X	Cser or Lsh
015	÷	±w		044	X≠Y	
016	ABS			045	RTN	End LBL C or LBL D calcs
017	STO0	w into R0		046	*LBLD	Start parallel calc
018	2			047	RCL3	Rnorm
019	Pi			048	÷	Y-scaled gi
020	x			049	GTO0	To common calc
021	RCL1	f0		050	*LBLA	Start calc (6.51)
022	x	ω0		051	x	
023	LSTX	f0		052	EEX	
024	X≠Y			053	2	100 removes the %
025	STO1	ω0 into R1		054	÷	1/δ as in (6.49)
026	X²			055	Pi	
027	STO2	ω0² into R2		056	X≠Y	
028	R↓	f0		057	÷	δπ
029	RTN	End f0 calc		058	eˣ	
				059	1	
				060	X≠Y	
				061	+	numerator of (6.51)
				062	LSTX	exp(δπ)
				063	1	
				064	-	denominator of (6.51)
				065	÷	
				066	RTN	End calc (6.51)

Program A6-4. Norton Transformations

See equations in Table 6.4.

Stack and key identification:

RY:	X_p				
RX:	X_s	LB→Pi	LB→T	LA→Pi	LA→T
Key:	A	B	C	D	E

Procedure: Put X_p (=L or $1/C$) and X_s into RY and RX. Press key A to initialize. Press key B or C or D or E. Display will then show the n^2 extreme fartherest from unity. Enter your choice of n^2 in the open range and press R/S. Display is X_a (=L_a or $1/C_a$). Roll down, see X_b; roll down, see X_c.

001	*LBLA	Store & Process Xs & Xp
002	STO1	Xs into R1
003	X⇄Y	
004	STO2	Xp into R2
005	÷	
006	1	
007	+	
008	ENT↑	
009	x	(1+Xs/Xp)²
010	SF1	Initialize
011	SF0	Flags (see table)
012	RTN	End store Xs & Xp
013	*LBLB	Start LB to Pi Calc
014	CF1	See flag table
015	*LBLC	Start Lb to T
016	CF0	See flag table
017	1/X	Min possible n²
018	R/S	Stop for user's n²
019	STO8	n² into R8
020	1/X	
021	STO3	1/n² into R3
022	GTO0	To common calculation
023	*LBLD	Start LA to Pi Calc
024	CF1	See flag table
025	*LBLE	Start LA to T Calc
026	R/S	Stop for user's n²
027	STO3	n² into R3
028	*LBL0	Common Calculation
029	√X	
030	STO4	n into R4
031	RCL1	Xs
032	x	
033	STO5	n·Xs
034	RCL4	n
035	1	
036	-	
037	STO7	n-1 into R7
038	RCL4	n
039	RCL2	Xp
040	x	

041	STO6	n·Xp into R6
042	F1?	
043	GTO9	To the T case
044	GTO4	To the Pi case
045	*LBL9	T Case
046	RCL7	n-1
047	x	Xc
048	RCL6	n·Xp
049	RCL1	Xs
050	RCL2	Xp
051	+	
052	RCL6	n·Xp
053	-	Xa
054	GTO5	To final transformation
055	*LBL4	Pi Case
056	RCL5	n·Xs
057	x	
058	RCL1	Xs
059	RCL7	n-1
060	RCL2	Xp
061	x	
062	-	
063	÷	
064	RCL5	n·Xs
065	RCL5	
066	RCL7	n-1
067	÷	Ans's now in stack
068	*LBL5	Final Transformations
069	F0?	
070	R/S	LA case OK as is
071	RCL8	n²
072	x	
073	R↑	
074	RCL8	
075	x	
076	R↑	
077	RCL8	
078	x	
079	X⇄Y	
080	R/S	Stop with LB ans's.

FLAG TABLE

Key:	B	C	D	E
FLG0	CLR	CLR	SET	SET
FLG1	CLR	SET	CLR	SET

Program A7-1. Bilinear Coefficients From Arbitrary Triples

See Section 7.1.1.

Stack input:

RT: w_i degrees

RZ: w_i magnitude

RY: Z_i degrees

RX: Z_i magnitude

Keys:

See a_i

X⌒Y	+	1/Z	*	Load and Go
A	B	C	D	E

Procedure: Press key E. When 1 appears, fill the stack as indicated above for $i = 1$, and press R/S. Do similarly when 2 and 3 appear. Calculator will stop in about 3 minutes. Then press 1, fa, and see a_1 magnitude; press A to see a_1 degrees, etc.

Example:

$w_1 = 0.1732 \; / -7.8675°$, $Z_1 = 0.1 \; / 30°$. $a_1 = 0.6 \; / 75°$.

$w_2 = 0.4473 \; / 129.5050°$, $Z_2 = 0.5 \; / 60°$. $a_2 = 0.18 \; / -23°$.

$w_3 = 0.5099 \; / -30.3244°$, $Z_3 = 1.1 \; / -10°$. $a_3 = 1.4 \; / 130°$.

001	*LBLa	See given ith coefficient	031	5	Z3	
002	6		032	6	w3	
003	+		033	8	R8	
004	STOI	Store indirect address	034	3	Z2	
005	GTO9	Rcl cmplx polar data	035	STOD	Sto sequence in RD	
006	*LBLE	Input wi & Zi on cue	036	GSB4	Add Z2·D31 into RO	
007	CLRG	(see table 7-2)	037	.	Det M opns sequence 3:	
008	P≷S		038	3	Z2	
009	CLRG		039	4	w2	
010	1		040	8	R8	
011	GSB1	Cue & Sto 1st input data	041	1	Z1	
012	2		042	2	w1	
013	GSB1	Cue & Sto 2nd input data	043	8	R8	
014	3		044	5	Z3	
015	GSB1	Cue & Sto 3rd input data	045	STOE	Sto sequence in RE	
016	CF0	Don't perform step 113	046	GSB4	Add Z3·D12 into RO	
017	.	Det M opns sequence 1:	047	GSBC	Polar invert	
018	5	Z3	048	GSB8	Sto inverse Det M in RO	
019	6	w3	049	SF0	To modify a1 sequences	
020	8	R8	050	RCLC	Rcl sequence 1	
021	3	Z2	051	GSB4	w1·D23	a1 calc in
022	4	w2	052	RCLD	Rcl sequence 2	steps 049-077
023	8	R8	053	GSB4	w2·D31	
024	1	Z1	054	RCLE	Rcl sequence 3	
025	STOC	Store sequence into RC	055	GSB4	w3·D12	
026	GSB4	Sto Z1·D23 in RO	056	GSB3	Mult by inverse det M	
027	.	|Det M opns sequence 2:	057	GSB8	Sto a1 in R7	
028	1	Z1	058	.	a3 sequence 1:	
029	2	w1	059	6	w3	
030	8	R8	060	4	w2	

See table 7-2.
Inverse Det M calc
in steps 016-048.

061	1	Z1
062	9	R9
063	GSB5	Partial a3 into R9
064	.	a3 sequence 2:
065	2	w1
066	6	w3
067	3	Z2
068	9	R9
069	GSB5	More a3 into R9
070	.	a3 sequence 3
071	4	w2
072	2	w1
073	5	Z3
074	9	R9
075	GSB5	Last of a3 into R9
076	GSB3	Multiply by inverse det M
077	GSB8	Sto a3 in R9
078	.	Entire a2 sequence:
079	3	Z2
080	4	w2
081	9	a3 Calc a2 in steps
082	8	R8 078-104
083	7	a1
084	3	Z2
085	8	R8
086	4	w2
087	8	R8
088	STOI	Sto sequence in RI
089	GSB0	Rcl Z2
090	GSB0	Rcl w2
091	GSBD	P2
092	GSB0	Rcl a3
093	GSBD	P2·a3
094	GSB7	Sto in R8
095	GSB0	Rcl a1
096	GSB0	Rcl Z2
097	GSBD	Polar multiply
098	CHS	Set up subtraction
099	GSB0	Rcl R8
100	GSBB	Polar addition
101	GSB0	Rcl w2
102	GSBB	Polar addition
103	GSB7	Sto a2 in R8
104	RTN	End program

105	*LBL4	Start det M or a1 opns
106	1	
107	7	
108	EEX	
109	CHS	
110	8	
111	X≷Y	
112	F0?	Only for a1 opns; see 049
113	+	Add to RX
114	STOI	Sto RX in RI
115	GSB0	Rcl digit 1 data
116	GSB0	Rcl digit 2 data
117	GSBD	Polar multiply
118	GSB7	Sto in digit 3 reg
119	GSB0	Rcl digit 4 data
120	GSB0	Rcl digit 5 data
121	GSBD	Polar multiply
122	*LBL6	Re-entry from LBL5
123	GSB0	Rcl next digit data
124	CHS	Set up subtraction
125	GSBB	Polar addition
126	GSB0	Rcl next digit data
127	GSBD	Polar multiply
128	GSB0	Rcl next digit data
129	GSBB	Polar addition
130	GSB8	Sto in last digit reg
131	RTN	End sequence of opns
132	*LBL5	Start a3 opns
133	STOI	Sto RX in RI
134	GSB0	Rcl digit 1 data
135	GTO6	Go use common opns
136	*LBLC	Polar cmplx invert
137	1/X	
138	X≷Y	
139	CHS	
140	*LBLA	Swap X & Y
141	X≷Y	
142	RTN	End invert and swap
143	*LBLD	Polar cmplx multiply
144	X≷Y	
145	R↓	
146	×	
147	R↓	
148	+	
149	R↑	
150	RTN	End cmplx multiply

```
151   *LBLB   Polar cmplx addition
152    →R
153    R↑
154    R↑
155    →R
156    X⇄Y
157    R↓
158    +
159    R↓
160    +
161    R↑
162    →P
163    RTN    End cmplx addition
164   *LBL0   Get ith digit
165    GSB2    Decode for digit
166   *LBL9   Rcl cmplx data pair
167    P⇄S
168    RCLi
169    P⇄S
170    RCLi
171    RTN    End cmplx data rcl
172   *LBL7   Get ith digit
173    GSB2    Decode for digit
174   *LBL8   Sto cmplx data
175    STOi
176    P⇄S
177    X⇄Y
178    STOi
179    P⇄S
180    X⇄Y
181    RTN    End cmplx data sto
182   *LBL2   Decode next digit in sequence
183    RCLI    Rcl sequence
184    FRC     Take fractional part
185     1
186     0
187     X     Move decimal one to right
188    STOI    Sto next digit and fraction
189    R↓     Flush RX from stack
190    RTN    End digit decode
191   *LBL3   Rcl inverse det M and multiply
192    P⇄S
193    RCL0
194    P⇄S
195    RCL0
196    GSBD
197    RTN    End mult by inverse det M
198   *LBL1   Cue and sto input data pairs
199    ISZI           RT: w deg
200    R/S            RZ: |w|
201    GSB8           RY: Z deg
202    R↓             RX: |Z|
203    R↓
204    ISZI
205    GSB8
206    RTN    End input
```

Program A7-2. Three-Port to Two-Port Conversion

See Equation (7.29) and Example 7.2.

Keys:

$S_3 \rightarrow$	$+$	$1/Z$	$*$	$\Gamma_{50}(Z^p)$	$Z^p(\Gamma_{50})$
a	b	c	d	e	RTN R/S

$Z_2^p \rightarrow$	S_{11}	S_{31}	S_{13}	S_{33}
A	B	C	D	E

Note: Key e and RTN R/S are inverse functions.

Input: Press CLRG, P⊃S, CLRG; then press fa. When 11 appears, key in θ_{11}^0, ↑, $|S_{11}|$, R/S; do the same for 21, etc. After all nine complex parameters are entered, save on a data card. Then input Z_2 in polar form: θ_2, ↑, $|Z_2|$, and press key A.

Output: Press key B; see θ_{11} and $|S_{11}|$ two-port parameters. Keys B, C, D, and E may be used in any order.

```
001    0     Start Zp(Γ50) Calculation    031    P⃗S
002   ENT↑                                 032   RCL5  |s22|
003    1                                   033   CHS   Set up subtraction
004   GSBb   Γ+1                           034   GSBb  Cmplx addition
005   STOD   Sto magnitude                 035   GSBc  Invert
006   X⃗Y                                    036   STO8  |(1/Γ2-s22)⁻¹|
007   STOE   Sto degrees                   037   P⃗S
008   X⃗Y                                    038   X⃗Y
009   CHS    Set up subtraction            039   STO8  (1/Γ2-s22)⁻¹ degrees
010    0                                   040   P⃗S
011   ENT↑                                 041   X⃗Y
012    2                                   042   RTN   End calc with Z2
013   GSBb   1-Γ                           043  *LBL2  Start stnd sequence calc
014   GSBc   Invert                        044   STOI  Store sequence
015   RCLE                                 045   P⃗S
016   RCLD   1+Γ                           046   RCL0  See step 039
017   GSBd   Cmplx multiplication          047   P⃗S
018    5                                   048   RCL0  See step 036
019    0                                   049   GSB0  Rcl digit 1 data
020    x     Z scale                       050   GSBd  Polar multiply
021   R/S    End Zp(Γ) Calculation         051   GSB0  Rcl digit 2 data
022  *LBLA   Start Calc with Z2            052   GSBd  Polar multiply
023   GSBe   Calc Γ2 from Z2               053   GSB0  Rcl digit 3 data
024   EEX                                  054   GSBb  Polar addition
025   CHS                                  055   X⃗Y
026    9                                   056   PRTX  Print degrees
027    +     Avoid divide by zero          057   X⃗Y
028   GSBc   1/Γ2                          058   PRTX  Print magnitude
029   P⃗S                                   059   RTN   End standard calculation
030   RCL5   s22 degrees
```

060	*LBL0	Decode next digit & rcl
061	RCLI	Rcl remaining sequence
062	FRC	Take fractional part
063	1	
064	0	
065	×	Move decimal 1 right
066	STOI	Sto digit and fraction
067	R↓	Flush stack RX
068	P⇄S	
069	RCLi	Rcl degrees
070	P⇄S	
071	RCLi	Rcl magnitude
072	RTN	End decode and rcl
073	*LBLc	Start polar cmplx invert
074	1/X	
075	X⇄Y	
076	CHS	
077	X⇄Y	
078	RTN	End invert
079	*LBLd	Start plor cmplx mult
080	X⇄Y	
081	R↓	
082	×	
083	R↓	
084	+	
085	R↑	
086	RTN	End mult
087	*LBLb	Start polar cmplx add
088	→R	
089	R↑	
090	R↑	
091	→R	
092	X⇄Y	
093	R↓	
094	+	
095	R↓	
096	+	
097	R↑	
098	→P	
099	RTN	End add

100	*LBLe	Start Γ50(Zp) Calc
101	5	
102	0	
103	÷	Zp/50
104	0	
105	ENT↑	
106	1	
107	GSBb	Cmplx addition
108	GSBc	Cmplx invert
109	2	
110	×	Similar to method
111	CHS	in eqn (2.2): -2/(z+1)
112	0	
113	ENT↑	
114	1	
115	GSBb	1-2/(z+1)
116	RTN	End Γ(Z) calc
117	*LBLa	Input 3x3 S Matrix
118	0	
119	STOI	Initialize RI
120	1	
121	1	
122	GSB1	s11 into R1
123	2	
124	1	
125	GSB1	s21 into R2
126	3	
127	1	
128	GSB1	s31 into R3
129	1	
130	2	
131	GSB1	s12 into R4
132	2	
133	2	
134	GSB1	s22 into R5
135	3	
136	2	
137	GSB1	s32 into R6
138	1	
139	3	
140	GSB1	s13 into R7
141	2	
142	3	
143	GSB1	s23 into R8
144	3	
145	3	
146	GSB1	s33 into R9
147	X⇄Y	
148	DSP4	
149	RTN	End 3x3 S Matrix Input

```
150   *LBL1  Cue & Sto 3x3 Elements
151   ISZI   Increment pointer
152   DSP0
153   R/S    Display cue 11, 12, etc.
154   STOi   Indirect store magnitude
155   P⇄S
156   X⇄Y
157   STOi   Indirect store degrees
158   P⇄S
159   RTN    End cue & store
160   *LBLB  Calc 2x2 S11
161    .     S11 opns sequence
162    2     s21
163    4     s12
164    1     s11
165   GTO2   To do this sequence
166   *LBLC  Calc 2x2 S31
167    .     S31 opns sequence
168    2     s21
169    6     s32
170    3     s31
171   GTO2   To do this sequence
172   *LBLD  Calc 2x2 S31
173    .     S31 opns sequence
174    8     s23
175    4     s12
176    7     s13
177   GTO2   To do this sequence
178   *LBLE  Calc 2x2 S33
179    .     S33 sequence
180    8     s23
181    6     s32
182    9     s33
183   GTO2   To do this sequence
```

Primary (magnitude) and secondary (degrees) register assignments;

R0	R1	R2	R3	R4	R5	R6	R7	R8	R9
See steps	s11	s21	s31	s12	s22	s32	s13	s23	s33
036 & 039									

Program A7-3. Impedance Mapping for a Scattering Two-Port Network

See Equations (7.36) and (7.40)–(7.42); see Example 7.4.

Keys:

$$S_2 \to$$
$$a$$

T, R, Z_c	+	$1/Z$	*	$\rho(Z_c^R, Z^R)$
A	B	C	D	E

Input: CLRG, P⊃S, then press fa. When 11 appears, key in θ_{11}, ↑, $|S_{11}|$, R/S, etc., until all four two-port scattering parameters have been entered.

Output: DSP4, press key A, and see θ_T, $|T|$, θ_R, $|R|$, X_c, and R_c. To calculate ρ (since X_c and R_c are already in the RY and RX registers, respectively), key in normalized load x and r, then press key E. The result is in polar form.

001	*LBLA	Start T,R,Zc Calculation	031	X⇄Y	
002	P⇄S		032	ST05	Δ degrees
003	RCL1	S11 degrees	033	RCL4	S22 degrees
004	P⇄S		034	P⇄S	
005	RCL1	$|S11|$	035	RCL4	$|S22|$
006	P⇄S		036	θ	
007	RCL4	S22 degrees	037	ENT↑	
008	P⇄S		038	1	
009	RCL4	$|S22|$	039	GSBB	1+S22
010	GSBD	Cmplx polar multiply	040	GSBC	Cmplx invert
011	ST05	$|S11 \cdot S22|$	041	ST06	$|1+S22|^{-1}$
012	P⇄S		042	P⇄S	
013	X⇄Y		043	X⇄Y	
014	ST05	S11·S22 degrees	044	ST06	$(1+S22)^{-1}$ degrees
015	RCL2	S21 degrees	045	RCL5	Δ degrees
016	P⇄S		046	P⇄S	
017	RCL2	$|S21|$	047	RCL5	$|Δ|$
018	P⇄S		048	CHS	Set up subtraction
019	RCL3	S12 degrees	049	GSB2	Calc a1
020	P⇄S		050	ST07	$|a1|$
021	RCL3	$|S12|$	051	X⇄Y	
022	GSBD	Cmplx multiply	052	P⇄S	
023	CHS	Set up subtraction	053	ST07	a1 degrees
024	P⇄S		054	RCL5	Δ degrees
025	RCL5	S11·S22 degrees	055	P⇄S	
026	P⇄S		056	RCL5	$|Δ|$
027	RCL5	$|S11 \cdot S22|$	057	GSB2	Calc a2
028	GSBB	Cmplx polar addition	058	ST08	$|a2|$
029	ST05	$|Δ|$	059	X⇄Y	
030	P⇄S		060	P⇄S	

061	STO8	a2 degrees
062	RCL4	S22 degrees
063	P⇄S	
064	RCL4	|S22|
065	CHS	Set up subtraction
066	0	
067	ENT↑	
068	1	
069	GSBB	1−S22
070	P⇄S	
071	RCL6	See step 044
072	P⇄S	
073	RCL6	See step 041
074	GSBD	Cmplx multiplication
075	STO9	|a3|
076	X⇄Y	
077	P⇄S	
078	STO9	a3 degrees
079	X⇄Y	
080	P⇄S	
081	→R	
082	2	
083	×	
084	STO0	$2 \Re e(a3)$
085	P⇄S	
086	RCL9	a3 degrees
087	CHS	Conjugate a3
088	P⇄S	
089	RCL9	|a3|
090	P⇄S	
091	RCL8	a2 degrees
092	P⇄S	
093	RCL8	|a2|
094	GSBD	a2·a3*
095	P⇄S	
096	RCL7	a1 degrees
097	P⇄S	
098	RCL7	|a1|
099	GSBB	Cmplx addition
100	RCL0	$2 \Re e(a3)$
101	÷	Equation (7.42)
102	P⇄S	
103	X⇄Y	
104	STO0	T degrees
105	PRTX	Print
106	P⇄S	
107	X⇄Y	
108	PRTX	Print
109	STO0	|T|
110	P⇄S	
111	RCL7	a1 degrees
112	P⇄S	
113	RCL7	|a1|
114	P⇄S	

115	RCL9	a3 degrees
116	P⇄S	
117	RCL9	|a3|
118	GSBC	Invert
119	GSBD	a1/a3
120	P⇄S	
121	RCL0	T degrees
122	P⇄S	
123	RCL0	|T|
124	CHS	Set up subtraction
125	GSBB	Equation (7.41)
126	X⇄Y	R degrees
127	PRTX	Print
128	X⇄Y	|R|
129	PRTX	Print
130	P⇄S	
131	RCL9	a3 degrees
132	CHS	Conjugate a3
133	P⇄S	
134	RCL9	|a3|
135	GSBC	Invert
136	→R	To rectangular
137	X⇄Y	
138	PRTX	xc
139	X⇄Y	
140	PRTX	rc
141	RTN	End program
142	*LBL2	Calc S11±Δ
143	P⇄S	
144	RCL1	S11 degrees
145	P⇄S	
146	RCL1	|S11|
147	GSBB	Cmplx addition
148	P⇄S	
149	RCL6	$(1+S22)^{-1}$ degrees
150	P⇄S	
151	RCL6	$\|1+S22\|^{-1}$
152	GSBD	Cmplx multiplication
153	RTN	End S11±Δ
154	*LBLC	Cmplx invert
155	1/X	
156	X⇄Y	
157	CHS	
158	X⇄Y	
159	RTN	End invert
160	*LBLD	Cmplx multiplication
161	X⇄Y	
162	R↓	
163	×	
164	R↓	
165	+	
166	R↑	
167	RTN	End multiplication

```
168  *LBLB  Cmplx addition
169  →R
170  R↑
171  R↑
172  →R
173  X⇄Y
174  R↓
175  +
176  R↓
177  +
178  R↑
179  →P
180  RTN    End addition
181  *LBLa  Input 2x2 s Matrix              Register Assignments
182  CLRG                                     Primary - magnitude
183  1                                        Secondary - degrees
184  1
185  GSB1   S11 into R1                  R0: 2Re(a3) & T
186  2
187  1                                   R1: S11
188  GSB1   S21 into R2                  R2: S21
189  1
190  2                                   R3: S12
191  GSB1   S12 into R3                  R4: S22
192  2
193  2                                   R5: Δ
194  GSB1   S22 into R4                  R6: (1+S22)⁻¹
195  RTN    End 2x2 input
196  *LBL1  Cue & store 2x2 elements     R7: a₁
197  ISZI   Increment pointer            R8: a₂
198  DSP0
199  R/S    Display cue 11, 12, etc.     R9: a₃
200  STOi   Indirect store magnitude
201  P⇄S
202  X⇄Y
203  STOi   Indirect store degrees
204  P⇄S
205  RTN    End cue and store
206  *LBLE  Start Calc Gen Refl Coeff. (7.16)
207  →P     Convert Z to polar
208  R↑
209  CHS    Conjugate Zc
210  R↑
211  STO0   Rc
212  →P     Zc to polar
213  GSBB   Polar addition: Zc*+Z
214  GSBC   Invert
215  RCL0   Rc
216  x
217  2
218  x
219  CHS    Set up subtraction
220  0      ρ=1-2·Rc/(Z+Zc*)
221  ENT↑
222  1
223  GSBB   Polar addition
224  RTN    End gen refl. coeff. calc.
```

Register Assignments correspond to the following:

R0: $2Re(a_3)$ & T
R1: $S11$
R2: $S21$
R3: $S12$
R4: $S22$
R5: Δ
R6: $(1+S22)^{-1}$
R7: a_1
R8: a_2
R9: a_3

Line 219-220: $\rho = 1 - 2 \cdot R_c/(Z + Z_c^*)$

Program A7-4. Maximally Efficient Gain Design

See Section 7.3.5 and Example 7.8.
Keys:

$$Y \rightarrow$$
a

Ans's	+	$1/Z$	*	$\Gamma_{50}(Z^p)$
A	B	C	D	E

Input: Enter θ_{11}, ↑, $|y_{11}|$, ↑, 11, fa (and 12, 21, and 22 entries similarly).
Output: Answers appear in the following order:

1. K (unloaded stability factor).
2. G_{ME} dB (maximally efficient gain).
3. θ_L degrees (load reflection angle).
4. $|\Gamma_L|$ (load reflection magnitude).
5. θ_s degrees (source reflection angle).
6. $|\Gamma_s|$ (source reflection magnitude).
7. K' (overall or loaded stability factor).

001	*LBLa	Input Y Parameters		031	STO5	g11 into R5
002	ENG	Set mho display		032	RCLE	y22 degrees
003	GSB0	Calc reg# from ij		033	RCLD	$|y22|$
004	R↓			034	→R	To rectangular
005	STOi	Indirect sto $|yij|$		035	STO6	g22 into R6
006	ISZI	Increment reg#		036	RCL5	g11
007	R↓			037	GSBi	Stab. factor K from (7.57)
008	STOi	Indirect sto yij deg		038	PRTX	
009	RTN	End input yij		039	RCLB	$|y21|$
010	*LBL0	Calc Reg# from ij		040	RCL3	$|y12|$
011				041	÷	
012	x			042	x	
013	2	See		043	LSTX	Calc (7.85) in
014	1	register		044	X²	steps 039–052
015	−	assignment		045	1	
016	STOI	table		046	−	
017	RTN	End		047	X⇄Y	
018	*LBLA	Start Main Program		048	1	
019	FIX			049	−	
020	RCL4	y12 degrees		050	2	
021	RCL3	$|y12|$		051	x	
022	RCLC	y21 degrees		052	÷	
023	RCLB	$|y21|$		053	LOG	G_{ME} in dB
024	GSBD	Cmplx polar multiply		054	1	
025	STO8	$|y12 \cdot y21|$		055	0	
026	→R	To rectangular		056	x	
027	STO7	Re(y12·y21)		057	PRTX	dB
028	RCL2	y11 degrees		058	RCLC	y21 degrees
029	RCL1	$|y11|$		059	RCLB	$|y21|$
030	→R	To rectangular		060	RCL4	y12 degrees

061	CHS	Conjugate y12
062	RCL3	|y12|
063	GSBB	
064	GSBC	Calc (7.82) in
065	RCL6	steps 058-079
066	x	
067	2	
068	x	
069	RCLC	
070	RCLB	
071	GSBD	
072	RCLE	
073	RCLD	|y22|
074	CHS	Set up subtraction
075	GSBB	Cmplx addition
076	STO0	|YL|
077	X≠Y	
078	STO9	YL degrees
079	X≠Y	
080	GSBC	ZL
081	GSBE	ΓL
082	X≠Y	
083	PRTX	Print ΓL degrees
084	X≠Y	
085	PRTX	|ΓL|
086	RCL9	YL degrees
087	RCL0	|YL|
088	GSBb	Calc related Yin
089	STOA	|Ys|
090	X≠Y	
091	CHS	Conjugate Yin
092	STOI	Ys degrees
093	X≠Y	
094	GSBC	Zs
095	GSBE	Γs
096	X≠Y	
097	PRTX	Print Γs degrees
098	X≠Y	
099	PRTX	Print |Γs|
100	RCL9	YL degrees
101	RCL0	|YL|
102	→R	To rectangular
103	RCL6	g22
104	+	GL+g22
105	RCLI	Ys degrees
106	RCLA	|Ys|
107	→R	To rectangular
108	X≠Y	
109	CLX	
110	RCL5	g11
111	+	Gs+g11
112	GSB1	Stab. factor K', (7.37)
113	PRTX	
114	RTN	End main program

115	*LBLb	Calc Yin by (7.49)
116	RCLE	y22 degrees
117	RCLD	|y22|
118	GSBB	Cmplx addition
119	GSBC	Cmplx invert
120	RCLC	y21 degrees
121	RCLB	|y21|
122	GSBD	Cmplx multiplication
123	RCL4	y21 degrees
124	RCL3	|y21|
125	GSBD	Cmplx multiplication
126	CHS	Set up subtraction
127	RCL2	y11 degrees
128	RCL1	|y11|
129	GSBB	Cpmlx addition
130	RTN	End Yin calc
131	*LBLC	Cmplx Invert
132	1/X	
133	X≠Y	
134	CHS	
135	X≠Y	
136	RTN	End invert
137	*LBLD	Cmplx Multiplication
138	X≠Y	
139	R↓	
140	x	
141	R↓	
142	+	
143	R↑	
144	RTN	End multiply
145	*LBLE	Cmplx Addition
146	→R	
147	R↑	
148	R↑	
149	→R	
150	X≠Y	
151	R↓	
152	+	
153	R↓	
154	+	
155	R↑	
156	→P	
157	RTN	End addition
158	*LBLE	Calc Γ50(Zp)
159	5	
160	0	
161	÷	z=Z/50
162	0	
163	ENT↑	
164	1	
165	GSBB	1+z
166	GSBC	Cmplx invert
167	2	
168	x	

```
169    CHS
170    0
171    ENT↑
172    1
173    GSBB   Γ=1-2/(1+z)
174    RTN    End calc Γ(Z)
175   *LBL1   Start Calc K or K'
176    x
177    2
178    x
179    RCL7
180    -
181    RCL8
182    ÷
183    RTN    End calc K
```

Register Assignments:

RO: $|YL|$

R1: $|y11|$

R2:   y11 degrees

R3: $|y21|$

R4:   y21 degrees

R5:   g11

R6:   g22

R7: $\text{Re}(y12 \cdot y21)$

R8: $|y12 \cdot y21|$

R9:   YL degrees

RA=R20: $|Ys|$

RB=R21: $|y21|$

RC=R22:   y21 degrees

RD=R23: $|y22|$

RE=R24:   y22 degrees

RI=R25:   Index & Ys degrees

Program A8-1. Bode Breakpoint Selectivity Estimate

See Equation (8.27).
Keys:

Singly	Doubly	$\overline{Q}_L$	F	NMI NCI
a	b	c	d	e
N	$\prod\limits_{1}^{N} Q_L \rightarrow$	ΠQ_L	f/f_0	dB
A	B	C	D	E

Input: Input N and press key A. Press fa or fb. Put NMI in RY and NMC in RX, and press fe. Press B and respond to prompts $1, 2, \ldots, N$ with Q_{L1}, Q_{L2}, etc.; or, input ΠQ_L or mean Q_L and press C or fc. Input F or f/f_0 and press fd or D, respectively. Press E and see dB attenuation for $dB \geqslant 20$. $F > 0$ and $f/f_0 > 1$ are required. Any of these can be changed individually. Registers A–E are unused.

Note: uses explicit arrangement, as in *HP Keynotes*, January, 1977, pp. 4–5. Keys A, C, D*, E, c, and d* solve for quantities if keyboard numbers are *not* pressed before the function key is pressed; otherwise, the function keys act as input. Keys B, a, b, and e are only single-purpose (input) function keys.

Example: See Example 8.1, Section 8.2.3.

Note: If A, C, D, E, c, or d don't run to completion when there was no input, just press again. Also, don't stop the program with R/S, because P⊃S is used.

*Iterates until round-off displayed is reached.

001	*LBLA	Input/Calc N		021	*LBL9	Entry from LBLs 2 or 9
002	STO1	N into R1		022	GSB7	dB+K
003	F3?	If data was entered		023	RCL2	F
004	R/S	then stop		024	GSB0	20Log(·)
005	GSB7	Db+K		025	RCL1	N
006	RCL6	ΠQL		026	x	
007	GSB0	20Log(·)		027	-	
008	-			028	GSB8	Inverter dB
009	GSB8	Inverter dB		029	-	
010	-			030	RCL7	20
011	RCL2	F		031	÷	
012	GSB0	20Log(·)		032	10ˣ	
013	÷			033	STO6	ΠQL into R6
014	STO1	N		034	F2?	If via LBL C
015	RTN	End calc N		035	RTN	then return
016	*LBLC	Input/Calc ΠQL		036	RCL1	N
017	SF2	See step 034		037	1/x	
018	STO6	Sto ΠQL into R6		038	Yˣ	Geo. mean QL
019	F3?	If data was entered		039	RTN	End calc ΠQL
020	R/S	the stop				

040	*LBLE	Input/Calc dB		094	*LBL3	Entry from LBL d
041	STO4	dB into R4		095	GSB7	dB+K
042	F3?	If data was entered		096	RCL7	20Log(·)
043	R/S	then stop		097	÷	
044	GSB8	Inverter dB		098	RCL6	ΠQL
045	RCL2	F		099	LOG	
046	GSB0	20Log(·)		100	-	
047	RCL1	N		101	STO2	Call this M: see note 1
048	x			102	P≷S	Start secant search
049	+			103	EEX	
050	RCL6	ΠQL		104	CHS	
051	GSB0	20Log(·)		105	4	
052	+			106	STO3	E-4 into S3
053	RCL5	K		107	1	
054	-			108	.	
055	STO4	dB		109	2	
056	RTN	End calc dB		110	STO1	1.2 into S1
057	*LBL0	Start Calc 20Log(·)		111	GSB2	Calc error function (note 1)
058	LOG			112	STO4	Error into S4
059	RCL7	20		113	RCL1	
060	x			114	RCL3	
061	RTN	End 20Log		115	+	1.2001
062	*LBL7	Sum dB+K		116	GSB2	Calc error function
063	RCL4	dB		117	STO2	
064	RCL5	K		118	RCL4	
065	+			119	-	Compare
066	RTN	End sum		120	RCL3	E-4
067	*LBLB	Input All QL's		121	÷	
068	RCL1	N		122	RCL2	
069	STOI	Store index		123	GTO4	Skip on first time
070	1			124	*LBL6	Re-enter search loop
071	DSP0			125	RCLI	Trial f/f0
072	STO6	1 into R6		126	GSB2	Calc error
073	+			127	STO2	
074	STO0	N+1 into R0		128	*LBL4	Entry from step 123
075	*LBL1	Loop re-entry point		129	RCLI	
076	RCL0	N+1		130	RCLI	This is a secant
077	RCLI	Index		131	STO1	search similar to
078	-			132	-	HP-67 Standard Pac
079	R/S	Stop for QLi input		133	RCL4	p.11-3 & p. L11-03.
080	ST×6	Form ΠQL in R6		134	RCL2	
081	DSZI	Decrement/test index		135	STO4	
082	GTO1	Loop if not done		136	-	
083	RCL6	ΠQL		137	÷	
084	DSP2			138	x	
085	RTN	End QL input		139	RCLI	
086	*LBLD	Input/Calc f/f0		140	x≷y	
087	SF2	See step 152		141	-	
088	STO3	f/f0 into R3		142	STOI	
089	GSB5	F		143	÷	
090	STO2	Sto in R2		144	RND	
091	RCL3	f/f0		145	X≠0?	
092	F3?	If data was entered		146	GTO6	Loop if not converged
093	R/S	then stop				

147	RCL1	Final f/f0	193	*LBLd	Input/Calc F
148	P≷S		194	ST02	F into R2
149	ST03	f/f0	195	X²	
150	GSB5	F	196	4	Calc f/f0
151	ST02		197	+	as in (6.87)
152	F2?	If via LBL D	198	√X	where w=F.
153	RCL3	then recall f/f0	199	RCL2	
154	RTN	End f/f0 calc	200	+	
155	*LBL2	Calc Error Function	201	2	
156	P≷S		202	÷	
157	ST0I	Save last f/f0	203	ST03	Sto f/f0 in R3
158	GSB5	F	204	RCL2	F
159	LOG		205	F3?	If data was entered
160	RCL1	N	206	R/S	then stop
161	x	NLogF	207	GT03	Iteratively calc f/f0 & F
162	RCLI	Trial f/f0	208	*LBLa	Singly-Terminated Selection
163	LOG		209	0	K=0 dB
164	GSB9	See step 178	210	ST05	into R5
165	+	Estimated inverter dB	211	RTN	End
166	RCL2	M	212	*LBLb	Doubly-Terminated Selection
167	-		213	6	K=6 dB
168	P≷S	Switch to S registers	214	ST05	into R5
169	RTN	End error calc	215	RTN	End
170	*LBL5	Calc F	216	*LBLe	Input Inverters Totals
171	ENT↑		217	2	
172	1/X		218	0	
173	-	F=f/f0-f0/f	219	ST07	Multiplier 20 into R7
174	RTN	End F calc	220	R↓	flush 20
175	*LBL8	Calc Inverter dB	221	ST09	NCI
176	RCL3	f/f0	222	R↓	
177	GSB0	20Log(·)	223	ST08	NMI
178	*LBL9	Local LBL, not as at 021	224	RTN	End
179	RCL8	NMI			
180	RCL9	NCI			
181	-				
182	x	(NMI-NCI)20Log(f/f0)			
183	RTN	End inverter dB			
184	*LBLc	Input/Calc Geo. Mean QL			
185	ST00				
186	RCL1	N			
187	Yˣ				
188	ST06	ΠQL if data was entered			
189	RCL0				
190	F3?	If data was entered			
191	R/S	then stop			
192	GT09	To geo. mean QL calc			

Note 1: Error function for secant search for f/f0 is

$$\text{Error} = N\text{LogF} + (NMI-NCI)\text{Log}(f/f0)$$
$$-(dB+K)/20 + \text{Log}\Pi QL$$

Register Assignment:

R0: Scratch	R6: ΠQL
R1: N	R7: 20
R2: F & scratch	R8: NMI
R3: f/f0	R9: NCI
R4: dB	RI: Indices, scratch
R5: K	

Program A8-2. Doubly Terminated Minimum-Loss Filters

See Section 8.4.5.

Keys:

STO F	L_0 dB	$\doteq L_s - L_0$	STO N	$L_s - L_0$
A	B	C	D	E

Input: Input $N \leqslant 6$ and press key D. Store unloaded Q_u in register 3. Input fractional frequency F and press key A. Input loaded Q_L and press key B; see L_0 dB. Press key C and see approximate relative dB loss (valid if greater than about 20 dB). Press key E and get exact relative loss (takes longer). To search for an approximate stopband loss, store a $\pm \Delta Q_L$ value in register I. Press key C and see approximate L_s; press R/S and see the Q_L that was just used. Press R/S again to increment the current Q_L value by the ΔQ_L stored in register I. Then recycle through keys C, R/S, and R/S to search for the desired L_s. A new ΔQ_L may be stored in RI at any time.

Example 1: $Q_u = 25$, $F = 0.3$, $Q_L = 5$:

N	1	2	3	4	5	6
$L_s - L_0$	1.6173	3.3804	3.4884	3.0701	3.5506	4.4478

Example 2: $N = 4$, $Q_u = 100$, $F = 0.1$:

$Q_L = 50$, $L_0 = 8.6859$ dB, $L_s - L_0 \doteq 41.2317$ dB, $L_s - L_0 = 41.0577$ dB.

Note: The midband-insertion-loss calculation assumes that the inverter(s) are adjusted for matched input resistance at the tune frequency.

001	*LBLA	Input F
002	STO5	into R5
003	RTN	End
004	*LBLC	Calc Rel. dB Loss
005	RCL4	QL
006	RCL5	F
007	x	
008	GSBd	20Log($\cdot$)
009	RCL1	N
010	x	
011	6	
012	−	dB per (8.99)
013	RCL7	L0 dB
014	−	
015	RTN	End rel. dB loss
016	RCL4	QL (after press'g R/S)
017	R/S	Stop to see QL
018	RCLI	User's ΔQL
019	+	Increment QL
020	*LBLB	Calc L0 dB per (8.103)
021	STO4	QL into R4
022	RCL3	Qu
023	÷	
024	STO6	QL/Qu into R6
025	RCL1	N
026	x	
027	.	
028	5	
029	e^x	
030	GSBd	4.3429

```
031    X
032    STO7  LO dB into R7
033    RTN   End LO dB calc
034   *LBLd  Calc 20Log(·)
035    LOG
036    2
037    0
038    X
039    RTN   End 20Log
040   *LBLE  Exact Calc Ls-LO
041    RCL1  N
042    STOI  Sto coefficient index
043    0
044    ENT↑
045    1     Initialize upper stack
046   *LBL0  Nesting re-entry
047    RCL4  QL
048    RCL5  F
049    X
050    RCL6  QL/Qu
051    →P    Polar (8.93)
052    R↓       Rectangular to
053    R↓       rectangular cmplx
054    →P       multiply in steps
055    X⇄Y      051-061.
056    R↓
057    X
058    R↓
059    +
060    R↑
061    →R
062    RCLI  # places dec shift
063    RCL2  Coefficient string
064    FRC
065    X⇄Y
066    Yˣ
067    RCL2
068    X
069    FRC
070    RCL2
071    FRC
072    ÷
073    INT   Table 8-9 ith coefficient
074    +     Add like in eqn (3.25)
075    DSZI  Decrement decimal shift
076    GTO0  Loop if not done
077    →P    To polar
078    2
079    ÷     See equation (8.97)
080    GSBd  20Log(·)
081    RCL7  LO dB
082    -
083    RTN   End exact rel dB

084   *LBLD  Input N
085    STO1  into R1
086    STOI  Store index
087    GSBi  Sto table 8-9 coefficients
088    RCL1  N
089    RTN   End input N
090   *LBL1  N=1 Coefficients
091    2
092    .
093    1
094    STO2
095    RTN   End
096   *LBL2  N=2 Coefficients
097    2
098    2
099    .
100    1
101    STO2
102    RTN   End
103   *LBL3  N=3 Coefficients
104    2
105    3
106    2
107    .
108    1
109    STO2
110    RTN   End
111   *LBL4  N=4 Coefficients
112    2
113    4
114    4
115    2
116    .
117    1
118    STO2
119    RTN   End
120   *LBL5  N=5 Coefficients
121    2
122    5
123    6
124    5
125    2
126    .
127    1
128    STO2
129    RTN   End
130   *LBL6  N=6 Coefficients
131    2
132    6
133    8
134    9
135    6
136    2
137    .
138    1
139    STO2
140    RTN   End
```

Program A8-3. Singly Terminated Minimum-Loss Filters

Keys:

STO F	L_0 dB	$\doteq L_s - L_0$	STO N	$L_s - L_0$
A	B	C	D	E

Input: Input $N \leqslant 6$ and press key D. Store unloaded Q_u in register 3. Input fractional frequency F and press key A. Input Q_L and press key B; see L_0 dB. Press key C and see approximate relative dB loss (valid if greater than about 20 dB). Press key E and get exact relative loss (takes longer). To search for and approximate stopband loss, store a $\pm \Delta Q_L$ value in register I. Press key C and see approximate L_s. Press R/S and see the Q_L that was just used. Press R/S again to increment the current Q_L value by the ΔQ_L stored in register I. Then recycle through keys C, R/S, and R/S to search for the desired L_s. A new ΔQ_L may be stored in RI at any time.

Example 1: $Q_u = 25$, $F = 0.3$, $Q_L = 5$:

N	1	2	3	4	5	6
$L_s - L_0$	4.8017	5.6109	4.0202	3.7603	5.3208	6.2320

Example 2: $N = 4$, $Q_u = 100$, $F = 0.1$,

$Q_L = 50$, $L_0 = 8.6859$ dB, $L_s - L_0 \doteq 47.2317$ dB, $L_s - L_0 = 46.7122$ dB.

Note: The midband-insertion-loss calculation assumes that the inverter(s) are adjusted for nominal input resistance at the tune frequency.

```
001  *LBLA Input F
002   ST05 into R5
003   RTN End
004  *LBLC Calc Rel. dB Loss
005   RCL4 QL
006   RCL5 F
007     x
008   GSBd 20Log(·)
009   RCL1 N
010     x  Similar to (8.99)
011   RCL7 L0 dB
012     -
013   RTN End rel. dB loss
014   RCL4 QL (after press'g R/S)
015   R/S Stop to see QL
```

```
016   RCLI User's ΔQL
017    +  Increment QL
018  *LBLB Calc L0 dB per (8.103
019   ST04 QL into R4
020   RCL3 Qu
021    ÷
022   ST06 QL/Qu into R6
023   RCL1 N
024    x
025    .
026    5
027    e^x
028   GSBd 4.3429
029    x
030   ST07 Store L0 into R7
```

031	RTN	End L0 dB calc		086	*LBL1	N=1 Coefficients
032	*LBLd	Calc 20Log(·)		087	1	
033	LOG			088	.	
034	2			089	1	
035	0			090	ST02	
036	x			091	RTN	End
037	RTN	End 20Log		092	*LBL2	N=2 Coefficients
038	*LBLE	Exact Calc Ls–L0		093	1	
039	RCL1	N		094	1	
040	ST0I	Sto coefficient index		095	.	
041	0			096	1	
042	ENT↑			097	ST02	
043	1	Initialize upper stack		098	RTN	End
044	*LBL0	Nesting re-entry		099	*LBL3	N=3 Coefficients
045	RCL4	QL		100	1	
046	RCL5	F		101	2	
047	x			102	1	
048	RCL6	QL/Qu		103	.	
049	→P	Polar (8.93)		104	1	
050	R↓			105	ST02	
051	R↓	Rectangular to		106	RTN	End
052	→P	rectangular cmplx		107	*LBL4	N=4 Coefficients
053	X⇄Y	multiply in steps		108	1	
054	R↓	049–059.		109	3	
055	x			110	2	
056	R↓			111	1	
057	+			112	.	
058	R↑			113	1	
059	→R			114	ST02	
060	RCLI	# places dec shift		115	RTN	End
061	RCL2	Coefficient string		116	*LBL5	N=5 Coefficients
062	FRC			117	1	
063	X⇄Y			118	4	
064	Yˣ			119	3	
065	RCL2			120	3	
066	x			121	1	
067	FRC			122	.	
068	RCL2			123	1	
069	FRC			124	ST02	
070	÷			125	RTN	End
071	INT	Table 8-10 ith coefficient		126	*LBL6	N=6 Coefficients
072	+	Add like in eqn (3.25)		127	1	
073	DSZI	Decrement decimal shift		128	5	
074	GT00	Loop if not done		129	4	
075	→P	To polar		130	6	
076	GSBd	20Log(·)		131	3	
077	RCL7	L0 dB		132	1	
078	–	See equation (8.109)		133	.	
079	RTN	End exact rel. dB		134	1	
080	*LBLD	Input N		135	ST02	
081	ST01	into R1		136	RTN	End
082	ST0I	Store index				
083	GSBi	Store table 8-10 coefficients				
084	RCL1	N				
085	RTN	End input N				

Program A9-1. Equal-Admittance-Stub Filters

See Section 9.1.1.

Keys:

$P_N(N, y)$ L dB
   a   b

$N, K \rightarrow$	$f_1, f_2 \rightarrow$	$f_1, f_2 \rightarrow$	$w \rightarrow$	$f/f_0 \rightarrow$
	See f_0	See w	See q	See q
A	B	C	D	E

Example: Input 4, ↑, 1.4, and press key A. Input 400, ↑, 1400, and press key B; see $f_0 = 900$. Press R/S; see 20.376 dB. Or, after key A, input 400, ↑, 1400, and press key C; see $w = 1.111$. Press R/S and see 20.376 dB. Or, after key A, input 1.111 and press key D; see $q = 0.766$. Press R/S and see 20.372 dB. Or, after key A, input 1400, ↑, 900, ÷, and press key E; see $q = 0.766$. Press R/S and see 20.376 dB.

```
001  *LBLA  Input N, K
002  STO2  K
003  X≷Y
004  STO1  N
005  X≷Y
006  RTN  End
007  *LBLB  Calc f0 & Other Params
008  GSB1  Calculate parameters
009  RCL6  f0
010  R/S  Stop (restart with R/S)
011  GTOb  To calc L dB
012  *LBLC  Calc w & Other Params
013  GSB1  Calculate parameters
014  RCL7  w
015  R/S  Stop (restart with R/S)
016  GTOb  To calc L dB
017  *LBL1  Calc All Parameters
018  STO5  f2
019  X≷Y
020  STO4  f1
021   +
022   2
023   ÷
024  STO6  Store f0 in R6
025  RCL5  f2
026  RCL4  f1
027   -
028   ABS
029  RCL6  f0
030   ÷   w
```

```
031  *LBLD  Calc q from w
032  STO7  w
033   2
034   ÷
035   1
036  X≷Y
037   -   f/f0=1-w/2
038  *LBLE  Calc q from f/f0
039   9
040   0
041   x
042   COS
043   ABS
044  STO3  q
045  RTN  End
046  *LBLb  Calculate L dB from (9.18)
047  RCL2  K
048   2
049   ÷
050   1
051   +
052  RCL3  q
053   x
054   2
055   x
056  RCL1  N
057  X≷Y
058  GSBa  P_N
059  RCL3  q
060   x
```

```
061   RCL2  K
062    x
063    X²
064    4
065    ÷
066    1
067   ENT↑
068   RCL3  q
069    X²
070    -
071    ÷
072    1
073    +
074   LOG
075    1
076    0
077    x    L dB
078   RTN   End
079   *LBLα P_N(y) Cheby Polynomial
080   STO0  y
081   X⇄Y   N
082    1
083   X=Y?  If N=1
084   RTN   P=1
085   X⇄Y
086    2
087   X=Y?  If N=2
088   GTO2  P=y
089    -
090   STOI  N-2
091   RCL0  P2=y
092    1    P1=1
093   *LBL0 Loop Re-entry for eqn (8.91)
094   X⇄Y   P_{K-1}
095   STO8
096   RCL0  y
097    x
098   X⇄Y   P_{K-2}  -
099    -    P_K
100   RCL8  P_{K-1}
101   DSZI  If not done
102   GTO0   loop back
103   X⇄Y   P_N
104   RTN   End
105   *LBL2 N=2 Case
106   RCL0  P=y
107   RTN   End
```

Register Assignments:

RI: Index

R0: y

R1: N

R2: K

R3: q

R4: f_1

R5: f_2

R6: f_0

R7: w

R8: Scratch

416

Appendix B

PET BASIC Programs

Program B2-1*A***.   Gauss–Jordan Solution of Real Equations with Screen Input**

```
10 REM GAUSS-JORDAN,LEY P.302.TRC9/79.
9010 DIM A(10,11)
9020 EP=1E-6
9040 PRINT"NUMBER OF EQUATIONS=";:INPUT N
9060 M=N+1
9070 FOR J=1 TO N
9072 PRINT"INPUT COL";J;"REAL COEFS:"
9074 FOR I=1 TO N
9076 PRINT"    ROW";I;:INPUT A(I,J)
9078 NEXT I
9080 NEXT J
9082 PRINT"INPUT REAL INDEP VARIABLES:"
9084 FOR I=1 TO N
9086 PRINT"    ROW";I;:INPUT A(I,M)
9088 NEXT I
9110 KK=0
9120 JJ=0
9130 FOR K=1 TO N
9140 JJ=KK+1
9150 LL=JJ
9160 KK=KK+1
9170 IF ABS(A(JJ,KK))-EP >0 THEN GOTO 9200
9180 JJ=JJ+1
9190 GOTO9170
9200 IF LL-JJ=0 THEN GOTO 9250
9210 FOR MM=1 TO M
9220 AT=A(LL,MM)
9230 A(LL,MM)=A(JJ,MM)
9240 A(JJ,MM)=AT:NEXT MM
9245 REM FORCE EQUA'S INTO DIAGONAL FORM
9250 FOR LJ=1 TO M
9260 J=M+1-LJ
9270 A(K,J)=A(K,J)/A(K,K):NEXT LJ
9280 FOR I=1 TO N
9300 FOR LJ=1 TO M
9310 J=M+1-LJ
9320 IF(I-K)=0 THEN GOTO 9340
9330 A(I,J)=A(I,J)-A(I,K)*A(K,J)
9340 NEXT LJ
9342 NEXT I
9343 NEXT K
9345 PRINT"*** THE VARIABLES FROM 1 TO N ARE:"
9350 FOR I=1 TO N
9370 PRINT"    #";I;A(I,M):NEXT I
9400 END
```

Program B2-1*B*. Gauss–Jordan Solution of Real Equations with DATA Input

```
10 REM GAUSS-JORDAN ,LEY P.302.TRC9/79.
15 PRINT"GAUSS-JORDAN EXAMPLE,LEY PP293&303"
21 DATA 1,0,-1,0,0,0,0,0,-1,0
22 DATA 1,0,0,0,0,0,0,2,2,0
23 DATA 1,0,0,-1,0,0,0,-3,0,0
24 DATA 0,0,1,0,0,0,0,0,-4,-4
25 DATA 0,0,0,1,0,0,-5,-5,0,0
26 DATA 0,0,-1,0,1,0,0,0,0,-6
27 DATA 0,0,0,0,1,-7,0,0,0,7
28 DATA 0,1,0,0,0,8,8,0,0,0
29 DATA 0,1,0,-1,0,0,-9,0,0,0
30 DATA 0,1,0,0,-1,-10,0,0,0,0
31 DATA 1,2,0,0,0,0,0,0,0,0
9010 DIM A(10,11)
9020 EP=1E-6
9030 N=10
9040 PRINT"NUMBER OF EQUATIONS=";N
9060 M=N+1
9070 FOR J=1 TO N
9074 FOR I=1 TO N
9076 READ A(I,J)
9078 NEXT I
9080 NEXT J
9084 FOR I=1 TO N
9086 READ A(I,M)
9088 NEXT I
9110 KK=0
9120 JJ=0
9130 FOR K=1 TO N
9140 JJ=KK+1
9150 LL=JJ
9160 KK=KK+1
9170 IF ABS(A(JJ,KK))-EP >0 THEN GOTO 9200
9180 JJ=JJ+1
9190 GOTO9170
9200 IF LL-JJ=0 THEN GOTO 9250
9210 FOR MM=1 TO M
9220 AT=A(LL,MM)
9230 A(LL,MM)=A(JJ,MM)
9240 A(JJ,MM)=AT:NEXT MM
9245 REM FORCE EQUA'S INTO DIAGONAL FORM
9250 FOR LJ=1 TO M
9260 J=M+1-LJ
9270 A(K,J)=A(K,J)/A(K,K):NEXT LJ
9280 FOR I=1 TO N
9300 FOR LJ=1 TO M
9310 J=M+1-LJ
9320 IF(I-K)=0 THEN GOTO 9340
9330 A(I,J)=A(I,J)-A(I,K)*A(K,J)
9340 NEXT LJ
9342 NEXT I
9343 NEXT K
9345 PRINT"*** THE VARIABLES FROM 1 TO N ARE:"
9350 FOR I=1 TO N
9370 PRINT"   #";I;A(I,M):NEXT I
9400 END
```

Program B2-2*A*. Gauss–Jordan Solution of Complex Equations with Screen Input

```
5 REM LEY P314 CMPLX-TO-REAL MATRICES.TRC9/79
15 DIM A(20,21),B(10,22)
40 PRINT"# OF COMPLEX EQUATIONS=";:INPUT NC
50 N=2*NC
70 FOR J=1 TO NC
71 K=2*J-1
72 PRINT"INPUT COL";J;"COMPLEX COEFS:"
74 FOR I=1 TO NC
76 PRINT"    ROW";I;"REAL,IMAG=";:INPUT B(I,K),B(I,K+1)
78 NEXT I
80 NEXT J
82 PRINT"INPUT COMPLEX INDEP VARIABLES:"
84 FOR I=1 TO NC
86 PRINT"    ROW";I;"REAL,IMAG=";:INPUT B(I,N+1),B(I,N+2)
88 NEXT I
100 MA=N-1
110 MB=N+1
120 MC=N
130 MD=N+2
135 REM SET UP ODD ROWS
140 IK=-1
150 FOR I=1 TO MA STEP 2
160 IK=IK+1
170 FOR J=1 TO MB STEP 2
180 LJ=I-IK
190 A(I,J)=B(LJ,J)
200 A(I,J+1)=-B(LJ,J+1)
210 NEXT J
220 NEXT I
225 REM SET UP EVEN ROWS
230 FOR I=2 TO MC STEP 2
240 FOR J=1 TO MB STEP 2
250 LJ=INT(I/2)
260 A(I,J)=B(LJ,J+1)
270 A(I,J+1)=B(LJ,J)
280 NEXT J
290 NEXT I
300 PRINT"*** SOLN VARS BY REAL THEN IMAG PARTS."
9000 REM GAUSS-JORDAN,LEY P.302.TRC9/79.
9020 EP=1E-6
9060 M=N+1
9110 KK=0
9120 JJ=0
9130 FOR K=1 TO N
9140 JJ=KK+1
9150 LL=JJ
9160 KK=KK+1
9170 IF ABS(A(JJ,KK))-EP >0 THEN GOTO 9200
9180 JJ=JJ+1
9190 GOTO9170
9200 IF LL-JJ=0 THEN GOTO 9250
9210 FOR MM=1 TO M
9220 AT=A(LL,MM)
9230 A(LL,MM)=A(JJ,MM)
9240 A(JJ,MM)=AT:NEXT MM
9245 REM FORCE EQUA'S INTO DIAGONAL FORM
9250 FOR LJ=1 TO M
9260 J=M+1-LJ
9270 A(K,J)=A(K,J)/A(K,K):NEXT LJ
9280 FOR I=1 TO N
9300 FOR LJ=1 TO M
9310 J=M+1-LJ
9320 IF(I-K)=0 THEN GOTO 9340
9330 A(I,J)=A(I,J)-A(I,K)*A(K,J)
9340 NEXT LJ
9342 NEXT I
9343 NEXT K
9345 PRINT"*** THE VARIABLES FROM 1 TO N ARE:"
9350 FOR I=1 TO N
9370 PRINT"   #";I;A(I,M):NEXT I
9400 END
```

Program B2-2*B*. Gauss–Jordan Solution of Complex Equations with DATA Input

```
5 REM LEY P314 CMPLX-TO-REAL MATRICES.TRC9/79
10 PRINT"CMPLX-TO-REAL NTWK MATRIX EX,LEY P316"
15 DIM A(20,21),B(10,22)
20 NC=6
21 DATA 1.25,1, 0,.5, 0,0, 0,0, 0,0, 0,0
22 DATA 0,.5, 2.5,4.2, -1,-2.2, 0,0, 0,0, 0,0
23 DATA 0,0, -1,-2.2, 3.6,5.3, -1.2,-1.6, 0,0, 0,0
24 DATA 0,0, 0,0, -1.2,-1.6, 1.7,2.27, 0,.8, -.5,-1.47
25 DATA 0,0, 0,0, 0,0, 0,.8, 2.7,1.5, 0,-2.3
26 DATA 0,0, 0,0, 0,0, -.5,-1.47, 0,-2.3, .5,1.27
27 DATA 50,86.6, 0,0, 0,0, 0,0, 0,0, 0,0
40 PRINT"# OF COMPLEX EQUATIONS=";NC
50 N=2*NC
70 FOR J=1 TO NC
71 K=2*J-1
74 FOR I=1 TO NC
76 READ B(I,K),B(I,K+1)
78 NEXT I
80 NEXT J
84 FOR I=1 TO NC
86 READ B(I,N+1),B(I,N+2)
88 NEXT I
100 MA=N-1
110 MB=N+1
120 MC=N
130 MD=N+2
135 REM SET UP ODD ROWS
140 IK=-1
150 FOR I=1 TO MA STEP 2
160 IK=IK+1
170 FOR J=1 TO MB STEP 2
180 LJ=I-IK
190 A(I,J)=B(LJ,J)
200 A(I,J+1)=-B(LJ,J+1)
210 NEXT J
220 NEXT I
225 REM SET UP EVEN ROWS
230 FOR I=2 TO MC STEP 2
240 FOR J=1 TO MB STEP 2
250 LJ=INT(I/2)
260 A(I,J)=B(LJ,J+1)
270 A(I,J+1)=B(LJ,J)
280 NEXT J
290 NEXT I
300 PRINT"*** SOLN VARS BY REAL THEN IMAG PARTS."
9000 REM GAUSS-JORDAN,LEY P.302.TRC9/79.
9020 EP=1E-6
9060 M=N+1
9110 KK=0
9120 JJ=0
9130 FOR K=1 TO N
9140 JJ=KK+1
9150 LL=JJ
9160 KK=KK+1
9170 IF ABS(A(JJ,KK))-EP >0 THEN GOTO 9200
9180 JJ=JJ+1
9190 GOTO9170
9200 IF LL-JJ=0 THEN GOTO 9250
9210 FOR MM=1 TO M
9220 AT=A(LL,MM)
9230 A(LL,MM)=A(JJ,MM)
9240 A(JJ,MM)=AT:NEXT MM
9245 REM FORCE EQUA'S INTO DIAGONAL FORM
9250 FOR LJ=1 TO M
9260 J=M+1-LJ
9270 A(K,J)=A(K,J)/A(K,K):NEXT LJ
9280 FOR I=1 TO N
9300 FOR LJ=1 TO M
9310 J=M+1-LJ
9320 IF(I-K)=0 THEN GOTO 9340
9330 A(I,J)=A(I,J)-A(I,K)*A(K,J)
9340 NEXT LJ
9342 NEXT I
9343 NEXT K
9345 PRINT"*** THE VARIABLES FROM 1 TO N ARE:"
9350 FOR I=1 TO N
9370 PRINT"    #";I;A(I,M):NEXT I
9400 END
```

420

Program B2-3. Romberg Integration

```
9000 REM ROMBERG, IBM SCI3, P298. TRC10/79
9010 EP=1E-5
9020 ND=11
9022 DIM AU(ND)
9025 PRINT"LOWER,UPPER LIMITS =";:INPUT XL,XU
9040 XX=XL:GOSUB10000:FL=FC
9050 XX=XU:GOSUB10000:FU=FC
9060 AU(1)=(FL+FU)/2
9070 H=XU-XL
9080 IF(ND-1)<=0 THEN GOTO9420
9090 IF H=0 THEN GOTO9430
9100 HH=H
9110 E=EP/ABS(H)
9120 D2=0
9130 P=1
9140 JJ=1
9150 FOR I=2 TO ND
9160 Y=AU(1)
9170 D1=D2
9180 HD=HH
9190 HH=HH/2
9200 P=P/2
9210 X=XL+HH
9220 SM=0
9230 FOR J=1 TO JJ
9240 XX=X:GOSUB10000
9250 SM=SM+FC
9260 X=X+HD
9270 NEXT J
9280 AU(I)=.5*AU(I-1)+P*SM
9290 Q=1
9300 JI=I-1
9310 FOR J=1 TO JI
9320 II=I-J
9330 Q=Q+Q
9340 Q=Q+Q
9350 AU(II)=AU(II+1)+(AU(II+1)-AU(II))/(Q-1)
9355 NEXT J
9360 D2=ABS(Y-AU(1))
9370 IF (I-5)<0 THEN GOTO9400
9380 IF (D2-E)<=0 THEN GOTO9430
9390 IF (D2-D1)>=0 THEN GOTO9460
9400 JJ=JJ+JJ
9410 NEXT I
9420 PRINT"CAN'T GET < 1E-5 ERROR IN ";ND-1;"BISECTIONS"
9430 Y=H*AU(1)
9440 PRINT"   VALUE OF INTEGRAL =";Y
9445 PRINT
9450 GOTO9025
9460 PRINT"ERROR > 1E-5 DUE TO ROUNDING"
9470 Y=H*Y
9480 GOTO9440
10000 REM INTEGRAND. FC(XX).
10010 FC=1/XX
10020 RETURN
15000 END
```

Program B2-4A. Polynomial Minimax Approximation of General Piecewise Linear Functions

```
10 REM VLACH P.232
20 DIM X(15),Y(15),SL(16),S(15),Q(15)
22 DIM F(15),C(15,15),A(15),B(15,15)
24 DIM QU(16),G(15)
27 PI=3.1415926
30 B(1,1)=1
40 B(1,2)=0
50 B(2,2)=1
60 FOR I=3 TO 15
70 B(1,I)=-B(1,I-2)
80 K=I-2
90 IF K-1 <= 0 GOTO 116
95 FOR J=2 TO K
100 B(J,I)= -B(J,I-2)+2*B(J-1,I-1)
110 NEXT J
116 B(I-1,I) = 2*B(I-2,I-1)
120 B(I,I) = 2*B(I-1,I-1)
130 NEXT I
140 PRINT"# OF GIVEN FUNCTION POINTS =";:INPUT M
150 PRINT"INPUT X(I),Y(I) FOR -1<X<+1:"
160 FOR I=1 TO M
170 PRINT"    I=";I;:INPUT X(I),Y(I)
180 NEXT I
190 PRINT"MIN,MAX POLY DEGREES=";:INPUT N1,N2
200 FOR I=1 TO M
210 Z=X(I)+1E-15
212 F(I)=ATN(SGN(Z)*SQR(ABS(1/(Z*Z)-1)))
214 IF F(I)<0 THEN F(I)=F(I)+PI
216 IF Z=-1 THEN F(I)=PI
220 NEXTI
230 SL(M+1)=0
240 SL(1)=0
250 QU(M+1)=0
260 QU(1)=0
270 L=M-1
280 FORI=1 TO L
290 SL(I+1)=(Y(I+1)-Y(I))/(X(I+1)-X(I))
300 QU(I+1)=Y(I)-SL(I+1)*X(I)
310 S(I)=(SL(I+1)-SL(I))/PI
320 Q(I)=2*(QU(I+1)-QU(I))/PI
330 NEXT I
340 S(M)=(SL(M+1)-SL(M))/PI
350 Q(M)=2*(QU(M+1)-QU(M))/PI
360 L=N2+1
370 FOR J=1 TO L
380 FOR I=1 TO M
390 C(J,I)=SIN(J*F(I))/J
400 NEXT I
405 NEXT J
410 SU=0
420 FOR I=1 TO M
430 SU=SU+S(I)*C(1,I)+Q(I)*F(I)/2
435 NEXT I
440 A(1)=SU
450 SU=0
460 FORI=1 TO M
470 SU=SU+S(I)*(C(2,I)+F(I))+Q(I)*C(1,I)
475 NEXT I
480 A(2)=SU
490 FOR J=3 TO L
500 SU=0
510 FORI=1 TO M
520 SU=SU+S(I)*(C(J,I)+C(J-2,I))+Q(I)*C(J-1,I)
530 NEXT I
540 A(J)=SU
550 NEXT J
555 PRINT
560 PRINT"CHEBYCHEV COEFFICIENTS:"
570 FOR J=1 TO L
580 PRINTJ,A(J)
585 NEXT J
590 N3=N1+1
600 N4=N2+1
610 FOR JO=N3 TO N4
615 PRINT
620 PRINT"POLYCOEFFS:"
630 FOR J=1 TO JO
640 SU=0
645 FOR I=J TO JO
650 SU=SU+A(I)*B(J,I)
660 NEXT I
670 G(J)=SU
680 PRINTJ,G(J)
690 NEXT J
695 PRINT
700 PRINT"  X   Y      APPROX    ERROR"
710 L=JO-1
720 FORJ=1 TO M
730 SU=G(JO)
740 FORK=1 TO L
750 I=L+1-K
760 SU=SU*X(J)+G(I)
770 NEXT K
780 DI=SU-Y(J)
790 PRINTX(J);Y(J);SU;DI
800 NEXT J
805 STOP
810 NEXT JO
820 GOTO140
830 END
```

Program B2-4*B*. Polynomial Minimax Approximation of Even Piecewise Linear Functions

```
10 REM VLACH P.232
20 DIM X(15),Y(15),SL(16),S(15),Q(15)
22 DIM F(15),C(15,15),A(15),B(15,15)
24 DIM QU(16),G(15)
27 PI=3.1415926
30 B(1,1)=1
40 B(1,2)=0
50 B(2,2)=1
60 FOR I=3 TO 15
70 B(1,I)=-B(1,I-2)
80 K=I-2
90 IF K-1 <= 0 GOTO 116
95 FOR J=2 TO K
100 B(J,I)= -B(J,I-2)+2*B(J-1,I-1)
110 NEXT J
116 B(I-1,I) = 2*B(I-2,I-1)
120 B(I,I) = 2*B(I-1,I-1)
130 NEXT I
140 PRINT"# OF GIVEN FUNCTION POINTS=";:INPUT M
150 PRINT"INPUT X(I),Y(I) FOR 0<X<+1:"
160 FOR I=1 TO M
170 PRINT"   I=";I;:INPUT X(I),Y(I)
180 NEXT I
190 PRINT"MIN,MAX POLY DEGREES=";:INPUT N1,N2
200 FOR I=1 TO M
210 Z=X(I)+1E-15
212 F(I)=ATN(SGN(Z)*SQR(ABS(1/(Z*Z)-1)))
214 IF F(I)<0 THEN F(I)=F(I)+PI
216 IF Z=-1 THEN F(I)=PI
220 NEXTI
230 SL(M+1)=0
240 SL(1)=0
250 QU(M+1)=0
260 QU(1)=0
270 L=M-1
280 FORI=1 TO L
290 SL(I+1)=(Y(I+1)-Y(I))/(X(I+1)-X(I))
300 QU(I+1)=Y(I)-SL(I+1)*X(I)
310 S(I)=(SL(I+1)-SL(I))/PI
320 Q(I)=2*(QU(I+1)-QU(I))/PI
330 NEXT I
340 S(M)=(SL(M+1)-SL(M))/PI
350 Q(M)=2*(QU(M+1)-QU(M))/PI
360 L=N2+1
370 FOR J=1 TO L
380 FOR I=1 TO M
390 C(J,I)=SIN(J*F(I))/J
400 NEXT I
405 NEXT J
406 FOR I=1 TO L
407 A(I)=0
408 NEXT I
410 SU=0
420 FOR I=1 TO M
430 SU=SU+S(I)*C(1,I)+Q(I)*F(I)/2
435 NEXT I
440 A(1)=SU*2
490 FOR J=3 TO L STEP 2
500 SU=0
510 FORI=1 TO M
520 SU=SU+S(I)*(C(J,I)+C(J-2,I))+Q(I)*C(J-1,I)
530 NEXT I
540 A(J)=SU*2
550 NEXT J
555 PRINT
560 PRINT"CHEBYCHEV COEFFICIENTS:"
570 FOR J=1 TO L
580 PRINTJ,A(J)
585 NEXT J
590 N3=N1+1
600 N4=N2+1
610 FOR JO=N3 TO N4
615 PRINT
620 PRINT"POLYCOEFFS:"
630 FOR J=1 TO JO
640 SU=0
645 FOR I=J TO JO
650 SU=SU+A(I)*B(J,I)
660 NEXT I
670 G(J)=SU
680 PRINTJ,G(J)
690 NEXT J
695 PRINT
700 PRINT"  X   Y      APPROX   ERROR"
710 L=JO-1
720 FORJ=1 TO M
730 SU=G(JO)
740 FORK=1 TO L
750 I=L+1-K
760 SU=SU*X(J)+G(I)
770 NEXT K
780 DI=SU-Y(J)
790 PRINTX(J);Y(J);SU;DI
800 NEXT J
805 STOP
810 NEXT JO
820 GOTO140
830 END
```

Program B2-4C. Polynomial Minimax Approximation of Odd Piecewise Linear Functions

```
10 REM VLACH P.232
20 DIM X(15),Y(15),SL(16),S(15),Q(15)
22 DIM F(15),C(15,15),A(15),B(15,15)
24 DIM QU(16),G(15)
27 PI=3.1415926
30 B(1,1)=1
40 B(1,2)=0
50 B(2,2)=1
60 FOR I=3 TO 15
70 B(1,I)=-B(1,I-2)
80 K=I-2
90 IF K-1 <= 0 GOTO 116
95 FOR J=2 TO K
100 B(J,I)= -B(J,I-2)+2*B(J-1,I-1)
110 NEXT J
116 B(I-1,I) = 2*B(I-2,I-1)
120 B(I,I) = 2*B(I-1,I-1)
130 NEXT I
140 PRINT"# OF GIVEN FUNCTION POINTS =";:INPUT M
150 PRINT"INPUT X(I),Y(I) FOR 0<X<+1:"
160 FOR I=1 TO M
170 PRINT"    I=";I;:INPUT X(I),Y(I)
180 NEXT I
190 PRINT"MIN,MAX POLY DEGREES=";:INPUT N1,N2
200 FOR I=1 TO M
210 Z=X(I)+1E-15
212 F(I)=ATN(SGN(Z)*SQR(ABS(1/(Z*Z)-1)))
214 IF F(I)<0 THEN F(I)=F(I)+PI
216 IF Z=-1 THEN F(I)=PI
220 NEXTI
230 SL(M+1)=0
240 SL(1)=0
250 QU(M+1)=0
260 QU(1)=0
270 L=M-1
280 FORI=1 TO L
290 SL(I+1)=(Y(I+1)-Y(I))/(X(I+1)-X(I))
300 QU(I+1)=Y(I)-SL(I+1)*X(I)
310 S(I)=(SL(I+1)-SL(I))/PI
320 Q(I)=2*(QU(I+1)-QU(I))/PI
330 NEXT I
340 S(M)=(SL(M+1)-SL(M))/PI
350 Q(M)=2*(QU(M+1)-QU(M))/PI
360 L=N2+1
370 FOR J=1 TO L
380 FOR I=1 TO M
390 C(J,I)=SIN(J*F(I))/J
400 NEXT I
405 NEXT J
406 FOR I=1 TO L
407 A(I)=0
408 NEXT I
450 SU=0
460 FORI=1 TO M
470 SU=SU+S(I)*(C(2,I)+F(I))+Q(I)*C(1,I)
475 NEXT I
480 A(2)=SU*2
490 FOR J=4 TO L STEP 2
500 SU=0
510 FORI=1 TO M
520 SU=SU+S(I)*(C(J,I)+C(J-2,I))+Q(I)*C(J-1,I)
530 NEXT I
540 A(J)=SU*2
550 NEXT J
555 PRINT
560 PRINT"CHEBYCHEV COEFFICIENTS:"
570 FOR J=1 TO L
580 PRINTJ,A(J)
585 NEXT J
590 N3=N1+1
600 N4=N2+1
610 FOR JO=N3 TO N4
615 PRINT
620 PRINT"POLYCOEFFS:"
630 FOR J=1 TO JO
640 SU=0
645 FOR I=J TO JO
650 SU=SU+A(I)*B(J,I)
660 NEXT I
670 G(J)=SU
680 PRINTJ,G(J)
690 NEXT J
695 PRINT
700 PRINT"  X  Y      APPROX    ERROR"
710 L=JO-1
720 FORJ=1 TO M
730 SU=G(JO)
740 FORK=1 TO L
750 I=L+1-K
760 SU=SU*X(J)+G(I)
770 NEXT K
780 DI=SU-Y(J)
790 PRINTX(J);Y(J);SU;DI
800 NEXT J
805 STOP
810 NEXT JO
820 GOTO140
830 END
```

Program B2-5. Levy's Matrix Coefficients*

```
10 REM LEVY FITT'G EQS 15-18.TRC9/79.
100 DIM OK(15),RK(15),IK(15),ZK(15)
110 PRINT"# OF FREQUENCIES =";:INPUT MN
115 MM=MN-1
120 FOR I=0 TO MM
130 PRINT"SAMPLE#";I:PRINT"   OMEGA,REAL,IMAG=";:INPUT OK(I),RK(I),IK(I)
150 ZK(I)=RK(I)*RK(I)+IK(I)*IK(I)
160 NEXT I
170 LA=0:SA=0:TA=0:UA=0
180 PRINT"H=";:INPUT HH
190 FOR K=0 TO MM
200 OH=OK(K)^HH
210 LA=LA+OH
220 SA=SA+OH*RK(K)
230 TA=TA+OH*IK(K)
240 UA=UA+OH*ZK(K)
250 NEXT K
260 PRINT"    LAMBDA";HH;"=";LA
270 PRINT"    S";HH;"=";SA
280 PRINT"    T";HH;"=";TA
290 PRINT"    U";HH;"=";UA
300 GOTO170
310 END
```

*See Equation (2.57).

Program B3-1.   Moore's Root Finder

```
10 REM  MOORE ROOT FINDER, TRC, 7/79.
100 DIM A(35),B(35),X(35),Y(35)
200 REM INPUT POLY DEG & CMPLX COEFS
210 PRINT"N=";:INPUT N
220 IF N>35 GOTO210
230 FOR K=0 TO N
240 PRINT"EXPONENT=";K;"     INPUT COEFFICIENT:"
250 PRINT"   REAL PART=";:INPUT A(K)
260 PRINT"   IMAG PART=";:INPUT B(K)
270 NEXT K
1000 REM REDUCED POLY RE-ENTRY
1010 IF N=1 GOTO5500
1020 YS=1:X(0)=1:Y(1)=1
1030 XS=.1:X(1)=.1
1040 Y(0)=0:L=0
1050 GOSUB3000
2000 REM NEW X,Y CORNER (LINEAR SEARCH)
2010 FS=F
2020 L=L+1
2030 M=0:UX=0:VX=0
2040 FOR K=1 TO N
2050 UX=UX+K*(A(K)*X(K-1)-B(K)*Y(K-1))
2060 VX=VX+K*(A(K)*Y(K-1)+B(K)*X(K-1))
2070 NEXT K
2080 PM=UX*UX+VX*VX
2090 DX=-(U*UX+V*VX)/PM
2100 DY=(U*VX-V*UX)/PM
2190 REM POST QRTRG CUTBACK RE-ENTRY
2195 M=M+1
2200 X(1)=XS+DX
2210 Y(1)=YS+DY
2220 GOSUB3000
2230 IF F>=FS GOTO4000
2240 IF ABS(DX)>1E-5 GOTO2260
2250 IF ABS(DY)<=1E-5 GOTO4500
2260 IF L>50 GOTO5200
2270 XS=X(1):YS=Y(1)
2280 GOTO2000
3000 REM CALC X(.),Y(.),U,V, & F
3010 X2=X(1)*2
3020 XY=X(1)*X(1)+Y(1)*Y(1)
3030 FOR K=2 TO N
3040 X(K)=X2*X(K-1)-XY*X(K-2)
3050 Y(K)=X2*Y(K-1)-XY*Y(K-2)
3060 NEXT K
3070 U=0:V=0
3080 FOR K=0 TO N
3090 U=U+A(K)*X(K)-B(K)*Y(K)
3100 V=V+A(K)*Y(K)+B(K)*X(K)
3110 NEXT K
3120 F=U*U+V*V
3130 RETURN
4000 REM FNCN INCRSD SO CUT BACK THE STEP
4010 IF M>10 GOTO4040
4020 DX=DX/4:DY=DY/4
4030 GOTO2190
4040 REM TEST FOR CONVERG > 10 CUTBACKS
4050 IF ABS(U)>1E-4 GOTO4070
4060 IF ABS(V)<=1E-4 GOTO4500
4070 PRINT"STEP SIZE TOO SMALL"
4080 STOP
4500 REM CONVERGED. PRNT ROOT, REMOVE FACTOR.
4510 GOSUB5000
4520 REM REMOVE LINEAR FACTOR
4530 K=N-1
4540 A(K)=A(K)+A(K+1)*X(1)-B(K+1)*Y(1)
4550 B(K)=B(K)+A(K+1)*Y(1)+B(K+1)*X(1)
4560 K=K-1
4570 IF K>=0 GOTO4540
4580 FOR K=1TON
4590 A(K-1)=A(K)
4600 B(K-1)=B(K)
4610 NEXT K
4620 N=N-1
4630 GOTO1000
5000 REM PRNT ROOT
5010 PRINT"A ROOT HAS"
5020 PRINT"   REAL PART=";X(1)
5030 PRINT"   IMAG PART=";Y(1)
5040 RETURN
5200 PRINT"NO ROOT FOUND"
5210 STOP
5500 REM CALC DEG=1 EQUATION ROOT
5510 XY=A(1)*A(1)+B(1)*B(1)
5520 X(1)=-(B(0)*B(1)+A(0)*A(1))/XY
5530 Y(1)=(A(0)*B(1)-A(1)*B(0))/XY
5540 GOSUB5000
5560 END
```

Flowchart for PET Program Roots

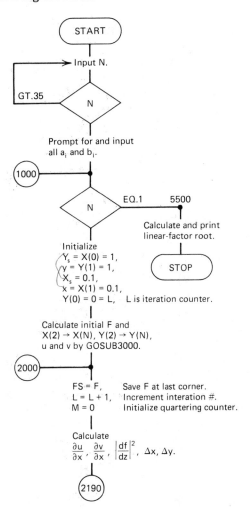

START

Input N.

GT.35

N

Prompt for and input
all a_i and b_i.

1000

N   EQ.1   5500

Calculate and print
linear-factor root.

STOP

Initialize
$Y_s = X(0) = 1$,
$y = Y(1) = 1$,
$X_s = 0.1$,
$x = X(1) = 0.1$,
$Y(0) = 0 = L$,   L is iteration counter.

Calculate initial F and
$X(2) \rightarrow X(N)$, $Y(2) \rightarrow Y(N)$,
u and v by GOSUB3000.

2000

$FS = F$,   Save F at last corner.
$L = L + 1$,   Increment interation #.
$M = 0$ Initialize quartering counter.

Calculate
$\dfrac{\partial u}{\partial x}$, $\dfrac{\partial v}{\partial x}$, $\left|\dfrac{df}{dz}\right|^2$, Δx, Δy.

2190

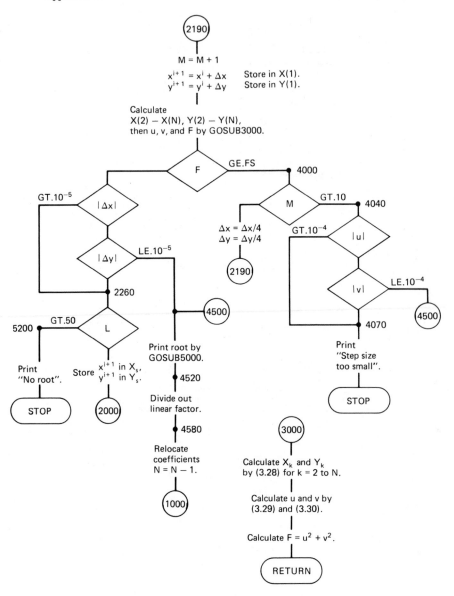

(2190)

$M = M + 1$

$x^{i+1} = x^i + \Delta x$ Store in X(1).
$y^{i+1} = y^i + \Delta y$ Store in Y(1).

Calculate
$X(2) - X(N)$, $Y(2) - Y(N)$,
then u, v, and F by GOSUB3000.

F — GE.FS — • 4000

$|\Delta x|$ GT.10^{-5}

M GT.10 — • 4040

$\Delta x = \Delta x/4$
$\Delta y = \Delta y/4$

(2190)

$|\Delta y|$ LE.10^{-5}

$|u|$ GT.10^{-4}

2260

$|v|$ LE.10^{-4}

(4500)

4070 (4500)

5200 • GT.50 L

Print
"Step size
too small".

Print
"No root".

Store x^{i+1} in X_s,
y^{i+1} in Y_s.

Print root by
GOSUB5000.

• 4520

STOP

STOP (2000)

Divide out
linear factor.

• 4580

(3000)

Relocate
coefficients
$N = N - 1$.

Calculate X_k and Y_k
by (3.28) for k = 2 to N.

(1000)

Calculate u and v by
(3.29) and (3.30).

Calculate $F = u^2 + v^2$.

RETURN

Program B3-2. Polynomials From Complex Zeros

```
10 REM POLYS FROM COMPLEX ZEROS.VLACH P214.TRC8/79.
100 PRINT"NUMBER OF ZEROS=";:INPUT N
110 N1=N+1
120 DIM A(N1),B(N1),G(N1),H(N1)
130 PRINT"INPUT THE ZEROS:"
140 FOR I=1 TO N
150 PRINT"ZERO#";I;"   REAL PART=";:INPUT A(I)
160 PRINT"           IMAG PART=";:INPUT B(I)
165 NEXT I
170 G(1)=1:G(2)=0:H(1)=0:H(2)=0
180 FOR J=1 TO N
190 G(J+1)=G(J)
200 H(J+1)=H(J)
210 IF(J-1)<=0 THEN GOTO290
220 FOR L=2 TO J
230 K=J-L+2
240 X=G(K-1)-A(J)*G(K)+B(J)*H(K)
250 Y=H(K-1)-A(J)*H(K)-B(J)*G(K)
260 G(K)=X
270 H(K)=Y
280 NEXT L
290 X=-G(1)*A(J)+H(1)*B(J)
300 H(1)=-H(1)*A(J)-G(1)*B(J)
310 G(1)=X
320 NEXT J
330 N=N+1
335 PRINT"POLYNOMIAL COEFFICIENTS ARE:"
340 FOR I=1 TO N
350 PRINT"EXPONENT=";I-1":"
360 PRINT"   REAL PART =";G(I)
370 PRINT"   IMAG PART =";H(I)
380 NEXT I
390 END
```

Program B3-3.   Polynomial Multiplication

```
10 REM POLYNOMIAL MULT.VLACH P.216,TRC8/79.
100 DIM A(21),B(21),G(41)
110 PRINT"POLY#1 DEGREE=";:INPUT N1
120 PRINT"INPUT REAL COEFFICIENTS:"
130 FOR J=1 TO N1+1
140 PRINT"    EXPONENT=";J-1;"COEF=";:INPUT A(J)
150 NEXT J
160 PRINT"POLY#2 DEGREE=";:INPUT N2
170 PRINT"INPUT REAL COEFFICIENTS:"
180 FOR J=1 TO N2+1
190 PRINT"    EXPONENT=";J-1;"COEF=";:INPUT B(J)
200 NEXT J
210 M=N1+N2+1
220 FOR I=1 TO M
230 G(I)=0
240 NEXT I
250 FOR J=1 TO N2+1
260 FOR I=1 TO N1+1
270 JT=I+J-1
280 G(JT)=G(JT)+A(I)*B(J)
290 NEXT I
300 NEXT J
310 PRINT"PRODUCT POLYNOMIAL IS:"
320 FOR J=1 TO M
330 PRINT"    EXPONENT=";J-1;"COEF=";G(J)
340 NEXT J
350 IF M>20 THEN GOTO400
360 FOR J=1 TO M
370 A(J)=G(J)
380 NEXT J
385 N1=M-1
390 GOTO160
400 END
```

Program B3-4. Polynomial Addition and Subtraction of Even, Odd, or All Parts

```
10 REM POLY ADD/SUB OF EV/ODD/ALL PARTS.TRC8/79.
90 DIM R(45)
100 PRINT"POLY#1 DEGREE=?";:INPUT N1
105 R(1)=N1+1
110 PRINT"INPUT REAL COEFFICIENTS:"
120 FOR I=1 TO N1+1
130 PRINT"    EXPONENT=";I-1;"COEF=";:INPUT R(I+4)
140 NEXT I
150 PRINT"POLY#2 DEGREE=";:INPUT N2
155 R(2)=N2+1
160 PRINT"INPUT REAL COEFFICIENTS:"
165 IB=R(1)+4
170 FOR I=1 TO N2+1
180 PRINT"    EXPONENT=";I-1;:INPUT R(IB+I)
190 NEXT I
200 MX=N1
210 IF N2>MX THEN MX=N2
220 R(4)=MX+1
230 PRINT"ADD OR SUBTRACT (1 OR -1)";:INPUT RI
240 PRINT"ODD,EVEN,OR ALL (-1,1,OR 0)";:INPUT RJ
250 PRINT"RESULT POLY IS:"
252 IA=RJ
254 RE=1
260 LM=R(4)
270 FOR I=1 TO LM
280 SM=0
290 RE=-RE
300 IF IA=0 THEN GOTO320
310 IF RE*RJ>0 THEN GOTO370
320 IF I>R(1) THEN GOTO340
330 SM=SM+R(4+I)
340 IF I>R(2) THEN GOTO360
350 SM=SM+RI*R(IB+I)
360 PRINT"    EXPONENT=";I-1;"COEF=";SM
370 NEXT I
380 GOTO230
390 END
```

Program B3-5. Continued Fraction Expansion

```
10 REM CONTIN FRAC EXPAN.VLACH,P.222.TRC8/79.
100 PRINT"INPUT DEGREE N=";:INPUT NN
110 N=NN+1
120 K=N+1
130 DIM A(K),Q(K)
140 A(K)=0
145 PRINT"IS 1ST ELEMENT CSH OR LSER (Y/N)";:INPUT A$
150 PRINT"INPUT RATIONAL POLY COEFFICIENTS:"
160 FOR I=1 TO N
162 K=I-1
164 IF A$="Y" THEN K=N-I
170 PRINT"EXPONENT=";K;:INPUT A(I)
180 NEXT I
200 PRINT"ELEMENT VALUES ARE:"
210 K=1
220 Q(K)=A(K)/A(K+1)
230 IF A$="Y" THEN PRINTK;":";Q(K)
240 I=K+2
250 IF (I-N)<=0 GOTO300
260 GOTO400
300 FOR L=I TO N STEP 2
310 A(L)=A(L)-Q(K)*A(L+1)
320 NEXT L
330 K=K+1
340 GOTO220
400 IF A$="Y" THEN GOTO500
420 FOR K=1 TO NN
430 PRINTK;":";1/Q(K)
440 NEXT K
500 END
```

Program B3-6. Long Division

```
10 REM LONG DIVISION,VLACH,P218.TRC8/79
100 DIM A(25),B(25),G(25)
110 FOR K=1 TO 25
120 A(K)=0
130 NEXT K
140 PRINT"NUMERATOR DEGREE=";:INPUT N1
150 PRINT"INPUT REAL NUMERATOR COEFFICIENTS:"
160 FOR I=1 TO N1+1
170 PRINT"    COEF#";I;:INPUT A(I)
180 NEXT I
190 PRINT"DENOMINATOR DEGREE=";:INPUT N2
200 PRINT"INPUT REAL DENOMINATOR COEFFICIENTS:"
210 FOR I=1 TO N2+1
220 PRINT"    COEF#";I;:INPUT B(I)
230 NEXT I
240 PRINT"QUOTIENT IS:"
250 K=N1+1
260 FOR J=1 TO K
270 G(J)=A(J)/B(1)
280 PRINT"    COEF#";J;"COEFFICIENT=";G(J)
290 IF (K-J)<=0 THEN GOTO999
300 FOR I=1 TO N2
310 A(J+I)=A(J+I)-G(J)*B(I+1)
320 NEXT I
330 NEXT J
999 END
```

432

Program B3-7. Partial Fraction Expansion

```
10 REM PARTIAL FRAC EXPAN,CT1/77P44.TRC8/79
100 PRINT"DENOMINATOR DEGREE=";:INPUT N
110 DIM AR(N),AI(N),PR(N),PI(N)
120 PRINT"INPUT REAL NUMERATOR COEFFICIENTS:"
130 FOR I=1 TO N
140 K=N+1-I
150 PRINT"EXPONENT=";I-1;:INPUT AR(K)
160 AI(K)=0
170 NEXT I
180 PRINT"INPUT DENOMINATOR ROOTS IN ORDER OF"
190 PRINT"ASCENDING MAGNITUDES:"
200 FOR I=1 TO N
210 PRINT"ROOT #";I
220 PRINT"   REAL PART=";:INPUT PR(I)
230 PRINT"   IMAG PART=";:INPUT PI(I)
240 NEXT I
250 EP=1.E-10
300 FOR I=1 TO N
310 I1=N-I+1
320 IF I1=1 GOTO400
330 FOR J=2 TO I1
340 AR(J)=AR(J)+PR(I)*AR(J-1)-PI(I)*AI(J-1)
350 AI(J)=AI(J)+PI(I)*AR(J-1)+PR(I)*AI(J-1)
360 NEXT J
400 FOR J=1 TO I
410 J1=N-J+1
420 IF J=1 GOTO460
430 IF(PR(J)-PR(J-1))^2+(PI(J)-PI(J-1))^2<=EP GOTO530
440 AR(I1)=AR(I1)-AR(J1+1)
450 AI(I1)=AI(I1)-AI(J1+1)
460 IF(PR(J)-PR(I))^2+(PI(J)-PI(I))^2<=EP GOTO600
470 YR=PR(J)-PR(I)
480 YI=PI(J)-PI(I)
490 Y2=YR*YR+YI*YI
500 AS=(AR(J1)*YR+AI(J1)*YI)/Y2
510 AI(J1)=(AI(J1)*YR-AR(J1)*YI)/Y2
515 AR(J1)=AS
520 GOTO570
530 DR=AR(J1)-AR(J1+1)
540 DI=AI(J1)-AI(J1+1)
550 AR(J1)=(DR*YR+DI*YI)/Y2
560 AI(J1)=(DI*YR-DR*YI)/Y2
570 NEXT J
600 NEXTI
610 PRINT"THE (REAL,IMAG) POLE RESIDUES IN"
620 PRINT"INPUT ORDER AND DESCENDING MULTIPLICITY:"
630 FOR I=1 TO N
635 K=N+1-I
640 PRINT"#";I;AR(K),AI(K)
650 NEXT I
660 END
```

Program B4-1. Level-0 Ladder Network Analysis

```
1010 PRINT"FREQ,INDUC,CAPAC UNITS=";:INPUT FU,LU,CU
1020 PRINT"LOAD RESISTANCE,REACTANCE=";:INPUT RL,XL
1030 PRINT"WATTS IN LOAD=";:INPUT WL
1040 DIM M(16),X(16),P(16)
1045 PI=3.1415926
1050 PRINT"INPUT RLC COMPONENTS AS TYPES
1060 PRINT"  1-RESISTOR, 2-INDUCTOR, 3-CAPACITOR."
1070 PRINT"  NEGATIVE 1,2, OR 3 MEANS"
1080 PRINT"    NULL PRIOR BRANCH."
1090 PRINT"LIST FROM LOAD END: TYPE,VALUE,Q"
1100 PRINT"    (TERMINATE WITH 0,0,0)"
1110 FOR N=1 TO 16
1120 PRINT"   ";N;
1130 INPUT M(N),X(N),P(N)
1140 IF P(N)=0 THEN P(N)=1E10
1150 P(N)=1/P(N)
1160 IF M(N)=0 GOTO1200
1170 NEXT N
1180 PRINT"MORE THAN 15 VALUES.":GOTO9999
1200 PRINT"FREQ=";:INPUT OM
1210 OM=2*PI*OM*FU
1220 BR=SQR(WL/RL):BI=0:DR=0:DI=0
1230 CR=RL:CI=XL
1240 PRINT"BR#  REAL      IMAGINARY"
1250 REM K=BRANCH #, N=COMPONENT POINTER
1260 K=0:N=0:F1=0
1300 K=K+1:N=N+1
1310 MK=M(N)
1315 IF F1>0 THEN MK=-MK
1317 F1=0
1370 GOSUB9900
1380 PRINT K;AR,AI
1381 IF MK=0 GOTO9955
1382 IF MK<0 GOTO9000
1385 ON MK GOSUB 9100,9200,9300
1390 GOTO1300
9000 REM NULL BRANCH
9010 CR=0:CI=0:N=N-1
9020 F1=1
9030 GOTO1300
9100 REM RESISTOR
9110 CR=X(N):CI=0
9120 IF K=INT(K/2)*2 THEN RETURN
9130 CR=1/CR:RETURN
9200 REM INDUCTOR
9210 CI=OM*X(N)*LU
9220 CR=CI*P(N)
9230 IF K=INT(K/2)*2 THEN RETURN
9240 DD=P(N)*P(N)+1
9250 CR=P(N)/DD/CI
9260 CI=-1/DD/CI
9270 RETURN
9300 REM CAPACITOR
9310 CI=OM*X(N)*CU
9320 CR=CI*P(N)
9330 IF K=INT(K/2)*2 GOTO9240
9340 RETURN
9900 REM COMPLEX LINEAR UPDATE
9910 AR=BR*CR-BI*CI+DR
9920 AI=BI*CR+BR*CI+DI
9930 DR=BR:DI=BI
9940 BR=AR:BI=AI
9950 RETURN
9955 REM CALC/PRINT ZIN=RIN+JXIN
9960 IF K=INT(K/2)*2 GOTO9970
9965 AR=BR:AI=BI:BR=DR:BI=DI:DR=AR:DI=AI
9970 AI=BR*BR+BI*BI
9975 AR=(DR*BR+DI*BI)/AI
9980 AI=(DI*BR-DR*BI)/AI
9985 PRINT"RIN=";AR;"XIN=";AI
9990 GOTO1200
9999 END
```

434

Flowchart for Ladder Analysis Program B4-1

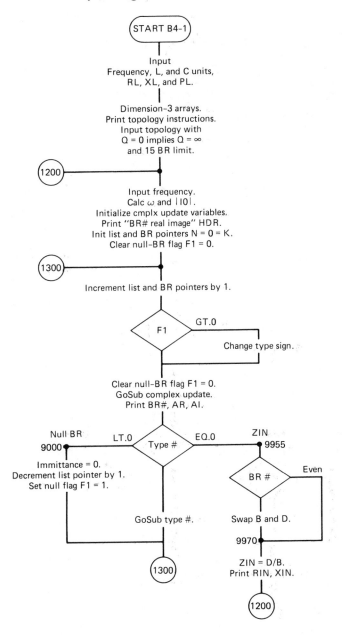

START B4-1

Input
Frequency, L, and C units,
RL, XL, and PL.

Dimension-3 arrays.
Print topology instructions.
Input topology with
Q = 0 implies Q = ∞
and 15 BR limit.

1200

Input frequency.
Calc ω and |I0|.
Initialize cmplx update variables.
Print "BR# real image" HDR.
Init list and BR pointers N = 0 = K.
Clear null-BR flag F1 = 0.

1300

Increment list and BR pointers by 1.

F1 — GT.0 — Change type sign.

Clear null-BR flag F1 = 0.
GoSub complex update.
Print BR#, AR, AI.

Null BR
9000 — LT.0 — Type # — EQ.0 — ZIN 9955

Immittance = 0.
Decrement list pointer by 1.
Set null flag F1 = 1.

BR # — Even

GoSub type #.

Swap B and D.

9970

1300

ZIN = D/B.
Print RIN, XIN.

1200

435

Program B4-2. Discrete Fourier Transform and Convolution

```
100 REM F-T & CONVO. LEYP269. TRC11/79.
105 PI=3.1415626
110 NF=27:NT=51
115 DEF FNT(X)=TAN(ATN(TAN(ATN(TAN(ATN(X))))))
120 DIM RE(NF),W(NF),H(NT),FI(NT),HF(NT)
125 NU=25
130 PRINT"CALC RE PT H(S)=1/(S+1)  VS ODD# FREQS:"
132 PRINT
135 TT=FNT(1):TT=FNT(1):TT=FNT(1):TT=FNT(1)
140 W=-.5
145 PRINT"RADIANS      RE H(JW)"
150 FOR I=1 TO NU
160 W=W+.5:W(I)=W
170 RE(I)=1/(1+W*W)
180 PRINTW(I),RE(I)
185 TT=FNT(1):TT=FNT(1):TT=FNT(1):TT=FNT(1)
190 NEXT I
200 REM CALC IMPULSE RESPONSE
210 DT=.1:TM=5
220 T=-DT
230 WI=W(NU)/(NU-1)
240 L1=INT(TM/DT+1)
242 TT=FNT(1):TT=FNT(1):TT=FNT(1):TT=FNT(1)
245 PRINT
250 PRINT"SECONDS    IMPULSE RESP"
260 FOR I=1 TO L1
270 T=T+DT
280 SU=0
290 L2=NU-2
300 FOR J=1 TO L2 STEP 2
310 XX=W(J)*T
320 SU=SU+RE(J)*COS(W(J)*T)+4*RE(J+1)*COS(W(J+1)*T)+RE(J+2)*COS(W(J+2)*T)
325 NEXT J
330 H(I)=(2/PI)*WI*SU/3
340 PRINT T,H(I)
350 NEXT I
400 REM INPUT A UNIT STEP & CONVOLVE
410 FOR I=1 TO L1
420 FI(I)=1
430 NEXT I
432 PRINT
434 TT=FNT(1):TT=FNT(1):TT=FNT(1):TT=FNT(1)
440 PRINT"FOR UNIT STEP EXCITATION -"
450 PRINT"SECONDS      OUTPUT"
460 T=0
470 L3=INT(TM/DT+2)
480 FOR I=4 TO L3 STEP 2
490 L4=I-1
500 FOR J=1 TO L4
510 K=I-J
520 HF(J)=H(K)*FI(J)
530 NEXT J
540 L5=L4-2
550 SU=0
560 FOR M=1 TO L5 STEP 2
570 SU=SU+HF(M)+4*HF(M+1)+HF(M+2)
580 NEXT M
590 V=DT*SU/3
600 T=T+2*DT
610 PRINT T,V
620 NEXT I
630 END
```

Program B5-1. The Fletcher–Reeves Optimizer*

```
5 REM ***** APPENDIX PROGRAM B5-1 *****
10 REM HARWELL VAO8A FLETCHER-REEVES
20 REM OPTIMIZER. SEE A.E.R.E. REPORT R-7073 (1972)
30 REM USER MUST FURNISH SUBROUTINE
40 REM FOR OBJ FNCN. BEGIN AT STMNT 1000.
50 REM IF Y%=0 RETURN F. IF Y%=1 RETURN F9.
60 REM UNUSED NAMES BEGIN WITH: BCHJLOPQRTUV
```

```
70 PRINT"# VARIABLES,N=";:INPUT N%
75 PRINT
90 DIM X(N%),G(N%),S(N%)
100 PRINT"INPUT STARTING VARIABLES X(I):"
110 FOR I=1 TO N%                          INPUT N,$\underline{x}^0$
120 PRINT"    ",I;
130 INPUT X(I)
140 NEXT I
145 PRINT
```

```
150 M%=100          Set max IFN=100, ITN=0, epsilon = 1.E-5,
160 E=.00001        and expect F decrease on first iteration
170 D9=-.1          to be 0.1*F
180 I9%=0
```

```
190 Y%=0
200 GOSUB1000        Calculate F, ∇F, and set IFN=1
210 I7%=1
```

```
215 D9=ABS(D9*F)     Expected change in F on iter#1 is 0.1*F
220 REM          Reentry point for resetting to S.D. 1st move
230 FOR I=1 TO N%
235 S(I)=0
240 NEXT I
```

```
250 G9=1            Set last gradient sqd norm to 1
260 FOR I5=1 TO N%  Loop to 850 for N search directions
265 PRINT
270 PRINT"ITN=";I9%,"IFN=";I7%
280 PRINT"F=";F
285 PRINT"  I      X(I)        G(I)" Print ITN,IFN,F,∇F
290 FOR K=1 TO N%
295 PRINT K;X(K);G(K)
300 NEXT K
310 REM
320 I9%=I9%+1        Increment iteration # (ITN)
```

```
330 G7=0
340 FOR K=1 TO N%
350 G7=G7+G(K)*G(K)   Calculate $\beta_i = \|g^i\|^2 / \|g^{i-1}\|^2$ and STOP if =0
360 NEXT K
370 Z=G7/G9
380 IF Z=0 GOTO900
390 FOR K=1 TO N%     New search direction:
395 S(K)=Z*S(K)-G(K)
400 NEXT K            $\underline{s}^i = \beta_i \underline{s}^{i-1} - \underline{g}^i$
```

```
410 G5=0
420 FOR K=1 TO N%     Compute slope in $\underline{s}^i$ direction
430 G5=G5+G(K)*S(K)      (directional derivative)
440 NEXT K
```

```
450 IF G5>=0 GOTO220   If slope is positive, reset to S.D.
460 G3=G5              Starting slope = last slope.
470 A=-2*D9/G5         Calculate initial α for current iteration.
480 D9=F               Temporary save; see line 840
490 REM        Reentry after cubic interpolation or extraploation on α.
500 IF I7%=M% GOTO900  If have calc'd F& F 100 times, then STOP.
```

*See flowchart in Appendix D.

```
510 I3%=0                              Set "converged" flag
520 FOR I=1 TO N%                      If change in any variable exceeds
530 IF ABS(A*S(I))>=E THEN I3%=1       tolerance epsilon, then lower flag.
535 X(I)=X(I)+A*S(I)                   Move to new x values.
540 NEXT I
```

```
545 Y%=1
550 GOSUB1000                          Calculate Fα,∇Fα, & increment IFN count.
560 I7%=I7%+1
```

```
570 G1=0
580 FOR K=1 TO N%
590 G1=G1+G(K)*S(K)
600 NEXT K
```

```
610 IF ABS(G1/G3)<=.1 GOTO800   If slope mag decr'd by 10, then α OK.
620 IF F9>=F GOTO710            If F increased, cubic interp on α.
630 IF G1>0 GOTO710            If pos slope, do cubic interp on α.
```

```
640 Z=4
650 IF G5<G1 THEN Z=G1/(G5-G1)
660 IF Z>4 THEN Z=4             Extrapolate α.
670 A=A*Z
```

```
680 F=F9
690 G5=G1                       Save F and slope values in "last value" bins,
700 GOTO490                     and go take next step in x.
```

```
710 REM
720 FOR K=1 TO N%
725 X(K)=X(K)-A*S(K)            Back up to last x location, and
730 NEXT K
740 IF I3%=0 GOTO870   If converged flag is set, go to termination.
750 Z=3*(F-F9)/A+G1+G5           do cubic interpolation on α.
760 W=SQR(1-G5/Z*G1/Z)*ABS(Z)
770 Z=1-(G1+W-Z)/(2*W+G1-G5)
780 A=A*Z                        Then change α to that new value.
790 GOTO490                      Now go take that step in x.
```

```
800 REM                   Branch point from line 620 when α is OK.
810 F=F9                       Save current F  in "last F" bin.
820 IF I3%=0 GOTO900   If converged flag set, go print out and STOP.
830 G9=G7                  Save current ‖g‖² in "last norm sqd" bin.
840 D9=D9-F  Save F change over last iteration for next start'g α calc.
850 NEXT I5            End the loop for N search directions.
860 GOTO220           Go to the "reset for S,D." branch point above.
```

```
870 REM        Calc F & g with present α; then print and STOP.
875 Y%=0
880 GOSUB1000
890 I7%=I7%+1
```

```
900 REM        Branch point from 380,820, or 890. Print results & STOP.
905 PRINT
910 PRINT"ITN=";I9%,"IFN=";I7%
920 PRINT"F=";F
925 PRINT" I      X(I)            G(I)"
930 FOR K=1 TO N%
935 PRINT K;X(K);G(K)
940 NEXT K
```

```
999 GOTO9999              STOP
```

```
1000 REM SAMPLE OBJ FNCN       Rosenbrock's Banana Function
1010 Q=100*(X(2)-X(1)*X(1))^2+(1-X(1))^2
1020 IF Y%=0 THEN F=Q
1030 IF Y%=1 THEN F9=Q
1040 G(1)=-400*(X(1)*X(2)-X(1)^3)-2*(1-X(1))
1050 G(2)=200*(X(2)-X(1)*X(1))
1060 RETURN
9999 END
```

Program B5-2. L-Section Optimization*

```
10 REM   L-SECTION OPTIMIZATION.
15 REM    BY TOM CUTHBERT
20 PRINT"# OF SWR GOAL VALUES=";:INPUT MZ
30 DIM FR(MZ),RL(MZ),XL(MZ)
40 PRINT"INPUT THOSE:"
50 FOR I=1 TO MZ
55 PRINT"FREQ(";I;"),RL,XL=";:INPUT FR(I),RL(I),XL(I)
60 NEXT I
72 N%=2
90 DIM X(N%),G(N%),S(N%)
100 PRINT"INPUT STARTING L VALUE";:INPUT X(1)
110 PRINT"INPUT STARTING C VALUE";:INPUT X(2)
120 PRINT"PTH DIFFERENCE: P=";:INPUT P

          USE LINES 150-940 FROM PROGRAM B5-1 OPTIMIZER

999 GOTO120
1000 REM   SWR-TYPE SQRD-ERROR OBJ FNCN
1005 C=0
1010 GOSUB2100
1015 UN=ER
1020 IF Y%=0 THEN F=ER
1030 IF Y%=1 THEN F9=ER
1040 FOR C=1 TO N%
1050 X(C)=X(C)*1.0001
1060 GOSUB2100
1070 X(C)=X(C)/1.0001
1080 G(C)=(ER-UN)/(1E-4*X(C))
1090 NEXT C
1100 RETURN
2100 REM   SUM OF SQRD-SWR ERROR OVER FREQS
2105 ER=0
2110 FOR U=1 TO MZ
2120 OM=FR(U)
2140 RL=RL(U)
2150 XL=XL(U)
2160 GOSUB3000
2165 IF C<1 THEN PRINT"           SWR(";U;") = ";SW
2170 ER=ER+SW^P
2180 NEXT U
2190 RETURN
3000 REM CALC L-SECTION INPUT SWR
3010 VA=1-OM*X(2)*(OM*X(1)+XL)
3020 VB=OM*X(2)*RL
3030 VC=RL
3040 VD=OM*X(1)+XL
3050 VR=SQR(((VC-VA)^2+(VD-VB)^2)/((VC+VA)^2+(VD+VB)^2))
3070 SW=(1+VR)/(1-VR)
3080 RETURN
3299 END
```

---

*See flowchart in Figure 5.28.

Program B6-1.   L, T, and Pi Matching

```
150 REM BY TOM CUTHBERT
160 PRINT"R1<R2    R2<R1   [-180<DEGREES<180]
1000 REM CALCULATION:
1010 PF=180/3.14159265
1020 PRINT
1025 X3=1
1030 PRINT"OHMS R1,R2=";:INPUT R1,R2
1031 R3=R1*R2:R5=SQR(R3)
1040 PRINT"L,T, OR PI:";:INPUT N$
1050 IF N$="L" GOTO2170
1060 PRINT"PHASE BETWEEN +/-180 DEGREES=";
1063 INPUTB:B=B/PF
1070 IF N$="T" GOTO2070
1080 R7=R3*SIN(B)
1085 R8=COS(B)
1090 X1=R7/(R2*R8-R5)
2000 PRINT"X1=";X1
2010 X2=R7/R5
2020 PRINT"X2=";X2
2030 IF X3=0 GOTO2499
2040 X3=R7/(R1*R8-R5)
2050 PRINT"X3=";X3
2060 GOTO2499
2070 R7=SIN(B)
2075 R8=COS(B)
2080 X1=(R5-R1*R8)/R7
2100 PRINT"X1=";X1
2110 X2=-R5/R7
2120 PRINT"X2=";X2
2130 IF X3=0 GOTO2499
2140 X3=(R5-R2*R8)/R7
2150 PRINT"X3=";X3
2160 GOTO2499
2170 X3=0
2172 B=R1/R2
2174 IF B<1 THEN B=1/B
2176 B=ATN(SQR(B-1))
2180 PRINT"DEGREES =";B*PF;:PRINT", IS IT + OR - ";:INPUT N$
2190 IF N$="-" THEN B=-B
2200 IF R1<R2 GOTO2230
2210 PRINT"L SECTION B:"
2220 GOTO1080
2230 PRINT"L SECTION A:"
2240 GOTO2070
2499 END
```

Flowchart for Matching Program B6-1

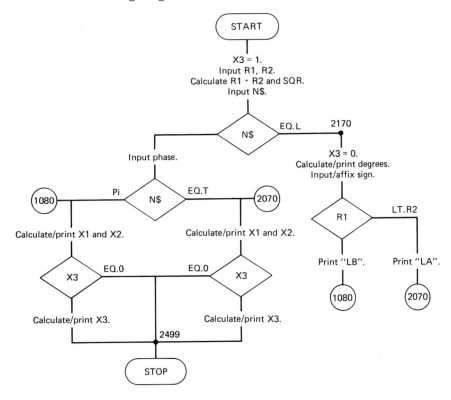

START

X3 = 1.
Input R1, R2.
Calculate R1 · R2 and SQR.
Input N$.

N$ — EQ.L — 2170

X3 = 0.
Calculate/print degrees.
Input/affix sign.

Input phase.

(1080) — Pi — N$ — EQ.T — (2070)

Calculate/print X1 and X2.   Calculate/print X1 and X2.

R1 — LT.R2

Print "LB".   Print "LA".

(1080)   (2070)

X3 — EQ.0   EQ.0 — X3

Calculate/print X3.   Calculate/print X3.

2499

STOP

Program B6-2.   Fano, Newton–Raphson Solution

```
100 DEF FNS(X)=(EXP(X)-EXP(-X))/2
110 DEF FNC(X)=(EXP(X)+EXP(-X))/2
120 DEF FNI(X)=LOG(X+SQR(X*X+1))
130 DEF FNG(X)=FNS(N*X)/FNC(N*X)/FNC(X)
140 DEF FNP(X)=(N-(FNS(N*X)*FNC(N*X)*FNS(X))/FNC(X))/(FNC(N*X)^2)/FNC(X)
150 DEF FNV(X)=(1+X)/(1-X)
160 PI=3.1415926
200 PRINT"N,QL,%BW=";:INPUT N,Q,BW
210 AL=Q*BW/100
300 REM: CALC INIT A,B BY CUTHBERT -
310 A=FNI((1.7*AL^.6+1)*(SIN(PI/2/N))/AL)
320 B=FNI((1.7*AL^.6-1)*(SIN(PI/2/N))/AL)
330 IF B<0 THEN B=.001
350 IT=0
400 REM: CALC FUNC, DELTA & NEW A,B
410 F1=FNS(A)-FNS(B)-(2/AL)*SIN(PI/2/N)
420 F2=FNG(A)-FNG(B)
430 J1=FNC(A)
440 J2=FNP(A)
450 J3=-FNC(B)
460 J4=-FNP(B)
470 JD=J1*J4-J2*J3
480 DA=-(J4*F1-J3*F2)/JD
490 DB=-(-J2*F1+J1*F2)/JD
500 FV=F1*F1+F2*F2
510 PRINT"ITER#,FUNC,A,B=";IT,FV,A,B
520 IF FV<1.E-9 GOTO600
530 IT=IT+1
540 A=A+DA
550 B=B+DB
560 GOTO400
600 PRINT"CONVERGED"
610 RH=FNC(N*B)/FNC(N*A)
620 RL=FNS(N*B)/FNS(N*A)
630 SH=FNV(RH)
640 SL=FNV(RL)
650 PRINT"SWR FROM";SL;"TO";SH
880 GOTO200
890 END
```

442

Program B6-3. Levy Matching to Resistive Source with g_i Prototype Values

```
10 REM LEVY RS WITH G(N+1) MODIFICATION.
20 REM 810116. TRC.
100 DEF FNS(X)=(EXP(X)-EXP(-X))/2
110 DEF FNC(X)=(EXP(X)+EXP(-X))/2
120 DEF FNI(X)=LOG(X+SQR(X*X+1))
130 DEF FNG(X)=FNS(N*X)/FNC(N*X)/FNC(X)
140 DEF FNP(X)=(N-(FNS(N*X)*FNC(N*X)*FNS(X))/FNC(X))/(FNC(N*X)^2)/FNC(X)
150 DEF FNV(X)=(1+X)/(1-X)
160 DEF FNN(R)=4*SIN((R-.5)*PN)*SIN((R+.5)*PN)
165 DEF FND(R)=X*X+Y*Y+SIN(R*PN)^2-2*X*Y*COS(R*PN)
170 DIM G(10)
180 PI=3.1415926
200 PRINT
205 PRINT"N,QL,%BW=";:INPUT N,Q,BW
210 AL=Q*BW/100
300 REM: CALC INIT A,B BY CUTHBERT -
310 A=FNI((1.7*AL^.6+1)*(SIN(PI/2/N))/AL)
320 B=FNI((1.7*AL^.6-1)*(SIN(PI/2/N))/AL)
330 IF B<0 THEN B=.001
350 IT=0
360 PN=PI/N
400 REM: CALC FUNC, DELTA & NEW A,B
410 F1=FNS(A)-FNS(B)-(2/AL)*SIN(PI/2/N)
420 F2=FNG(A)-FNG(B)
430 J1=FNC(A)
440 J2=FNP(A)
450 J3=-FNC(B)
460 J4=-FNP(B)
470 JD=J1*J4-J2*J3
480 DA=-(J4*F1-J3*F2)/JD
490 DB=-(-J2*F1+J1*F2)/JD
500 FV=F1*F1+F2*F2
520 IF FV<1.E-9 GOTO600
530 IT=IT+1
540 A=A+DA
550 B=B+DB
560 GOTO400
600 REM:CALC REFLECTION & SWR EXTREMES
602 X=FNS(A)
604 Y=FNS(B)
610 RH=FNC(N*B)/FNC(N*A)
620 RL=FNS(N*B)/FNS(N*A)
630 SH=FNV(RH)
640 SL=FNV(RL)
650 PRINT"SWR FROM";SL;"TO";SH
720 IF N>10 GOTO200
800 REM:CALC ELEMENT VALUES
810 G(1)=AL
820 FOR R=1 TO N-1
830 G(R+1)=FNN(R)/FND(R)/G(R)
840 NEXT R
850 FOR R=1 TO N
860 PRINT"G(";R;")=";G(R)
870 NEXT R
874 GS=2/G(N)*SIN(PI/2/N)/(X+Y)
876 PRINT"G(";N+1;")=";GS
880 GOTO200
890 END
```

Program B6-4. Romberg Integration of Two Fano Gain-Bandwidth Integrals

```
10 REM LP PB MATCH'G.CAS5/79P319.TRC10/79.
20 ND=11
22 DIM AU(ND)
80 PI=3.1415926
90 DEF FNE(X)=-ATN(X/SQR(-X*X+1))+PI/2
100 DEF FNC(X)=(EXP(X)+EXP(-X))/2
110 DEF FNI(X)=LOG(X+SQR(X*X-1))
120 PRINT"#L&C(EVEN#)=";:INPUT N
130 PRINT"OMEGA LOW,HIGH=";:INPUT WA,WB
140 PRINT"RESISTANCE RATIO (>1)=";:INPUT R
150 PRINT"# REACTANCES IN LOAD (1 OR 2)=";:INPUT NR
155 PRINT
160 PRINT"MAX PB DB=";:INPUT LA
170 SU=EXP(LA/10*LOG(10))
180 A=(WB*WB-WA*WA)/2
190 WO=SQR((WA*WA+WB*WB)/2)
200 EC=(FNC(N/2*(FNI(WO*WO/A))))^2
210 DC=(R+1)*(R+1)/R/4
220 EE=(SU-DC)/(1-EC)
230 DL=SU-EE-1
250 XU=WO+(WB-WA)*2^(10/N)
260 XL=0
9000 REM ROMBERG,IBM SCI3,P298.TRC10/79
9010 EP=1E-3
9040 XX=XL:GOSUB10000:FL=FC
9050 XX=XU:GOSUB10000:FU=FC
9060 AU(1)=(FL+FU)/2
9070 H=XU-XL
9080 IF(ND-1)<=0 THEN GOTO9420
9090 IF H=0 THEN GOTO9430
9100 HH=H
9110 E=EP/ABS(H)
9120 D2=0
9130 P=1
9140 JJ=1
9150 FOR I=2 TO ND
9160 Y=AU(1)
9170 D1=D2
9180 HD=HH
9190 HH=HH/2
9200 P=P/2
9210 X=XL+HH
9220 SM=0
9230 FOR J=1 TO JJ
9240 XX=X:GOSUB10000
9250 SM=SM+FC
9260 X=X+HD
9270 NEXT J
9280 AU(I)=.5*AU(I-1)+P*SM
9290 Q=1
9300 JI=I-1
9310 FOR J=1 TO JI
9320 II=I-J
9330 Q=Q+Q
9340 Q=Q+Q
9350 AU(II)=AU(II+1)+(AU(II+1)-AU(II))/(Q-1)
9355 NEXT J
9360 D2=ABS(Y-AU(1))
9370 IF (I-5)<0 THEN GOTO9400
9380 IF (D2-E)<=0 THEN GOTO9430
9390 IF (D2-D1)>=0 THEN GOTO9460
9400 JJ=JJ+JJ
9410 NEXT I
9420 PRINT"CAN'T GET < 1E-3 ERROR IN ";ND-1;"BISECTIONS"
9430 Y=H*AU(1)
9440 PRINT"    VALUE OF INTEGRAL =";Y
9445 PRINT
9450 GOTO160
9460 PRINT"ERROR > 1E-5 DUE TO ROUNDING"
9470 Y=H*Y
9480 GOTO9440
10000 REM GAIN-BW INTEGRANDS
10010 WP=ABS((XX*XX-WO*WO)/A)
10020 IF XX<WA OR XX>WB THEN GOTO10050
10030 AG=EE*(COS(N/2*FNE(WP)))^2
10040 GOTO10060
10050 AG=EE*(FNC(N/2*FNI(WP)))^2
10060 FC=LOG(SQR((1+DL+AG)/(DL+AG)))
10070 IF NR=2 THEN FC=FC*XX*XX
10080 RETURN
15000 END
```

Program B6-5. Hilbert Minimum Reactance Calculated From Piecewise Linear Resistance

```
50 REM: HILBERT IMPEDANCE CALCULATION
70 PRINT"# BREAKPOINTS";:INPUT L%
72 N%=L%-1
75 PRINT
90 DIM X(L%),OK(L%),AJ(L%),BJ(L%)
91 PI=3.1415926
92 PRINT"1ST BREAK FREQ MUST BE 0."
94 PRINT"INPUT ANY # FOR LAST EXCURSION"
96 PRINT"(PROGRAM RECOMPUTES IT)."
100 PRINT"INPUT EXCURSIONS & BREAK FREQS:"
110 FOR I=1 TO L%
120 PRINT"R(";I;"),W(";I;")=)";:INPUT X(I),OK(I)
130 NEXT I
140 OK(1)=0
150 PRINT"FREQUENCY=";:INPUT OM
160 GOSUB3000
170 PRINT"(RQ,XQ)=";RQ,XQ
180 PRINT
190 GOTO150
3000 REM: CALC ZQ=(RQ,XQ)
3010 X(L%)=0
3020 FOR J=1 TO N%
3030 X(L%)=X(L%)-X(J)
3040 NEXT J
3050 AJ(1)=1
3060 FOR J=2 TO L%
3070 AJ(J)=0
3080 IF OM<=OK(J-1) GOTO3130
3090 IF OM>=OK(J) GOTO3120
3100 AJ(J)=(OM-OK(J-1))/(OK(J)-OK(J-1))
3110 GOTO3130
3120 AJ(J)=1
3130 NEXT J
3140 BJ(1)=0
3150 DEF FNB(WB)=WB*((V+1)*LOG(V+1)+(V-1)*LOG(ABS(V-1))-2*V*LOG(V))
3160 FOR J=2 TO L%
3170 V=OM/OK(J)+1E-9
3175 BJ(J)=FNB(OK(J))
3180 IF OK(J-1)>0 THEN V=OM/OK(J-1)+1E-9
3190 BJ(J)=(BJ(J)-FNB(OK(J-1)))/PI/(OK(J)-OK(J-1))
3200 NEXT J
3210 RQ=1E-9:XQ=0
3220 FOR J=1 TO L%
3230 RQ=RQ+X(J)*AJ(J)
3240 XQ=XQ+X(J)*BJ(J)
3250 NEXT J
3260 RETURN
3299 END
```

445

Program B6-6. Carlin Resistance Excursion Optimization with Independent Source Resistance

```
9 REM CARMINNORN
10 REM CARLIN RESISTANCE EXCURSION OPTIMIZATION WITH DERIVATIVES
12 REM EXCUR RN IS DEPENDENT & RO INDEP HERE.
15 REM BY TOM CUTHBERT 801209.
20 PRINT"# OF MEASURED ZL VALUES=";:INPUT MZ
30 DIM FR(MZ),RL(MZ),XL(MZ),GO(MZ)
35 PI=3.1415926
40 PRINT"INPUT FREQ,ZL,&GAIN<=1:"
50 FOR I=1 TO MZ
55 PRINT"FREQ(";I;"),RL,XL,GAIN=";:INPUT FR(I),RL(I),XL(I),GO(I)
60 NEXT I
70 PRINT"# BREAKPOINTS INCLUDING 0 RADIANS";:INPUT L%
72 N%=L%-1
90 DIM X(L%),OK(L%),AJ(L%),BJ(L%),G(N%),S(N%)
92 PRINT"    FIRST BREAK FREQUENCY MUST BE ZERO."
94 PRINT"    USE ANY NUMBER FOR LAST EXCURSION."
96 PRINT"    (PROGRAM RECOMPUTES IT)."
100 PRINT"INPUT EXCURSIONS & BREAK FREQS:"
110 FOR I=1 TO L%
120 PRINT"EXCUR(";I;"),OMEGA(";I;")=";:INPUT X(I),OK(I)
130 NEXT I
140 OK(1)=0

        USE LINES 150-940 FROM PROGRAM B5-1 OPTIMIZER

999 GOTO92
1000 REM SQRD ERROR OBJ FNCN & GRADIENT
1010 ER=0
1020 FOR J=1 TO N%
1030 G(J)=0
1040 NEXT J
1050 REM SAMPLE FREQ LOOP
1060 FOR U =1 TO MZ
1070 OM=FR(U)
1080 GOSUB3000
1090 RL=RL(U)
1100 XL=XL(U)
1110 LA=(RL+RQ)*(RL+RQ)+(XL+XQ)*(XL+XQ)
1120 TX=4*RL/LA
1130 GA=GO(U)
1135 ES=TX*RQ/GA-1
1140 ER=ER+ES*ES
1150 E2=2*ES
1160 TX=TX/LA
1170 TR=(LA-2*RQ*(RL+RQ))*TX
1180 TX=-2*RQ*(XL+XQ)*TX
1190 REM GRADIENT LOOP AT EACH FREQ
1200 FOR J=1 TO N%
1210 G(J)=G(J)+E2*(TR*(AJ(J)-AJ(L%))+TX*(BJ(J)-BJ(L%)))/GA
1220 NEXT J
1230 NEXT U
1240 IFY%=0 THEN F=ER
1250 IFY%=1 THEN F9=ER
1260 RETURN
3000 REM  CALC ZQ=(RQ,XQ) BY LIN INTERP & HILBERT
3010 X(L%)=0
3020 FOR J=1 TO N%
3030 X(L%)=X(L%)-X(J)
3040 NEXT J
3050 AJ(1)=1
3060 FOR J=2 TO L%
3070 AJ(J)=0
3080 IF OM<=OK(J-1) GOTO3130
3090 IF OM>=OK(J) GOTO3120
3100 AJ(J)=(OM-OK(J-1))/(OK(J)-OK(J-1))
3110 GOTO3130
3120 AJ(J)=1
3130 NEXT J
3140 BJ(1)=0
3150 DEF FNB(WB)=WB*((V+1)*LOG(V+1)+(V-1)*LOG(ABS(V-1))-2*V*LOG(V))
3160 FOR J=2 TO L%
3170 V=OM/OK(J)+1E-9
3175 BJ(J)=FNB(OK(J))
3180 IF OK(J-1)>0 THEN V=OM/OK(J-1)+1E-9
3190 BJ(J)=(BJ(J)-FNB(OK(J-1)))/PI/(OK(J)-OK(J-1))
3200 NEXT J
3210 RQ=1E-9:XQ=0
3220 FOR J=1 TO L%
3230 RQ=RQ+X(J)*AJ(J)
3240 XQ=XQ+X(J)*BJ(J)
3250 NEXT J
3260 RETURN
3299 END
```

Program B9-1. Elliptic Filter Pole/Zero and Loss Calculations

```
1000 PRINT"ELLIP LP FLTR LOSS. DANIELS P.79"
1050 DIM X(25),XZ(25),A(250),O(250)
1060 PI=3.1415926:C1=PI/2
1100 PRINT"AMAX DB, AMIN DB,FB,FH=":INPUT AX,AN,FB,FH
1110 EE=10^(AX/10)-1:L=SQR((10^(AN/10)-1)/EE):XL=FH/FB
1130 K=1/L:GOSUB5000
1140 KL=KK
1160 K=SQR(1-1/L/L):GOSUB5000
1170 L1=KK
1410 K=SQR(1-1/XL/XL):GOSUB5000
1420 K1=KK
1430 K=1/XL:GOSUB5000
1440 N=INT(L1/KL*KK/K1)+1
1450 PRINT"DEGREE=";N
1460 NI=0:RN=1
1470 IFN=2*INT(N/2) THENNI=1
1480 PRINT"LOSS POLE FREQUENCIES:"
1500 FOR I=1 TO INT(N/2):GOSUB7000:NEXT I
1600 X=1:GOSUB9300
1700 C=1/RN
1800 PRINT"F=";:INPUT F:GOSUB6000
1900 GOTO1800
3000 REM CALCSN(U,K)
3100 Q=EXP(-PI*K1/KK)
3200 V=PI/2*U/KK:SN=0:J=0
3300 W=Q^(J+.5):SN=SN+W*SIN((2*J+1)*V)/(1-W*W):J=J+1
3400 IFW>1E-7 GOTO3300
3500 SN=SN*2*PI/K/KK
3600 RETURN
5000 REM CALC COMPLETE ELLIP INTEGRAL
5100 YY=K:XX=SQR(1-K*K):GOSUB8000:A(0)=ZZ:O(0)=PI/2:P=1:I=0
5200 X=2/(1+SIN(A(I)))-1:Y=SIN(A(I))*SIN(O(I))
5300 YY=SQR(1-X*X):XX=X:GOSUB8000:A(I+1)=ZZ
5400 YY=Y:XX=SQR(1-Y*Y):GOSUB8000:O(I+1)=(O(I)+ZZ)/2
5500 E=1-A(I+1)*2/PI:I=I+1
5600 IFE>1E-7 GOTO5200
5700 FOR J=1 TO I:P=P*(1+COS(A(J))):NEXT J
5800 X=PI/4+O(I)/2:KK=LOG(SIN(X)/COS(X))*P
5900 RETURN
6000 REM CALC LOSS
6100 X=F/FB
6200 RN=C*X^(1-NI)
6300 GOSUB9300
6400 A=10*LOG(1+EE*RN*RN)/LOG(10)
6500 PRINT"    DB LOSS=";A
6600 RETURN
7000 REM CALC POLES AND ZEROS
7100 U=(2*I-NI)*KK/N:GOSUB3000
7200 XZ(I)=SN:X(I)=XL/SN
7300 F(I)=FB*X(I):PRINT F(I)
7400 RETURN
8000 REM ATAN2
8001 S=SQR(XX*XX+YY*YY)
8002 IF XX<0 GOTO8009
8003 IF XX>0 GOTO8008
8004 IF YY=0 GOTO8012
8005 IF YY>0 GOTO8007
8006 YY=-C1:GOTO8013
8007 YY=C1:GOTO8013
8008 YY=ATN(YY/XX):GOTO8013
8009 IF YY>=0 GOTO8011
8010 YY=ATN(YY/XX)-PI:GOTO8013
8011 YY=ATN(YY/XX)+PI:GOTO8013
8012 YY=0
8013 ZZ=YY:RETURN
9300 FOR I=1 TO INT(N/2)
9400 RN=RN*(X*X-XZ(I)^2)/(X*X-X(I)^2)
9500 NEXT I
9600 RETURN
9999 END
```

Program B9-2. Symmetric Elliptic Filters

```
10 PRINT"SYMMETRICAL ELLIPTIC FLTR,C&S12/78,1009"
1010 DIM B(16),C(16),D(16),E(15),F(30)
1020 DN=LOG(10)/10:PI=3.1415926
2010 PRINT"STBND EDGE (KHZ)=";:INPUT FS
2020 PRINT"PSBND EDGE (KHZ)=";:INPUT FP
2030 IF ABS(FS-FP)<=0 GOTO2010
2040 PRINT"NUMBER OF PEAKS(1-15)=";:INPUT N
2050 IF N<=0 GOTO2010
2060 M=2*N+1
2080 FC=SQR(FS*FP)
2090 R=FC+FC
2100 FOR K=1TO2
2110 S=FS+FP
2120 FOR J=1TO6
2130 P=SQR(S*R)
2140 S=(S+R)/2
2150 IF1E8*(S-P)<S GOTO2170
2160 R=P:NEXT J
2170 IF K>=2 GOTO2200
2180 Q=M/S
2190 R=ABS(FS-FP):NEXT K
2200 Q=Q*S
2210 S=EXP(-PI/Q)
2220 Y=S
2230 PRINT"CRITICAL Q=";Q/(4*(1-S)*S^N)
2250 PRINT"STBND REJECTION (DB)=";:INPUT S
2260 IF S<=0 GOTO2010
2270 S=EXP(S*DN/2)
2280 R=EXP(PI*Q)
2290 P=(LOG(1+(S*S-1)/(R/4+1/R)^2))/DN
2300 PRINT"PSBND RIPPLE (DB)=";P
2310 R=R/(2*(S+SQR(S*S-1)))
2320 R=LOG(R+SQR(R*R+1))/(2*Q)
2330 R=SIN(R)/COS(R)
2340 W=R
2350 PRINT"3 DB (KHZ) ABOUT =";FP+(FS-FP)/(1+FC/(FP*R*R))
2360 PRINT"NOMINAL OHMS RESISTANCE=";:INPUTR
2370 IF R<=0 GOTO2040
2390 Z=Y:E(N)=W:W=W*W
2400 FOR J=1TOM-1
2410 F(J)=1:NEXT J
2420 K=1
2430 FOR J=1TO1024
2440 F(K)=F(K)*(1-Z)/(1+Z)
2450 IF K<M-1 GOTO2500
2460 Z=Z*Y
2470 X=((1-Z)/(1+Z))^2
2480 E(N)=E(N)*(W+X)/(1+W*X)
2490 K=0
2500 Z=Z*Y
2510 IF Z<.25E-18 GOTO2530
2520 K=K+1:NEXT J
2530 FOR J=1TON
2540 F(J)=F(J)*F(M-J)
2550 F(M-J)=F(J):NEXT J
3010 FOR J=1TON
3020 D(J)=F(2*J)*(1-F(J)^4)/F(J)
3030 B(J)=E(N)*F(J):NEXT J
3040 C(1)=1/B(N)
3050 FOR J=1TON-1
3060 C(J+1)=(C(J)-B(N-J))/(1+C(J)*B(N-J))
3070 E(N-J)=E(N+1-J)+E(N)*D(J)/(1+B(J)*B(J)):NEXT J
4010 FOR J=1TON
4020 B(J)=(((1+C(J)*C(J))*E(J)/D(J)-C(J)/F(J))/2
4030 C(J)=C(J)*F(J)
4040 D(J)=F(J)*F(J):NEXT J
4050 B(N+1)=B(N):C(N+1)=C(N):D(N+1)=D(N)
5010 IF N=1 GOTO6020
```

448

```
5020 L=1
5030 FOR K=L+2TON+1 STEP2
5040 FOR J=LTOK-2 STEP2
5050 Y=C(J)-C(K)
5060 Z=1/(Y/(B(J)*(D(K)-D(J)))-1)
5070 B(K)=(B(K)-B(J))*Z*Z-B(J)*(1+Z+Z)
5080 C(K)=Y*Z:NEXTJ
5081 NEXT K
5082 IFL=2GOTO6010
5083 L=2:GOTO5030
6010 S=B(N)/B(N+1)-1
6020 Q=.0005/(PI*FC)
6030 P=Q*R:Q=Q/R
6040 IF FS<FP GOTO6150
6060 PRINT"      ** LOW-PASS FILTER **"
6070 FOR J=1TON
6080 C(J)=Q*C(J)
6090 D(J)=Q*B(J)*D(J)
6100 B(J)=P/B(J)
6110 F(J)=FC/F(J):NEXT J
6120 C(N+1)=Q*C(N+1)
6130 GOTO6230
6150 PRINT"      ** HIGH-PASS FILTER **"
6160 FOR J=1TON
6170 C(J)=Q/C(J)
6180 D(J)=Q/(B(J)*D(J))
6190 B(J)=P*B(J)
6200 F(J)=FC*F(J):NEXT J
6210 C(N+1)=Q/C(N+1)
6230 PRINT"    KHZ          FARAD          HENRY"
6240 FOR J=1TONSTEP2
6250 PRINTTAB(12);C(J)
6260 PRINTF(J);J;D(J);B(J):NEXT J
6270 PRINTTAB(12);C(N+1)
6280 IF N=1 THEN STOP
6290 L=(INT((N+1)/2))*2:K=M-1-L
6300 FOR J=L+2TOM-1STEP2
6310 PRINTF(K);K;D(K);B(K)
6320 PRINTTAB(12);C(K)
6330 K=K-2:NEXT J
6340 PRINT"PRECISION TEST:";S
7020 STOP
```

Program B9-3. Antimetric Elliptic Filters

```
20 PRINT"ANTIMET ELLIP FLTR,CT12/78,1008"
1030 DIM B(16),C(16),D(16),E(30),F(16),R(15),S(15),DB(16),TB(16)
1040 DN=LOG(10)/10:PI=3.1415926
1050 PRINT
1060 PRINT"REJECTION,RIPPLE(DB),1/2-DEG(2-15),TYPE(A,B,OR C):"
1070 INPUT AS,AP,M,T$
1080 IF AS<=AP THEN STOP
1090 N=2*M
2010 ES=EXP(DN*AS)-1
2020 EP=EXP(DN*AP)-1
2030 V=SQR(ES/EP)+SQR(ES/EP-1)
2040 U=PI*PI/(2*LOG(V+V))
2050 V=V/(SQR(ES)+SQR(ES+1))
2060 W=U*LOG(V+SQR(V*V+1))/PI
2070 W=SIN(W)/COS(W):AO=W:W=W*W
2080 Y=EXP(-U):Z=Y:K=M-1
2090 FOR J=1TON
2100 E(J)=1:NEXT J
2110 FORJ=1TO1024
2120 IF K<>M GOTO2150
2130 X=((1-Z)/(1+Z))^2
2140 AO=AO*(W+X)/(1+W*X)
2150 E(K)=E(K)*(1-Z)/(1+Z)
2160 Z=Z*Y:IF Z<.25E-18 GOTO2180
2165 K=K-1
2170 IFK=0 THEN K=N
2175 NEXTJ
2180 E(M)=0:E(N)=E(N)*E(N)
2190 PRINT"U=";U;"AO=";AO;"EP=";E(N)
2200 FOR J=1TOM-1
2210 E(J)=-E(J)*E(N-J)
2220 PRINT"E=";-E(J)
2230 E(N-J)=-E(J):NEXT J
2250 X=SQR(AO*AO+1/(AO*AO)+E(N)*E(N)+1/(E(N)*E(N)))
2260 FOR J=1TOM-1STEP2:K=(J+1)/2
2270 Y=AO*E(J):Y=Y+1/Y
2280 Z=E(N)*E(J)
2290 R(K)=E(M-J)*(1/Z-Z)/Y:S(K)=-X/Y
2300 PRINT"RE=";R(K);"SE=";S(K)
2310 R(M-K+1)=R(K)
2320 S(M-K+1)=-S(K):NEXT J
2330 IF K+K=M GOTO3010
2340 R(K+1)=-AO:S(K+1)=0
2350 PRINT"RE=";-AO
3010 IT=2:IF T$="A" THEN IT=1
3020 E8=-E(1):IF T$="A" THEN E8=E(N)
3030 EO=E(N):IF T$="C" THENEO=-E(1)
3040 FP=SQR((E(N)+EO)/(1+E(N)*E8))
3050 FS=SQR((1+E(N)*EO)/(E(N)+E8))
3060 D(1)=0
3065 FOR J=ITTOM
3070 D(J)=(E(2*J-1)+E8)/(1+E(2*J-1)*EO)
3080 F(J)=SQR(1/D(J)):NEXT J
3100 SR=0:TQ=0:TO=0:B(1)=0:I=1
3110 FOR J=1TOM
3120 W=(AO^2+E(2*J-1)^2)/(1+(AO*E(2*J-1))^2)
3130 X=(1+EO*E8)*S(J)+EO+E8*W
3140 Y=EO^2+2*EO*S(J)+W
3150 Z=1+2*E8*S(J)+E8^2*W
3160 U=SQR(Y/Z):V=X/Z
3170 R(J)=SQR((U-V)/2):S(J)=SQR((U+V)/2)
3180 PRINT"RF=";-R(J)/FP;"SF=";S(J)/FP
3200 SR=SR+R(J)/U
3210 I=-I:W=I*R(J)/S(J)
3220 TQ=(TQ+W)/(1-TQ*W)
3230 IF T$<>"A" GOTO3270
3240 U=(F(2)-S(J))/R(J):V=(F(2)+S(J))/R(J)
3250 W=I*(V-U)/(1+U*V)
```

450

```
3260 TO=(TO+W)/(1-TO*W)
3270 B(1)=B(1)+R(J):NEXT J
4010 IF T$="A" THEN TO=TO/(1+SQR(1+TO*TO))
4020 FOR K=ITTOM
4030 DB(K)=0:TB(K)=TO:I=1
4040 FOR J=1TOM
4050 DB(K)=DB(K)+1/(R(J)+(F(K)-S(J))^2/R(J))+1/(R(J)+(F(K)+S(J))^2/R(J))
4070 I=-I:W=(F(K)-I*S(J))/R(J)
4080 TB(K)=(TB(K)+W)/(1-TB(K)*W):NEXT J:NEXT K
5010 D(M+1)=D(M):F(M+1)=F(M):DB(M+1)=DB(M):TB(M+1)=TB(M):C(1)=0
5020 FOR J=1TOM+1-IT STEP2
5030 TB(M+1-J)=-1/TB(M+1-J):NEXT J
5040 FOR J=ITTOM+1
5050 B(J)=(1+TB(J)^2)*DB(J)/(4*D(J))-TB(J)*F(J)/2
5060 C(J)=TB(J)/F(J):NEXT J
6010 FOR L=1TO2
6020 FOR K=L+2TOM+1 STEP2
6030 FOR J=LTOK-2STEP2
6040 U=C(J)-C(K)
6050 V=1/(U/(B(J)*(D(K)-D(J)))-1)
6060 B(K)=(B(K)-B(J))*V*V-B(J)*(V+V+1)
6070 C(K)=U*V:NEXTJ:NEXTK:NEXTL
7010 W=1:IFT$<>"C" THEN W=((1-TQ*TO)/(TQ+TO))^2
7020 FORJ=1TOM+1 STEP2:B(J)=B(J)*W
7030 C(J)=C(J)*W:NEXT J
7040 PRINT"LD RESIS=";W;" (";1/W;" )"
7050 PRINT"          L(C)          C(L)          PEAK"
7070 IF T$<>"A" THENPRINT" 1                ";FP/B(1)
7085 V=0
7090 FOR J=ITTOM:V=V+C(J)
7100 PRINTJ;FP*C(J);FP/B(J);F(J)/FP:NEXTJ
7110 PRINTM+1;FP*C(M+1);"STPBD EDGE=";FS/FP
7130 PRINT"TESTS";B(M)/B(M+1)-1;(W+1)*SR-V-C(M+1)
7140 STOP
7150 END
```

Appendix C

Derivation of the Fletcher–Reeves Scalar Multiplier

Given Equation (5.54) and $\beta_1 = 0$, find β_i, $i = 2, 3, \ldots, n$. By (5.52),

$$0 = (\mathbf{s}^i)^T \mathbf{A}\mathbf{s}^{i-1} = (-\mathbf{g}^i + \beta_i \mathbf{s}^{i-1})^T \mathbf{A}\mathbf{s}^{i-1}$$

$$= -\mathbf{g}^{iT} \mathbf{A}\mathbf{s}^{i-1} + \beta_i (\mathbf{s}^{i-1})^T \mathbf{A}\mathbf{s}^{i-1},$$

$$\therefore \beta_i = \frac{(\mathbf{g}^i)^T \mathbf{A}\mathbf{s}^{i-1}}{(\mathbf{s}^{i-1})^T \mathbf{A}\mathbf{s}^{i-1}}.$$

From (5.48),

$$\mathbf{A}\mathbf{s}^{i-1} = \frac{\mathbf{g}^i - \mathbf{g}^{i-1}}{\alpha_{i-1}},$$

$$\therefore \beta_i = \frac{(\mathbf{g}^i)^T (\mathbf{g}^i - \mathbf{g}^{i-1})}{(\mathbf{s}^{i-1})^T (\mathbf{g}^i - \mathbf{g}^{i-1})} = \frac{(\mathbf{g}^i)^T \mathbf{g}^i - 0^*}{\underbrace{(\mathbf{s}^{i-1})^T \mathbf{g}^i}_{0 \text{ by tangency}} - (\mathbf{s}^{i-1})^T \mathbf{g}^{i-1}},$$

$$\beta_i = \frac{(\mathbf{g}^i)^T \mathbf{g}^i}{(\mathbf{g}^{i-1})^T \mathbf{g}^{i-1}},$$

since

$$(\mathbf{s}^{i-1})^T \mathbf{g}^{i-1} = (-\mathbf{g}^{i-1} + \beta_{i-1} \mathbf{s}^{i-2})^T \mathbf{g}^{i-1}$$

$$= -(\mathbf{g}^{i-1})^T \mathbf{g}^{i-1} + \beta_{i-1} \underbrace{(\mathbf{s}^{i-2})^T \mathbf{g}^{i-1}}_{0 \text{ by orthogonality}}.$$

---

*It turns out in conjugate direction line searches to successive line minima that all $\mathbf{g}^i$ are orthogonal (see Aoki, 1971, p. 121).

Appendix D

*Linear Search Flowchart**

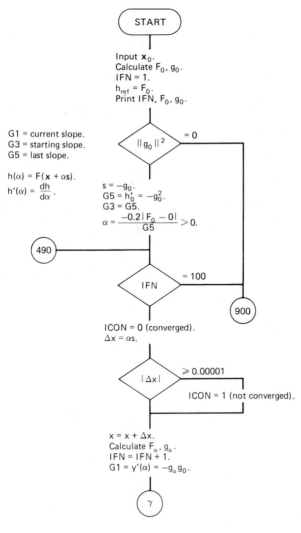

START

Input x_0.
Calculate F_0, g_0.
IFN = 1.
$h_{ref} = F_0$.
Print IFN, F_0, g_0.

G1 = current slope.
G3 = starting slope.
G5 = last slope.

$h(\alpha) = F(x + \alpha s)$.
$h'(\alpha) = \dfrac{dh}{d\alpha}$.

$\|g_0\|^2$ = 0

$s = -g_0$.
$G5 = h'_0 = -g_0^2$.
$G3 = G5$.
$\alpha = \dfrac{-0.2|F_0 - 0|}{G5} > 0$.

490

IFN = 100

900

ICON = 0 (converged).
$\Delta x = \alpha s$.

$|\Delta x|$ ⩾ 0.00001

ICON = 1 (not converged).

$x = x + \Delta x$.
Calculate F_α, g_α.
IFN = IFN + 1.
$G1 = y'(\alpha) = -g_\alpha g_0$.

γ

*For one variable.

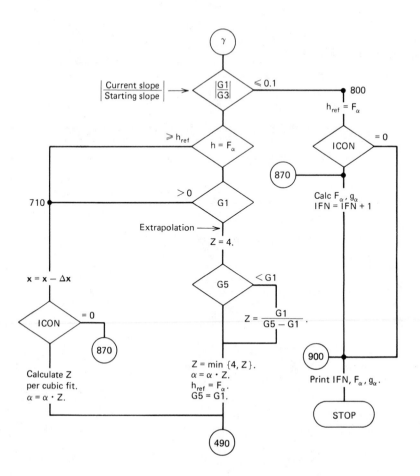

$$\gamma$$

$$\left| \frac{\text{Current slope}}{\text{Starting slope}} \right| \longrightarrow \left| \frac{G1}{G3} \right| \quad \leqslant 0.1 \qquad \bullet \; 800$$

$$h_{ref} = F_\alpha$$

$$\geqslant h_{ref} \qquad h = F_\alpha$$

ICON $\qquad = 0$

$$710 \qquad \qquad > 0 \qquad G1 \qquad \qquad 870 \qquad \text{Calc } F_\alpha, g_\alpha$$
$$IFN = IFN + 1$$

Extrapolation $\longrightarrow$ $\quad Z = 4.$

$$x = x - \Delta x$$

$$G5 \qquad < G1$$

ICON $\qquad = 0$

$$Z = \frac{G1}{G5 - G1}.$$

870

$$900$$

Calculate Z
per cubic fit.
$\alpha = \alpha \cdot Z.$

$$Z = \min \{4, Z\}.$$
$$\alpha = \alpha \cdot Z.$$
$$h_{ref} = F_\alpha.$$
$$G5 = G1.$$

Print IFN, F_α, g_α.

STOP

490

Appendix E _____

Defined Complex Constants for Amplifier Scattering Analysis*

The scattering-parameter determinant is

$$\Delta = S_{11}S_{22} - S_{21}S_{12}, \tag{E.1}$$

The stability factor is

$$K = \frac{1 + |\Delta|^2 - |S_{11}|^2 - |S_{22}|^2}{2|S_{21}S_{12}|}. \tag{E.2}$$

Two arbitrary constants are

$$B_1 = 1 + |S_{11}|^2 - |S_{22}|^2 - |\Delta|^2 = g_{L\,max}^{-1} + D_1, \tag{E3}$$

$$B_2 = 1 + |S_{22}|^2 - |S_{11}|^2 - |\Delta|^2 = g_{s\,max}^{-1} + D_2, \tag{E4}$$

where the unilateral input and output gain factors are

$$g_{s\,max} = \frac{1}{1 - |S_{11}|^2}, \tag{E.5}$$

$$g_{L\,max} = \frac{1}{1 - |S_{22}|^2}. \tag{E.6}$$

Four other commonly recurring constants are

$$D_1 = |S_{11}|^2 - |\Delta|^2, \tag{E.7}$$

$$D_2 = |S_{22}|^2 - |\Delta|^2, \tag{E.8}$$

$$C_1 = S_{11} - \Delta S_{22}^*, \tag{E.9}$$

$$C_2 = S_{22} - \Delta S_{11}^*. \tag{E.10}$$

*See Sections 7.4 and 7.5.

The maximum possible efficiency (see note below) is

$$\eta_{max} = \frac{|S_{21}|}{|S_{12}|}\left(K \pm \sqrt{K^2 - 1}\right). \tag{E.11}$$

The source and load conjugate-image reflection coefficients (see note below) are

$$r_{ms} = C_1^* \left[\frac{B_1 \pm \sqrt{B_1^2 - 4|C_1|^2}}{2|C_1|^2}\right], \tag{E.12}$$

$$r_{mL} = C_2^* \left[\frac{B_2 \pm \sqrt{B_2^2 - 4|C_2|^2}}{2|C_2|^2}\right]. \tag{E.13}$$

The maximum 50-ohm transducer gain is

$$g_0 = |S_{21}|^2. \tag{E.14}$$

The center of the input-plane stability circle is at

$$r_{s1} = \frac{C_1^*}{D_1}. \tag{E.15}$$

The center of the output-plane stability circle is at

$$r_{s2} = \frac{C_2^*}{D_2}. \tag{E.16}$$

The radius of the input-plane stability circle is

$$\rho_{s1} = \frac{|S_{12}S_{21}|}{D_1}. \tag{E.17}$$

The radius of the output-plane stability circle is

$$\rho_{s2} = \frac{|S_{12}S_{21}|}{D_2}. \tag{E.18}$$

The unilateral figure of merit is

$$u = \frac{|S_{11}S_{12}S_{21}S_{22}|}{|(1 - |S_{11}|^2)(1 - |S_{22}|^2)|}. \tag{E.19}$$

The maximum unilateral transducer gain is

$$G_{Tumax} = \frac{|S_{21}|^2}{|(1 - |S_{11}|^2)(1 - |S_{22}|^2)|}. \tag{E.20}$$

Note: Use a plus sign when $B_1 < 0$, and similarly for B_2 in (E.13).

Appendix F

Doubly Terminated Minimum-Loss Selectivity

Doubly Terminated Minimum-Loss Passband Selectivity for N = 4

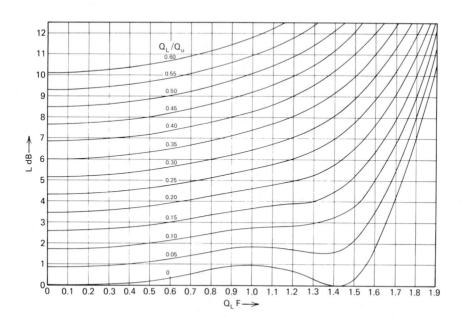

Doubly Terminated Minimum-Loss Stopband Selectivity for $N = 4$

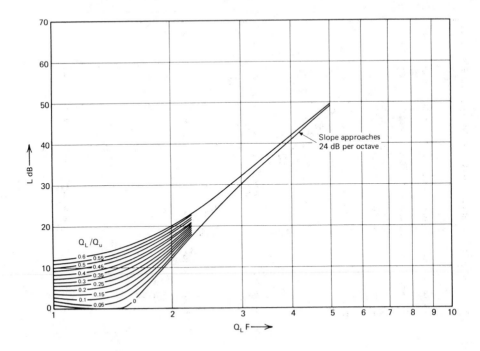

Appendix G

Direct-Coupled Filter Design Equations

These equations are grouped according to the design steps indicated in the flowchart in Figure 8.28 (Chapter Eight).

G.1. Response Shapes

$$F = \frac{f_b}{f_0} - \frac{f_0}{f_b} = \frac{f_b - f_a}{f_0} \doteq 2\left(\frac{f_b}{f_0} - 1\right) \quad \text{if} \quad F < 0.1; \quad f_0^2 = f_a f_b. \quad \text{(G.1)}$$

$$\bar{f} = \frac{f}{f_0} = \left[1 + \left(\frac{F}{2}\right)^2\right]^{1/2} \pm \frac{F}{2}. \quad \text{(G.2)}$$

$$Q_L = \frac{R_p}{X_p} = \frac{VA}{W}, \quad \overline{Q}_{LK} = \frac{Q_{LK}}{Q_{LN}}, \quad K = 1, 2, \ldots, N. \quad \text{(G.3)}$$

$$\sinh^{-1}x = \ln\left[x + (x^2 + 1)^{1/2}\right], \quad \cosh^{-1}x = \ln\left[x + (x^2 - 1)^{1/2}\right], \quad x > 1. \quad \text{(G.4)}$$

$$\theta = \frac{\pi}{2N}, \quad \phi = 2(N - r) - 1, \quad \Psi = 2(N - r) + 1, \quad \xi = N - r,$$

$$A_2 = \sin^2(\xi\theta), \quad A_3 = \cos^2(\xi\theta), \quad r = N - 1, N - 2, \ldots, 1. \quad \text{(G.5)}$$

$$\varepsilon = (10^{L_p/10} - 1)^{1/2}, \quad d_1 Q_{LN} = \begin{cases} 0, & G_s = 0, \\ \dfrac{1}{\overline{Q}_{L1}}, & G_s \neq 0. \end{cases} \quad \text{(G.6)}$$

G.1.1. Overcoupled Shape

$$T_N(x) = \begin{cases} \cos(N\cos^{-1}x), & |x| < 1, \\ \cosh(N\cosh^{-1}x), & |x| \geqslant 1. \end{cases} \tag{G.7}$$

$$L(f) = 10\log_{10}\left(1 + \varepsilon^2 T_N^2 \frac{F}{F_p}\right) \text{dB}. \tag{G.8}$$

$$S_N = \sinh\frac{\left[\sinh^{-1}(1/\varepsilon)\right]}{N}, \qquad B_1 = \left[\frac{S_N}{1 + d_1 Q_{LN}}\right]^2. \tag{G.9}$$

$$g_{N+1} = \begin{cases} 1, & \text{odd } N, \\ \coth^2\dfrac{\ln\left[\coth(L_p/17.37)\right]}{4}, & \text{even } N. \end{cases} \tag{G.10}$$

$$\overline{Q}_{L1} = g_{N+1}, \qquad \overline{Q}_{LN} = 1. \tag{G.11}$$

$$Q_{LN}F_p = \frac{\sin\theta}{\sqrt{B_1}}. \tag{G.12}$$

$$\overline{Q}_{Lr}\overline{Q}_{L(r+1)} = \frac{(\sin\phi\theta)(\sin\psi\theta)B_1/\sin^2\theta}{\left[A_3 + (d_1 Q_{LN})^2 A_2\right]B_1 + A_2 A_3}. \tag{G.13}$$

G.1.2. Maximally Flat Shape

$$L(f) = 10\log_{10}\left[1 + \varepsilon^2\left(\frac{F}{F_p}\right)^{2N}\right] \text{dB}. \tag{G.14}$$

$$\overline{Q}_{L1} = 1 = \overline{Q}_{LN}. \tag{G.15}$$

$$\overline{Q}_{Lr}\overline{Q}_{L(r+1)} = \frac{(\sin\phi\theta)\sin\psi\theta}{\left[A_3 + (d_1 Q_{LN})^2 A_2\right]\sin^2\theta}. \tag{G.16}$$

$$Q_{LN}F_p\varepsilon^{-1/N} = (1 + d_1 Q_{LN})\sin\theta. \tag{G.17}$$

G.1.3. Undercoupled Shape

$$L_d < 10\log_{10}\left[1 + \sinh^2(0.8814N)\right]. \tag{G.18}$$

$$q = \left(10^{L_d/10} - 1\right)^{1/2}. \tag{G.19}$$

$$k = \frac{q}{\sinh(0.8814N)}. \tag{G.20}$$

$$F_d = \frac{F_p}{\sinh\left\{\left[\sinh^{-1}(\varepsilon/k)\right]/N\right\}}. \tag{G.21}$$

$$L(f) = 10 \log_{10}\left[1 + k^2 \sinh^2\left(N \sinh^{-1}\frac{F}{F_d}\right)\right]. \tag{G.22}$$

$$C_N = \cosh\frac{\cosh^{-1}\{[\sinh(0.8814N)]/q\}}{N}. \tag{G.23}$$

$$\text{Error} = \overline{Q}_{LN} - 1 = \text{function of } \overline{Q}_{L1}. \tag{G.24}$$

$$A_1 = \left(\frac{C_N}{1 + d_1 Q_{LN}}\right)^2. \tag{G.25}$$

$$\overline{Q}_{Lr}\overline{Q}_{L(r+1)} = \frac{(\sin\phi\theta)(\sin\psi\theta)A_1/\sin^2\theta}{\left[A_3 + (d_1 Q_{LN})^2 A_2\right]A_1 - A_2 A_3}. \tag{G.26}$$

Note: For even N, find $N/2$ error roots yielding $\overline{Q}_{L1} < 1$. Then there are $N/2$ roots using $\overline{Q}_{L1} \leftarrow 1/\overline{Q}_{L1}$. For odd N, use one error root $\overline{Q}_{L1} = 1$, then find $(N-1)/2$ roots yielding $\overline{Q}_{L1} < 1$. Then there are $(N-1)/2$ roots using $\overline{Q}_{L1} \leftarrow 1/\overline{Q}_{L1}$.

$$Q_{LN}F_d = \frac{\sin\theta}{\sqrt{A_1}}. \tag{G.27}$$

G.1.4. Doubly Terminated Minimum-Loss Shape

$$y = (-1)^N X, \qquad X = Q_L\left(\frac{1}{Q_u} + jF\right). \tag{G.28}$$

$$P_K(y) = yP_{K-1}(y) - P_{K-2}(y), \qquad P_1 = 1, \qquad P_2 = y, \qquad K = 3,\ldots,N. \tag{G.29}$$

$$L(f) = 10 \log_{10}|(X+2)P_N(X) + (-1)^N P_{N-1}(X)|^2 - 6.0206 \text{ dB}. \tag{G.30}$$

$$\overline{Q}_{LK} = 1, \qquad K = 1,\ldots,N. \tag{G.31}$$

G.2. Physical Data

$$\underline{L} \leqslant L \leqslant \overline{L}, \qquad \underline{C} \leqslant C \leqslant \overline{C}, \qquad \hat{X}_{ij} = \omega_0\frac{(\overline{L}_{ij} + \underline{L}_{ij})}{2} \quad \text{or} \quad \frac{(\overline{C} + \underline{C})}{2\omega_0\overline{C}\,\underline{C}}, \tag{G.32}$$

or, for traps, $\hat{X}_{ij} = \omega_0|(\overline{L}_{ij} + \underline{L}_{ij})/2[1 - (\omega_0/\omega_n)^2]|$.

$$Q_{uK} = \frac{1}{\omega L_K G_u} = \frac{\omega L_{ij}}{R_u}. \tag{G.33}$$

G.3.   Passband

$$Q_{LN} = \frac{(Q_{LN}F_x)}{F_x}, \qquad x = \begin{cases} p, & \text{not undercoupled,} \\ d, & \text{undercoupled.} \end{cases} \qquad \text{(G.34)}$$

G.3.1.   Narrow-Band Approximation

$$Z_{ij} = (R_{ii}R_{jj})^{1/2}, \qquad j = i+1, \qquad i = 1, \ldots, N-1 \text{ (prototype).} \quad \text{(G.35)}$$

$$Z_{ij}(\omega) = \begin{cases} \dfrac{\omega}{\omega_0}\omega_0 L_{ij} & \text{for inverter L,} \\[2mm] \dfrac{\omega_0/\omega}{\omega_0 C_{ij}} & \text{for inverter C,} \\[2mm] \left| \dfrac{(\omega_0 L_{ij})(\omega/\omega_0)}{1-(\omega/\omega_n)^2} \right|, \\[2mm] C_{ij} = \dfrac{1}{\omega_n^2 L_{ij}}, \end{cases} \Bigg\} \text{LC trap at } \omega = \omega_n. \qquad \text{(G.36)}$$

G.3.2.   Loss Effects

Note: For overcoupled, even N, add

$$L_p = 10\log_{10}(1+\varepsilon^2)\,\text{dB}. \qquad \text{(G.37)}$$

G.3.3.   Stopbands

$$\bar{f}_s = \frac{f_s}{f_0}, \qquad F_s = \bar{f}_s - \bar{f}_s^{-1}, \qquad \bar{f}_K = \frac{f_K}{f_0}. \qquad \text{(G.38)}$$

$$L_s(f) \doteq -6. + DB1 + DB2 + DB3 + DB4 + DB5 \geqslant 20\,\text{dB}. \qquad \text{(G.39)}$$

Note: For overcoupled, even N, add $10\log_{10}g_{N+1}$.

Calculate each of the following subheadings in reverse order, starting with G.4.4 and working backward through G.4.1. Repeat for each $L_s(f_s)$, but use $L_s(f_K)$ for each trap.

G.4.1.   Loaded-Q Product

$$\Pi Q_L \overset{\Delta}{=} Q_{L1}Q_{L2}\cdots Q_{LN}. \qquad \text{(G.40)}$$

$$Q_{LN} = \left(\frac{\Pi Q_L}{\Pi\bar{Q}_{LN}}\right)^{1/N}. \qquad \text{(G.41)}$$

$$\Pi Q_L = \text{antilog}_{10}\frac{DB1}{20}. \qquad \text{(G.42)}$$

$$DB1 = L_K + 6 - DB2 - DB3 - DB4 - DB5. \qquad \text{(G.43)}$$

Note: For overcoupled, even N, subtract $10\log_{10}g_{N+1}$.

G.4.2. Resonator Asymptote Slopes

$$DB2 = N20 \log_{10} F. \tag{G.44}$$

G.4.3. L- and C-Inverter and L-Section Slopes

$$DB3 = (NLINV - NCINV)20 \log_{10} \bar{f}_s, \tag{G.45}$$

where NLINV is the number of L inverters including traps resonant at higher frequencies, and NCINV is the number of C inverters including traps resonant at lower frequencies.

$$DB4 = 10 \log_{10} \frac{(A - eCD)^2 + C^2}{A(1 + B^2)} \tag{G.46a}$$

for each terminal L section, where

$$e = \begin{cases} +1 & \text{for top-coupling L,} \\ -1 & \text{for top-coupling C,} \end{cases} \tag{G.46b}$$

and

$$A = 1 + Q_0^2, \tag{G.46c}$$

$$B = Q_{L1} F_s, \tag{G.46d}$$

$$C = B + eQ_0 \bar{f}_s^{-e}, \tag{G.46e}$$

$$D = Q_0 \bar{f}_s^e, \tag{G.46f}$$

$$Q_0 = \left(\frac{R_{11}}{R_g - 1} \right)^{1/2} \quad \text{or} \quad Q_0 = \left(\frac{R_{NN}}{R_L - 1} \right)^{1/2}. \tag{G.46g}$$

$$DB4 \doteq e20 \log_{10} \bar{f}_s \quad \text{if} \quad Q_0 > 3. \tag{G.47}$$

Note: Sum this expression for each L-section, since Q_0 is unknown beforehand (a conservative estimate).

G.4.4. Trap-Inverter Effects

$$DB5 = \sum_{n=1}^{NTRAPS} 20 \log_{10} \left| \frac{\bar{f}_n^2 - 1}{1 - (\bar{f}_n / \bar{f}_K)^2} \right|. \tag{G.48}$$

Note: The term f_n represents the null resonance frequency of each trap. For traps where $f_n < f_K$, use (G.48) as stated. For traps where $f_n > f_K$, use inverted variables in (G.48): $\bar{f}_n \leftarrow 1/f_n$, $\bar{f}_n \leftarrow 1/\bar{f}_K$.

$$L_K \doteq \tfrac{1}{2} \left\{ L_s + \left[L_s^2 + 4(0.585 \, ma)^2 \right]^{1/2} \right\}, \tag{G.49}$$

where

$$m \overset{\Delta}{=} 6(N + NLINV - NCINV). \tag{G.50}$$

Note: See nomenclature in (G.45) and count L sections as inverters. Classify traps at a frequency just beyond null resonance away from the passband.

If	$\bar{f}_K$	a	
$\bar{f}_s < 2$	$= (\bar{f}_s)^{1.585}$	$= 3.322 \log_{10} \bar{f}_s$	
$\bar{f}_s \geqslant 2$	$= 1.5 \bar{f}_s$	$= 1$	(G.51)

G.5. Q Effects

$$Q_{LN} = \max(\text{stopband } Q_{LN}\text{'s}), \quad \text{except}$$

if F_p is more important *and*   (G.52)

passband Q_{LN} is greater than the stopband Q_{LN}'s,

then $Q_{LN} = $ passband Q_{LN} [see (G.34) and (G.41)].

$$\hat{Q}_K \overset{\Delta}{=} \frac{Q_{LK}}{Q_{uK}}, \qquad K = 1, \ldots, N. \tag{G.53}$$

$$L_0 \doteq \begin{cases} -10 \log_{10} \prod_{K=1}^{N} (1 - \hat{Q}_K) \, dB, & \text{except that} \\ 4.34 N \hat{Q}_L \, dB \text{ for minimum-loss shape.} \end{cases} \tag{G.54}$$

Note: For overcoupled, even N, add L_p.

G.6. Design Limitations

$$\Pi Q_L \geqslant 3^N. \tag{G.55}$$

$$L_0 \leqslant 2N \, dB. \tag{G.56}$$

G.7. Minimum Shunt Inductances

$$L_K = \underline{L}_K, \qquad K = 1, \ldots, N. \tag{G.57}$$

G.8. Prototype Ohmic Values

$$R_{KK} = Q_{LK} \omega_0 L_K, \qquad K = 1, \ldots, N, \tag{G.58}$$

except for overcoupled even N, design to $R_{NN} = Q_{LN}\omega_0 L_N/g_{N+1}$, but use $R_{NN} = Q_{LN}\omega_0 L_N$ as modified in Section G.11.
Note: R_{11} and/or R_{NN} may be dependent.

$$X_{ij} = (R_{ii}R_{jj})^{1/2}, \qquad j = i+1, \qquad i = 1, \ldots, N-1. \qquad (G.59)$$

$$X_{g1} = \left[R_g(R_{11} - R_g) \right]^{1/2}, \qquad X_{NL} = \left[R_L(R_{NN} - R_L) \right]^{1/2}, \qquad (G.60)$$

or, for $N > 2$ and inductive X_{g1} and X_{12} and $L_1 \to \infty$,

$$X_{g1} = Q_{sg}R_g, \qquad Q_{sg} = g\left(\frac{R_{22}}{R_g}, Q_{L1} \right), \qquad R_{11} = R_g\left(1 + Q_{sg}^2 \right). \qquad (G.61)$$

For $N > 2$ and inductive X_{NL} and $X_{N,N-1}$ and $L_N \to \infty$,

$$X_{NL} = Q_{SL}R_L, \qquad Q_{SL} = g\left(\frac{R_{N-1,N-1}}{R_L}, Q_{LN} \right), \qquad R_{NN} = R_L\left(1 + Q_{SL}^2 \right),$$

$$(G.62)$$

where

$$g(x,y) \overset{\Delta}{=} \left| \frac{xy - |x(y^2+1) - 1|^{1/2}}{x^2 - 1} \right|. \qquad (G.63)$$

G.9. Component Acceptability

$$\min E \overset{\Delta}{=} \left[\left(\hat{X}_{g1}^2 - X_{g1}^2 \right)^2 + \cdots + \left(\hat{X}_{k,k+1}^2 - X_{k,k+1}^2 \right)^2 + \cdots + \left(\hat{X}_{NL}^2 - X_{NL}^2 \right)^2 \right] \cdot 10^{-12},$$

$$(G.64)$$

so that finally

$$\underline{L}_i \leqslant L_i \leqslant \overline{L}_i, \qquad i = 1, \ldots, N. \qquad (G.65)$$

Note:

1. See (G.32) and (G.59)–(G.62).
2. $1 \leqslant k \leqslant (N-1)$.
3. Adjust prototype shunt-L values for minimization; constraint is on shunt-L values after combining with inverters and, perhaps, an L section.

G.10.   Shunt Inductance Adjustment

$$K_L = L_K \text{ units (example: } 10^{-9} \text{ for nH).} \tag{G.66}$$

$$\frac{\partial E}{\partial L_1} = \begin{cases} 0 & \text{if } \hat{X}_{g1} = 0 \text{ or } L_1 \to \infty, \\ -2K_L 10^{-9}\omega_0 Q_{L1}\left[R_g(\hat{X}_{g1}^2 - X_{g1}^2) + R_{22}(\hat{X}_{12}^2 - X_{12}^2) \right]. \end{cases} \tag{G.67}$$

$$\frac{\partial E}{\partial L_j} = \begin{cases} 0 & \text{if } L_1 \to \infty \text{ and } N = 2, \\ -2K_L 10^{-9}\omega_0 Q_{Lj}\left[R_{j-1,\,j-1}(\hat{X}_{j-1,j}^2 - X_{j-1,j}^2) + R_{j+1,j+1}(\hat{X}_{j,j+1}^2 - X_{j,j+1}^2) \right], \end{cases}$$
$$2 \leqslant j \leqslant (N-1). \tag{G.68}$$

$$\frac{\partial E}{\partial L_N} = \begin{cases} 0 & \text{if } \hat{X}_{NL} = 0 \text{ or } L_1 \to \infty, \\ -2K_L 10^{-9}\omega_0 Q_{LN}\left[R_{N-1,N-1}(\hat{X}_{N-1,N}^2 - X_{N-1,N}^2) + R_L(\hat{X}_{NL}^2 - X_{NL}^2) \right]. \end{cases}$$
$$\tag{G.69}$$

G.11.   Final Component Values

$$\hat{R} = \frac{R_{11}(\text{lossy})}{R_{11}(\text{lossless})} = \frac{Y_N + Y_{N-1}}{X_N + X_{N-1}}. \tag{G.70}$$

K	Y_K	X_K
1	1.	$\hat{Q}_{L1}$
2	$\hat{Q}_{L2}$	$\hat{Q}_{L1}\hat{Q}_{L2} + 1$
3 ⎫ ⋮ ⎬ N ⎭	$Y_K = \hat{Q}_{LK}Y_{K-1} + Y_{K-2}$ $X_K = \hat{Q}_{LK}X_{K-1} + X_{K-2}$	

$$\tag{G.71}$$

Note: For overcoupled, even N, use $\hat{Q}_{LK}$ in reverse order.

N	$R_{11}(\text{lossy})$	$R_{NN}(\text{lossy})$
Even	$= \hat{R}^{1/2}R_{11}(\text{lossless})$	$= \hat{R}^{1/2}R_{NN}(\text{lossless})*$
Odd	$= \hat{R}^{1/2}R_{11}(\text{lossless})$	$= R_{NN}(\text{lossless})/\hat{R}^{1/2}$ (G.72)

*For overcoupled, even N, divide by g_{N+1}.

Note: If only one or no end coupling is employed, apply (G.70) directly in the end coupling or an inverter, respectively.

$$X_{g1} = \left(\frac{R_{11} - 2R_g}{2Q_u}\right) + \left[R_g(R_{11} - R_g) + \left(\frac{R_{11} - 2R_g}{2Q_u}\right)^2\right]^{1/2},$$
$$Q_u \gg 1. \tag{G.73}$$

$$X_{p1} = \frac{R_{11}}{\left[R_{11}/(R_g + X_{g1}/Q_u) - 1\right]^{1/2}}; \quad \text{parallels } C_1. \tag{G.74}$$

$$X_{NL} = \left(\frac{R_{NN} - 2R_L}{2Q_u}\right) + \left[R_L(R_{NN} - R_L) + \left(\frac{R_{NN} - 2R_L}{2Q_u}\right)^2\right]^{1/2},$$
$$Q_u \gg 1. \tag{G.75}$$

$$X_{PN} = \frac{R_{NN}}{\left[R_{NN}/(R_L + X_{NL}/Q_u) - 1\right]^{1/2}}, \quad \text{parallels } C_N. \tag{G.76}$$

$$C = \frac{B_C}{\omega}, \quad L = \frac{X_L}{\omega}, \quad C = \frac{1}{\omega_r L} \quad \text{(for resonance).} \tag{G.77}$$

$$L_{ij} = X_{ij} \frac{|1 - (1/\bar{f}_n)^2|}{\omega_0} \quad \text{(for traps).} \tag{G.78}$$

$$C_a + C_b = C_T, \quad \frac{1}{L_a} + \frac{1}{L_b} = \frac{1}{L_T}. \tag{G.79}$$

G.12. Performance and Sensitivity Analysis

G.12.1. *Transfer Sensitivities to Load SWR*

$$S_{21}(\Gamma_L) \doteq S_{21}(0)\left[1 + S_{22}(0) \cdot \Gamma_L\right], \quad |\Gamma_L| \ll 1. \tag{G.80}$$

$$\Delta|S_{21}| \doteq \pm 20 \log_{10}(1 + |S_{22}| \cdot |\Gamma_L|) \text{ dB maximum.} \tag{G.81}$$

$$\Delta\phi_{21} \doteq \pm 57.2958 |S_{22}| \cdot |\Gamma_L| \text{ degrees maximum.} \tag{G.82}$$

$$|\Gamma_L| = \frac{SWR_L - 1}{SWR_L + 1}. \tag{G.83}$$

$$S_v^T \triangleq \frac{\partial T}{\partial v} \cdot \frac{v}{T} \doteq \frac{\Delta T/T}{\Delta v/v}. \tag{G.84}$$

G.12.2. Sensitivities to Resonators

$$S_{C_K}^{Z_{11}} = \pm jQ_{LK}, \qquad S_{Z_{ij}}^{Z_{11}} = \pm 2. \tag{G.85}$$

$$\frac{df_0}{f_0} = -\frac{1}{2}\left(\frac{dL}{L} + \frac{dC}{C}\right). \tag{G.86}$$

G.13. Design Adjustment

$$\theta_{g1} = \cos^{-1}\left(\frac{R_g}{R_{11}}\right)^{1/2}, \qquad \theta_{NL} = \cos^{-1}\left(\frac{R_L}{R_{NN}}\right)^{1/2}. \tag{G.87}$$

$$V_1\,dB = 20\log\frac{V_1}{V_g} = 10\log\frac{R_{11}}{R_g}. \tag{G.88}$$

Appendix H

Zverev's Tables of Equivalent Three- and Four-Element Networks

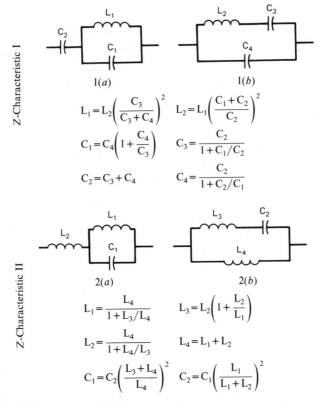

From Zverev (1967).

3(a) 3(b)

$$L_0 = L_2 \qquad L_2 = L_0$$

$$C_0 = \frac{C_2 C_3}{C_2 + C_3} \qquad C_2 = C_1 + C_0$$

$$L_1 = L_3\left(1 + \frac{C_3}{C_2}\right)^2 \qquad L_3 = L_1\left(\frac{C_1}{C_1 + C_0}\right)^2$$

$$C_1 = \frac{C_2}{1 + C_3/C_2} \qquad C_3 = C_0\left(\frac{1 + C_0}{C_1}\right)$$

4(a) 4(b)

$$L_0 = L_2 + L_3 \qquad L_2 = \frac{L_1 L_0}{L_1 + L_0}$$

$$L_1 = L_2\left(1 + \frac{L_2}{L_3}\right) \qquad L_3 = \frac{L_0^2}{L_1 + L_0}$$

$$C_0 = C_2 \qquad C_2 = C_0$$

$$C_1 = C_3\left(\frac{L_3}{L_2 + L_3}\right)^2 \qquad C_3 = C_1\left(\frac{L_1 + L_0}{L_0}\right)^2$$

5(a) 5(b)

$$L_1 = W \quad L_2 = X \qquad L_4 = \frac{L_1 L_2}{L_1} + L_2$$

$$C_1 = Y \quad C_2 = Z \qquad L_3 = \frac{(L_1^2 C_1 + L_2^2 C_2)^2}{(L_1 + L_2)(L_1 C_1 - L_2 C_2)^2}$$

$$A = L_4 L_3 C_4 C_3 \qquad C_4 = \frac{C_1 C_2 (L_1 + L_2)^2}{L_1^2 C_1} + L_2^2 C_2$$

$$B = L_4 C_4 + L_3 C_3 + L_4 C_3 \qquad C_3 = \frac{(L_1 C_1 - L_2 C_2)^2}{L_1^2 C_1} + L_2^2 C_2$$

$$E = L_3 C_4 C_3$$

$$D = C_4 + C_3$$

Z-Characteristic III

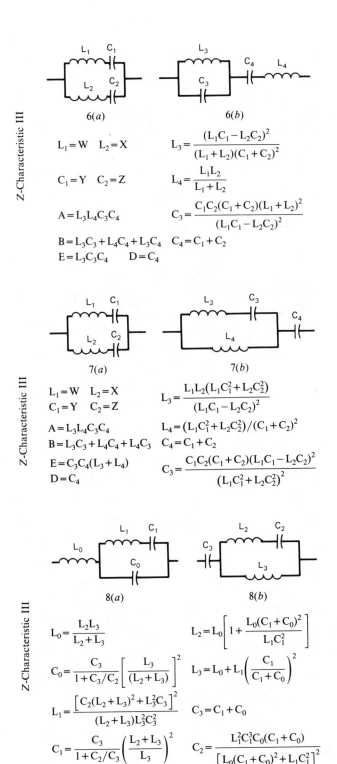

6(a) 6(b)

$L_1 = W \quad L_2 = X$

$C_1 = Y \quad C_2 = Z$

$A = L_3 L_4 C_3 C_4$

$B = L_3 C_3 + L_4 C_4 + L_3 C_4$
$E = L_3 C_3 C_4 \qquad D = C_4$

$L_3 = \dfrac{(L_1 C_1 - L_2 C_2)^2}{(L_1 + L_2)(C_1 + C_2)^2}$

$L_4 = \dfrac{L_1 L_2}{L_1 + L_2}$

$C_3 = \dfrac{C_1 C_2 (C_1 + C_2)(L_1 + L_2)^2}{(L_1 C_1 - L_2 C_2)^2}$

$C_4 = C_1 + C_2$

Z-Characteristic III

7(a) 7(b)

$L_1 = W \quad L_2 = X$
$C_1 = Y \quad C_2 = Z$
$A = L_3 L_4 C_3 C_4$
$B = L_3 C_3 + L_4 C_4 + L_4 C_3$
$E = C_3 C_4 (L_3 + L_4)$
$D = C_4$

$L_3 = \dfrac{L_1 L_2 (L_1 C_1^2 + L_2 C_2^2)}{(L_1 C_1 - L_2 C_2)^2}$

$L_4 = (L_1 C_1^2 + L_2 C_2^2)/(C_1 + C_2)^2$

$C_4 = C_1 + C_2$

$C_3 = \dfrac{C_1 C_2 (C_1 + C_2)(L_1 C_1 - L_2 C_2)^2}{(L_1 C_1^2 + L_2 C_2^2)^2}$

Z-Characteristic III

8(a) 8(b)

$L_0 = \dfrac{L_2 L_3}{L_2 + L_3}$

$C_0 = \dfrac{C_3}{1 + C_3/C_2} \left[\dfrac{L_3}{(L_2 + L_3)} \right]^2$

$L_1 = \dfrac{\left[C_2 (L_2 + L_3)^2 + L_3^2 C_3 \right]^2}{(L_2 + L_3) L_3^2 C_3^2}$

$C_1 = \dfrac{C_3}{1 + C_2/C_3} \left(\dfrac{L_2 + L_3}{L_3} \right)^2$

$L_2 = L_0 \left[1 + \dfrac{L_0 (C_1 + C_0)^2}{L_1 C_1^2} \right]$

$L_3 = L_0 + L_1 \left(\dfrac{C_1}{C_1 + C_0} \right)^2$

$C_3 = C_1 + C_0$

$C_2 = \dfrac{L_1^2 C_1^3 C_0 (C_1 + C_0)}{\left[L_0 (C_1 + C_0)^2 + L_1 C_1^2 \right]^2}$

471

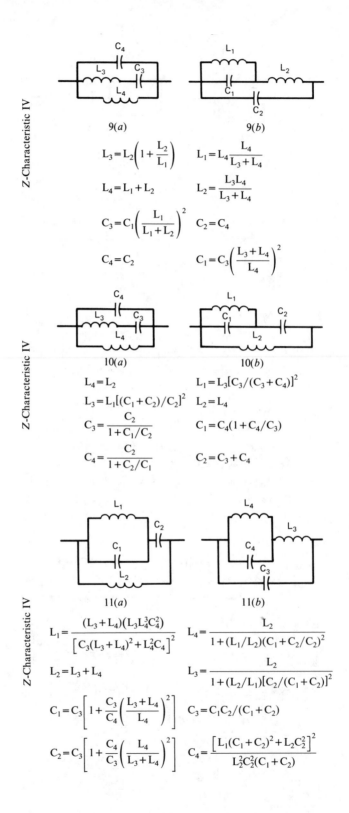

9(a)

9(b)

$$L_3 = L_2\left(1 + \frac{L_2}{L_1}\right)$$

$$L_1 = L_4\frac{L_4}{L_3 + L_4}$$

$$L_4 = L_1 + L_2$$

$$L_2 = \frac{L_3 L_4}{L_3 + L_4}$$

$$C_3 = C_1\left(\frac{L_1}{L_1 + L_2}\right)^2 \quad C_2 = C_4$$

$$C_4 = C_2$$

$$C_1 = C_3\left(\frac{L_3 + L_4}{L_4}\right)^2$$

10(a)

10(b)

$$L_4 = L_2$$

$$L_1 = L_3[C_3/(C_3 + C_4)]^2$$

$$L_3 = L_1[(C_1 + C_2)/C_2]^2 \quad L_2 = L_4$$

$$C_3 = \frac{C_2}{1 + C_1/C_2}$$

$$C_1 = C_4(1 + C_4/C_3)$$

$$C_4 = \frac{C_2}{1 + C_2/C_1}$$

$$C_2 = C_3 + C_4$$

11(a)

11(b)

$$L_1 = \frac{(L_3 + L_4)(L_3 L_4^3 C_4^2)}{\left[C_3(L_3 + L_4)^2 + L_4^2 C_4\right]^2}$$

$$L_4 = \frac{L_2}{1 + (L_1/L_2)(C_1 + C_2/C_2)^2}$$

$$L_2 = L_3 + L_4$$

$$L_3 = \frac{L_2}{1 + (L_2/L_1)[C_2/(C_1 + C_2)]^2}$$

$$C_1 = C_3\left[1 + \frac{C_3}{C_4}\left(\frac{L_3 + L_4}{L_4}\right)^2\right]$$

$$C_3 = C_1 C_2/(C_1 + C_2)$$

$$C_2 = C_3\left[1 + \frac{C_4}{C_3}\left(\frac{L_4}{L_3 + L_4}\right)^2\right]$$

$$C_4 = \frac{\left[L_1(C_1 + C_2)^2 + L_2 C_2^2\right]^2}{L_2^2 C_2^2 (C_1 + C_2)}$$

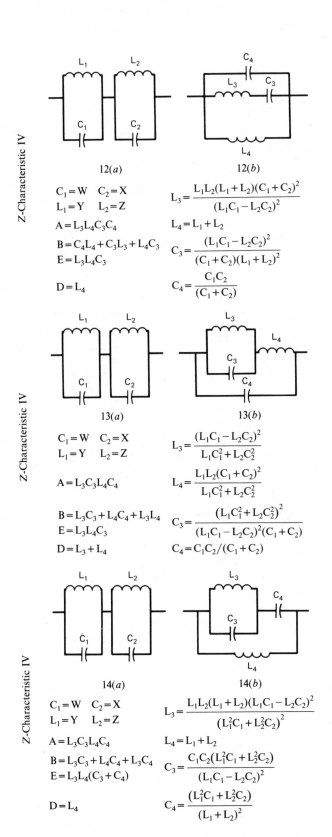

Z-Characteristic IV

12(a)   **12(b)**

$C_1 = W \quad C_2 = X$
$L_1 = Y \quad L_2 = Z$

$A = L_3 L_4 C_3 C_4$

$B = C_4 L_4 + C_3 L_3 + L_4 C_3$
$E = L_3 L_4 C_3$

$D = L_4$

$L_3 = \dfrac{L_1 L_2 (L_1 + L_2)(C_1 + C_2)^2}{(L_1 C_1 - L_2 C_2)^2}$

$L_4 = L_1 + L_2$

$C_3 = \dfrac{(L_1 C_1 - L_2 C_2)^2}{(C_1 + C_2)(L_1 + L_2)^2}$

$C_4 = \dfrac{C_1 C_2}{(C_1 + C_2)}$

Z-Characteristic IV

13(a)   **13(b)**

$C_1 = W \quad C_2 = X$
$L_1 = Y \quad L_2 = Z$

$A = L_3 C_3 L_4 C_4$

$B = L_3 C_3 + L_4 C_4 + L_3 L_4$
$E = L_3 L_4 C_3$
$D = L_3 + L_4$

$L_3 = \dfrac{(L_1 C_1 - L_2 C_2)^2}{L_1 C_1^2 + L_2 C_2^2}$

$L_4 = \dfrac{L_1 L_2 (C_1 + C_2)^2}{L_1 C_1^2 + L_2 C_2^2}$

$C_3 = \dfrac{(L_1 C_1^2 + L_2 C_2^2)^2}{(L_1 C_1 - L_2 C_2)^2 (C_1 + C_2)}$

$C_4 = C_1 C_2 / (C_1 + C_2)$

Z-Characteristic IV

14(a)   **14(b)**

$C_1 = W \quad C_2 = X$
$L_1 = Y \quad L_2 = Z$

$A = L_3 C_3 L_4 C_4$

$B = L_3 C_3 + L_4 C_4 + L_3 C_4$
$E = L_3 L_4 (C_3 + C_4)$

$D = L_4$

$L_3 = \dfrac{L_1 L_2 (L_1 + L_2)(L_1 C_1 - L_2 C_2)^2}{(L_1^2 C_1 + L_2^2 C_2)^2}$

$L_4 = L_1 + L_2$

$C_3 = \dfrac{C_1 C_2 (L_1^2 C_1 + L_2^2 C_2)}{(L_1 C_1 - L_2 C_2)^2}$

$C_4 = \dfrac{(L_1^2 C_1 + L_2^2 C_2)}{(L_1 + L_2)^2}$

473

Impedance Characteristics

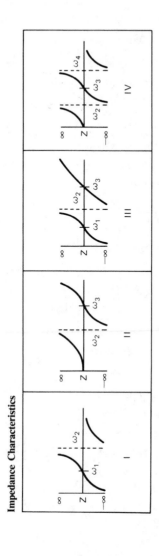

Impedance Resonance Frequencies

Z-Char	Fig.	ω_1^2	ω_2^2	ω_3^2	ω_4^2
I	1a	$1/L_1(C_1+C_2)$	$1/L_1C_1$		
I	1b	$1/L_2C_3$	$(C_3+C_4)/L_2C_3C_4$		
II	2a		$1/L_1C_1$	$(L_1+L_2)/L_1L_2C_1$	
II	2b		$1/L_1C_1$	$1/L_3C_2$	
III	5a	$1/L_2C_2$	$1/(L_3+L_4)C_2$	$1/L_1C_1$	
III	6b	$2/Q$	$(C_1+C_2)/(L_1+L_2)C_1C_2$	$Q/2L_3L_4C_3C_4$	
III	7b	$2/S$	$1/L_3C_3$	$S/2L_3L_4C_3C_4$	
III	5b	$2/S$	$1/(L_3+L_4)C_3$	$S/2L_3L_4C_3C_4$	
IV	12a		$(C_3+C_4)/L_3C_3C_4$	$(L_1+L_2)/L_1L_2(C_1+C_2)$	$1/L_1C_1$
IV	12b		$1/L_2C_2$	$1/L_4C_4$	$S/2L_4C_4L_3C_3$
IV	13b		$2/S$	$(L_3+L_4)/C_3L_3L_4$	$S/2L_4C_4L_3C_3$
IV	14b		$2/S$	$1/L_3(C_3+C_4)$	$S/2L_4C_4L_3C_3$

Simplified Notations

The equations describing the element values of some of the networks can be clarified by the use of more simplified notations. This has been done in the charts by substitution of symbols for some of the more complicated relationships. These symbols represent the following expressions:

1. The coefficients A, B, D, and E represent various combinations of L and C. They can be determined from the applicable equations on the chart.

2. $W = A(A - P^2)/P(AD - PE)$,
   $X = (A - P^2)/(E - PD)$,
   $Y = (AD - PE)/(A - P^2)$,
   $Z = P(E - PD)/(A - P^2)$,
   where $P = (B + \sqrt{B^2 - 4A})/2$.

3. $Q = L_3C_3 + L_3C_4 + L_4C_4 + \sqrt{(L_3C_3 + L_3C_4 + L_4C_4)^2 - 4L_3L_4C_3C_4}$.

4. $S = L_3C_3 + L_4C_3 + L_4C_4 + \sqrt{(L_3C_3 + L_4C_4 + L_4C_3)^2 - 4L_3L_4C_3C_4}$.

References

Aaron, M. R. (1956). The use of least squares in system design. *IEEE Trans. Circuit Theory*, December, pp. 224–231.

Abramowitz, M., and I. A. Stegun (1972). *Handbook of Mathematical Functions with Formulas, Graphs, and Mathematical Tables*. Washington, D.C.: U.S. Government Printing Office.

Acton, F. S. (1970). *Numerical Methods That Work*. New York: Harper & Row.

Amstutz, P. (1978). Elliptic approximation and elliptic filter design on small computers. *IEEE Trans. Circuits Syst.*, December, pp. 1001–1011.

Aoki, M. (1971). *Introduction to Optimization*. New York: Macmillan.

Attikiouzel, J. (1979). The use of orthogonal polynomials in all-pole filter design. *Trans. Inst. Eng. Australia*, October, pp. 80–87.

Bandler, J. W. (1973). In *Modern Filter Theory and Design* (G. C. Temes and S. K. Mitra, eds.). New York: Wiley, Chap. 6.

Bandler, J. W., and R. E. Seviora (1970). Current trends in network optimization. *IEEE Trans. Microwave Theory Tech.*, December, pp. 1159–1170.

Bauer, F. L., H. Rutishauser, and E. Stiefel (1963). New aspects in numerical quadrature. *Proc. Symp. Appl. Math. XV, Am. Math. Soc.*, pp. 199–218.

Beale, E. M. L. (1972). In *Numerical Methods for Non-Linear Optimization* (F. A. Lootsma, ed.). New York: Academic, pp. 39–43.

Beatty, R. W., and D. W. Kerns (1964). Relationships between different kinds of network parameters, not assuming reciprocity or equality of the waveguide or transmission line characteristic impedances. *Proc. IEEE*, January, p. 84.

Besser, L., and S. Swenson (1977). Take the hassle out of FET amplifier design. *Microwave Syst. News*, September, pp. 97–105.

Beveridge, G. S., and R. S. Schechter (1970). *Optimization: Theory and Practice*. New York: McGraw-Hill.

Blinchikoff, H. J., and A. I. Zverev (1976). *Filtering in the Time and Frequency Domain*. New York: Wiley.

Bode, H. W. (1945). *Network Analysis and Feedback Amplifier Design*. New York: Van Nostrand, p. 319.

Bodway, G. E. (1967). Two-port power flow analysis using generalized scattering parameters. *Microwave J.*, May, pp. 61–69.

—— (1968). Circuit design and characterization of transistors by means of three-port scattering parameters. *Microwave J.*, May, pp. 55–63.

476

Box, M. J., D. Davies, and W. H. Swann (1969). *Non-linear Optimization Techniques*. Edinburgh: Oliver & Boyd.

Bramham, B. (1961). A convenient transformer for matching coaxial lines. *Electron. Eng.*, January, pp. 42–44.

Branin, F. H., Jr. (1973). Network sensitivity and noise analysis simplified. *IEEE Trans. Circuit Theory*, May, pp. 285–288.

Carlin, H. J. (1977). A new approach to gain-bandwidth problems. *IEEE Trans. Circuits Syst.*, April, pp. 170–175.

Carlin, H. J., and J. J. Komiak (1979). A new method of broadband equalization applied to microwave amplifiers. *IEEE Trans. Microwave Theory Tech.*, February, pp. 93–99.

Chen, W. K. (1977). Explicit formulas for the synthesis of optimum broadband impedance-matching networks. *IEEE Trans. Circuits Syst.*, April, pp. 157–169.

Chin, F. Y., and K. Steiglitz (1977). An $O(N^2)$ algorithm for partial fraction expansion. *IEEE Trans. Circuits Syst.*, January, pp. 42–45.

Churchill, R. V. (1960). *Complex Variables and Applications*. New York: McGraw-Hill.

Coan, J. S. (1978). *Basic Basic*. Rochelle Park, N.J.: Hayden.

Cohn, S. B. (1957). Direct-coupled resonator filters. *Proc. IRE*, February, pp. 187–196.

COMPACT Engineering (1982). *COMPACT User Manual*. Comsat General Integrated Systems, 1131 San Antonio Road, Palo Alto, Calif. 94303.

Cottee, R. M., and W. T. Joines (1979). Synthesis of lumped and distributed networks for impedance matching of complex loads. *IEEE Trans. Circuits Syst.*, May, pp. 316–329.

Cristal, E. G. (1964). Coupled rods between ground planes. *IEEE Trans. Microwave Theory Tech.*, July, pp. 428–439.

——— (1968). Data for partially decoupled round rods between parallel ground planes. *IEEE Trans. Microwave Theory Tech.*, May, pp. 311–314.

——— (1969). Microwave synthesis techniques. Menlo Park, Ca.: Stanford Res. Inst. Tech. Rep. ECOM-0088F; ASTIA AD-861 887.

Cuthbert, T. R. (1980). Filter load effects, absorption, and loaded-Q design. Southern Methodist University, University Microfilm.

Daniels, R. W. (1974). *Approximation Methods for Electronic Filter Design*. New York: McGraw-Hill.

Davidon, W. C. (1959). Variable metric method for minimization. USAEC Report ANL-5990 (Rev), November.

Day, P. I. (1975). Transmission line transformation between arbitrary impedances using the Smith chart. *IEEE Trans. Microwave Theory Tech.*, September, pp. 772–773.

Dejka, W. J., and D. C. McCall (1969). Mathematical programming. San Diego, Ca.: USNELC Technical Note TN-1487, pp. 30–33.

Dishal, M. (1949). Design of dissipative band-pass filters producing desired exact amplitude-frequency characteristics. *Proc. IRE*, September, pp. 1050–1069.

——— (1951). Alignment and adjustment of synchronously tuned multiple-resonant-circuit filters. *Proc. IRE*, November, pp. 1448–1455.

——— (1953). Concerning the minimum number of resonators and the minimum unloaded Q needed in a filter. *IRE Trans. Vehic. Comm.*, June, pp. 85–117.

Dixon, L. C. W. (1971). In *Numerical Methods for Nonlinear Optimization* (F.A. Lootsma, ed.). New York: Academic, pp. 149–170.

Dwight, H. B. (1961). *Mathematical Tables of Elementary and Some Higher Mathematical Functions*. New York: Dover.

Fano, R. M. (1950). Theoretical limitations on the broadband matching of arbitrary impedances. *J. Franklin Inst.*, February, pp. 139–154.

Fiacco, A. V., and G. P. McCormick (1968). *Nonlinear Programming and Sequential Unconstrained Minimization Techniques*. New York: Wiley.

Fidler, J. K. (1976). Network sensitivity calculation. *IEEE Trans. Circuits Syst.*, September, pp. 567–571.

Fidler, J. K., and C. Nightingale (1978). *Computer Aided Circuit Design*. New York: Wiley.

Fisk, J. R. (1970). How to use the Smith chart. *Ham Radio*, November, pp. 92–101.

Fletcher, R. (1972a). In *Numerical Methods for Unconstrained Optimization* (W. Murray, ed.). New York: Academic, pp. 73–86.

——— (1972b). A FORTRAN subroutine for minimization by the method of conjugate gradients. Harwell, Berkshire, England: Atomic Energy Research Establishment Report No. AERE-R7073.

Fletcher, R., and C. M. Reeves (1964). Function minimization by conjugate gradients. *Computer J.*, 7:149–154.

Froehner, W. H. (1967). Quick amplifier design with scattering parameters. *Electronics*, October 16, pp. 100–109.

Geffe, P. R. (1963). *Simplified Modern Filter Design*. New York: J. F. Rider.

——— (1979). Microcomputer-aided filter design. *IEEE Circuits Syst. Mag.* March, pp. 5–8.

Gill, P. E., and W. Murray (1974). *Numerical Methods for Constrained Optimization*. New York: Academic.

Green, E. (1954). *Amplitude-Frequency Characteristics of Ladder Networks*. Chelmsford, Essex, England: Marconi House.

Guillemin, E. A. (1957). *Synthesis of Passive Networks*. New York: Wiley.

Ha, T. T., and T. H. Dao (1979). Application of Takahasi's results to broad-band matching for microwave amplifiers. *IEEE Trans. Circuits Syst.*, November, pp. 970–973.

Hadley, G. (1964). *Nonlinear and Dynamic Programming*. Reading, Mass.: Addison-Wesley.

Hamid, M. A., and M. M. Yunik (1967). On the design of stepped transmission-line transformers. *IEEE Trans. Microwave Theory Tech.*, September, pp. 528–529.

Hamming, R. W. (1973). *Numerical Methods for Scientists and Engineers*, 2nd ed. New York: McGraw-Hill.

Hewlett–Packard Co. (1968). S-parameters...circuit analysis and design. Application Note 95, September.

——— (1972). S-parameter design. Application Note 154, April.

——— (1976a). *General Utilities and Test Routines*, Part No. 09815–10001, pp. 47–64.

——— (1976b). *HP-67/97 EE Pac*. Part No. 00097–90057.

——— (1980). *Programming Tips*. Fort Collins, Col. 2-9 and 4-4.

Hindin, H. J. (1968). Insertion loss vs normalized frequency curves for Mumford's maximum flat filters. *IEEE Trans. Microwave Theory Tech.*, May, pp. 316–318.

Hindin, H. J., and J. J. Taub (1967). Design of TEM equal stub admittance filters. *IEEE Trans. Microwave Theory Tech.*, September, pp. 525–528.

Hooke, R., and T. A. Jeeves (1961). Direct search solution of numerical and statistical problems. *J. ACM*, April, pp. 212–229.

Hyde, W. L. (1966). Welcome from the Optical Society. In *Recent Advances In Optimization Techniques* (A. Lair and T. P. Vogl, eds.). New York: Wiley.

IBM (1968). *System 360 Scientific Subroutine Package (360A-CM-03X), Version III: Programmer's Manual*. White Plains, N.Y.: IBM Document No. H20-0205-3.

Jahnke, E., and F. Emde (1945). *Tables of Functions with Formulae and Curves*, 4th ed. New York: Dover.

Jasik, H. (1961). *Antenna Engineering Handbook*. New York: McGraw-Hill.

Johnson, L. W., and R. D. Riess (1977). *Numerical Analysis*. Reading, Mass.: Addison-Wesley.

Jong, M. T., and K. S. Shanmugam (1977). Determination of a transfer function from amplitude and frequency response data. *Int. J. Control*, **25**:941–948.

Kajfez, D. (1975). Numerical determination of two-port parameters from measured unrestricted data. *IEEE Trans. Instrum. Meas.*, March, pp. 4–11.

—— (1980). Computer aided analysis of noise in lossy microstrip filters. *IEEE Trans. Microwave Theory Tech.*, June, pp. 671–672.

Kokotovic, P., and D. D. Siljak (1964). Automatic analog solution of algebraic equations and plotting of root loci by generalized Mitrovic method. *IEEE Trans. Appl. Ind.*, 83:324–328.

Komiak, J. J., and H. J. Carlin (1977). Real frequency design of broad band microwave amplifiers. *Proc. 6th Biann. Cornell Electr. Eng. Conf.*, August, pp. 65–75.

Kotzebue, K. L. (1976). A quasi-linear approach to the design of microwave transistor power amplifiers. *IEEE Trans. Microwave Theory Tech.*, December, pp. 975–978.

—— (1979). Microwave design with potentially unstable FET's. *IEEE Trans. Microwave Theory Tech.*, January, pp. 1–3.

Ku, W. H. (1967). Design of transistor amplifiers using the scattering parameters of active two-ports. *Proc. 1st Asilomar Conf. on Circuit Theory*, November, Pacific Grove, Calif., pp. 100–111.

Kuester, J. L., and J. H. Mize (1973). *Optimization Techniques with Fortran*. New York: McGraw-Hill.

Kurokawa, K. (1965). Power waves and the scattering matrix. *IEEE Trans. Microwave Theory Tech.*, March, pp. 194–202.

Lago, G. and L. M. Benningfield (1979). *Circuit and System Theory*. New York: Wiley.

Lasdon, L. S., and A. D. Waren (1967). Mathematical programming for optimal design. *Electro-Technology*, November, pp. 55–70.

Levy, E. C. (1959). Complex-curve fitting. *IRE Trans. Autom. Control*, May, pp. 37–43.

Levy, R. (1964). Explicit formulas for Chebyshev impedance-matching networks. *Proc. IEE*, June, pp. 1099–1106.

—— (1973). A generalized design technique for practical distributed reciprocal ladder networks. *IEEE Trans. Microwave Theory Tech.*, August, pp. 519–526.

Ley, B. J. (1970). *Computer Aided Analysis and Design for Electrical Engineers*. New York: Holt, Rinehart & Winston.

Lin, C. C., and Y. Tokad (1968). On the element values of mid-series and mid-shunt low-pass LC ladder networks. *IEEE Trans. Circuit Theory*, December, pp. 349–353.

Lind, L. F. (1978). Accurate cascade synthesis. *IEEE Trans. Circuits Syst.*, December, pp. 1012–1014.

Linvill, J. G., and J. F. Gibbons (1961). *Transistors and Active Circuits*. New York: McGraw-Hill.

Lootsma, F. A. (1972). *Numerical Methods for Non-linear Optimization*. New York: Academic, pp. 149–170.

McCalla, T. R. (1967). *Introduction to Numerical Methods and* FORTRAN *Programming*. New York: Wiley.

Manaktala, V. K. (1972). Optimization of electrical networks using nonlinear programming. *J. Franklin Inst.*, May, pp. 313–324.

Mattaei, G. L., Leo Young, and E. M. T. Jones (1964). *Microwave Filters, Impedance-Matching Networks, and Coupling Structures*. New York: McGraw-Hill.

Mellor, D. J. (1975). Computer-aided synthesis of matching networks for microwave amplifiers. Stanford Electronics Laboratories, University Microfilm No. 75-21887.

Milligan, T. A. (1976). Transmission line transformation between arbitrary impedances. *IEEE Trans. Microwave Theory Tech.*, March, p. 159.

Minnis, B. J. (1981). Printed circuit coupled-line filters for bandwidths up to and greater than an octave. *IEEE Trans. Microwave Theory Tech.*, March, pp. 215–222.

Moad, M. F. (1970). A sequential method of network analysis. *IEEE Trans. Circuit Theory*, February, pp. 99–104.

Moore, J. B. (1967). A convergent algorithm for solving polynomial equations. *J. ACM*, April, pp. 311–315.

Muller, D. E. (1956). A method for solving algebraic equations using an automatic computer. *MTAC*, 10:208–215.

Mumford, W. W. (1965). Tables of stub admittances for maximally flat filters using shorted quarter-wave stubs. *IEEE Trans. Microwave Theory Tech.*, September, pp. 695–696.

Murdock, B. K. (1979). *Handbook of Electronic Design and Analysis Procedures Using Programmable Calculators*. New York: Van Nostrand Reinhold.

Noble, B. (1969). *Applied Linear Algebra*. Englewood Cliffs, N. J.: Prentice-Hall.

Penfield, P., R. Spence, and S. Duinker (1970). *Tellegen's Theorem and Electrical Networks*. Cambridge, Mass.: MIT Press, Research Mongraph 58.

Przedpelski, A. (1980). Simple transmission line matching circuits. *r. f. design*, Summer, pp. 27–32.

Ralston, A. (1965). *A First Course in Numerical Analysis*. New York: McGraw-Hill.

Ramo, S., and J. R. Whinnery (1953). *Fields and Waves in Modern Radio*. New York: Wiley.

Roberts, S. (1946). Conjugate-image impedances. *Proc. IRE*, April, pp. 198P–204P.

Rollett, J. M. (1962). Stability and power gain invariants of linear two-ports. *IEEE Trans. Circuit Theory*, March, pp. 29–32.

Rosemarin, D., and M. Chorev (1977). Easy-to-plot graphs show power-gain tradeoffs. *Microwaves*, December, pp. 178–183.

Saal, R. (1979). *Handbook of Filter Design*. Berlin: AEG-Telefunken.

Saal, R. and E. Ulbrich (1958). On the design of filters by synthesis. *IEEE Trans. Circuit Theory*, December, pp. 284–327.

Sanathanan, C. K., and J. Koerner (1963). Transfer function synthesis as a ratio of two complex polynomials. *IRE Trans. Autom. Control*, January, pp. 56–58.

Singhal, K., and J. Vlach (1973). Accuracy and speed of real and complex interpolation. *Computing*, 11:147–158.

Seshu, S., and N. Balabanian (1959). *Linear Network Analysis*. New York: Wiley.

Spence, R. (1970). *Linear Active Networks*. New York: Wiley.

Storch, L. (1954). The transmission matrix of N alike cascaded networks. *AIEE Trans.*, January, pp. 616–618.

Szentirmai, G. (1977). FILSYN—a general purpose filter synthesis program. *IEEE Proc.*, October, pp. 1443–1458.

Taub, J. J. (1963). Design of minimum-loss band-pass filters. *Microwave J.*, November, pp. 67–76.

——— (1968). Equal-element elliptic function filters. *IEEE Trans. Circuit Theory*, December, pp. 478–481.

Taub, J. J., and H. J. Hindin (1964). Minimum insertion loss microwave filters. *Microwave J.*, August, pp. 41–45.

Temes, G. C., and D. Y. F. Zai (1969). Least pth approximations. *IEEE Trans. Circuit Theory*, May, pp. 235–237.

Temes, G. C., and S. K. Mitra (1973). *Modern Filter Theory and Design*. New York: Wiley.

Terman, F. E. (1943). *Radio Engineers' Handbook*. New York: McGraw-Hill.

Traub, J. F. (1964). *Interactive Methods of Solution of Equations*. Englewood Cliffs, N.J.: Prentice-Hall.

Turnbull, H. W. (1952). *Theory of Equations*. New York: Interscience.

Van Valkenburg, M. E. (1960). *Introduction to Modern Network Synthesis*. New York: Wiley.

Vendelin, G. D., W. Alexander, and D. Mock (1974). Computer analysis of RF circuits with generalized Smith charts. *Electronics*, March 21, pp. 102–109.

Vincent, G. A. (1969). Impedance transformation without a transformer. *Frequency Technol.*, September, pp. 15–21.

Vlach, J. (1967). Chebyshev coefficients for piecewise linear functions. *IEEE Proc.*, April, pp. 572–574.

Vlach, J. (1969). *Computerized Approximation and Synthesis of Linear Networks*. New York: Wiley.

Weinberg, L., and P. Slepian (1960). Takahasi's results on Tchebycheff and Butterworth ladder networks. *IEEE Trans. Circuit Theory*, June, pp. 88–101.

Wenzel, R. J. (1964). Exact design of TEM microwave networks using quarter-wave lines. *IEEE Trans. Microwave Theory Tech.*, January, pp. 94–111.

Woods, D. (1976). Reappraisal of the unconditional stability criteria for active 2-port networks in terms of *S* parameters. *IEEE Trans. Circuits Syst.*, February, pp. 73–81.

Zehentner, J. (1980). Analysis and synthesis of coupled microstrip lines by polynomials. *Microwave J.*, May, pp. 95–98, 110.

Zverev, A. I. (1967). *Handbook of Filter Synthesis*. New York: Wiley.

Author Index

Subject Index